PALEOZOOLOGY AND PALEOENVIRONMENTS

Paleozoology and Paleoenvironments outlines the reconstruction of ancient climates, floras, and habitats on the basis of animal fossil remains recovered from archaeological and paleontological sites. In addition to outlining the ecological fundamentals and analytical assumptions attending such analyses, Tyler Faith and Lee Lyman describe and critically evaluate many of the varied analytical techniques that have been applied to paleozoological remains for the purpose of paleoenvironmental reconstruction. These techniques range from analyses based on the presence or abundance of species in a fossil assemblage to those based on taxon-free ecological characterizations. All techniques are illustrated using faunal data from archaeological or paleontological contexts.

Aimed at students and professionals, this volume will serve as a fundamental resource for courses in zooarchaeology, paleontology, and paleoecology.

J. Tyler Faith is Curator of Archaeology at the Natural History Museum of Utah and Assistant Professor of Anthropology at the University of Utah. His research addresses the relationships between Quaternary mammal communities, environmental change, and human–environment interactions, with an emphasis on eastern and southern Africa. This is his first book.

R. Lee Lyman is Emeritus Professor of Anthropology at the University of Missouri, Columbia. A scholar of late Quaternary paleozoology and human prehistory of the Pacific Northwest United States, he is author of *Vertebrate Taphonomy* (Cambridge, 1994), *Quantitative Paleozoology* (Cambridge, 2008), and *Theodore E. White and the Development of Zooarchaeology in North America* (2016).

PALEOZOOLOGY AND PALEOENVIRONMENTS

FUNDAMENTALS, ASSUMPTIONS, TECHNIQUES

J. TYLER FAITH

University of Utah, Salt Lake City

and

R. LEE LYMAN

University of Missouri, Columbia

CAMBRIDGE
UNIVERSITY PRESS

DAMAGED

CAMBRIDGE
UNIVERSITY PRESS

University Printing House, Cambridge CB2 8BS, United Kingdom

One Liberty Plaza, 20th Floor, New York, NY 10006, USA

477 Williamstown Road, Port Melbourne, VIC 3207, Australia

314–321, 3rd Floor, Plot 3, Splendor Forum, Jasola District Centre, New Delhi – 110025, India

79 Anson Road, #06-04/06, Singapore 079906

Cambridge University Press is part of the University of Cambridge.

It furthers the University's mission by disseminating knowledge in the pursuit of education, learning, and research at the highest international levels of excellence.

www.cambridge.org
Information on this title: www.cambridge.org/9781108480352
DOI: 10.1017/9781108648608

First published 2019

Printed in the United Kingdom by TJ International Ltd, Padstow, Cornwall

A catalogue record for this publication is available from the British Library.

Library of Congress Cataloging-in-Publication Data
Names: Faith, J. Tyler, author. | Lyman, R. Lee, author.
Title: Paleozoology and paleoenvironments : fundamentals, assumptions, techniques / J. Tyler Faith, R. Lee Lyman
Description: Cambridge ; New York : Cambridge University Press, 2019. | Includes bibliographical references and index.
Identifiers: LCCN 2018039941 | ISBN 9781108480352 (hardback) | ISBN 9781108727327 (pbk.)
Subjects: LCSH: Paleontology. | Paleoecology.
Classification: LCC QE761.F35 2019 | DDC 560–dc23
LC record available at https://lccn.loc.gov/2018039941

ISBN 978-1-108-48035-2 Hardback
ISBN 978-1-108-72732-7 Paperback

CONTENTS

FIGURES

TABLES

ACKNOWLEDGMENTS

This volume is the outcome of years of reflection and discussion with our mentors, collaborators, colleagues, and friends, and we sincerely thank the many people who contributed directly or indirectly to the ideas presented here. We are particularly indebted to those who responded to our questions, comments, and requests while writing this volume: Andrew Barr provided a brief but important discussion concerning measurements in ecomorphology; Larisa DeSantis sent us dental microwear photosimulations; Andrew Du supplemented our limited coding skills and offered insightful comments on taxonomic diversity; Christine Janis and John Damuth sent us raw data concerning precipitation estimates for the Great Plains; Julien Louys clarified points concerning a methodological technique; Dan Peppe directed us to some useful paleobotanical literature; Steve Wolverton provided raw data concerning the size of deer from Texas. René Bobe and an anonymous reader inspired useful fine-tuning in later drafts of this volume. A portion of Faith's contribution to this volume was supported by an Australian Research Council DECRA fellowship at the University of Queensland.

ONE

WHY A BOOK ON PALEOENVIRONMENTAL RECONSTRUCTION FROM FAUNAL REMAINS?

Once *fossils* were recognized for what they are – ancient remains of organisms – early earth scientists attempted to ascertain what those remains might reveal about the past. Fossils represented ancient life, but what else might they signify? Georges Cuvier is often regarded as the first scientific paleontologist (Rudwick 1985, 1997); he initiated a now long-standing tradition of research focusing on the morphology of past animals, including their functional anatomy (Haber 1959; Rainger 1981). After the publication of Darwin's (1859) *On the Origin of Species*, the morphological tradition (Rainger 1981) included efforts to decipher the transitions between major groups of animals such that by the 1880s "paleontology was an evolutionary science" (Hall 2002:649). Paleontology's major applied value was in facilitating the exploitation of the earth's mineral resources, leading it to be typically associated with academic departments of geological sciences rather than biological sciences (Gould 1977; Rainger 1985). By the middle of the twentieth century, however, that somewhat singular focus began to broaden.

Ancient remains of animals and plants had been parts of living organisms. Those organisms must have had species-specific ecologies, just as now-living organisms do. It was a relatively easy (though perhaps a surprisingly long-time-coming) step to deciphering the paleoecological signal of fossils in order to reconstruct past environments, *habitats*, and climates. Although not without precedent in the nineteenth century (e.g., Dawkins 1871; Lartet 1875), early modern work began with Everett C. Olson in the 1940s (Rainger 1997). The focus of much paleontological research at that time was on interpreting fossil morphology and writing evolutionary history (beyond the basic geological work of biostratigraphy and mineral exploitation). Olson broadened that traditional focus to include study of the paleoecology of particular organisms and the biological and ecological interrelationships of communities of organisms, providing a framework for reconstructing terrestrial paleoenvironments through time. His relatively unique insights to past landscapes and biota were a catalyst for paleontologists, many of whom turned at least some of their attention to paleoecology.

A major landmark in the development of paleoecology was the 1957 publication of the *Treatise on Marine Ecology and Paleoecology* (Hedgpeth and Ladd 1957), followed a few years later by *Approaches to Paleoecology* (Imbrie and Newell 1964). These volumes focused on marine ecosystems, but interest in terrestrial paleoecology was blossoming at the same time. In 1961, F. Clark Howell and François Bourlière organized a symposium aiming to "integrate the results of increasingly numerous field studies bearing on the biological-behavioral evolution of the higher primates (especially hominids) with other field studies in the paleoecology and the recent mammalian ecology of sub-Saharan Africa" (Howell and Bourlière 1963:v). The proceedings were published a few years later in *African Ecology and Human Evolution* (Howell and Bourlière 1963); in the pages of that volume the contributors (paleontologists, archaeologists, geologists, and zoologists) acknowledged that paleoenvironments were the context in which humans evolved biologically and culturally. Knowing something about those ancient environments would, it was believed (and still is), facilitate understanding of humankind's deep history. This idea was echoed a year later by North American archaeologists in Hester and Schoenwetter's (1964) monograph *The Reconstruction of Past Environments*. Old World archaeologists had exploited a particular value of paleoenvironmental research based on faunal remains some years earlier, specifically thinking about what paleoenvironmental fluctuations might tell us about the possible necessity of the origins of agriculture (Reed and Braidwood 1960). Australian paleontologists held a symposium in 1978, the proceedings of which were published that same year under the title *Biology and Quaternary Environments* (Walker and Guppy 1978). The main purpose of that symposium was to review and evaluate in explicit terms the operating principles and requisite assumptions that underpin paleoenvironmental reconstruction based on faunal remains such that they did not become so implicit as to facilitate complacency among researchers. Clearly, a florescence of interest in paleoenvironmental reconstruction based on faunal remains took place during the middle of the twentieth century.

The literature concerning the paleoenvironmental implications of prehistoric faunal remains (much of it is cited in this volume) has grown considerably in the past several decades. Much of that literature is made up of case studies, that is, analyses of particular faunas that address aspects of the paleoecology and paleoenvironments in which the represented faunas participated and lived. Reading that literature can reveal the analytical techniques and requisite interpretive assumptions that attend such analyses. However, one must read numerous articles to be exposed to the myriad techniques that have been used and the various assumptions that, when stated explicitly, underscore the sometimes tenuous nature of interpretations.

We have both analyzed ancient faunas with the goal of deciphering their paleoecological and paleoenvironmental implications, having been taught some of the basics of doing so by our academic advisors and teachers, and learning more by reading much of the pertinent literature. During the course of our research, we perceived a major lacuna in the extant literature – there was no textbook that described the ecological basics, the analytical assumptions, and the numerous analytical techniques one might use to reconstruct past environments on the basis of a *collection* of ancient faunal remains. Our students' inquiries about the topic underscored this gap in the literature; there was no single title to which we could refer them. Producing a list of a dozen or fewer titles to which we could direct them might have sufficed, but that raised the question of which titles to include and which to exclude. Writing (what turned out to be) this book would fill this significant gap in the literature. And we perceived another reason to write the book, a reason we identify shortly.

In the remainder of this chapter, we continue our sketch of the history of *paleozoology* with a focus on the emergence of paleoecological research and paleoenvironmental reconstruction. Our intention is to provide some background to modern research in this area. The sketch is brief because our intentions in writing this volume are not to provide a lengthy history of this field of inquiry. We provide the history that we do because we believe that knowing something of the background of one's chosen line of research can enhance understanding of why practitioners today ask the research questions they do and seek answers to those questions in the ways they do. It is for this reason we include historical tidbits throughout the volume. Following our outline of the history of paleoenvironmental reconstruction in this chapter, we describe the structure of the remainder of the volume. Near the end of this chapter, we identify the other reason we decided to write this book, one that is of equal if not more significance than that of filling a gap in the literature.

A BIT MORE HISTORY

Old World archaeologists had long studied faunal remains recovered from archaeological deposits in order to facilitate building cultural chronologies (O'Connor 2007). It was the association of undisputed stone tools with remains of animal species believed to have gone extinct at the end of the Pleistocene ice age that persuaded the nineteenth-century scientific community that human ancestors had been present well prior to the biblically documented creation (Grayson 1983a; Van Riper 1993). Similarly, it was the association of remains of an extinct form of bison (*Bison antiquus* [at the time the specimens were attributed to *B. taylori*, a form no longer considered valid]) with stone artifacts in New Mexico (USA) that convinced local archaeologists that people had been in North America near the end of the Pleistocene (Meltzer 2006). Use of ancient faunal remains for these kinds of biostratigraphic (stratigraphic correlation and age assessment) purposes had been around for some time in geology (Rudwick 1996). For instance, Charles Lyell (1833), sometimes referred to as the father of modern geology, had used the fossil record to build a chronology of geological eras (Rudwick 1978; see also Lyman and O'Brien 2000).

In short, the remains of ancient animals have a deep history of analytical use in the earth sciences, including archaeology, but using them to decipher ancient environmental conditions did not really emerge in any consistent or formalized way until, as we noted above, the middle of the twentieth century (Grayson 1981; Lundelius 1998; Semken 1983). Nearly four decades ago Grayson (1981:28) noted that critical examinations of "the principles and processes of paleoenvironmental reconstruction using archaeological vertebrates are quite rare." North American paleontologists later provided some critical discussion (Churcher and Wilson 1990; Graham and Mead 1987; Graham and Semken 1987; Harris 1963; Semken and Graham 1987), but we note that those discussions are seldom if ever cited by archaeologists interested in paleoenvironments. The still very useful discussion produced by Australian paleontologists had appeared a few years before Grayson's analysis (Walker and Guppy 1978). Paleozoologist Peter Andrews' (1995, 1996) excellent summaries written from an Old World perspective did not appear until the end of the twentieth century. The Australian volume seems not to be well known, but the latter items do receive some attention (in the form of citations), likely as a result of Andrews' (1990) landmark book *Owls, Caves and Fossils*. That volume is a *tour de force* review of how various species of raptor accumulate and modify remains of their animal prey, how those raptors might skew the paleoenvironmental signal of an ancient fauna, and how to contend with such things when analyzing an ancient fauna.

As indicated above, part of the reason for the expansion of interest in paleoenvironments is that it is those environments that served as the context of evolution of today's biotas. Therefore, knowledge of paleoenvironments as the evolutionary context, especially that of human biological and cultural evolution, has come to be seen as mandatory to writing and understanding evolutionary history (e.g., Behrensmeyer 2006; Kingston 2007). Archaeological and anthropological concerns over paleoenvironments grew as interest in the driving forces of hominin evolution expanded coincident with the discovery of more and more fossils of human ancestors (e.g., compare select chapters in Coppens et al. [1976] with the entire volume of Bobe et al. [2007]). Determination of what those ancient environments were like has become commonplace, particularly in the Old World where the majority of human biological and cultural evolution took place. Several excellent texts are available that cover many of the kinds of data and analytical techniques that can be used to reveal aspects of paleoenvironments (e.g., Bradley 1985, 2015; Dodd and Stanton 1990; Lowe and Walker 1997, 2015). Barring Dodd and Stanton (1990), who deal primarily with marine invertebrates, there is, from our admittedly biased viewpoint, surprisingly little discussion in those volumes on how ancient (terrestrial) faunal remains might be studied and analyzed, despite the fact that most researchers realize that paleozoological specimens are sometimes all that is available and that paleozoological data can be used to supplement those other data.

VOLUME STRUCTURE

A major purpose of this volume is to summarize many of the varied analytical techniques that have been applied to paleozoological remains for the purpose of reconstructing past environmental conditions, including climatic parameters, habitat characteristics, etc. Our discussion and analytical examples focus on mammals, as remains of other kinds of animals – birds, fish, shellfish, reptiles – tend to be less abundant in the terrestrial paleozoological record. This does not mean non-mammalian remains are not valuable when it comes to their paleoenvironmental implications. The study of shellfish is, in fact, where much modern paleoecology began (e.g., Hedgpeth and Ladd 1957). Nevertheless, our expertise tends to be with mammalian remains, so that is where the bulk of our attention is directed. Even so, much of what we say throughout the volume pertains to all taxa of organisms, plant or animal. We do reference some of the pertinent literature with respect to insects, shellfish, herpetofauna, and birds, though perhaps not surprisingly, the literature on mammals outweighs that on those other kinds of animals, ignoring for the moment the truly ancient paleontological record wherein invertebrate faunas dominate ecosystems.

Given our intention (and hope) that this volume serve as a text, it is necessary that we outline some basic ecological and biogeographic principles that underpin paleoenvironmental reconstruction based on zoological remains. We describe these principles in Chapter 2, where we also emphasize that taking courses in *ecology* and *biogeography*, reading widely on the subjects, or both are highly recommended.

Analytical methods in virtually any field of inquiry rest on one or more assumptions. Many of these are implicit and few are typically discussed in textbooks. We have, for example, been somewhat baffled by how few of the requisite assumptions to paleoenvironmental reconstruction have been identified in the paleoecological literature, and even more baffled by how seldom the mentioned assumptions are critically evaluated (e.g., Dodd and Stanton 1981; Lowe and Walker 1997; Wing et al. 1992). Even though the assumptions underpinning paleoenvironmental reconstructions based on zooarchaeological remains (e.g., Findley 1964; Harris 1963; Lundelius 1964; Redding 1978; Yalden 2001) and on paleontological remains (e.g., Walker and Guppy 1978) have been previously outlined, none of these earlier discussions have discussed all key assumptions, nor have those assumptions been critically evaluated from a modern ecological or taphonomic perspective. We think it exceptionally important to do just that within the context of this volume. Therefore, in Chapter 3 we identify and critically discuss ten assumptions underpinning the analysis and limitations to the interpretation of taxa represented by a set of faunal remains in terms of their paleoenvironmental implications. We focus on the assumptions and limitations that result from analytical dependence on taxonomic identifications of faunal remains; such analyses typically concern biogeography, taxonomic presences and abundances, and skeletal morphometry. Assumptions necessary to paleoenvironmental reconstruction based on analytical methods not directly resting on taxonomic identifications are covered as necessary in other chapters.

In Chapter 4 we describe several previously published mammalian faunas that are used in exemplary analyses in later chapters. Although we do not always use these same faunas and often refer to others to illustrate certain things, subjecting the same faunas as often as possible to the different analytical techniques discussed in later chapters accomplishes two things. First, it ensures that detected differences in analytical results must be a function of the analytical techniques that are used. Second, use of the same faunas throughout means the reader can focus on variability in analytical techniques rather than on variability in faunas and their paleoenvironmental implications.

One of the most challenging aspects of this volume, and a major reason for wanting to put it together, was to produce a summary of the diverse analytical methods that have been developed over the years to decipher the paleoenvironmental meaning of ancient animal remains. Several excellent introductory overviews of analytical methods applicable to paleozoological remains are available (e.g., Andrews 1996; Graham and Semken 1987; Reed 2013; Reed et al. 2013). There are also texts that discuss how to analyze and interpret paleobiological remains in terms of their paleoenvironmental implications (e.g., Bradley 1985, 2015; Dodd and Stanton 1990; Lowe and Walker 1997, 2015), but these tend to focus most closely on geological data, marine invertebrates, and botanical remains such as pollen while terrestrial faunal remains receive minimal discussion. And while there are several excellent volumes that are made up of case studies of individual faunas (e.g., Bobe et al. 2007; Graham et al. 1987), a text that focuses on zoologically pertinent ecology and analytical assumptions and techniques does not exist.

We organize the myriad techniques for analyzing faunal remains in terms of their paleoenvironmental implications into what we believe is a sensible framework, even if the boundaries between them are not hard and fast. In Chapter 5 we describe some of the most basic techniques, ones that depend on the identity of the taxa that are found and their ecological and biogeographic predilections; these techniques rely on what are often called presence/absence data. Analytical techniques that focus on taxonomic abundances are the subject of Chapter 6. Techniques discussed in Chapter 7 extend into the realm of taxon-free methods, in which taxa and communities are characterized according to ecological or morphological variables. The topics we cover in that chapter include such things as community structure analysis, *ecomorphology* and *ecometrics*, and paleodietary reconstruction. Taxonomic *diversity* (*richness*, *evenness*, and *heterogeneity*) is sometimes considered a taxon-free metric, but because it poses a distinct set of analytical challenges we address it separately in Chapter 8. In Chapter 9 we address the suite of techniques – relying on taxonomic presences, abundances, and taxon-free characterizations – that are designed to provide numerical estimates of paleoenvironmental variables (e.g., temperature, precipitation). And lastly, Chapter 10 discusses the use of size *clines* and ecogeographic rules, including the well-known *Bergmann's rule*, in paleoenvironmental reconstruction.

It has become increasingly clear over the past three or four decades that the taphonomic histories of assemblages of faunal remains can variously mute or skew the paleoecological implications of those assemblages (e.g., Behrensmeyer et al. 2000, 2007a; Fernández-Jalvo et al. 2011; Lyman 1994; Soligo 2002; Soligo and Andrews 2005; Turvey and Cooper 2009). An appreciation of taphonomic processes enhances our confidence that faunal signals of potential paleoenvironmental significance are meaningful, or at least reminds us to be sufficiently cautious with our interpretations. In this volume we do not, however, spend much time trying to disentangle taphonomic processes, especially for case studies where our main goal is to demonstrate how a particular analytical technique works; to do so would result in tremendous lengthening of the discussion. But we do call attention to taphonomic issues relevant to the exemplary case studies and that are also likely to cause problems with future applications of the techniques. A topic raised several times in Chapter 3 also merits attention here: any paleoenvironment reconstructed on the basis of faunal remains (or any other data source) should be evaluated against independent data, such as palynological or geomorphic data. Means to understand and overcome the limitations of paleoecological analyses of ancient faunas have grown markedly in the past several decades, and these are mentioned throughout the volume as occasions arise.

There are increasing efforts to forecast future environments in the face of what seems to be increasingly anthropogenically (humanly) driven climate change (e.g., Blois et al. 2013; Williams et al. 2013). The record of ancient environmental and climatic changes revealed by paleozoological data can be used to test models of the future and also provide insight to, for instance, how individual species of animals as well as communities thereof may respond to those changes. As testimony to this, we note the emergence since about 1990 of what is today known as conservation paleobiology (Dietl et al. 2015), the paleontological equivalent to what archaeologists sometimes refer to as applied *zooarchaeology* (Lyman 1996). The paleozoological record, whether paleontological or zooarchaeological, comprises an archive of experimental results concerning how biota respond to environmental change (Barnosky et al. 2017; Dietl et al. 2015; Sandweiss and Kelley 2012, respectively). As such, that record is being consulted more and more often as conservation biologists and others seek to predict the future biological health of planet earth. As custodians of a large portion of the paleozoological record, archaeologists, paleoecologists, and paleobiologists have, we believe, along with a growing number of other scientists, a moral and ethical responsibility to not only protect but utilize that record for humankind's benefit (e.g., Dietl and Flessa 2009; Dietl et al. 2012; Faith 2012b; Lyman and Cannon 2004; Wolverton and Lyman 2012; Wolverton et al. 2016). By "utilization" we emphatically do not mean collection of ancient animal remains only to learn about the past. Rather we mean one should use those remains for just such purposes, but also and equally importantly, use those remains to help conservation biologists learn about what might happen in the future as humankind's influences on ecological processes and biotic and abiotic resources take on ever greater onerous implications. If, for example, anthropogenically driven global warming continues (e.g., Barnosky 2009), what might be the effects on plants and animals? Numerous instances of climatic warming occurred in the past; the end of the Pleistocene ice age about 11,700 years ago is an obvious recent example, but there are also important examples much deeper in time (e.g., Gingerich 2006). Knowing how biota responded to those shifts in climatic regimes would suggest what we could expect, and also how we might work to avoid (or adapt to) such changes if they would seem to adversely impact the ecological goods and services humans depend on for survival. We return to this issue in the final chapter (Chapter 11) of this volume where we describe some real-world examples.

The applied aspects of paleoenvironmental reconstruction are, to us as parents, extremely important. There are now numerous efforts to forecast changes in biota that may occur as a result of future (anthropogenically driven) climatic change (e.g., Hijmans and Graham 2006; Parmesan 2006; Williams and Jackson 2007; Yackulic et al. 2011). That large and growing body of literature contains several nuanced discussions of the limitations of using biological data as proxies for environmental variables that are directly pertinent to using paleozoological remains to reconstruct or "hindcast," the modelers say, paleoenvironments (e.g., Belyea 2007; Birks et al. 2010; Huntley 2012; Varela et al. 2011). That literature also includes rather nifty studies of what early urbanization can do to local faunas (e.g., Weissbrod et al. 2014); urbanization has long been perceived as a threat to *biodiversity*, and the paleozoological (particularly the zooarchaeological) record demonstrates the threat has a deep history (see also Boivin et al. 2016). Finally, we note the argument has recently been put forth that the basic subject of paleoecology should be used to facilitate teaching students and the public, too, about modern ecological topics such as global warming, the loss of biodiversity, ecological sustainability and resilience, and disruption of ecosystem services (e.g., Raper and Zander 2009). In our admittedly biased opinion, we think this is a swell idea!

WHAT WE DO NOT DO

In this volume we do not spend significant time discussing the quantification of faunal remains; that topic has been covered in detail elsewhere (e.g., Grayson 1984b; Lyman 2008b). Instead, we acknowledge that the topic is in fact contentious, but for sake of simplicity throughout the volume we presume the favored unit of quantification for measuring taxonomic abundances is not a significant issue, although in the minds of some it is (e.g., Domínguez-Rodrigo 2012; Giovas 2009; Lyman 2008b; Morin et al. 2017a, 2017b; Nikita 2014; Thomas and Mannino 2017; Turvey and Blackburn 2011). For analyses that require quantification of taxonomic abundances, we rely exclusively on the number of identified specimens (NISP) of a taxon, where a specimen is a bone, shell, or tooth or fragment thereof, or the minimum number of individuals (MNI), though others have used alternative measures such as biomass (e.g., Pokines 1998; Staff et al. 1985).

We discuss the geochemical approaches (e.g., stable isotopes) used to reconstruct paleoenvironments from faunal remains in Chapter 7, but we do not spend any time discussing the biogeochemical foundations of these techniques or the appropriate laboratory methods. To do so would require one or more additional chapters, which we feel is unnecessary given the numerous books that deal with these issues (e.g., Allègre 2008; Faure and Mensing 2004; Hoefs 2015; Sharp 2007), as well as volumes and reviews written for archaeological and paleontological audiences (e.g., Ambrose and Katzenberg 2000; Lee-Thorp 2008; Lee-Thorp and Sponheimer 2006, 2013; Pate 1994; Sandford 1993; Schoeninger 1995). Just as one should learn the basics of faunal analysis (e.g., identification and quantification – topics we do not cover here) from experienced paleozoologists, those interested in geochemistry should seek training and laboratory experience from experienced geochemists.

We also do not cover aspects of paleoecology beyond those topics relevant to the reconstruction of past environments – considered here to include things like ancient climate, floras, habitats, biomes, and the like. As a discipline, paleoecology is concerned with the study of interactions of ancient organisms with each other and with their environment. Although a major focus of paleoecological research deals with paleoenvironments, many paleoecologists are also interested in a broader range of topics, including, for example, the assembly and disassembly of biotic communities, food webs and trophic linkages, and evolutionary processes. We deal only with paleoenvironmental reconstruction here, and this is why we have entitled this book *Paleozoology and Paleoenvironments* rather than *Paleozoology and Paleoecology*.

FINAL COMMENTS

Although we have already used them above, there are several terms appearing throughout the volume that we need to define at the outset. We use the term *fossil* to denote any ancient remain of an animal, regardless of the specimen's age or fossilization (mineralization) condition. We use the term *assemblage* to denote a collection of faunal remains whose aggregation is the result of an analytical decision. That decision might be to aggregate (analytically) all fossils recovered from a depositional unit such as a geological stratum, or to aggregate all the remains of a particular taxon recovered from a particular multi-stratum site. We emphasize that the aggregation, or choosing the spatio-temporal-taxonomic boundaries of an assemblage is an analytical decision, sometimes facilitated by stratigraphic boundaries.

The remains of ancient animals – bones, teeth, shells, dermal structures – have been and are regularly recovered from two general kinds of deposits. Archaeological deposits are those that include human artifacts such as arrowheads and fragments of pottery. Faunal remains from these deposits are often referred to as zooarchaeological remains. The oldest reported zooarchaeological remains are about 3.4 million years old and are from Africa; in the Americas the oldest generally accepted (there is ongoing debate over the validity of proposed more ancient remains) zooarchaeological remains are about 14,000 years old. Paleontological deposits are those that do not include human artifacts and can be of virtually any age. The included fossils are typically referred to as faunal remains or paleontological remains. It is not always possible to determine whether a particular assemblage of faunal remains is archaeological or paleontological, in part because some faunal remains thought to have been accumulated and deposited by hominins have no associated artifacts. And in many cases, substantial portions of the faunal remains from archaeological deposits are accumulated by geological processes (e.g., fluvial action) or other taphonomic agents (e.g., raptors, carnivores). Because for our purposes the distinction is not always pertinent, we refer to any set of ancient faunal remains, zooarchaeological or paleontological, as paleozoological. Other terms will be defined when first encountered in the text. For convenience, we have compiled the specialized terminology in this book into a Glossary of key terms, each of which is italicized at first appearance.

We attempt to be objective in presenting the numerous analytical techniques that have been proposed, and not be too judgmental. We do, however, point out what we believe to be particular weaknesses of some techniques, and strengths of others. Finally, although we reference much of the pertinent literature, particularly examples of paleoenvironmental analyses of ancient faunas, the list of references we cite is in no sense complete, but we have tried to include literature from all major continents. Our combined linguistic expertise is English, so that puts a limit on what we have read while writing this volume. Faith has focused his research attention on southern and eastern Africa, and Lyman on western North America. That likely puts another constraint on the literature with which we are familiar. Something that became increasingly apparent while writing this book is that many of the analytical techniques commonly used by paleozoologists working in the Old World differ from those used by paleozoologists working in the New World, even though our goals (paleoenvironmental reconstruction) are the same. There is no good reason why these geographic traditions should remain distinct, and the divergence likely reflects the simple fact that paleozoologists working in any particular part of the world are most familiar with the literature from that part of the world. Our effort to include literature from all continents reflects our hope to facilitate cross-over between these traditions and to include literature that might be familiar to anyone who reads the volume. We look forward to hearing from colleagues who wish to identify our weaknesses, both in literature cited and in our reasoning.

TWO

FUNDAMENTALS OF ECOLOGY AND BIOGEOGRAPHY

This book is about paleoenvironmental reconstruction founded on analysis and interpretation of evidence extracted from faunal remains. Despite its multiple syllables, paleoenvironmental reconstruction is something of a shorthand term for determining or (more typically) estimating what ancient climates, floras, habitats, ecologies, and the like were, if and how they differed from the present, and how, why, and when they changed. This means knowledge of biogeographic and ecological fundamentals is required. We outline many of these basics here, and a few more in Chapter 3 that we think fit better within discussion of the various analytical assumptions that underpin paleoenvironmental reconstruction. Do not expect to learn all there is to know about these topics in this and the following chapter, but we hope you will glean sufficient information to facilitate understanding the remainder of the book.

The development of ecology as a field of inquiry influenced the development of paleoecology – of which paleoenvironmental reconstruction is an important objective – because the former was the source of several key interpretive concepts embedded within the latter (Terry 2009). It is perhaps not surprising, then, that one of the so-called founders of modern twentieth-century ecology, Eugene P. Odum (1971:159), stated the most fundamental assumption (number 2 below) required by those seeking to reconstruct past environments on the basis of faunal (or any taxonomically identified biotic) remains: "The basic assumptions of paleoecology are: (1) that the operation of ecological principles has been essentially the same throughout various geological periods and (2) that the *ecology of a fossil may be inferred from what is known about equivalent or related species now living*" (emphasis added). We begin this chapter with a few words about the history of biogeography and ecology, two intimately interrelated fields of biological inquiry, before turning to what we take to be basic ecological principles, concepts, and processes.

The historical sketch helps explain why paleoecologists approach their field of interest in the ways that they do. The fundamental concepts and theses of ecology and biogeography underpin paleoenvironmental reconstructions based on faunal remains. Our intent is not to summarize the entirety of these two fields (that cannot be done in a single volume), but rather to introduce those parts of ecological thinking that are necessary to paleoenvironmental reconstruction. We focus on those concepts, terms, and theories that inform analysis and interpretation of ancient animal remains. Given the depth and breadth of the ecological and biogeographic literature currently available, we suggest that should a particular topic we cover here strike you as requiring additional study, you begin with reading some of the recent literature regarding that topic that we cite (see also Real and Brown 1991 and F. A. Smith et al. 2014 for compilations of classic studies). Another good place to broaden knowledge of many topics is in selected articles in the *Annual Review of Ecology, Evolution and Systematics* volumes. As well, having ready to hand a text on biogeography (e.g., Lomolino et al. 2010) and basic ecology (e.g., Begon et al. 2006; Whittaker 1975) is a good idea. And for our student readers, especially those in archaeology programs where ecology and biogeography are not traditionally part of the curriculum, taking courses in these subjects will prove useful.

Something we do not do in this chapter is summarize the geographic ranges and ecological predilections of individual animal species. For many continents or countries, there are surveys of the mammals (often), birds (less often), and other vertebrates and invertebrates found there today or during the very recent past. For instance, a classic reference on historic-era biogeographic records of North American mammals is Hall (1981). Kingdon et al. (2013) provide much the same for African mammals, and similar references are available across the continents (e.g., Francis 2008; Gardner 2007; Patton et al. 2015). For more information on the distribution, ecology, and habitat preferences of some species of mammal, there are regional guides (e.g., Jones et al. 1985; Skinner and Chimimba 2005; Verts and Carraway 1998). Many of the larger, charismatic mammalian species have one or more books devoted to their biology and ecology (e.g., Byers 1997; O'Gara and Yoakum 2004). All of these reference materials will be of value to reading the paleoenvironmental signal of the species identified in a collection of faunal remains. Indeed, for many of the paleoenvironmental techniques outlined later in this volume, familiarity with the species involved is required. And even if it is not needed for some types of analysis (e.g., taxonomic diversity), it is always important to know about the species under study. We have found that the tedium (in our opinion – others claim to find the process therapeutic!) of identifying bones and teeth can be alleviated by reading about the species that have been identified. This will facilitate analysis and interpretation once all of the faunal remains making up a collection have been identified.

HISTORICAL SKETCH

Our sketch of the history of biogeography is short because it need not be long (see Lomolino et al. 2010 for a detailed overview). Perhaps the most important thing to recognize is that early biogeographers sought explanations for distributions in the species-specific ecologies of organisms, and that fed the development of ecology (Lomolino et al. 2010). Key biogeographic concepts and terms pertinent to this volume – e.g., *disjunct distributions*, *sweepstakes dispersal*, *vicariance event* – will be introduced and defined as necessary. For quick reference, see the Glossary at the end of this volume.

Biogeography

Biogeography, the study of the geographical distribution of organisms, had its beginnings in the age of exploration about 250 years ago. Early European naturalists not only collected and catalogued specimens of plants and animals previously unknown to western science, they also identified similarities and differences between biotas in different areas of the world. In the mid-eighteenth century George-Louis Leclerc, Comte de Buffon observed that isolated regions with similar environments tended to have taxonomically distinct (but not always ecologically distinct; see Chapter 7) assemblages of plants and animals; this became the first principle of biogeography and is known as *Buffon's law* (Lomolino et al. 2010; Posadas et al. 2006). Near the end of the eighteenth century, botanist Johann Reinhold Forster outlined one of the first systems of worldwide biotic regions along latitudinal gradients. This was followed in the early nineteenth century by the observation by Alexander von Humboldt (the father of phytogeography) that Forster's zones could be detected at more local *scales* along elevational gradients (Jackson 2009a, 2009b). Importantly, from the perspective of paleoenvironmental analysis, they and others noted that floras tended to be strongly associated with climate (e.g., Holdridge 1947; Peel et al. 2007).

The prevailing view of the mid-eighteenth century that both the earth and its organisms had been static (since the biblically documented divine creation) – a view championed by such luminaries as Carolus Linnaeus – began to weaken by the early nineteenth century. Georges Cuvier (the father of paleontology) demonstrated that individual species became extinct and new species appeared at different times in the past. Charles Lyell later provided evidence that the earth is both ancient and dynamic; sea levels fluctuated, mountain chains rose and fell, and the distribution of species, some of which are now extinct, changed through time. Although he was right about this, Lyell and many of his contemporaries were wrong in their belief that species were immutable and that a divine creator was responsible for the emergence of new species.

Armed with a growing appreciation of earth's antiquity, nineteenth-century naturalists later focused on trying to account for the origin, spread, and diversification of plants and animals. Charles Darwin's observations on natural history facilitated overturning the static view of biogeography as well as the immutability of species. A major contribution to biogeography and catalyst for abandonment of biogeographic stability was made by Alfred Russell Wallace (Fichman 1977; Whittaker et al. 2013). A couple of years before Darwin and Wallace proposed natural selection as a major evolutionary force, Philip Lutley Sclater suggested there were multiple areas of (divine) creation and that these could be recognized by distinct biotas; on that basis he proposed a system of biogeographic regions that is still used (if in modified form) today (Olson et al. 2001). Sclater focused on birds. Wallace, often considered the father of zoogeography because he developed many of the basic concepts and principles of the field, turned his attention to multiple groups of terrestrial animals to assemble the first global map of biogeographic regions (also still used today; Holt et al. 2013). In the context of this volume it is relevant to note that Wallace also suggested the fossil record provided evidence of past distributions and dispersals of organisms, and climatic events were a major influence on those dispersals and distributions. It was also during the nineteenth century that several well-known ecogeographic rules were proposed, including Bergmann's rule and *Allen's rule*, which we describe in Chapter 10.

A remaining source of debate among nineteenth-century naturalists was explanation of how species dispersed and later became isolated from their sources of origin. By now it had become apparent that just as species were not immutable, neither were their geographic ranges. But how did remote islands become colonized? Why are there taxonomic affinities, for instance, between southern hemisphere biotas on different continents? How could such *disjunct* (discontinuous) *distributions* be explained? Observations of island biotas by Darwin suggested to him that long-distance dispersal across biogeographic barriers, sometimes due to completely random processes (e.g., dispersal of seeds by birds), was responsible. Such random, long-distance colonization events are now called *sweepstakes dispersals* (after Simpson 1940). Others (including Lyell during part of his career) proposed the existence of massive transoceanic land bridges later broken up by geological and climatic mechanisms. While evidence for the types of land bridges envisioned by nineteenth-century naturalists never materialized, we know now that this idea is partially correct in the sense that earth's continents were once connected and that sea-level fluctuations can bridge some islands and even continents (e.g., North America and Asia across Beringia), meaning that long-distance dispersals are not always required. It is now recognized that long-distance dispersals and *vicariance events* – tectonic or climate-driven events that cause populations to become geographically fragmented – are both important biogeographic processes (Lomolino et al. 2010).

By the middle of the twentieth century, major books that attempted to synthesize biogeography were available (e.g., Darlington 1957), and syntheses of biogeographic concepts important to evolution had been outlined (e.g., Simpson 1940, 1952). Biogeography had come to focus on geographic distributions as largely ecologically controlled or historically influenced, but it was soon recognized that both kinds of variables were important to greater or lesser degrees depending on the species under consideration and local conditions (both abiotic and biotic) (Endler 1982). This is something to keep in mind in this volume – both historical and ecological factors may be important when trying to decipher the paleoenvironmental significance of ancient animal remains.

Ecology

"Why were all species not found everywhere?" was an inductively generated question based on an empirical generalization (e.g., Buffon's law). Environmental variables were obvious possible causes (e.g., Grinnell 1924, 1928), but identifying the particular one(s) that limited a species' distribution was difficult. Barriers such as mountain ranges between lowlands, wide rivers, and seas were another. Eventually, inductive pattern recognition suggested possible relationships between some variables but not others, and resulted in more focused research questions. As is often the case in natural history endeavors, familiarity with the phenomena under study eventually began to reveal patterns that seemed to have explanatory potential. Figuring out why organisms had the geographic distributions they did eventually became a, if not *the*, foundational basis for paleoenvironmental reconstruction. To demonstrate that this is in fact pertinent, let us outline an example that is well known in North America.

An early study on the distribution of organisms – both plants and animals – in North America resulted from the work of C. Hart Merriam for the US Department of Agriculture's Economic Division of Ornithology and Mammalogy (Phillips et al. 1989; Sterling 1989). Merriam was interested in determining the environmental parameters of particular areas and the organisms that occurred in those areas. Once a correspondence was found, he could recommend which commercially important plant and animal taxa could be successfully farmed in those areas. Merriam recognized patterns in the combinations of taxa that seemed to recur within particular climatic regimes (Merriam 1890, 1892, 1894, 1895, 1898). It was on the bases of these corresponding distributions that he developed his famous *life zone* concept (e.g., Daubenmire 1938; Hoffmeister 1964; Phillips et al. 1989) that dominated much early twentieth-century North American biogeographic thinking. Merriam identified what he took to be a more or less universal relationship between climate, particularly temperature, and the distributions of individual species. Although aspects of the concept of life zones as originally envisioned are now out of favor (Whitaker 1975), the concept influenced some early efforts to reconstruct paleoenvironments on the basis of faunal remains (e.g., Miller 1937; Simpson 1947; White 1954). The life zone concept is still used occasionally by North American vertebrate zooarchaeologists in such efforts (e.g., Hughes 2009; Walker 2007).

In Merriam's (e.g., 1890, 1894, 1898) view, particular species of plants and animals were found at particular altitudes on warmer (and thus drier) south-facing slopes of mountains and occupied slightly lower altitudes on cooler (and thus moister) north-facing slopes given their exposure to less direct sunlight than those on south-facing slopes (Figure 2.1). This notion was translated to geography's horizontal dimension when Merriam (1895, 1898) drafted a map of North America showing more or less latitudinally delineated life zones. Merriam (1894:233, 234) stated what he believed to be biogeographic laws: "animals and plants are restricted in northward distribution by the total quantity of heat during the season of growth and reproduction [and] are restricted in southward distribution by the mean temperature of a brief period covering the hottest part of the year." In Merriam's view, temperature minima and maxima, not seasonal or annual averages, were biogeographically and ecologically important.

At the time Merriam made his remarks, ecology was becoming a rigorous science, and it soon became apparent that numerous variables, not just temperature, influenced the distribution of plants and animals. Precipitation, substrate, vegetation, elevation, competitors, predators, and the like all had a role to play (e.g., Grinnell 1917a, 1924, 1928). The life zone concept eventually evolved into that of *biomes*, or particular combinations of kinds of plants and animals found over large and multiple areas as a result of similar climatic regimes (e.g., Clements 1936; Clements and Shelford 1939; Whittaker 1975) (Figure 2.2). Biomes and similar conceptual entities were initially thought to be immutable and to have shifted north and south as units coincident with warmer and cooler climates, respectively. The latter notion of biota as immutable entities influenced much biogeographic and paleoenvironmental reconstruction (e.g., Blair 1958; Martin 1958).

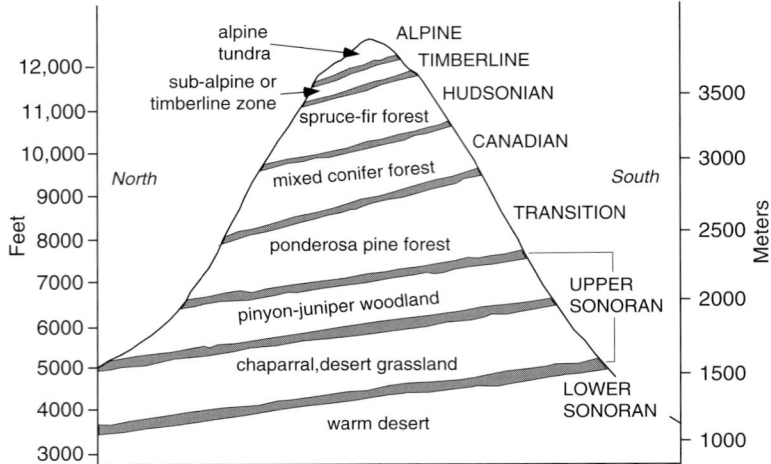

2.1. A model of Merriam's life zones superimposed on the San Francisco Peaks, Arizona. Life-zone names are capitalized. Note the shift in elevation from north-facing to south-facing slopes. Gray bands are community boundaries and are meant to signify the somewhat gradual transition from one life zone to another, or the ecotone/edge effect. Description of life zones: Alpine = alpine meadows or tundra; lichen, grass; above treeline; 33–40 inches annual precipitation; Hudsonian = spruce forest; Englemann spruce, alpine fir, bristlecone pine; 30–35 inches annual precipitation; Canadian = fir forest; Douglas fir, quaking aspen; 25–30 inches annual precipitation; Transition = open woodland; ponderosa pine; 18–26 inches annual precipitation; Upper Sonoran = desert steppe or chaparral; sagebrush, scrub oak, Colorado pinyon, Utah juniper; 8–20 inches annual precipitation; Lower Sonoran = low, hot desert; creosote bush, Joshua tree; ≤10 inches annual precipitation.

Early in the twentieth century, animal ecologist Charles Elton (1927:5) stated that "each different kind of habitat contains a characteristic set of animals. We call these animal associations, or better, animal communities [because] they are not mere assemblages of species living together, but form close-knit communities or societies comparable to our own." Elton championed the idea that communities could be viewed as an interconnected organic unit in which each animal is closely linked with other animals with which it lives and interacts. Linkages can be conceived as vertical or trophic level linkages (e.g., predator–prey interactions), or as horizontal linkages such as interspecific competition (e.g., Southwood 1987). Paleozoologists have long attempted to determine the taxonomic makeup – both taxonomic presences and abundances – of prehistoric faunal communities (e.g., Damuth 1982; Fagerstrom 1964; Hoffman 1979; Shotwell 1955, 1958). These efforts must presume that communities in some real way exist; they must exist if they are to be discovered and their makeup described. Given that defining a modern *community* today can be difficult given the different sorts of boundaries possible (Figure 2.3) and the varied geographic scales over which they might be recognized, it is not surprising that determining when a prehistoric faunal

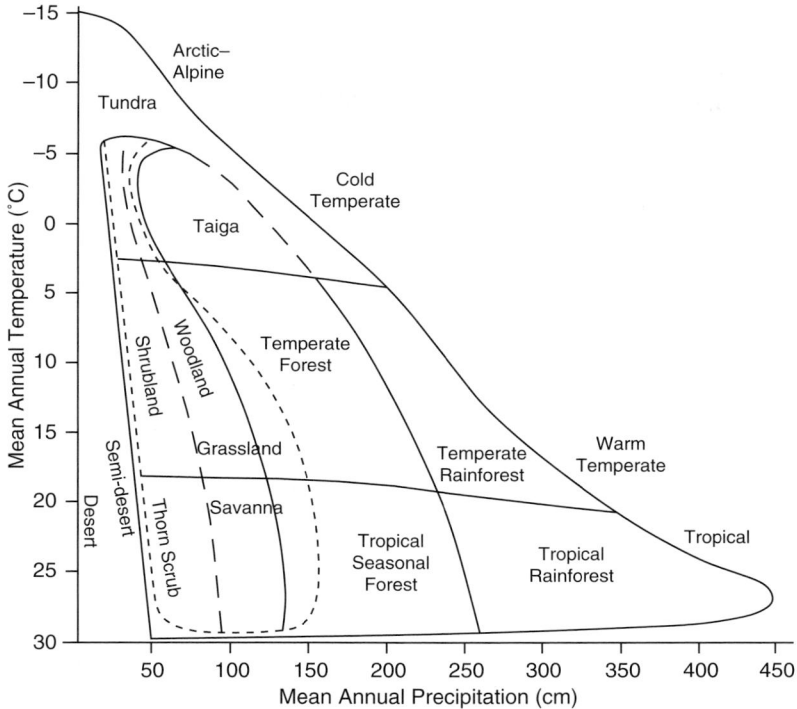

2.2. A model of world biome types in relation to mean annual precipitation and mean annual temperature. Boundaries between biome types are approximate. The fine dashed line encloses a range of environments in which either grassland or one of the woody-plant dominated biomes may be the prevailing vegetation. Redrawn after Whittaker (1975).

assemblage represents one community as opposed to a partial one or a mixture of several can be difficult.

The so-called *"cliseral shift hypothesis"* (Graham 1976, 1979), which suggests that ancient communities were immutable and organism-like, was a direct result of the reasoning of plant ecologist Frederic E. Clements (1936; Clements and Shelford 1939), who advocated the notion that biological units such as communities and biomes were immutable and organism-like because the component species of plants and animals were believed to be ecologically mutually interdependent (McIntosh 1998). Thus, boundaries between communities and biotic provinces were abrupt and clear because those boundaries were composites of the multiple range edges of their component species (Figure 2.3A). Clements also believed that climate was largely responsible for the cohesive nature of biotic communities (particularly for plants), noting that "[t]he inherent unity of [plant communities] rests upon the fact that it is not merely the response to a particular climate, but is at the same time the expression and indicator of it" (Clements 1936:254). Especially relevant to paleo-environmental reconstruction, the Clementsian viewpoint implies that ancient

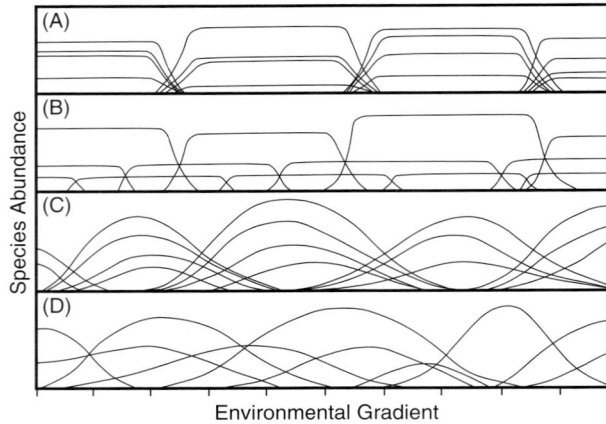

2.3. Models of variability in the abruptness of community boundaries along an environmental gradient. Each curve represents the frequency distribution of a species along the environmental gradient: (A) boundaries between communities are abrupt and self-evident; (B) boundaries between communities are diffuse and not apparent; (C) boundaries between communities are diffuse but evident; (D) boundaries between communities are diffuse and not apparent. Redrawn after Whittaker (1975:114).

communities will respond to past climate changes as intact units through synchronized range shifts of all component species (e.g., Graham 1979). We now know that this is not always so.

Clementsian ecology was largely replaced by Henry A. Gleason's "individualistic concept" during the early middle of the twentieth century (Gleason 1926; see McIntosh 1975, 1990, 1995). The Gleasonian model of communities and biotic provinces holds that each species of organism responds individually to environmental variables given that each species has its own unique *ecological tolerance* limits (see below). Moving upslope on a real mountain through Merriam's modeled life zones, one observes not abrupt ecological boundaries defined by the edges of ranges of multiple species, but rather a more or less continuous scattering of those range edges (Figure 2.3B, 2.3D). This model often (but not always) provides a more parsimonious account of so-called nonanalog communities observed in the fossil record (e.g., Graham 2005; Williams et al. 2013; see also Cole 1995). These are communities that include species living together in the past that do not have overlapping distributions today (i.e., they are *allopatric*), or that include combinations of taxonomic abundances unknown in modern communities (we discuss non-analog communities a bit more later in this chapter). Importantly, the *individualistic hypothesis* does not mean that one cannot sometimes perceive and specify variously clear to fuzzy boundaries between communities (Figure 2.3C), or that particular species never have significant interdependencies (e.g., Blois et al. 2013). Ultimately, though, the Gleasonian individualistic model is more flexible and accounts for more of the empirical record, both modern and prehistoric, than the Clementsian community-as-organism model.

Summary

Perhaps the key theme of our historical sketch is the replacement of the view that species and communities were immutable with one that recognizes a much more dynamic history. Species evolve and adapt, their geographic ranges shift, and they become extinct. And because communities are composed of species with individual ecological tolerances, the communities we observe today need not be the same as those that existed in the past or will exist in the future. Importantly, since the time of Lyell, the history of species and communities is thought to have been influenced at least in part by the *environment* and how it changed through time. These insights are precisely what make paleo-environmental reconstruction a worthwhile but also a challenging pursuit. If communities responded to environmental changes as super-organisms, then all that would be required of us is to identify which community is represented in the fossil record and reason from there what the ancient environment would have been; we now know that reality is more complex. But before we can delve into this complexity and begin reconstructing the environment from paleozoological remains, we must first turn our attention to those ecological concepts and principles that are essential to the process.

ECOLOGICAL TOLERANCES

Early in the twentieth century, ecologist Victor Shelford (1913, 1931) proposed what is today sometimes referred to as "Shelford's *law of tolerance*." Shelford (1931:462) noted that ecological "factors usually control animals in ways manifested by presence, absence and abundance. Commonly a series of factors or conditions are essential to [a species'] presence and to any given degree [its] abundance. The factors must necessarily operate in a manner suggested by *Liebig's law of minimum*." Liebig's law holds that the ecological variable necessary to an organism's occurrence that is least abundant in an area will limit the organism's abundance in that area; other necessary resources may be significantly more abundant than necessary, but the least available resource will be the limiting one. Shelford indicated that if a critical variable was too abundant, it too could limit an organism's abundance. This "law of tolerance," as Shelford referred to it, is typically modeled as shown in Figure 2.4.

Any ecological variable can be plotted on the x-axis for any species represented on the y-axis of Figure 2.4 (see Stonehouse [1997] for an introduction to climatic variables that influence animals). Typical x-axis variables include temperature, precipitation, frost-free days, biomass of food sources,

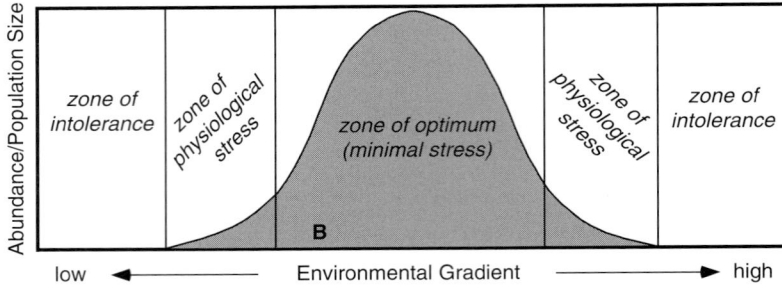

2.4. Ecological tolerance curve. Any environmental variable can be plotted on the horizontal axis. The gray shaded area represents the population frequency distribution of one species along the environmental gradient of the horizontal axis. When interpreting the presence or abundance of a species in a prehistoric assemblage, the analyst typically assumes the individuals represented fall near the middle of the tolerance curve (see Figure 2.5). See the text for discussion of "B."

frequency of competitors, frequency of predators, or any of a plethora of other variables that an organism requires or may have to contend with. The filled-in graph is referred to as an organism's *niche* (Lynch and Gabriel 1987; Pianka 1978) (see further discussion of the niche concept below). Figure 2.4 is also a model of how environmental variables influence an organism's presence/absence and abundance across an environmental gradient (e.g., cool to warm temperature). The values of an environmental variable at the extremes of the *tolerance curve* are referred to as *limiting factors* (e.g., King and Graham 1981; Raup and Stanley 1971). Whether or not values of any given environmental variable on the x-axis are known for a particular taxon is another matter. Those values are often unknown because any organism must adapt simultaneously to a plethora of (but typically not all) environmental variables, and the multitude of environmental variables that could be plotted on the horizontal axis of Figure 2.4 are not completely independent of one another (Kearney and Porter 2009; Sexton et al. 2009; Wisz et al. 2013). Paleozoologists regularly suggest that because the ecological tolerances of organisms are poorly known, the geographic distributions of species can be used to derive an approximation of those tolerances (see Chapter 3, assumption 3). Similar reasoning has permeated much of modern ecological modeling and the niche concept, a topic that we turn to later in this chapter.

Another part of the difficulty of knowing the values of any particular environmental variable for a species in Figure 2.4 concerns the fact that the variable controlling the abundance of a taxon in one area may not be the same variable that is controlling that taxon's abundance in another area. This is one of several subsidiary principles of the law of tolerance (from Odum 1971:107–108). Others are: (1) a species may have a wide range of tolerance for one environmental variable but a narrow range for another variable; (2) species with wide tolerance ranges (referred to as *eurytopic*) for most or all environmental variables tend to have larger geographic distributions than those with narrow tolerances (referred to as *stenotopic*) for many variables; (3) environmental variables may interact through the physiology of a species such that, say, if conditions of one environmental variable are suboptimal for that species, tolerances may be suboptimal with respect to other variables as well; (4) one or several environmental variables may be more important than others with respect to limiting where a species can live; and (5) "the period of reproduction is usually a critical period when environmental factors are most likely to be limiting" (Odum 1971:108).

Turning to the fossil record, paleozoologist Arthur Harris (1985:9) indicated:

> the basic task of the paleoecologist is to decipher which of the potential [environmental] parameters were operative in a particular place during a particular span of time for the organisms under study. Correct identification of the parameters governing distribution of those organisms places definite constraints on what the nature of that environment could have been; the more parameters identified, the stronger the constraints and the more exactly defined the paleoenvironment.

There are various ways to determine those parameters and we discuss them in more detail in later chapters. It suffices to note here that a relationship between a particular environmental variable, such as a gradient from low to high annual rainfall or from open to closed habitats, and the distribution or abundance of a species is usually taken as evidence that the environmental variable was the "operative" one in the past. As we explore in more detail in Chapter 3, this assumes a species has the same ecological tolerances today as it had in the past, that most fundamental assumption we identified (using Odum's [1971] words) at the beginning of this chapter.

The breadth of a species' "tolerance curve" (Lynch and Gabriel 1987:283) is the "response of a species' total fitness over an environmental gradient." Within a population of a species, each individual organism's unique genotype results in a slightly different location and breadth of tolerance curve along the environmental gradient than other individuals, recognizing that each unique phenotype has a more or less unique *adaptation* because it is a result of interaction between the environment and the genotype (Figure 2.5). The tolerance curve for a population (or all populations) of a species is, then, a composite of the particular tolerances of individuals (Lynch and Gabriel 1987:299), much like a summed probability distribution of multiple radiocarbon dates. Intraspecific variability is masked. This insight reveals a potential interpretive pitfall when it comes to paleoenvironmental reconstruction. Our reconstructions are founded on the overall tolerances of all individuals making up a species. This means we interpret the presence of a species in a prehistoric assemblage as indicating environmental conditions near the middle (or perhaps the minima, or maxima) of the tolerance curve, even though the individuals represented

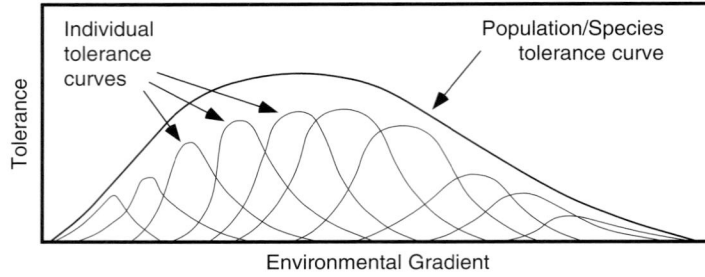

2.5. A model of how tolerances of individual organisms of a species produce a fundamental niche or a species-level tolerance curve, here labeled "Population/Species" (see Figure 2.4). When interpreting the presence or abundance of a species in a prehistoric assemblage, the analyst typically assumes the individuals represented fall near the middle of the population's tolerance curve. This model reveals that those individuals might fall near the ends of the curve.

in the assemblage of paleozoological remains could be ones with tolerances near the ends of the curve (Walker 1978). Paleozoologist Elaine Anderson (1968:52) provided an explicit comment on this potentiality when she noted that we presume each individual of a species provides an indication of that species' "typical" ecological tolerances, but then goes on to caution that the represented individuals "may have been [ecologically] marginal and atypical." Fortunately, she also indicates a way around this problem:

> When a number of species having the same habitat preferences are found together in a deposit, it is likely that this represents the habitat found at or near the deposit. The chances of an atypical habitat decrease as the number of forms having similar habitat requirements increases. It is unlikely that a large segment of the fauna would be living atypically.
>
> (Anderson 1968:52)

We will have several occasions in the remainder of our discussion to call upon multiple as opposed to single taxa.

The model of a species' ecological tolerances presented in Figure 2.4 suggests that if we knew precisely what the environmental or climatic envelope (defined by tolerance limits) was for each species, having a list of species in an assemblage (and perhaps their abundances) would readily reveal rather clearly the paleoenvironment in which the represented fauna lived. This idea is central to some of the methods we discuss in Chapter 5, including the *mutual climate range* and *area of sympatry* techniques. An approximation of a multidimensional *fundamental niche* (see below) might be modeled like the two-dimensional model in Figure 2.6. But precise knowledge of the values to put on the two environmental gradients and how those values influence one another and the resulting shape of the population density curve typically is unavailable. The result is that paleozoologists often discuss the environmental preferences of different species

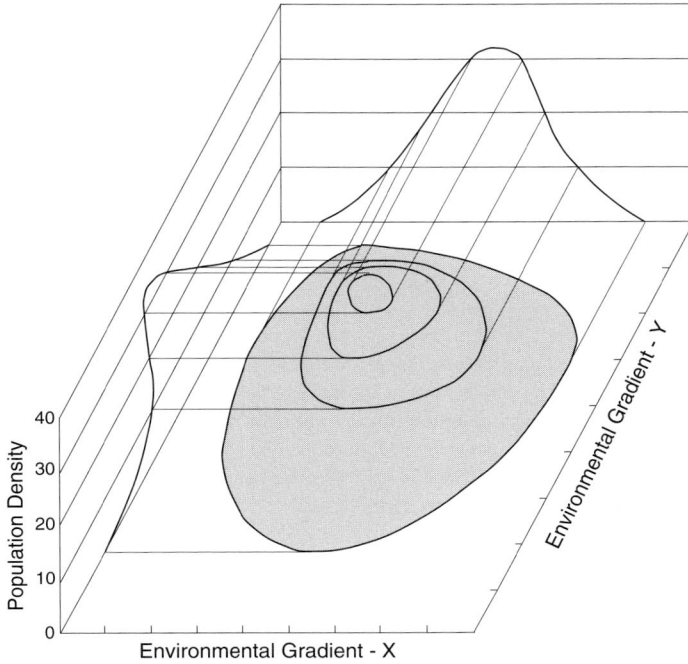

2.6. A model of a two-dimensional or two environmental gradient niche for a species population. The population responds to the environmental gradients by decreasing in density in all directions away from the population peak. A Hutchinsonian niche is a multidimensional hypervolume of environmental gradients. Compare with Figure 2.4. Redrawn after Whittaker (1975:123).

in relative terms: species A prefers mesic habitats relative to the more arid habitats preferred by species B.

To help visualize what we mean by relative environmental differences and similarities of species, consider the model in Figure 2.7. We illustrate the model with a moisture gradient, but as in other models of ecological tolerances we describe in this volume (Figures 2.4 and 2.6), any environmental parameter could be used, say temperature or biome type (Figure 2.2). We have included both stenotopic and eurytopic species in the model. As graphed, species 3 and 4 prefer wet habitats whereas species 5, 7, and 8 prefer dry habitats. What about species 1 and 2? Well, relative to species 3 and 4, species 1 and 2 are indicative of xeric environments, but relative to species 5, 7, and 8, species 1 and 2 would be interpreted as indicating mesic habitats. In later chapters the perceptive reader will note that a particular species of mouse found in North America, the western harvest mouse (*Reithrodontomys megalotis*), is interpreted by Grayson (2000b) to represent mesic habitats at one site, but Semken (1980) interprets that same mouse to be indicative of xeric habitats at another. Our point is not that these researchers disagree about the ecological tolerances of western harvest mice; they likely agree on those tolerances. Rather, our point is

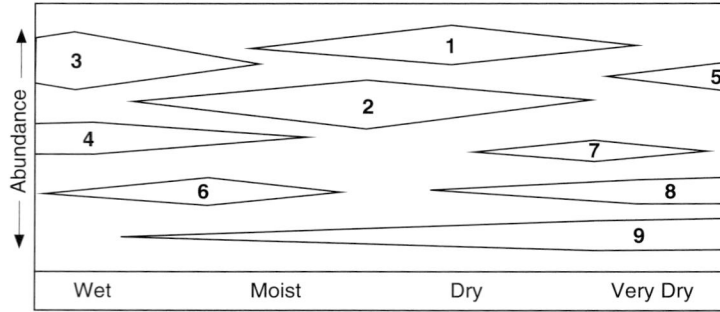

2.7. A model of presences and abundances of nine species along an environmental gradient of moisture illustrating that ecological tolerances are relative between species. Each numbered diamond plots the presence of a species along the environmental gradient (horizontal); diamond thickness represents the abundance of the species (vertical). Species 1 and 2 prefer xeric habitats relative to species 3 and 4, but species 1 and 2 prefer mesic habitats relative to species 5, 7, and 8. Species 7 is the most stenotopic and species 9 is the most eurytopic on this environmental gradient. Modified after Whittaker (1975:132).

that Grayson and Semken would agree with one another as to the paleoenvironmental implications of the western harvest mouse, given the other species with which it is associated in each of the faunas. These two paleozoologists are inferring paleoenvironmental conditions in relative terms – what one (or more) species suggests about past environments relative to another species (or two). For both analysts, western harvest mouse is species 1 in Figure 2.7; the difference is that Grayson is comparing western harvest mouse with species 5, 7, and 8, whereas Semken is comparing it with species 3 and 4. That is what we mean by *relative tolerances*.

ENVIRONMENTS AND NICHES

The *environment* is "the sum total of all physical and biological factors impinging on a particular organismic unit" (Pianka 1978:2). Physical factors include such abiotic things as weather, climate, geology, solar radiation, and the like; biotic factors include the plants and animals present. The environment is hyperdimensional; it consists of all those variables external to the organism. It is complex and varies along both spatial (size) and temporal scales (we discuss *scale* in more detail later in this chapter). With respect to paleoenvironmental reconstruction, the term environment is usually restricted to the natural environment – climate, weather, flora, fauna, topography, geography, geology, etc., and their types, frequencies, amplitudes, distributions in geographic space and time (or structure), and the like. The paleoenvironment is the natural context of the organism or community of organisms under study. Given how the term is used by paleozoologists who reconstruct paleoenvironments, a (paleo)*habitat* is a description of a kind of a spatio-temporally bounded place where an organism lived; vegetation is usually the emphasized environmental variable.

Ecologists and biogeographers today generally agree that three categories of variables influence the distribution of a species. These are (1) the distribution of biotic factors (B) that allow a species to exist, such as the availability of forage and the number and type of competitors, predators, parasites, pathogens, and so on; (2) the abiotic factors (A), such as climate and elevation, to which the species must adapt; and (3) the mobility (M) or dispersal capabilities of the species, including its vagility (Soberón 2007; Soberón and Peterson 2005). Together these categories are sometimes referred to as comprising the BAM model (Soberón and Peterson 2005). The included variables are important ones to keep in mind because the variables most often targeted in paleoenvironmental reconstruction (e.g., climate, vegetation) need not be the only ones that influence where a species is found. The total variables are also important because they act in concert to determine what we perceive as the ecological niche of a species.

A few years after Merriam developed his life zone concept and applied it in North America (Figure 2.1), ecologist Joseph Grinnell (e.g., 1914, 1917a, 1917b) suggested a species' niche was determined by the habitat in which it lived and included its behavioral adaptations to that habitat. Ecologist Charles S. Elton (1927) adopted a more functional view of the interaction of a species with its environment. An Eltonian niche concerns both a species' response to and its influence on the environment in which it lives. Ecologist G. Evelyn Hutchinson (1957, 1978) discussed variables comprising ecological niches and the relation of those variables to areas of distribution of species (e.g., Colwell and Rangel 2009; Soberón 2007). The Hutchinsonian niche is a multidimensional hypervolume of environmental conditions defining the requirements for a species to exist. Figure 2.6 illustrates a two-dimensional Hutchinsonian niche of a species.

The Hutchinsonian niche has now been refined to include two kinds. A species' *fundamental niche* concerns the particular combination of environmental variables in which that species can occur; the species' *realized niche* incorporates the effects of factors that preclude that species' occurrence such that it only occurs in a portion of its fundamental niche (Graham 2005; Jackson and Overpeck 2000; Williams and Jackson 2007). The former concerns where a species *could* occur whereas the latter concerns where a species *actually does* occur, thus a species' realized niche tends to be geographically smaller than its fundamental niche (Soberón and Arroyo-Peña 2017). Factors that preclude a species' presence in an area include competitors, predators, and biogeographic barriers. The distinction between fundamental and realized niches is important to paleoenvironmental reconstruction because what we perceive to be the ecology of a species is based on its realized niche. To infer paleoenvironmental conditions we must assume that the species' realized niche is a robust reflection of that species' fundamental niche.

The relationship of a species' realized and fundamental niches has become increasingly important as efforts to model niches for purposes of conservation biology have become more sophisticated. Greater computing power combined with modeling expertise and growing concerns over accurately predicting how future anthropogenically driven climatic change may influence biota (Dawson et al. 2011) has prompted numerous researchers to construct *ecological niche models* (Morin and Lechowicz 2008; Qiao et al. 2015; Soberón and Nakamura 2009), *species distribution models* (Austin 2007; Elith and Leathwick 2009), and *bioclimatic envelope models* (Araújo and Peterson 2012; Polly and Eronen 2011). We will use the generic term *biological distribution models* to refer to all three kinds of model. Coarsely put (Peterson and Soberón 2012), biological distribution models incorporate as many environmental variables as possible into a species' ecological tolerance model (Figures 2.4 and 2.6) in order to predict where that species might survive on a changing environmental landscape (e.g., Mota-Vargas and Rojas-Soto 2016). The result is that we are learning more about fundamental and realized niches of various species and how the two differ, and that increased knowledge will ultimately benefit those interested in paleoenvironmental reconstruction, paleobiogeography, and paleoecology in general (Nogués-Bravo 2009; Varela et al. 2011).

Biological distribution models are now occasionally tested with paleobiological data for which environmental parameters are known (e.g., Dietl et al. 2015; McGuire and Davis 2014; Rowe and Terry 2014; Terry et al. 2011). Much of this work has thus far involved establishing correlations and stopped short of identifying causal linkages between a species' presence/absence (taxonomic abundances are seldom considered) and an environmental variable (Belyea 2007; Hampe 2004; Kearney and Porter 2009; Qiao et al. 2015; Sexton et al. 2009). This seems less an issue in paleozoology where causal linkages have been postulated and tested (see Chapter 9).

It has become increasingly clear since Hutchinson's (1957) influential conceptualization that a species' niche may be static over long temporal periods or it may evolve (see review and references in Holt 2009). This makes modeling a modern niche difficult, particularly if that niche model is to be used to reconstruct paleoenvironments. Another challenge is that in most (if not all) cases, the fundamental niche of a species extends beyond the limits of environmental parameters observed in the present (Figure 2.8). If novel environments were to arise (e.g., altered seasonality due to *insolation* change), then the realized niche of a species will differ relative to what we observe today (e.g.,

2.8. A model of species' niches, non-analog faunas, and species assemblage change over time as a result of environmental change. Two different environments (bold ovals) exist at two different times. The fundamental niches (ecological tolerances) of three species (non-bold ovals) are shown and do not change from time 1 to time 2. Co-occurrences of species can occur when their fundamental niches overlap with one another and with the environment at a particular time. Thus, species 1 and 3 co-occur at time 1 without species 2, and species 2 and 3 co-occur at time 2 without species 1. Species 1 and 2 cannot co-occur because their fundamental niches do not overlap. Modified after Williams and Jackson (2007:476).

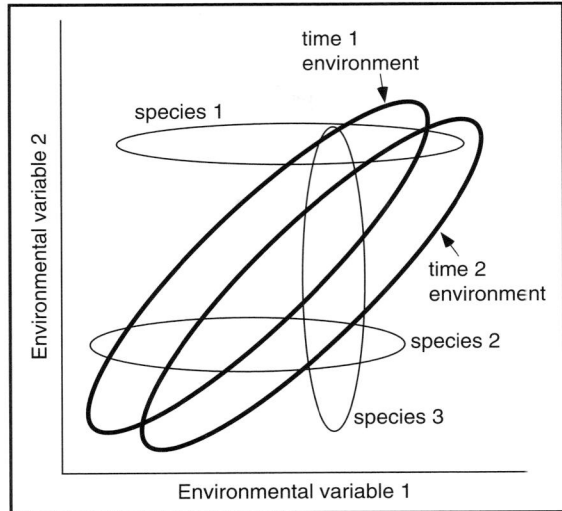

Williams and Jackson 2007). As indicated in Figure 2.8, this process can give rise to what Quaternary paleozoologists in particular refer to as *disharmonious* or *non-analog faunas* (Graham 2005), which occurred with some regularity in the past. These faunas are made up of species that are stratigraphically and temporally associated (e.g., Semken et al. 2010) and are inferred to represent biological communities of species that were in the past biogeographically *sympatric* or coexisted but which today are *allopatric* or do not co-occur in an area or habitat. The association of "tropical" and "temperate" species was the subject of much discussion and debate among nineteenth-century paleontologists in Europe (see discussion in Chapter 5), especially when it became increasingly apparent that they did not represent stratigraphically mixed or time-averaged remains from different climatic eras.

Zeuner (1936) considered the implications of the apparent association of temperate and tropical species in the early twentieth century. Early modern work on the same issue was also provided by paleozoologist Claude Hibbard's (1960) study of late Cenozoic North American faunas. Echoing the arguments of previous generations (e.g., Lartet 1867, 1875), multiple records of such non-analog faunas in North America lead to the hypothesis that past climates were at least sometimes more seasonally equable than modern climates (e.g., Graham and Mead 1987). Summer temperatures were cooler, allowing northern species to occupy more southern latitudes than at present, and winter temperatures were warmer, allowing southern species to live north of their modern range. Such reasoning is not inconsistent with Merriam's views on the role of temperature in mediating species' ranges, though it is worth noting North America's *no-analog* plant communities have been interpreted as reflecting increased seasonality relative to the present (e.g., Williams and Jackson 2007).

Non-analog is the preferred label rather than disharmonious because the fact that the species occurred together suggests there was no ecological disharmony (Semken 1988); no-analog has been used to label floras without modern analogs and is conceived differently than non-analog faunas (Graham 2005) because the former are typically recognized by atypical taxonomic abundances as opposed to associations of now-allopatric taxa. As noted above, Gleason's (1926) *individualistic hypothesis* is preferred today and explains why non-analog communities are found in the past. In short, each individual species has unique ecological tolerances and fundamental niches, is ecologically independent (more or less) of all other species with which it might come in contact and interact, and disperses biogeographically independent of other species. Thus it is not surprising that unique climatic regimes in the past would produce unique faunal communities. This important aspect of non-analog faunas (and novel ecosystems; see below) is illustrated in Figure 2.8, which shows how species may exist under environmental conditions that do not exist today but did exist in the past and may exist in the future (Jackson and Overpeck 2000; Williams and Jackson 2007). This of course renders modeling species-specific bioclimatic envelopes, niches, and distributions tenuous when only modern data for a species are used. However, as our knowledge of critical variables making up a species' niche increases, models should become more accurate and thus of value to those seeking to reconstruct past environments.

Interestingly, without explicit recognition that they were labeling a concept similar to non-analog faunas, botanists proposed the term *novel ecosystem* (Chapin and Starfield 1997) when considering the effects of modern, anthropogenically driven global warming on arctic plant communities. Paleobotanists had acknowledged that Pleistocene plant communities were often novel in species composition relative to modern communities (e.g., M. B. Davis 1976, 1981). Conservationists quickly picked up on the concept of novel ecosystems (e.g., Hobbs et al. 2009, 2013; Williams and Jackson 2007), but have seldom acknowledged the deep roots of these ideas (e.g., Lartet 1867, 1875).

Acknowledgment of humankind's growing impacts on ecosystems has prompted the emergence of what might be thought of as a new conservation biology. Whereas conservationists traditionally sought to recreate past floras, faunas, and ecosystems (e.g., pre-industrial), today they acknowledge such recreations are increasingly impossible (Hobbs et al. 2014). Therefore, the anticipated appearance of new or "novel" ecosystems has become the focus of conservation efforts in the hope that ecosystem services (e.g., carbon sequestration, sufficient fresh water) requisite to survival of humans (particularly, but obviously not solely) can, through informed management, be maintained (e.g.,

Hobbs et al. 2009, 2013, 2014). Fortunately, paleobiologists who have argued for an applied (conservation-oriented) paleozoology have picked up on this new thinking (e.g., Barnosky et al. 2017). For instance, the concepts of non-analog faunas and novel ecosystems are sufficiently similar that we anticipate we can learn much about future novel ecosystems from close study of prehistoric non-analog faunas (e.g., Williams et al. 2013). A big part of that learning process will involve implementation of analytical techniques we describe in later chapters of this book.

ECOLOGICAL SUCCESSION AND CLIMAX

"A succession is an ecocline in time" (Whittaker 1975:171). Communities do not just instantaneously come into existence from nothing. They grow and develop. If we start with a bare piece of real estate, such as the biotically barren landscape around a recently erupted volcano, early colonizing weedy plant species show up first, then a few animals, then later-colonizing plants and animals, and so on. As the plants and animals modify the landscape, still other plants and animals immigrate, perhaps simply joining the community, per-haps replacing some members thereof (see Wood and del Moral [1987] for a real-world example). These changes in or development of the community are called a *succession*. Processes of succession can be external to the community (e.g., addition of sediment) or internal (e.g., early colonizing plants decom-posing and adding nutrients to the sediment, allowing colonization by other plant species). Successions can be observed, given sufficient time (e.g., Wood and del Moral 1987). Eventually, a community is believed to reach maturity and stop developing, at which time it is referred to as a *climax*. This idea was encapsulated by Clements (1936:256), who remarked that "stabilization is the universal tendency of all vegetation under the ruling climate." The *climax com-munity* is supposed to be developmentally static, but because organisms appear, grow, and die, and storms, fires, and erosion do not simply no longer happen, stability is unlikely even over short-term ecological timescales (Whittaker 1975). Stasis may seem to occur at some large scale – a stand of trees is in a location for decades – but that does not mean things are not happening at finer scales (Delcourt et al. 1983).

Without belaboring this part of the discussion, our central point here is that successional changes are not necessarily environmentally driven in the sense of being a response to (usually climate-driven) environmental change that is of interest to us. Rather, presuming sufficient temporal *resolution* in the paleozoological record, we do not want to confuse ecological succession with other types of (extrinsic) paleoenvironmental change. This may be a more pernicious problem than one might initially suspect. Ecologists now realize that communities, habitats, and ecosystems are undergoing constant change and are seldom static at all scales for very long. Thus, ecological succession and perturbations to particular successions are incessant, if varying in magnitude over time. The *time averaging* that much of the paleozoological record reflects (see Chapter 3, assumption 9) might mute the ecological signal of a succession, leaving signals of paleoenvironmental shifts to be monitored. It will, of course, be an analytical issue to determine whether a shift in paleofaunas was the result of succession or extrinsic (e.g., climatically driven) environmental change.

WHAT IS A SPECIES TO DO WHEN THE ENVIRONMENT CHANGES?

What happens when the environment changes, say, winter temperatures get colder than the temperature minimum to which an organism is adapted? The taxon represented by the bell-shaped curve in Figure 2.4 (see also Figure 2.6) has at least three options (Gauthreaux 1980; Gienapp et al. 2008; Lister 1997; Vrba 1985, 1992). It can relocate, that is, geographically track (relocate to) the environmental conditions to which it is adapted; this is sometimes called *habitat tracking*. Alternatively, the local population can die out in the area where the temperature has dropped (extirpation). Or the local population can adapt, that is, the population may evolve as a result of natural selection, or individuals may phenotypically adjust to those new conditions and thus the geographic range of the species does not shift (see the Adaptation or Adjustment section). These potential responses need not be mutually exclusive (e.g., M. B. Davis and Shaw 2001; M. B. Davis et al. 2005).

Relocating to a new geographic area does not require genotypic or phenotypic change as long as the newly occupied environment is sufficiently similar to the old conditions of the abandoned area. Of course, organisms may relocate without prompting (ecologists today say "forcing") by environmental change, perhaps even owing to purely stochastic processes. As with a species' range coming to include new areas, loss of old areas resulting from extirpation of a local population may happen as a result of myriad processes other than climatic change (e.g., drought, fire, disease). These possibilities exacerbate the difficulties of assigning high-resolution values to the x-axis of the tolerance curve for a particular species (Figure 2.4).

If winter temperature minima decrease a degree or two in a particular area, we have essentially relocated the position on the tolerance graph we are considering. Consider point "B" on the x-axis of Figure 2.4. Point "B" represents the drop in winter temperature from the optimal value corresponding to the peak in the population curve. Notice that the upper edge of the bell-shaped curve in this horizontal position is low, indicating fewer individuals because that lower temperature is not the "optimal" one. The implication is that populations of a species can respond to environmental change by altering the frequency of individuals, increasing when environments get better (i.e., carrying capacity is higher), decreasing when environments become more stressful and less optimal (i.e., carrying capacity is lower). Yet, as noted, organisms can increase or decrease in abundance for unclear reasons, perhaps even for non-"environmental" reasons.

The three possible responses to environmental change – move, adapt, or die (coarsely put) – indicate a species may survive by tracking favorable environmental conditions. It need not adapt. The retention of ancestral ecological traits is referred to as *niche conservatism*, a phenonmenon that underpins the key assumption to paleoenvironmental reconstruction, a topic we explore in some detail in Chapter 3 (assumption 1). It suffices here to point out that the key assumption – the one listed as number 2 by Odum (1971) and described in the introduction to this chapter – states that a species represented by fossils is assumed to have had the same ecological tolerances in the past that it has today. If we could not assume, or preferably demonstrate, the validity of this assumption, paleoenvironmental reconstruction would not be possible.

TOP-DOWN OR BOTTOM-UP ECOLOGY

During much of the early history of ecology the preference was to conceive of the composition and structure of an animal community as strongly influenced by available resources, particularly vegetation (e.g., Elton 1927). The abundance or availability of resources at the lowest level or tier of the *trophic pyramid* (Figure 2.9) was believed to influence abundances (and kinds) of organisms at higher levels (Leroux and Loreau 2015). This is today referred to as a *bottom-up model* (Hunter and Price 1992). The preference for bottom-up ecological models began to change in the second half of the twentieth century when it was suggested that predators (e.g., carnivorous mammals) could, by, say, their abundances, influence the abundance of herbivores, which in turn influenced the abundance and composition of vegetation (e.g., Estes 1996; Hairston et al. 1960). The reintroduction of wolves to Yellowstone National Park is an excellent example of this (D. W. Smith et al. 2003). These *top-down models* resulted in the notions of *keystone species* (e.g., Simberloff 1998) and *trophic cascade effects*. The ultimate keystone species today is humans as reflected in the concept of the Anthropocene (e.g., Ruddiman 2013); modern manifestations of cascade effects in archaeology are found in both niche construction theory (Laland and O'Brien 2011; Odling-Smee et al. 2003) and models of prey depression used by zooarchaeologists (e.g., Broughton and Cannon 2010; Lupo 2007). Recent ecological research indicates how both bottom-up and top-down processes typically operate, and at times one, the other, or neither is dominant (Hanley and La Pierre 2015).

Sun

energy

energy flow

Top Carnivores/ Tertiary Consumers

Decomposers

(Small) Carnivores/ Secondary Consumers

recycled nutrients

bottom-up processes

top-down processes

Herbivores/Primary Consumers

Photosynthesis/Primary Producers

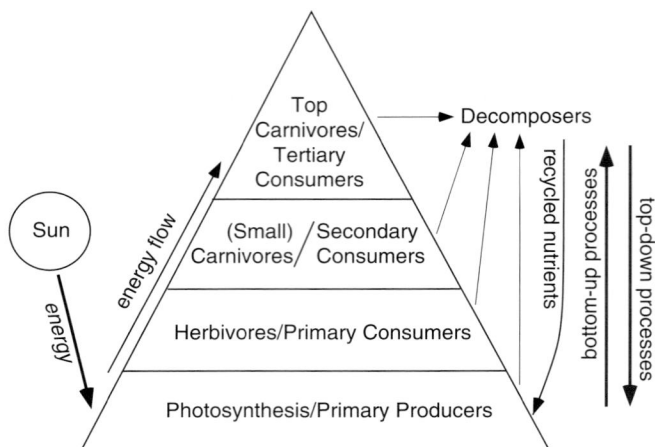

2.9. A model of the trophic pyramid including the flow of energy and recycling of nutrients through decomposition at all levels of the pyramid. Bottom-up processes work from the sun (climate) upwards in the direction of energy flow; top-down processes involve changes at high levels (e.g., an increase in tertiary consumers) such that lower levels are influenced.

What does the distinction of bottom-up and top-down ecological processes have to do with paleoenvironmental reconstruction? We think there are two reasons to be aware of the distinction. First, much of the literature concerning paleoenvironmental reconstruction suggests that plant remains are a better source of information on ancient environments than animal remains because the distributions and abundances of animals are more strongly influenced by their habitats or the flora in which they live than by climatic variables such as temperature or precipitation (e.g., Andrews 1996; Avery 1982, 1990, 2007; Bökönyi 1982; Chaline 1977; Churcher and Wilson 1990; Flannery 1967; Grayson 1977; Huntley 2012; King and Graham 1981; Reed et al. 2013; Tchernov 1975; White 2008; Woodcock 1992). This is not a new idea. Clements(1936), for instance, remarked that "plants constitute the fixed matrix of the biome in direct connection with the climate, while the animals bear a dual relation, to plants as well as to climate." Climate change influences types and abundances of vegetation that in turn influence types and abundances of herbivores that feed on it (and on up the trophic pyramid; see below). The appropriate analytical protocol should, it is therefore reasoned, involve reconstruction of past vegetation on the basis of the fauna, and then reconstruct past climate based on the (reconstructed) flora. This recommendation would seem to be at least partially a reflection of the bottom-up model of ecosystem structuring.

At least some species of animals are sometimes strongly influenced directly by climate (e.g., Birch 1957; Grayson and Delpech 2005; Heisler et al. 2014 [and references therein]; Polly and Eronen 2011). Nevertheless, we find some of the major early luminaries who reconstructed paleoenvironments based on animal remains stating unequivocally that "Climate can affect individual mammals directly by lethal extremes of temperature or drought, but its effects on species or genera of mammals through geologic time are probably indirect, acting through the soils and vegetation of the habitat" (Hibbard et al. 1965:515). Here, Hibbard et al. cite biologist William H. Burt (1958), at the time a colleague of Hibbard's at the University of Michigan. Burt (1958:141) had stated that he was "of the opinion that climatic conditions, as regards the movements of mammals over long periods of time, are important in an indirect way, as they affect vegetation and soils, rather than in a direct one … Temperature tolerance in mammals is primarily physiological, and most of them can withstand great changes if supplied with food." He made this statement as a counter to paleontologist George G. Simpson's (1947:685) suggestion that climate, particularly temperature (perhaps following Merriam's notion of life zones), is "not the only, but probably the most important, selective factor" influencing faunal interchanges between Eurasia and North America. The common conception in the middle twentieth century was that influences predominantly flowed from climate to vegetation to fauna, a bottom-up perspective (Figure 2.9). To reconstruct past climates was thus a two-step analytical process: reason from faunal remains (against the flow of ecological influence) to vegetation, and then, in step two, reason from the vegetation (again against the flow of bottom-up ecological influence) to the climate. This model is still generally followed by researchers who attempt to build paleohabitat or paleoclimate models (e.g., Avery 1990; Reed et al. 2013), but it has also been shown that the middle step may not be necessary (e.g., Polly and Eronen 2011). One can sometimes reason from animals directly to climate once a climatic envelope for each species is determined, although we still have to worry about the extent to which such envelopes reflect the role of climate in mediating vegetation structure (see Chapter 9).

The second reason to be aware of the distinction between bottom-up and top-down ecological processes is that the latter include anthropogenic processes, both modern industrial ones and prehistoric ones. The significance of this observation is that those interested in reconstructing paleoenvironments must not confuse top-down (e.g., anthropogenic) driven change from bottom-up (e.g., climatic) driven change in paleofaunas. Zooarchaeologists long ago recognized this potentiality when they worried about the "*cultural filter*" (Reed 1963; Reed and Braidwood 1960), a concept labeling the process thought by many to jeopardize if not invalidate attempts to reconstruct paleoenvironments from zooarchaeological remains (e.g., Daly 1969; Holbrook 1975, 1982a, 1982b; Reed 1963). More recently, Semken and Graham (1987:476) astutely observed that the "cultural filter" is indeed a potential problem, but it is no more of a problem than any taphonomic process that results in a collection of faunal remains that, because of biased accumulation or preservation or both, provides an inaccurate environmental signal (see also Semken 1983). Despite the fact that we are much more taphonomically sophisticated today than a few decades ago, the decipherment of which changes in prehistoric faunas reflect environmental (bottom-up) changes and which represent shifts in human behaviors or other top-down processes (e.g., appearance or disappearance of a top carnivore) remains a significant but not always insurmountable challenge (Gandiwa 2013; Parmesan 2006; Pierce et al. 2012). We mention some of the ways to analytically contend with the cultural filter and other top-down processes that may skew the paleoenvironmental signal of a prehistoric fauna elsewhere in this volume.

Resolution and Scale

What resolution might we expect of our paleoenvironmental reconstructions? This depends on a handful of analytical, ecological, and taphonomic factors, only some of which we have control over. Analytical factors that we can control include the methods we employ and how we view the ecology of the species under consideration. As we outline in Chapters 5–10, the paleoenvironmental techniques at our disposal provide varying levels of resolution ranging from biome-scale classifications (e.g., tundra, tropical forest) to finer-scale reconstructions of local habitats or climates (e.g., estimates of mean annual temperature). The level of resolution depends in part on how we (or the methods we use) view the ecological preferences of various species. If habitats occupied by species are conceived in very general terms (e.g., woodland, grassland), the resulting reconstructions will be of similar resolution (Andrews et al. 1979; Gustafson 1972; Kingston 2007; Nesbit Evans et al. 1981; Wilson 1973). For extinct or poorly known species, this may be the best that we can (usually) hope for. But if the habitat preferences of a species are conceived in more specific terms (e.g., montane grasslands with light shrub cover), then we can potentially provide reconstructions at finer resolution.

Even when we have control over analytical decisions, we are at the same time constrained by the ecology of the species in our fossil collections. Stenotopic species should allow for finer-scale reconstructions than their eurytopic counterparts because the habitat preferences of the former can be more clearly defined. Because they perceived microvertebrates to be good reflections of microenvironments, including microclimates, Churcher and Wilson (1990:70) were hopeful that "it should be possible to apply transfer functions [to faunal remains] comparable to those used by palynologists in the not-too-distant future." A *transfer function* is an established statistical relationship between paleobiotic data and (usually) climate variables such as seasonal temperature or annual precipitation (Sachs et al. 1977; Webb and Clark 1977; see Chapter 9). At the same time that Churcher and Wilson (1990) expressed optimism about the future, Avery (1990:410) was skeptical: "Attempts to derive more precise or quantitative environmental information from micromammalian data are fraught with difficulties. This is due partly to the present state of knowledge and partly to the nature of the species involved. Mammals, even micromammals, have fairly wide physical tolerance levels and are more or less adaptable" (see also Tchernov 1975; Wells 1978). Lundelius (1974:142) had noted some years earlier that most animals have "behavior patterns that enable them to avoid many environmental extremes." Walker (1978) would agree and supplement such cautions by noting the potentially mutable relationships between organisms and their environments. Still others expressed major doubt about the utility of mammalian remains for paleoenvironmental reconstruction, noting that because mammals were homeothermic, they could withstand major variability in climatic parameters, particularly temperature (e.g., van der Meulen and Daams 1992).

Despite the skepticism, Churcher and Wilson's (1990) predictions have begun to come true. Although a few early attempts were made to derive interval scale climatic reconstructions from faunal remains (e.g., Dulian 1975; Graham 1984; Hokr 1951; Roberts 1970), the construction of ecological niche models, species distribution models, and bioclimatic envelope models mentioned above seems to hold great promise for providing interval-scale resolution in paleoenvironmental reconstructions (Nogues-Bravo 2009). In fact, paleozoologists themselves have built bioclimatic envelope models (Polly and Eronen 2011) for particular species and what might be broadly construed as transfer function types of models for those species. We discuss the latter in some detail in Chapter 9.

The taphonomic issues we must contend with are those that influence the spatial and temporal scales of resolution provided by our fossil samples (e.g., Behrensmeyer et al. 2000; Kidwell and Behrensmeyer 1993). These are constraints imposed by the nature of the fossil record and by the different types of fossil evidence under consideration. For instance, some paleodietary indicators (e.g., *dental microwear*; see Chapter 7) record the diet of an individual over a period of several weeks. That individual might be part of a fossil deposit that accumulated in a geological instant (e.g., burial in a mudflow) or over decades to tens of thousands of years. Linking evidence from the fossil record to long-term environmental changes requires understanding these issues. Here, it is important to offer a couple of observations on spatial and temporal scales of resolution.

Spatial and Temporal Scale

Although interest in variability of scale of ecological patterns and processes had been around since earlier in the twentieth century, serious concern over such variability emerged with force in the 1980s (Schneider 2001; Wu 2007). At about the same time, paleoecologists began to discuss scale, particularly in terms of spatial and temporal scales, by which they meant magnitude of area and length of temporal duration (Delcourt et al. 1983; Delcourt and Delcourt 1988). Ecologists quickly came to explicitly recognize that space and time were the fundamental dimensions along which scale varied (Levin 1992; Nekola and White 1999; Sayre 2005; Wu 2007). This recognition was facilitated by the use of "space–time diagrams" in which ecological processes were plotted against geographic area and temporal duration (Schneider 2001). Ecologists also quickly adopted two aspects of ecological scales (Sayre 2005; Wu 2007). *Extent* concerns the total size of the area or total duration of time under study (e.g., 10 × 10 m, or 10 × 10 km); *grain* concerns the sampling strategy, including size and frequency of samples. Extent and grain "are analogous to the overall size of a sieve and its mesh size, respectively" (Wiens 1989:387).

Grain influences *resolution*; think of it as the DPI in a photograph. We cannot perceive patterns at a scale below the grain size because larger individual grains include more heterogeneity; enlarging extent usually also enlarges grain size (Wiens 1989). And, typically, with increased spatial scale, time scale also increases (Behrensmeyer et al. 2000; Delcourt et al. 1983; Delcourt and Delcourt 1988; Landres 1992; Wiens 1989). Many ecologists have come to acknowledge that the scale used is an analytical choice; intrinsic ecological scales may exist, but it is likely they are more or less continuous and we must choose a scale that is appropriate to the research questions and process(es) of interest (Levin 1992; Sayre 2005). To understand phenomena, we must first identify patterns within those phenomena; once we have patterns, causal mechanisms can be postulated and tested (Levin 1992). In performing such tests, we must be aware that patterns might vary in scale from their causal mechanisms (Delcourt et al. 1983). "A successful research design first defines the scale at which the phenomenon of interest can be observed, and then selects methods of analysis appropriate to resolving ecological pattern and process at that spatial-temporal scale" (Delcourt and Delcourt 1988:25). It should be obvious that different processes operate at different spatial and temporal scales. Individual organisms, for instance, interact on an hour-to-hour or day-to-day scale; community dynamics likely take longer, perhaps on a season-to-season scale.

One way to conceptualize the importance of spatial (and temporal) scale is provided by paleontologist Stephen J. Gould (1990:24): "Phenomena unfold on their own appropriate scales of space and time and may be invisible in our myopic world of dimensions assessed by comparison with human height and times metered by human life spans." With respect to biological phenomena, one can conceive several scales of inclusiveness (Figure 2.10; see Delcourt and Delcourt [1988] and Delcourt et al. [1983] for similar models regarding plants). Changes in phenomena at one scale, say in a population's membership, may or may not influence phenomena at another scale, say at the community level. Earlier in this chapter we mentioned the concept of a biological community

2.10. A model of covariation of spatial and temporal scales of biological phenomena and processes. Modified after Landres (1992).

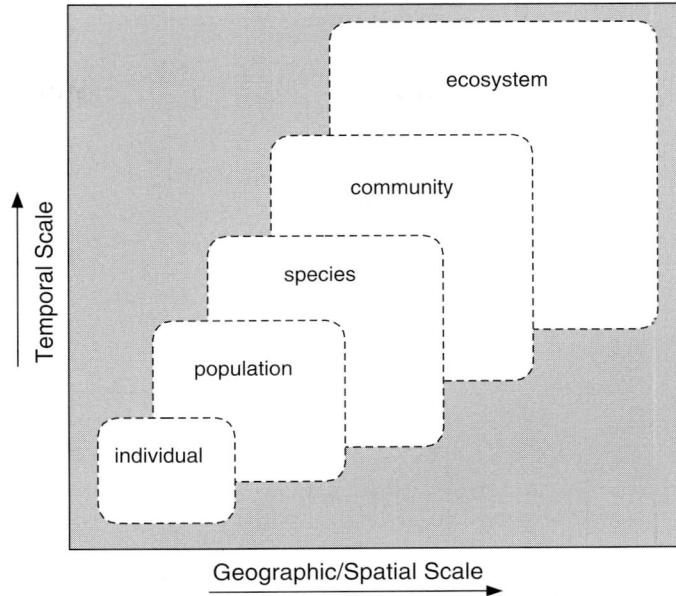

and described why they can be difficult to identify despite direct observation of modern landscapes (Figure 2.3). We also indicated it can be difficult to know when a collection of ancient faunal remains represents a partial or complete community, or several communities. A community is, nevertheless, a spatial unit, however its boundaries in space and time might be specified.

Paleoecologists have explicitly recognized the importance of scale in terrestrial and marine ecosystems (e.g., Behrensmeyer et al. 2000, 2007a; Delcourt and Delcourt 1988; Delcourt et al. 1983; Jablonski and Sepkoski 1996; Kidwell and Behrensmeyer 1993; Schoonmaker 1998). Behrensmeyer et al. (2000) plotted the approximate spatial and temporal scales of fossil samples (plants and animals) from different continental depositional environments. We provide a modified version of their figure here in Figure 2.11, emphasizing those environments that typically provide vertebrate fossil remains and adding cave depositional environments. It is useful to compare this with a similar figure provided by Clark (1985), who plotted the spatio-temporal scales of weather events such as hurricanes and small-scale climatic events such as El Niño events on a single graph. We modified that graph a bit and added the terminal Pleistocene–earliest Holocene Younger Dryas event, and the late Holocene Little Ice Age (Figure 2.12). The latter two are for illustration purposes only as we did not determine their spatio-temporal extents as rigorously as Clark (1985) did for the events he plotted. Note that the amount of time and space averaging provided by fossil samples (Figure 2.11) need not overlap with all climatic (Figure 2.12) processes potentially of interest, and though not illustrated

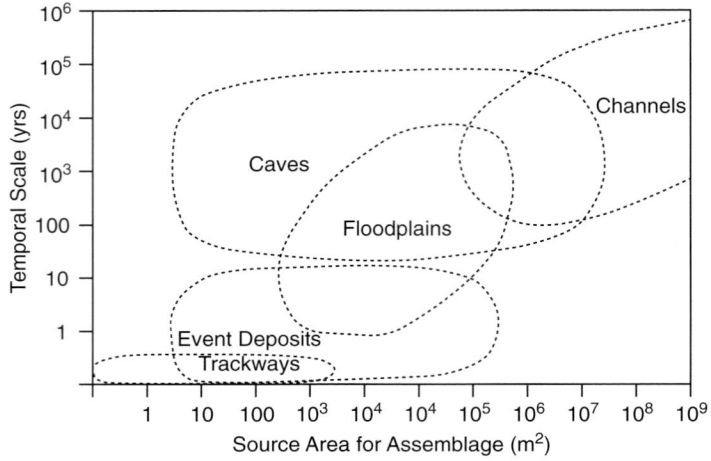

2.11. A model of the approximate spatial and temporal scales provided by typical fossil assemblages of terrestrial vertebrates from different depositional settings. Adapted from Behrensmeyer et al. (2000).

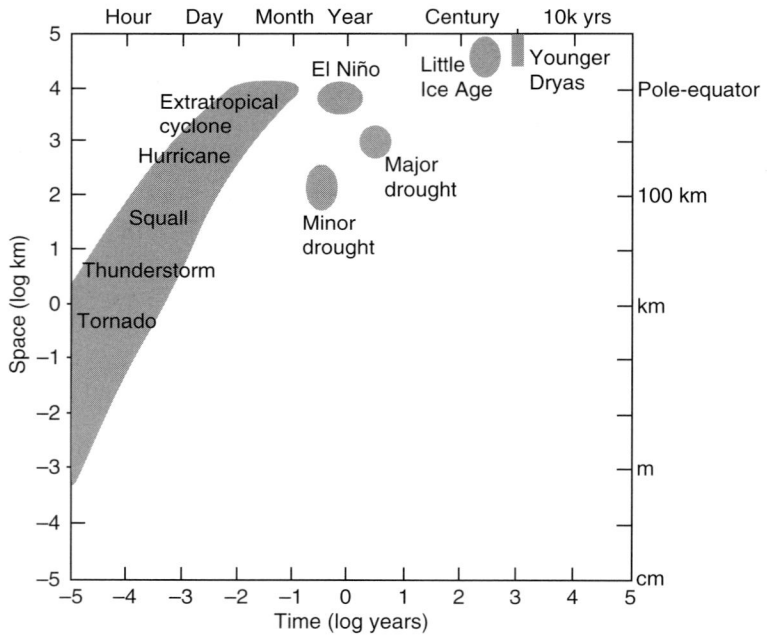

2.12. A model of the spatial and temporal scales of selected weather and climatic episodes. Modified after Clark (1985).

here, the same applies to various ecological or environmental processes. As discussed in further detail in Chapter 3, the resolution of the fossil record determines the resolution of environmental processes that we can detect. It is possible to increase the amount of fossil spatio-temporal averaging in our fossil samples by aggregating assemblages from different sites or times, but it is (usually) impossible to achieve finer resolution. For instance, a vertebrate fossil deposit that accumulated over a period of a hundred years – excellent temporal resolution by most standards – cannot be parsed out into finer temporal scales of days, years, or decades (unless we directly date each specimen, a very destructive and costly undertaking).

A final aspect of ecological scale concerns the difference between synecology and autecology. *Synecology* involves the ecological study of a group of organisms, species, or communities and their relationships to their environment. *Autecology* is the study of an individual organism or species and its relationship to its environment. As implied in Figure 2.5, an individual organism has a slightly different relationship to its environment than its conspecifics, and the same can be said for each individual species and, say, its congeners. As we shift up in scale from an individual organism to a species to a community of multiple species, the pertinent ecological processes also shift in scale. As will become clear in later chapters, depending on whether our analysis involves study of a single species or a community of multiple species, the resulting insight to past environments likely will vary in scale as well.

To reiterate, there is no universally correct spatio-temporal scale (Levin 1992; Sayre 2005). Rather, an appropriate scale is the one indicated by the research question being asked or hypothesis being tested. Whether or not that scale can in fact be obtained depends of course on the resolution of that portion of the paleozoological record under study (e.g., Behrensmeyer et al. 2000; Roy et al. 1996). Resolution is what allows us to detect and track various kinds of paleoenvironmental change. We turn next to what those kinds of change might look like, noting that issues of scale must be part of the categorization, as will become clear.

Kinds of Paleoenvironmental Change

We are not concerned here with sorting environmental changes into categories like "warming," "aridification," and "more grassland or more open habitat." Rather, we outline several forms of change that are spatio-temporal scale dependent, which is not to preclude environmental stasis as a possibility. But as we show, even what seems to be stasis might include change that is irrelevant to our research questions. For instance, day-to-day fluctuation in weather patterns or seasonal variation in climatic parameters such as mean temperature may not be pertinent (ignoring for the moment that they may not have left a perceptible signature among faunal remains) (Figure 2.12). What sorts of change might be pertinent?

Change implies the passage of time. Our interest is in environmental variables such as density of trees on the landscape or annual precipitation. The

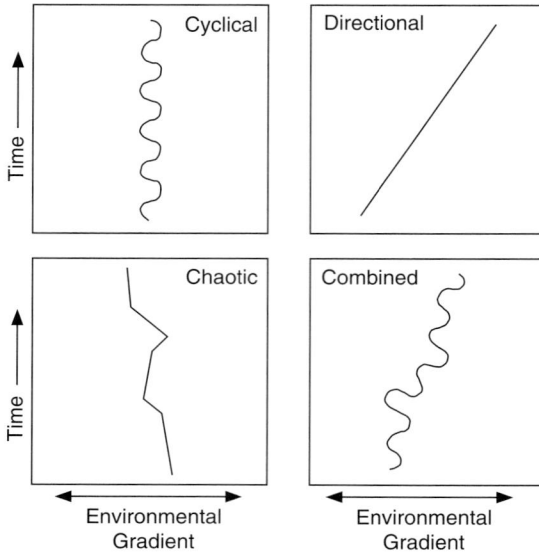

2.13. Categories of environmental change along space and time. Modified after Landres (1992).

environmental variables find expression across space. Thus our two dimensions of variability (and resolution) are space and time. Change can be cyclical at virtually any spatio-temporal scale; some change occurs at regular intervals (Figure 2.13) (much of this paragraph is derived from Landres [1992]). An obvious example involves the change in seasons year after year in temperate latitudes, and the same applies to longer-term phenomena such as the glacial–interglacial cycles of the Quaternary. Directional change involves a relatively long-term shift (increase or decrease) in an environmental parameter. An example of this is the rise in global temperatures that occurred from the Last Glacial Maximum to the onset of the Holocene (barring the Younger Dryas cooling event). Chaotic change involves random or stochastic events such as earthquakes, volcanic eruptions, and catastrophic floods or fires. Combinations of any two or all three of these can result in a very dynamic environmental history (Figure 2.13).

Ecologists, particularly restoration ecologists, today speak of the *historical* or *natural range of variation* evident in ecosystems (e.g., Wiens et al. 2012a). In doing so they explicitly acknowledge that habitats, communities, and ecosystems are not literally static but rather dynamic, undergoing all sorts of changes of multiple kinds and scales. In documenting the historical range of variation, ecologists do not seek natural benchmarks or reference conditions they hope to recreate; that is an unattainable goal. Rather, they hope to discover the ecological processes that structure the landscape and create patterns in order to predict what might happen ecologically in the future, given humankind's increasing population and resource needs, and to suggest what should be done to decrease the loss of biodiversity and ecological services necessary to the

2.14. Environmental fluctuation over time highlighting dynamic equilibrium around an average (panel A), flux around an average yet a long-term change (panel B), and (panel C) tipping points and thresholds being crossed resulting in a relatively abrupt change from one stable state (in dynamic equilibrium) to another. Modified after Wiens et al. (2012b).

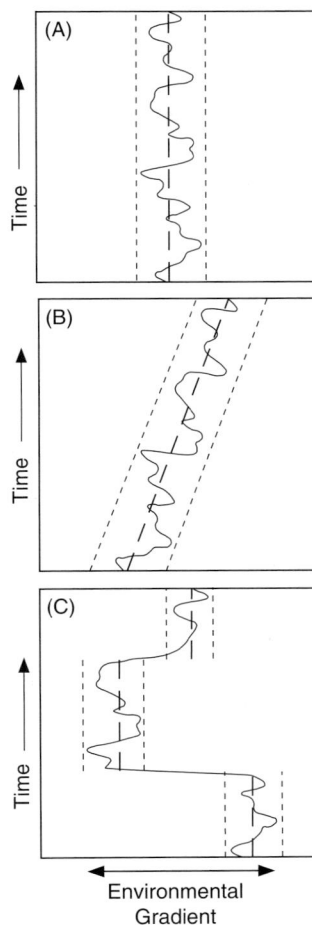

survival of humankind. We will have more to say about this in the final chapter of this volume. In the context of this section of the volume, knowing about the historical range of variation in environments underscores that change can occur at all scales of space and time.

Figure 2.13 presents several kinds of change, but a clear indication of scale is not included. Figure 2.14 begins to clarify the scale of change by including the notion of historical (or natural) range of variation (Wiens et al. 2012b:67). The latter adds an explicit indication of the average conditions and also the amount of fluctuation around that average in dynamic (realistic) ecological systems. The "cyclical" change of Figure 2.13 is not that different from the stasis of Figure 2.14A. The "directional" change of Figure 2.13 has more detail (greater resolution) in Figure 2.14B. And the "chaotic" and "combined" categories of change in Figure 2.13 have something of a different character in Figure 2.14C that we will return to momentarily. Here it is important to note that there is still little indication of scale, either extent or grain, in Figure 2.14. How do we know that Figure 2.14B is not simply a small part of Figure 2.14A writ large? We do not. Scale, as we noted above, is an analytical choice, and thus demands an indication of magnitude along either the temporal or the environmental gradient axis.

Let us return to Figure 2.14C as there is another point to make about it, and it will underscore the importance of our analytical choice of scale. Figure 2.14A could be said to model an ecological steady state, one that is in dynamic equilibrium; fluctuation is incessant, but things do not really change over the long time span (extent) measured. Not so in Figure 2.14C; not only is change incessant, but at least three temporally sequent steady states (or periods of dynamic equilibrium) occur over the time span represented. An ecological threshold has been crossed during the change from one steady state to another; an *ecological threshold* is defined as an abrupt change in ecosystem quality, property, or phenomenon, or a case in which small changes in an environmental driver produce large responses in ecosystems (Groffman et al. 2006). A related concept involves what is referred to as a *tipping point*, a particular moment in time when a small change can have long-term consequences for a system (Lenton et al. 2008), or a time when a system shifts radically and potentially irreversibly into a different state (Brook et al. 2013). Interest in such *regime shifts* (as the results are sometimes referred to, e.g., Hughes et al. [2013]), has increased in the past ten to fifteen years because of anticipation that human-created ecological change may soon (and suddenly and without warning) jeopardize human well-being (Scheffer et al. 2012). With respect to paleoenvironmental reconstruction, the notions of tipping point and regime shift are captured in Figure 2.14C as are the notions of dynamic equilibrium and steady state. Depending on the rate and magnitude of environmental change, such phenomena may be detectable in the paleozoological record, presuming of course that we have sampled at the correct scale (both extent and grain) and taphonomic processes have not completely obscured the paleozoological signals.

Change is incessant, and it can occur in a variety of ways across numerous spatial and temporal scales. Environmental parameters vary and change across space and time; environmental parameters often (but not always) prompt or force responses among organisms (e.g., move, adapt, die); responses can involve changes in species, communities, ecosystems, or some combination thereof (Delcourt et al. 1983). A final point about scale concerns the part of the paleozoological record that is under study, the assemblage. In many cases, the assemblage of fossils has been collected from a spatially limited geological deposit typically referred to as a *stratum*, or a more or less distinct sedimentary unit of approximately contemporaneous deposition (for sake of discussion, we will ignore a deposit made up of multiple strata). The faunal remains in that stratum were deposited there after having been accumulated by one or more processes (e.g., fluvial action) or agents (e.g., predators). These are well-known aspects of *taphonomy* (Gifford 1981; Gifford-Gonzalez 1991a; Lyman 1994), but the related point we wish to make has seldom been explicitly stated.

Every fossil assemblage derives from what might be thought of as a *sampling* or *accumulation radius* around the location of the deposit that produced the fossils (Behrensmeyer et al. 2000). Shotwell (1955, 1958) was concerned with just this issue, though not in these terms, when he proposed a (now known to be flawed) analytical technique for distinguishing species representing the proximal (nearby) community around the deposit from species representing distal (distant) communities. Identifying the agents and processes of accumulation responsible for a fossil assemblage will allow estimation of the accumulation radius represented by the assemblage. Some owl species, for instance, that create distinctive taphonomic signatures have known foraging radii (Andrews 1990); identifying a particular taphonomic signature implies a particular accumulating agent that in turn suggests an accumulation radius. Turning to zooarchaeological assemblages, for example, Faith and Behrensmeyer (2013:617) drew from ethnographic observations indicating that southern African hunter-gatherers typically forage ~8–12 km from their residential sites to argue that "the ungulate species identified in [South African zooarchaeological assemblages] represent animals that lived within a similar distance of the archaeological sites." The larger the area sampled by the accumulator(s), the more likely a larger number of microhabitats have contributed animal remains, and vice versa. The same applies to the source area of fossil deposits accumulated by geological processes (Figure 2.11).

ADAPTATION (EVOLUTIONARY GENETIC) OR ADJUSTMENT (PHENOTYPIC PLASTICITY)

Paleozoologists acknowledge that an animal species can adapt to environmental change. However, some evolutionary biologists argue that such statements are at best inaccurate and at worst simply wrong. They are deemed inaccurate because we do not *know* that the species adapted by the process of natural selection had changed its genotype. Or they may be wrong because the species did not adapt ("evolve") via genetic change but instead merely adjusted to the new environment by altering its phenotype (not its genotype). The latter is possible because of what is known as phenotypic or *ecophenotypic plasticity* (e.g., Jass et al. 2015; Lister 1997; Réale et al. 2003). Temporal shifts in the mean size of a species, for example, might be interpreted to be an adaptation but in reality be merely an expression of ecophenotypic plasticity – an ecological adjustment. Altering behaviors is another way to adjust phenotypically to environmental change (Lister 1997; Lundelius 1964; Redding 1978).

In an explicit paleozoological study of whether a prehistoric change in phenotype was a result of genetic (adaptive) change or merely an expression of phenotypic plasticity (adjustment), Hadly (1997) compared two skeletal (phenotypic) traits of a species across different subspecies. She argued that the trait that did not covary among modern subspecies but did covary with environmental conditions represented a phenotypically plastic trait, whereas the trait that did covary among modern subspecies but not with environmental conditions represented a genetically controlled trait. Thus temporal change in the former trait within the species over prehistoric time reflected an adjustment to environmental change. She later extracted aDNA from some of the same fossils to evaluate her earlier conclusions and found that her interpretations were correct (Hadly et al. 1998, 2004; see also Jass et al. 2015).

Distinguishing whether the individuals making up a population modified their behaviors, or their phenotypes or genotypes in subsequent generations, may be difficult to determine in the fossil record (Lister 1997), but study of aDNA might resolve some instances (e.g., examples in Stewart 2009). Whether or not paleozoologists need to rigorously maintain the distinction between ecophenotypically plastic adjustments and genetic evolutionary change seems to not be as important as one might think at first blush. Conservation biologists, for instance, who worry about the modern vulnerability of species to anthropogenically caused environmental change, define "adaptive capacity" as "the capacity of a species or constituent populations to cope with climate change by persisting in situ, by shifting to more suitable local microhabitats, or by migrating to more suitable regions," and note that adaptive capacity depends in part on phenotypic plasticity, genetic diversity, evolutionary rates, and other variables (Dawson et al. 2011:53). The distinction between phenotypic adjustment and genetic evolution is, however, sufficiently important that paleozoologists interested in reconstructing paleoenvironments need to be aware of it.

CONCLUSION

We introduce in this chapter some basic ecological concepts that underpin paleoenvironmental reconstruction and that must be kept in mind when undertaking such analyses. We do not pretend to think we have provided all the ecological knowledge that might be required for a rigorous analysis, hence our recommendation early in the chapter that a good text on the ecology of species and communities and one on biogeography would be good things to have on hand for consultation. And we have cited some of the basic ecology and biogeography literature in this chapter as a form of introduction to the relevant literature.

Perhaps the key things to remember from this chapter are that each species has a more or less unique set of ecological tolerances (Figures 2.4 and 2.6) and while these are absolute in the real world and represent a species' fundamental niche, various factors influence species such that they occupy a smaller realized niche. An additional complication is that we often only know in an ordinal-scale, low-resolution sort of way the values of the environmental gradients describing a species' niche. Thus paleozoologists must typically (though not always) fall back on such interspecific relative statements – species A is arid adapted relative to species B which occupies more mesic habitats.

Given what we have covered in this chapter, it is now time to turn from modern or neoecology to paleoecology in the following sense: What are the bases for applying what we know of neoecological principles to the paleozoological record? In particular, what are the necessary assumptions upon which paleoenvironmental reconstruction on the basis of paleozoological remains rests? We identify these assumptions and discuss their implications at some length in the following chapter.

THREE

ANALYTICAL ASSUMPTIONS

Many analytical techniques used by archaeologists, paleontologists, and other scientists require that the researcher make some assumptions about how the part of the world under investigation works. These necessary assumptions are not always stated explicitly in a discipline's literature, making it difficult for a novice to grasp analytical nuances and sometimes resulting in experts disagreeing about analytical fundamentals or interpretations. The typical literature on paleoenvironmental reconstruction is no different. In an effort to rectify this situation, Lyman (2017) recently summarized many of the assumptions requisite to paleoenvironmental reconstruction attending the analysis and interpretation of the taxonomic identities of the animals in a fossil collection. This chapter repeats much of that discussion, and also identifies assumptions necessary to virtually any analysis of animal fossils and interpretation of their paleoenvironmental implications.

One of the nearly (if not completely) universal assumptions that scientists make concerns *uniformitarianism*. Earth scientists recognize this as "the present is the key to the past" (Gould 1965, 1987; Rudwick 1971; Simpson 1970), but things are more complex than that (Inkpen 2008; Romano 2015). A great deal of very interesting literature covers this assumption in some detail, and we cannot hope to summarize it all here. In our view, Cleland (2001, 2011) provides an excellent discussion of uniformitarianism from the perspective of the philosophy of science that is pertinent to archaeology, paleontology, and paleoenvironmental reconstruction. It suffices here to say that uniformitarianism is the label for the assumption that the laws of nature (how earth and natural processes work) are invariant in time and space. As we noted in Chapter 2, ecologist Eugene Odum (1971:159) stated the principle well when he said the basic assumption of paleoecology is that "the operation of ecological principles has been essentially the same throughout the various geological periods." In case the importance of this assumption is not clear, we note that the analytical utility of everything said in Chapter 2 rests on uniformitarianism.

Another way to highlight the significance of Odum's (1971) key ecological assumption is to outline what we take to be the fundamental generic approach to paleoenvironmental reconstruction. First, the researcher searches for and (hopefully eventually) identifies a correlation between a biological variable and an environmental variable in the modern world. The former might be the geographic distribution of a species (its realized niche) and the latter a kind of habitat such as an open woodland or a climatic parameter such as maximum summer temperature (either one approximating, it is assumed, the species' fundamental niche). Second, the spatial correlation of a species and the environmental variable is in a sense (not literally) converted to a temporal correlation such that when the species is identified in a fossil collection, it is inferred that the value of the environmental variable for that species was the same in the past as observed today, which in turn allows the researcher to say something about ("reconstruct") the past environment. The implicit assumption allowing the inference is that the correlation between the two variables held in all times and all places of the past – this is uniformitarianism. This assumption is what allows us to derive transfer functions, to construct bioclimatic envelope models, to infer paleoenvironmental conditions and the like (see Chapter 2).

Paleozoologists have referred to this most basic and fundamental assumption of paleoecology as "taxonomic uniformitarianism" (Bottjer et al. 1995), "physiological uniformitarianism" (Tiffney 2008), and simply as the principle of uniformitarianism (Ager 1979; Behrensmeyer et al. 2007a; Birks et al. 2010; Demirel et al. 2011; Gifford 1981; Graham and Semken 1987; Guthrie 1968a; Johnson 1960; Kingston 2007; McCown 1961; Rymer 1978; Scott 1963; Wilson 1968). In the context of this book, it is important to note that Lawrence (1971:602) refers to this assumption critically as "transferred ecology." What Lawrence is concerned about is clarified by Behrensmeyer et al. (2007a:13–14) who point out that making this assumption "limits our ability to see how the ecology of the past was different from that of today." Thus the requisite identification of any correlation between a zoological variable and an environmental variable as the first step of paleoenvironmental reconstruction may unnecessarily restrict or limit our interpretations to the point of making incorrect inferences about past environments. We discuss ways to circumvent such pitfalls later in this chapter and in subsequent chapters (see especially Chapter 7).

Assumptions required by analytical techniques described in later chapters are our concern here. These assumptions were compiled from the perspective of analyses based on taxonomic identifications – what we take to be the most basic family of analytical techniques (Chapters 5–6) – though several are also relevant to other types of techniques (additional assumptions specific to other techniques are discussed as needed in other chapters). Many analytical methods in use today depend upon taxonomic identifications, yet the assumptions necessary for analysis and interpretation are, as noted above, seldom explicitly acknowledged in the recent literature (e.g., Bañuls-Cardona et al. 2014; Emery and Thornton 2014; Gómez Cano et al. 2014; López-García et al. 2015a, 2015b; Matthews et al. 2011; Socha 2014; Widga 2013). There are several incomplete lists of the assumptions in the older paleozoological literature (Findley 1964; Harris 1963; Lundelius 1964). The assumptions described below are a compilation of these, plus a few additional ones. If this volume is to be used effectively as a textbook or a guide to how to do particular sorts of analyses, we are of the unshakable opinion that those who follow the guide must understand not only how to do those analyses but why those analyses are philosophically allowed. Knowing the necessary underpinning assumptions allows us to highlight where an argument is strong and where it is weak. As well, the fear of becoming complacent would seem to be sufficient reason to devote some page space to the assumptions, particularly from a taphonomically, analytically, and ecologically up-to-date perspective.

In this chapter we identify key assumptions, discuss each assumption's strengths and weaknesses, and identify relevant ecological principles. Analytical means to overcome or circumvent weaknesses are described. Although the literature cited is incomplete, the number of titles cited per assumption provides a coarse indication of the scale of perceived importance ascribed to each assumption relative to the others. Most of the cited literature concerns vertebrate fauna in general and mammals in particular, but some references concern mollusks, insects, or plants. It is thus important to underscore that all of the assumptions mentioned in this chapter, while phrased in terms of animal remains, are equally pertinent to plant remains.

ASSUMPTION 1: UNIFORMITARIANISM

The key assumption to interpreting the remains of any organism, plant or animal, in terms of their paleoenvironmental implications, is that the identified species, if it exists today, had the same ecological tolerances in the past that it has today (Andrews 1995; Avery 1982; Brown 1908; Cleland 1966; Coope 1986; Craig 1961; Dalquest 1965; Demirel et al. 2011; Frazier 1977; Grayson 1976, 1981; Guilday 1971; Hibbard 1955; Hibbard et al. 1965; Hokr 1951; Holbrook 1975, 1980; Holbrook and Mackey 1976; Lawrence 1971; Lozek 1986; Lundelius 1964,

1983; Miller 1937; Patton 1963; Polly and Eronen 2011; Redding 1978; Romer 1961; Rymer 1978; Schultz 1967, 1969; Semken 1966; Taylor 1965; Tiffney 2008; Walker 1978; Zeuner 1961). In other words, the ecological tolerances of extant species are assumed to have not evolved or changed over time (Belyea 2007; Chaline 1977; Cheatum and Allen 1964; Findley 1964; Johnson 1960; Levinson 1985; Scott 1963; Yalden 2001).

Many researchers acknowledge that this assumption could be invalid because all species have evolved to one degree or another. Because of the likelihood that the ecological tolerances of a species have evolved, analysts often consider more than one species during analysis to "reduce the probability of error owing to an evolutionary change in tolerances of one species" (Lundelius 1964:26; see also Avery 1988; Bobe and Eck 2001; Dalquest 1965; Dorf 1959; Grayson 1981; Guilday 1962; Johnson 1960; Redding 1978; Romer 1961; Semken 1966; Walker 1978; Wells 1978; Winkler and Gose 2003; Zeuner 1961). The use of multiple taxa is seen as a better means of reconstructing past environments than using one taxon. The reasoning behind the recommended consultation of multiple species is remarkably straightforward, and that is probably why it is seldom explicit in the literature. It is unlikely that every one of the multiple taxa consulted will have evolved in precisely the same way to changing environments. For instance, not all cold-adapted taxa will respond the same way to warming climates. It is improbable (though not altogether impossible) that all of them will have (or obtain through mutation or breeding) the necessary alleles that result in each taxon becoming adapted to warmer climates. Increasing the improbability that all species in a prehistoric fauna evolved in like manner, it is likely the plethora of selective processes will not be the same for all represented species given what we know of ecological tolerances. Only some of the taxa will adapt as a result of natural selection altering their ecological tolerances; some taxa will alter their distributions (or abundances) and others will be locally extirpated. Nevertheless, any paleoenvironmental reconstruction based on, say, mammal remains should be corroborated as rigorously as possible by consultation of independent biological (e.g., plant, insect, avian, and molluskan remains) and geological data.

That evolutionary change and altered adaptations or ecophenotypic adjustments present a not insignificant analytical challenge is found in the observation that the chronologically older the faunal collection examined, the more tenuous the key assumption becomes (Craig 1961; Faith 2011a; Guilday 1962; Guthrie 1968a; Reed 2007; Romer 1961; Scott 1963; Smith 1919). The tenuousness becomes greater because with greater age the greater the probability that evolutionary change in a species' tolerances has occurred (Guthrie 1968a; Lundelius 1964, 1985; Walker 1978). Because of the possibility of evolutionary change in adaptations and ecological tolerances, and thus the potentially mutable relationship between a species and its tolerances, environmental reconstructions are said by some to be imperfect or tenuous (Avery 2007; Walker 1978). An alternative way to think about paleoenvironmental reconstructions is not that they are imperfect but rather that they are of low resolution, or ordinal scale (greater than–less than) rather than ratio scale (environment A has 10 cm more annual precipitation than environment B) (Birks et al. 2010). Although some might bemoan such low resolution, it is difficult to justify not undertaking an analysis of any particular aspect of the past simply because our insights are not as focused or as high resolution as we might hope. As later chapters indicate, numerous analytical techniques we discuss produce ordinal-scale inferences, and that is often sufficient and is certainly better than no inference at all. Some techniques are advertised as providing high-resolution ratio-scale inferences, but whether this is possible is something we address in Chapter 9.

In the absence of specific information on the ecological tolerances of species, some analysts have determined the habitats preferentially occupied by the species represented in a collection and then developed a variety of metrics that can be loosely categorized as *habitat metrics* (e.g., Andrews et al. 1979; Fernández-Jalvo et al. 1998; Hernández-Fernández and Peláez-Campomanes 2005; López-García et al. 2014; Matthews et al. 2005; Nesbit Evans et al. 1981; D. N. Reed 2007; K. E. Reed 2008; Su and Harrison 2007). These approaches (described in Chapter 7) appear analytically elegant because they seem to shift the resolution of paleoenvironmental reconstructions from relatively coarse life-zone or biome-like interpretations to seemingly higher-resolution tropical forest versus steppe/savanna versus open woodland versus grassland interpretations given how the indices are calculated. Do not be misled by the mathematics involved in the calculations (Belyea 2007); the paleohabitat reconstructions are akin to life-zone type interpretations and both are ordinal scale. "Terms such as woodland or savanna are open to interpretation and have come to encompass a wide range of potential habitats. The vagueness of the terminology is, in part, tolerated as the low resolution of the fossil data typically does not furnish specifics that allow us to go beyond these generalized terms" (Kingston 2007:32). We cover habitat metric techniques in more detail in Chapter 7. Here it is relevant to point out that habitat metrics rest on a slightly modified version of the first assumption – that the habitat preferences of an extant species were the same in the past as they are today. This first key assumption does not appear in any habitat metric study with which we are familiar.

A growing body of research suggests that the key assumption of no shift having occurred in a species' ecological tolerances over time, known as *niche conservatism*, is true for at least some taxa (e.g., DeSantis et al. 2012b; DiMichele et al. 2004; Martínez-Meyer et al. 2004; Wake et al. 2009; Wiens et al. 2010). This is good news but at the same time we must not overlook the fact that some species likely have evolved and their ecological tolerances have likely shifted. Robust tests of apparent cases of niche conservatism are still under development (Nogués-Bravo 2009). For now, we must be careful to not over-interpret data suggestive of niche conservatism (Peterson 2011) or to presume that paleoenvironmental reconstructions are interval or ratio scale. Best analytical practice in paleoenvironmental reconstruction is to consult as many independent lines of evidence as possible. Convergence of those varied kinds of evidence toward the same reconstruction would suggest not only a correct reconstruction but also niche conservatism among the monitored species.

ASSUMPTION 2: NEAREST LIVING RELATIVE

Extinct species are assumed to have had ecological tolerances similar to those of their closest living relative (Allan 1948; Andrews 1996; Avery 1982; Bobe and Eck 2001; Dorf 1959; Findley 1964; George 1958; Gidley and Gazin 1938; Guilday 1971; Guthrie 1982; Harris 1985; Hibbard 1944, 1958, 1960, 1963; Kingston 2007; D. N. Reed 2007; K. E. Reed 2013; Rhodes 1984; Romer 1961; Schultz 1967, 1969; Semken 1966; Simpson 1953; Stephens 1960; Wing et al. 1992; Woodcock 1992; Woodring 1951; Zeuner 1936, 1961). This assumption is sometimes referred to as "taxonomic uniformitarianism" (Reed et al. 2013) or "taxonomic analogy" (Bobe and Eck 2001). In paleoclimatology, this assumption is referred to as the "*nearest living relative*" or NLR method (Mosbrugger 2009); it involves choosing the taxon that is the nearest living relative of the extinct taxon and using the former as an ecological analog for the latter.

Choosing its closest living relative as an appropriate ecological analog for an extinct species assumes the two species, because they are (or should be) members of the same genus, tribe, family, or clade, not only share lots of genes but have in common the same or similar ecological tolerances. The more distant the genetic relationship between the extinct and extant (modern analog) taxon, the more tenuous the assumption becomes because as genetic relationship grows distant, the greater the chance that similarity in ecological tolerances decreases (Jehl 1966; Reed et al. 2013). Wing et al. (1992:6) suggest "logically, analogy with close living relatives is a derivative of functional morphology; species have been placed in the same higher taxon because they have similar morphological traits, and species with similar morphological traits are assumed to have similar ecological characteristics." Mosbrugger (2009) cautions the major hurdle to using the NLR method is determination of exactly how close an extinct taxon is to its modern analog taxon – at the species, genus, or higher taxonomic level.

Some early paleontologists worried that we could never know the ecological tolerances of extinct taxa and, because of that, they argued extinct taxa were of limited if any analytical value when reconstructing past environments (e.g.,

Ager 1979; Andrews 1990, 1995; Dalquest 1965; Hester 1964; Hokr 1951; Ladd 1959; Levinson 1985; Lundelius 1972, 1974, 1976, 1985; Simpson 1947; Tchernov 1975). Combined with a lack of trust in the validity of using an extant relative as an ecological analog, it might seem that extinct taxa provide minimal insight to past environments. That extinct species had particular ecological tolerances just as extant species do is undeniable; this is fundamental ecology and biology (and the assumption of uniformitarianism). To learn something about the ecological tolerances of an extinct species, analysts have suggested two alternatives. One involves the assumption that similar phenotypic features signify similar adaptations, such as the familiar one of low-crowned teeth in *browsers* versus high-crowned teeth in *grazers* (Bobe and Eck 2001; Gould 1970; Lundelius 1964; Simpson 1953; Yalden 2001). In a slight variation of this alternative, Semken (1966) suggested that paleoecologists assume that the closer the morphometry of ancient animal remains is to bones and teeth of modern species with known ecological tolerances, the closer the ecological tolerances of the ancient animal to the modern animal (see also Tiffney 2008; Wells 1978; Whittington 1964; Wing et al. 1992). In short, then, the corollary assumption to our assumption 2 is that morphometrically similar functional anatomical traits denote similar ecologies (e.g., Herm 1972; Lundelius 1964; Nelson and Semken 1970; Semken 1966; Tiffney 2008; Wells 1978; Whittington 1964; Wing et al. 1992; Zeuner 1936, 1961).

The second alternative to assuming that an extinct taxon and its closest living relative have similar ecological tolerances is to determine the ecological tolerances of extant taxa that are stratigraphically and temporally associated with the extinct taxon (e.g., Case 1936; Cheatum and Allen 1964; Harris 1985; Miller 1937; Romer 1961; Simpson 1947; Wells 1978). Co-occurrence of the two (extinct and extant) suggests they had similar ecological tolerances (e.g., Hughes 2009; Walker 1978; but see assumption 3). This analytical strategy rests on the first assumption, that extant taxa have not changed their ecological tolerances. But using the same strategy of study of all associated extant taxa, it has been shown that occasionally the first assumption is false for some extant taxa (e.g., Mead and Spaulding 1995; Owen et al. 2000). Ecological tolerances or niches of extant taxa have evolved; niche conservatism does not apply to these taxa. Today, there are also computer-based models of past climates that can be used to estimate the climatic parameters to which an extinct species was adapted (e.g., McDonald and Bryson 2010).

Other alternatives to assumption 2 include study of ecomorphological skeletal traits (e.g., Faith et al. 2012; Plummer et al. 2008), *tooth wear* (e.g., Faith 2011a; Pinto-Llona 2013; Rivals et al. 2011), and stable isotopes in bones and teeth (e.g., Cerling et al. 2015; Eastham et al. 2016). These techniques, which we cover in more detail in Chapter 7, potentially allow paleozoologists to learn quite a bit about the ecology of an extinct species, ameliorating some of the

concerns about the use of extinct species in paleoenvironmental reconstruction. Importantly, any of the alternatives to the second assumption could reveal
much about past environments. They may even suggest a particular extinct
taxon had different ecological tolerances and habitat preferences than its
closest living relative (Lundelius 1985; Miller 1937). For example, the western
North American extinct noble marten (*Martes americana nobilis*) was originally thought to have ecological tolerances similar to the extant conspecific
pine/American marten (*M. a. caurina*). Study of extant fauna stratigraphically
associated with the noble marten shows that its ecological tolerances were
in fact unique among members of this species (Hughes 2009; Lyman 2011).
And in a different context, analysis of tooth wear of bovids and equids from
the Middle Pleistocene site of Elandsfontein in South Africa showed that
many species had dietary preferences distinct from those of their nearest living
relatives (Stynder 2009).

ASSUMPTION 3: ECOLOGICAL TOLERANCES

It is assumed each species has a set of ecological tolerances more or less unique
from all other species (Avery 2007; Cheatum and Allen 1964; Cleland 1966;
Guilday 1962; Harris 1963; Lundelius 1964; Moine et al. 2002), and we know
those tolerances sufficiently well for every species to predict with near cer-
tainty where, in terms of habitats and climates, a particular species will occur
(Findley 1964; Harris 1963, 1985; Kingston 2007; Redding 1978; Rhodes
1984; Romer 1961; Woodcock 1992; Yalden 2001). Cabrera's law of ecological
incompatibility holds that only one species can occupy a particular niche in a
given area at a given time (Simpson 1936:413, 1937:64). There can be more than
one species of a genus in a fauna but they need to occupy different niches so
as to avoid interspecific competition. Thus, some authors believe that multiple
related species in a collection means environmental variety (Simpson 1936),
or multiple microhabitats blended to reveal regional environment (Semken
1966; Wilson 1973), or multiple genotypes signifying multiple tolerance levels
(Walker 1978; see Figure 2.5).

   Some researchers believe we know very little about the ecological tolerances
of many extant taxa (Anderson 1968; Andrews 1995; Avery 1988; Calaby 1971;
Cheatum and Allen 1964; Guilday 1962; Harris 1963; Huntley 2012; Lundelius
1967, 1985; McCown 1961; Taylor 1965; Wells 1978). An analytical alternative
when details of ecological tolerances and limiting factors of extant taxa are
poorly known is to consult the modern distribution of each particular taxon
as a low-resolution clue to their ecological tolerances. (Think of a geographic
version of Figure 2.7.) It has, however, been argued that inferring a prehistoric
biome on this basis is error prone (Mares and Willig 1994), though it has been
shown that such is not necessarily the case (Andrews 2006; Le Fur et al. 2011).

If known, the biome or life zone rather than the specific habitat(s) in which a species commonly occurs or is abundant is used as a coarse, low-resolution indication of the paleoenvironment (e.g., Gidley and Gazin 1938; Gustafson 1972; Hughes 2009). Ranges of multiple taxa are thought to give greater resolution because each taxon is conceived as ecologically independent of all others (the individualistic hypothesis; see Chapter 2) and thus each provides an independent indication of biome, life zone, habitat, or environment (Moine et al. 2002).

Any time the geographic range of a taxon is used as a proxy for the ecological tolerances of an organism, the analyst is assuming that the organism's distribution is controlled by its ecological tolerances (Lundelius 1983); that is, the analyst assumes a species' distribution reflects its fundamental as opposed to its realized niche. Some suggest, however, that a species' geographic range may not be a good proxy for ecological tolerances (Dalquest 1965; Guilday 1971; Guthrie 1968a; Lundelius 1983; Moine et al. 2002; Smith 1957; Walker 1978; Woodcock 1992). Analytical use of a species' geographic range is nevertheless not unusual (Birks et al. 2010; Churcher and Wilson 1990; Davies et al. 2009; Escarguel et al. 2011; Gidley and Gazin 1938; Guilday 1971; Hoffmann and Jones 1970; Slaughter 1967). It is, for instance, something of an ecological truism (though not a rigid law) that taxa with large ranges tend to have wider (less restrictive) ecological tolerances than taxa with small ranges (Cheatum and Allen 1964; Davies et al. 2009; Lundelius 1964). The presumption is that a correlation between the distribution of a species and some environmental/climatic variable represents a causal relationship (Birks et al. 2010; Lundelius 1983); this is the implicit assumption behind the biological distribution models described in Chapter 2. Some researchers indicate the causal mechanisms producing the relationship between an environmental variable and a species' distribution need not be known in order to use the correlation as an interpretive algorithm (Avery 2007; Lawrence 1971; see also Kearney and Porter 2009; Polly and Eronen 2011). Gifford (1981:391) is a rare exception and argues it is critically important to seek "causal relationships" between ecological variables when it comes to deciphering paleoecologies. In her view, "the search for regular and ecologically relevant linkages between static attributes of the fossil record and their dynamic causes and associations is the key to understanding the prehistoric evidence." We agree.

Much that we know about biogeographic ranges is descriptive (Avery 1982; Kearney and Porter 2009; Lavergne et al. 2010; Redding 1978; Scott 1963); causal relationships between distributions and environmental variables are unknown (see overviews by Holt 2003; Louthan et al. 2015). Further, published surveys and distribution maps for species are not always thorough or accurate, respectively (e.g., Grayson et al. 1996). As a result, the paleozoologist may need to trap animals around the deposits producing the fossils to determine the local environment and the animal species present (e.g., Grayson 1983b; Holbrook 1975; Walker 1982). Such data also provide a check on what we know of the ecological tolerances of species and their biogeographic ranges. (Be sure to determine the regulations regarding collecting animals in your research area and obtain all required permits before collecting live animals or dead ones such as road-killed individuals.)

ASSUMPTION 4: TAXONOMIC PRESENCE/ABSENCE

The natural or autochthonous occurrence of a taxon in an assemblage is assumed to indicate the existence of a set of local environmental conditions under which that taxon could exist at the time of its occurrence (Harris 1963; Lundelius 1964; Moine et al. 2002; Redding 1978; Tchernov 1982). This assumption is distinguished from assumption 1 because analysts sometimes interpret the absence of a taxon from an assemblage as indicative of the presence of environmental conditions that are outside the absent taxon's ecological tolerances (see also Chapter 5), the absent taxon did not have access to the area (for whatever reason), or the absent taxon did not have sufficient time to immigrate to the area (Harris 1963; Walker 1978).

Although some analysts have interpreted the absence of a taxon in the manners indicated, they were aware of the potential taphonomic pitfalls of doing so. Simpson (1937) was among the first to point out that the absence of remains of a taxon does not necessarily mean that that taxon was not present in the area. Paleontologist Theodore White (1954) provided a detailed consideration of how taxonomic absences should be treated. He first noted it had often been stated the absence of remains of a taxon from a deposit did not necessarily indicate the taxon was not present in the area when the deposit was laid down. To accept the notion that the absence of a taxon from a paleozoological assemblage signified absence of the taxon from a prehistoric fauna was problematic. One should instead, White (1954:429) reasoned, consider (1) if a missing taxon was present in areas near the site under investigation or only remote areas; (2) if the missing taxon was present in geologically earlier, later, or both kinds of deposits; (3) the abundance of the missing taxon where it was found, the implication being that rare taxa were present but rare on the ancient landscape and thus might not often be found in the paleozoological record; (4) whether remains of taxonomically related taxa were present; (5) whether the depositional conditions were favorable to preservation of the remains of the missing taxon; (6) the thoroughness with which the deposits supposedly lacking the remains of the missing taxon had been examined; and (7) whether or not the local climate at the time the missing taxon's remains might have been deposited was conducive to occupancy of the area by the missing taxon. These all were then, and are still today, important things to consider should one wish to interpret the negative evidence for a taxon as implying a particular kind of paleoenvironment.

Other analysts have subsequently made the point that the absence of evidence of a taxon is not necessarily evidence of that taxon's absence (e.g., Ervynck 1999; Grayson 1981, 1983b; Guthrie 1968b; Lyman 2008a). There are several reasons a taxon's remains might not occur in a fossil assemblage. The remains of a taxon may not be in an assemblage because the taxon was not in the area at the time the remains comprising the assemblage were accumulated and deposited. This is of course necessary to know if the absence of a taxon is to serve as the basis for inferring this taxon's requisite environmental conditions were not present. The taxon may, however, have been present in the area at the time but its remains were not accumulated and deposited in site sediments (e.g., Grayson 1981). Alternatively, the remains were accumulated and deposited but were not preserved. Or, if they were accumulated and deposited and were preserved (in an identifiable state), perhaps they were not found given recovery methods such as screens with large mesh (e.g., Lyman 2012c) or perhaps they were not correctly identified as belonging to that taxon (see assumption 6). A few analysts have suggested analytical techniques to determine when the absence of remains of a species can be taken as evidence that the species was in fact absent (Burnham 2008; Lyman 1995; Nowak et al. 2000), but these techniques have not been generally adopted because they rest on tenuous assumptions of their own. Paleontologists, on the other hand, often use similar techniques to determine what they refer to as *first appearance dates* (FADs) and *last appearance dates* (LADs) of fossil species. These techniques incorporate statistical estimates of chronological distributions based on both observed stratigraphic ranges and measures of data quality (e.g., number of specimens, sampling intensity) (e.g., Alroy 2000; Barry et al. 2002; Bobe and Leakey 2009; Marshall 1990).

We currently lack robust techniques for determination of which possible reason for the absence of remains of a taxon applies in any given case. Thus it has been suggested that analysis must be asymmetrical in the sense that interpretive emphasis is placed on taxonomic presences rather than absences (Ervynck 1999; Grayson 1981, 1983b; Lyman 2008a; but see Pinto et al. 2016 for a recent exception). Finally, recalling assumption 3, the analyst should not overlook the fact that we still do not know in detail the biogeographic histories or even the modern ranges of many species, in which case our understanding of the ecological reasons why a taxon might be absent from a site (provided there is an ecological reason) is only going to be as good as our understanding of the ecological reasons why that taxon might be present at a site.

A final comment about assumption 4 concerns its seeming dependence on a more deeply implicit assumption – that the presence of a taxon's remains are indicative of the local environment at the time of accumulation and deposition because they are (assumed to be) of local, not extralocal or extratemporal, origin. This is a taphonomic issue and is addressed in more detail under assumptions 5 and 7.

ASSUMPTION 5: FAUNAL COMPOSITION

Changes in faunal composition (species represented, species abundances) are assumed to reflect changes in environmental conditions (Avery 1990, 2007; Graham 1985a; Graham and Semken 1987; Grayson 1976, 1977, 1979a; Harris 1963; Lawrence 1971; Tchernov 1968). Although the point had been made earlier, Grayson (1981, 1983b) pointed out explicitly that the taxa in an assemblage could be treated as attributes of a fauna – their presence in or absence from the assemblage is the subject of study – or as variables – their frequencies relative to one another are the subject of study (we discuss analytical techniques for each in Chapters 5 and 6, respectively). Because the latter provides more information, it is more powerful analytically; changes in relative abundances of taxa are more sensitive to small-scale environmental flux than the presence–absence of taxa (Graham and Semken 1987). But to blindly believe taxonomic abundances are universally better than presence–absence data (particularly noting that one should not place great interpretive weight on the absence of taxa, as documented under assumption 4) ignores all the well-known problems of determining those abundances (Grayson 1984b; Lundelius 1964; Lyman 2008b; McCown 1961). Some analysts have taken those problems to be so great that they recommend against using the abundances of remains as a proxy for abundances on the landscape (e.g., Holbrook 1975; Redding 1978). Fortunately, we have learned assemblages are best quantified, if quantified at all, as NISP, MNI, or biomass depending on their taphonomic histories and depositional contexts (e.g., Badgley 1986; Guilday 1969; Terry 2009; Thomas and Mannino 2017). Regardless of the quantitative unit one uses, the data can be analyzed using interval- or ratio-scale statistics (as we do in later chapters), but there are good reasons to interpret results in ordinal-scale terms. For one, the quantitative data themselves are often demonstrably only ordinal scale (Grayson 1984b; Lyman 2008b). For another, recall the implications of Figure 2.5 – that each individual of a species has more or less unique environmental tolerances and we must presume our fossils represent individuals with average or modal tolerances of the species.

Once taxonomic abundances have been determined, the next issue concerns deciphering what shifts in those abundances signify. Darwin (1859:68) suggested the "amount of food for each species of course gives the extreme limit to which each can increase," and some modern researchers agree and add the abundance of food is in turn controlled by climate (White

2008). As noted in Chapter 2, this bottom-up perspective is likely what was assumed or implied by the suggestion that animals are an indirect reflection of climate whereas plants are a (relatively more) direct reflection of climate. But there are several possible reasons for changes in faunal composition and taxonomic abundances other than food availability or climate (Avery 1982; Belyea 2007; Grayson 1979a; Guilday 1969; Guthrie 1968b; McCown 1961). The appearance or disappearance of a competitor (Belyea 2007; Guilday 1971), cyclical or random flux in abundance (Guthrie 1968b), ecological succession (Avery 1982), historical contingency (Fukami 2015; Svenning et al. 2015), or other processes may cause flux in taxonomic abundances on the landscape. And, not necessarily contradicting his observation regarding the influence of food availability on the abundance of a species, Darwin (1859:69) stated that if a species was "favoured by any slight change of climate, [it] will increase in numbers."

The more frequently expressed worry regarding shifts in taxonomic abundances in the paleozoological record concerns the possibility that those shifts resulted from a change in the mechanism of bone accumulation or preservation that had nothing to do with environmental change (Behrensmeyer and Hook 1992; Behrensmeyer et al. 2007a; Guilday 1962; McCown 1961; Redding 1978; Turvey and Cooper 2009). For instance, different predators exploit different habitats that contain different prey species. Grayson (1981) demonstrated the assumption that changes in taxa or their abundances represent changes in environment is not necessarily true by finding varied abundances of rodent remains in collections of modern pellets egested by different species of owl (see Chapter 6). Andrews (1990) described similar observations. This does not mean all changes in the taxonomic composition of faunal assemblages or the abundances of taxa in those assemblages are suspect. A host of *fidelity studies* – *actualistic* assessments of "the quantitative faithfulness of the [fossil] record of morphs, age classes, species richness, species abundance, trophic structure, etc. to the original biological signals" (Behrensmeyer et al. 2000:120) – suggest that at least sometimes abundances of taxa in collections of faunal remains accurately reflect the environmental context in which those remains have been accumulated (e.g., Behrensmeyer et al. 2003; Cutler et al. 1999; Hadly 1999; Lyman 2012d; Lyman and Lyman 2003; Miller 2011; Miller et al. 2014; Terry 2010a, 2010b; Western and Behrensmeyer 2009).

How does one know if a particular prehistoric assemblage of faunal remains accurately reflects the paleoenvironmental conditions within which that assemblage was accumulated and deposited? How do we know if the prehistoric fauna is a representative sample of the fauna on the landscape at the time the former was formed? In short, we do not know and cannot know these things. Detailed taphonomic analysis is the usual suggested means of estimating the confidence that should be placed in the environmental fidelity of a prehistoric faunal collection (e.g., Andrews 1990; Domínguez-Rodrigo and Musiba 2010; Terry 2009; Wing et al. 1992). This suggestion presumes that we can decipher taphonomic data sufficiently well to analytically distinguish taphonomic or other non-environmental causes of faunal change from environmental causes. Only recently has our taphonomic knowledge become sufficient that we can often (though not always) analytically contend with various biases. Some of the more interesting studies along these lines have used large data sets to simulate the effects of impoverished samples and of mixed samples on interpretations (e.g., Le Fur et al. 2011); the former tend to often provide fairly accurate signals whereas the latter do not (Andrews 2006; contra Mares and Willig 1994).

But what does one do when faced with, say, a set of assemblages that reveal change in taxa represented or fluctuations in taxonomic abundances over time? One approach is to compare only assemblages that are isotaphonomic (Wilson 2001), those assemblages that have been subjected to the same taphonomic processes and biasing agents. This vetting of data is meant to eliminate variability that is the result of taphonomic processes and retain variability that is the result of environmental change. That might be more reasonable than attempting to strip away the taphonomic overprint (e.g., Lawrence 1968, 1971) in some cases. Another, similar alternative is to compare assemblages to distinguish those that are taphonomically skewed from those that are not (e.g., Alemseged et al. 2007). Or one might array the taxonomic abundances against one or several taphonomic variables that tell us something about potential biases (e.g., skeletal part representation, bone modifications) to determine whether changes in the former are tracking (and thus presumably are related to) changes in the latter (e.g., Bobe and Behrensmeyer 2004; de Ruiter et al. 2008; Faith 2013a). Finally, the morphometry of ancient animal remains is independent of taxonomic abundances (Gould 1970), so co-variation of the two can be taken as mutually reinforcing.

And it should be obvious at this point that, as many researchers we have cited encourage and as we have noted before, multiple lines of independent evidence of paleoenvironments should be consulted. Thus one might compare the paleoenvironmental implications of large mammals with those of small mammals on the assumption that it is unlikely that the two categories of remains underwent similar taphonomic histories. Or one might compare the paleoenvironmental signal provided by mammal remains with that provided by mollusk remains, palynological analysis, and geomorphology and sedimentology. If all lines of evidence suggest the same paleoenvironmental conditions, then because it is unlikely that all these kinds of evidence were skewed so as to provide the same false paleoenvironmental signal, the analyst can have confidence in the results.

ASSUMPTION 6: TAXONOMIC IDENTIFICATION

If interpretations depend on the taxonomic identifications of remains, one must assume faunal remains have been correctly identified to taxon (Avery 2007; Behrensmeyer et al. 2007a; Bell et al. 2010; Calaby 1971; Coope 1986; Findley 1964; Gustafson 1972; Harris 1963; Rhodes 1984; Stewart 2005; Walker 1978; Woodcock 1992; Zeuner 1961). Identification to subspecies typically is impossible, and even identification to the species level is not always possible (Graham and Semken 1987). Finer taxonomic distinction typically means greater resolution because of narrower ecological tolerances (Avery 2007; Findley 1964). Subspecies have smaller geographic ranges than the species to which they belong, implying narrower ecological tolerances of the former; species have smaller geographic distributions than the genus to which they belong, implying narrower ecological tolerances; and so on. Do not misunderstand; identifications to only the genus level (or even higher taxonomic levels) may be useful for paleoecological inferences though it is likely that resolution will be coarse relative to inferences based on species-level identifications (George 2012). And while taxonomic identification is now facilitated by study of ancient DNA (e.g., Orlando and Cooper 2014; Rull 2012) and peptide fingerprinting (Buckley 2018; Buckley et al. 2009; Richter et al. 2011), these methods are destructive, they may be cost prohibitive, and they depend on the preservational condition of the remains as well as the existence of suitable reference libraries. The standard, typically used taxonomic identification procedure is our concern here. Because it is seldom discussed in detail, it is worthwhile to consider the usual identification protocol.

Simpson referred to the identification procedure as the "comparative method" – hence the oft-applied label of "comparative collection" to a collection of modern skeletons of known taxonomic identity – and noted that it required the assumption "that the bones of different [taxa] have characteristic forms, more or less constant for any one [taxon]" (Simpson 1942:144). The paleozoologist typically makes a taxonomic identification by placing two homologous bones – one the modern comparative specimen of known taxonomy, the other the prehistoric taxonomically unknown – next to one another "and looking" (Simpson 1942:145). Given the assumption that each taxon's bones have characteristic forms that "reflect similar underlying genotypes" (Tiffney 2008:136), if the two bones look alike, the unknown is "identified" as the same taxon as the known. If the bones look different, the prehistoric unknown is "identified" as being from a taxon other than the modern known with which it is being compared. As Bell et al. (2010:33) put it in their important review of the taxonomic identification process, "the primary operational method for identifying [biological] remains is based on gross morphological [and metric] resemblance."

Zooarchaeologist Barbara Lawrence (1973:397) emphasized several decades ago that taxonomic "identification is the foundation on which all subsequent analysis rests. Yet, more often than not, site reports devote very little space to discussing the criteria used for making identifications. Somehow it is assumed that differences between species are so well understood that no two workers would identify the same fragment differently" (see also Brothwell and Jones 1978). Given the particularly critical nature of accurate taxonomic identifications, and Lawrence's explicit comment about it, there is surprisingly little discussion of this fundamental first analytical step in the zooarchaeological literature in general (but see Driver 1992, 2011; Gobalet 2001; Lyman 2010a; Rea 1986; Wolverton 2013). Not so with paleontological reports. Examples of published morphometric criteria are not unusual in the paleontological literature (e.g., Barnosky 2004; Grayson 1983b; Guilday et al. 1964, 1978; Schultz 2010) but rare in the zooarchaeological literature except when a particular paleoecologically significant taxon is the focus of analysis (e.g., Grayson 1977; Lyman 2012b). Paleontological reports typically (but not always) devote numerous pages to describing the morphometric anatomical features of individual fossils that are believed to be taxonomically diagnostic. This allows the foundational data of paleoenvironmental interpretations to be examined, tested, and either used, modified, or rejected by other analysts. It is for this reason that some researchers argue that particularly important identifications must be well documented in publications (e.g., Calaby 1971; Lyman 2002). Posting the morphometric criteria used to make taxonomic identifications as part of the supplemental online material for a published study is now possible for many journals and this possibility should be exploited.

A major hurdle to taxonomic identification is that because extant species are "often differentiated based on traits (pelage color) not normally preserved in the fossil record, [the] paleozoologist must establish osteological criteria that are species specific" (Graham and Semken 1987:3; see also Avery 2007; Cheatum and Allen 1964). Describing those criteria in publications is important because they are sometimes "not statistically significant, not applicable beyond local regions, or not agreed upon by specialists" (Graham and Semken 1987:3). If the supposed taxonomically diagnostic criteria are not described, published identifications cannot be independently evaluated (Semken and Graham 1987).

Some researchers suggest identifying the remains of small-bodied mammals such as insectivores and rodents is difficult and often restricted to cranial and dental remains. Recent advances in identifying dental remains of the genus *Microtus*, a taxon with numerous species which is widespread across North America and also found in the Old World, not only demonstrate this but underscore how difficult it is to identify even dental remains of some rodent groups (compare Semken and Wallace [2002] with McGuire [2011]). Postcranial remains of this taxon are typically not studied because they cannot often be identified to genus let alone species. The same is often true for other speciose families for which postcranial morphological differences between taxa are extremely subtle (e.g., Bovidae).

As anyone who has worked with a large collection of remains that includes numerous taxa of similar morphometry knows, the identification process is infrequently simple. Walker (1978:259) summarized the issue well:

> The [taxonomic] level to which a fossil can be identified often depends on the part which is fossilized and our technical capacity for recording details of morphology … Practically every attribute of a potential or actual fossil of a species has some variance; exactness of match will be greatest where there are several attributes on which to base comparisons, each of which has small variance.

The paleozoologist must be cognizant of intrataxonomic variability and intertaxonomic variability, have knowledge of which morphometric attributes or features of each skeletal element are taxonomically diagnostic, and have a feel for how similar skeletal specimens must be to confidently identify them as from the same taxon. To learn these things requires long hours of study of comparative collections of skeletons of known taxonomy (Bell et al. 2010; Findley 1964; Lyman 2010a). As if that were not enough of a hurdle to this fundamental analytical step, nearly constant revision of the biological taxonomy (which populations should be lumped into one species, which species should be placed in one genus, etc.) exacerbates the difficulty (Avery 2007; Lundelius 1985).

Although there are several detailed studies of the taxonomically diagnostic osteological criteria that distinguish certain difficult to distinguish taxa (e.g., Balkwill and Cumbaa 1992; Castaños et al. 2014; Emslie 1982; Grayson 1977; Hargrave and Emslie 1979; Jacobson 2003, 2004; Monchot and Gendron 2010; Peters and Brink 1992), such studies are also rare relative to the diversity of species that might be represented in a prehistoric fauna. The in-place alternative is to consult paleozoological case studies that include descriptions of taxonomically diagnostic morphometric features used to identify fossils. This can at once be tedious but also extremely educational and, for those of us who relish such things, quite interesting.

ASSUMPTION 7: SAMPLE SIZE AND TAPHONOMY

The researcher must assume there are minimal sample size effects and minimal taphonomic skewing of the sample(s) under study (Harris 1963; Lundelius 1964; Redding 1978). Paleozoologists have long worried about the adequacy of their fossil samples for measuring numerous variables (e.g., Gamble 1978; Gould 1970; Guilday 1962; Guthrie 1968b; Jamniczky et al. 2003, 2008; Morlan

1984; Simpson 1937; Travouillon et al. 2011; Wolff 1975) and also about how taphonomic histories of prehistoric faunal assemblages might skew or obscure a paleoecological signal (e.g., Cheatum and Allen 1964; Dodson 1973; George 1958; Gidley and Gazin 1938; Guilday 1962, 1969; Guthrie 1968a, 1968b; Harris 1963; B. Lawrence 1968; D. R. Lawrence 1971; Lundelius 1964; McCown 1961; Simpson 1953; White 1953, 1956; Wilson 1968; Wolff 1973). We deal with both issues in some detail in Chapter 8, and touch on them elsewhere throughout subsequent chapters where relevant. Here it suffices to mention just a few things about both.

How large must a sample of faunal remains be to be statistically representative of the fauna on the landscape from which the remains originated (ignoring for the moment taphonomic issues)? García–Alix et al. (2008) indicate sample adequacy is a slippery slope; Daams et al. (1999) suggest one needs an NISP ≥100 per assemblage to have representative samples, whereas Casanovas–Vilar and Agusti (2007) suggest an NISP ≥50 per assemblage is sufficient (see also Travouillon et al. 2007). Although one could hope for such a magic bullet, no such thing exists. Sampling theory indicates that the greater the heterogeneity in the population being sampled – the more kinds of things and the more equitably things are distributed across those kinds – the larger a sample must be to correctly estimate population parameters such as the number of kinds of things (richness), how individuals are distributed across those kinds (evenness), or some combination of the two (heterogeneity) (see Chapter 8 for more thorough discussion of these variables). Thus any suggestion of a universally adequate sample size is, in our view, ill-advised.

Concern about taphonomic skewing of fossil samples continues today (e.g., Fernández-Jalvo and Andrews 2016). A well-known early consideration of taphonomic biasing is Shotwell's (1955) effort to distinguish taxa that were members of the *proximal* (geographically local relative to the paleozoological deposit) *community* from those taxa that had been transported into the location of the fossil collection and thus represented what he called the *distal* (distant, non-local) *community*(ies). Although some analysts recommended use of Shotwell's technique (e.g., Raup and Stanley 1971) and some used it (e.g., Estes and Berberian 1970), subsequent consideration of that technique showed it to be flawed (Dodson 1973; Grayson 1978; Wolff 1973); it did not do what it purported to do. For one thing, Shotwell's method was dependent on sample size and actually ignored many taphonomic processes. Further, although it was meant to measure skeletal completeness of represented carcasses, it ignored frequencies of each of the particular skeletal elements (mandibles, humeri, tibiae, etc.). Recently developed measures have addressed precisely this deficiency (Mannion and Upchurch 2010).

The level of concern regarding potential taphonomic biasing of the paleo-environmental signal of a fossil fauna seems to have increased markedly over the past decade or two, if the recent literature is any indication (e.g., Allison and Bottjer 2011; Domínguez-Rodrigo and Musiba 2010; Fernández-Jalvo and Andrews 2016; O'Connor 2005; Pickering et al. 2007; Rogers et al. 2007). In the 1960s the hope of paleozoologists was to be able to strip away the taphonomic overprint from assemblages such that any skewing or obfuscation of the environmental or anthropogenic signal of an assemblage was removed (Lawrence 1968, 1971; Lyman 1994). Because such seems to often be impossible, three techniques to contend with possible taphonomic skewing are in use today. These are the same techniques noted in our discussion of assumption 5 concerning how one might deal with changes in taxonomic presences and abundances. To briefly reiterate, compare assemblages that are *isotaphonomic*; plot the value of the variable of interest (e.g., taxonomic richness) against a measure of taphonomic modification; finally, compare the environmental signal or variable of interest with independent paleoenvironmental data.

ASSUMPTION 8: SMALL BODIES OR LARGE BODIES

It is often assumed small-bodied mammals such as most rodents and insectivores (by insectivore we mean the now defunct taxonomic order that includes shrews, moles, hedgehogs, and the like) are better ecological indicators than large-bodied mammals (Andrews 1990; Avery 1988, 1990, 2007; Bate 1937; Calaby 1971; Churcher and Wilson 1990; Falk and Semken 1998; Graham 1976, 1981, 1985b, 1991; Guthrie 1968b, 1982; Harris 1985; Hibbard 1955; Jaeger and Wesselman 1976; Mares and Willig 1994; McCown 1961; Morlan 1984; Redding 1978; Reed 2007; Semken 1966; Smith 1957; Stephens 1960; Yalden 2001). This assumption is exemplified in (and likely gained traction from) a statement authored by five individuals who, at the time, were the North American leaders in paleoenvironmental reconstruction based on paleozoological remains.

> Smaller mammals, with a narrower range of habitat and individual home range than the larger carnivores or herbivores, are influenced more by the microclimate in which they live than by regional climatic change (Burt 1958). Hence it is reasonable and a justification of uniformitarian interpretations that the ecological conclusions drawn from rodents and insectivores are similar to those drawn from the associated lower vertebrates, mollusks, and plants.
>
> (Hibbard et al. 1965:515)

Interestingly, the size threshold of how small is small is seldom acknowledged, even though Brothwell and Jones (1978) noted early on that it was unclear how small a "small mammal" was. Avery (1988, 1990, 2007) indicates "micromammals" are ≤150 g body mass but provides no warrant for that threshold. The usual definitive criterion is taxonomic; rodents and insectivores are considered to be small bodied, which of course overlooks the fact that some rodents such as North American beaver (*Castor canadensis*) are not particularly small, having an adult average body weight of 20–22 kg.

Assumption 8 is usually warranted by calling attention to properties of a small-bodied mammal. Small-bodied species are less able to migrate from an environmentally unfavorable area to a favorable one (recall from Chapter 2 the options an organism has in the face of environmental change) (Churcher and Wilson 1990; Falk and Semken 1998; Graham 1976, 1981, 1985b; Graham and Semken 1987; Harris 1985; Smith 1957); small-bodied taxa generally have relatively narrow environmental tolerances and are adapted to local (as opposed to regional) environments (Avery 2007; Bate 1937; Chaline 1977; Falk and Semken 1998; Graham 1985b; Graham and Semken 1987; Guthrie 1968b; Harris 1963, 1985; Holbrook 1975; Jaeger and Wesselman 1976; Morlan 1984; Redding 1978; Stephens 1960); small-bodied taxa have relatively small home ranges (Falk and Semken 1998; Guthrie 1968b; Smith 1957); remains of small-bodied taxa are abundant in some contexts (Avery 1990, 2007; Graham 1976, 1985b; Guilday 1971); remains of small-bodied taxa are easy to identify (Graham 1976, 1985b; Graham and Semken 1987; but see Morlan 1984); small-bodied taxa are unlikely to have served as food resources for humans and thus have not been transported far from their habitat of origin by human hunters (Avery 2007; Holbrook 1975, 1977, 1982a; Holbrook and Mackey 1976; Olsen 1982; Reed 2007).

It has been noted this assumption does not hold universally or across all taxa at all times and places. Guilday (1967) and White (1953, 1956) point out ecological tolerances are wide for some taxa and narrow for others, regardless of body size, thus before accepting this assumption at face value the analyst must know the ecological tolerances of the taxa under study in some detail. Otherwise, as noted earlier, use the modern range of a taxon as a coarse indication of the breadth of that taxon's ecological tolerances.

Small-bodied mammals are better for reconstructing paleoenvironments only when the small mammals in an assemblage have narrow tolerances relative to the large mammals in the assemblage (Lang and Harris 1984). In addition, small-bodied species tend to adapt to microenvironmental variables such as whether a hillside is north- or south-facing, whereas large-bodied species tend to be more influenced by larger-scale variables such as vegetation zones or biomes (Wilson 1973). For this reason Guilday (1967:124) observed, "Small mammals, because of their size and smaller home ranges, can survive in microhabitats barred to larger forms. They are not so drastically affected by a given climatic change as a large mammal. An environmental change, say from a park-savanna to steppe, would eliminate the large browsing forms from the fauna but would have less effect upon herbivorous rodents." Stegner (2015) reiterated this point when she noted local edaphic conditions create microhabitats that may buffer the influences of regional climatic shifts on small mammals.

Biologists have observed the responses of living mammals to climate change tend to be mediated by body size and behavioral patterns (e.g., nocturnal or diurnal) such that large-bodied species respond more often than small-bodied species, perhaps because microhabitat variability provides sufficient areas for the latter but insufficient area for the former (McCain and King 2014). In sum, although it is sometimes assumed that small-bodied species are better indicators of environmental change than large-bodied species, this assumption may be untrue in many cases. We suspect it is at least partially for this reason that Andrews (1990) implies sole reliance on microfauna can hinder efforts to determine paleoenvironments. One could, as noted above, compare the paleo-environmental implications of small-bodied mammals with those of large-bodied mammals, keeping in mind that observed similarities and differences may be telling us more about spatial scale (smaller for small-bodied species and larger for large-bodied species) than whether the inferences from the former are somehow better than those from the latter. Knowing the ecological tolerances of the species represented in an assemblage would seem a good way to evaluate the validity of assumption 8. And once again, the analyst should compare reconstructions based on as many independent lines of evidence as possible.

ASSUMPTION 9: TEMPORAL RESOLUTION

It is assumed the temporal resolution of faunal assemblages under study is sufficient to monitor the magnitude and rate of change in the environmental variable of interest (habitat composition or structure, temperature, effective moisture, etc.) (Behrensmeyer et al. 2007a; Birks et al. 2010; Graham and Semken 1987; Lundelius 1985; Morlan 1984; Wells 1978). The necessary degree of chronological resolution is dictated by two requirements of reconstruction. First, we need tight chronological control of assemblages so that they can be placed in their correct temporal order (Behrensmeyer et al. 2007a; Wells 1978) or temporally correlated (Lundelius 1985). The necessity of temporal ordering should be obvious; concluding there was environmental stasis or change demands we know the age and duration of periods represented by the faunal assemblages. The necessity for temporal correlation concerns two distinct scales. At the coarser scale, tight chronological control may facilitate recognizing if one assemblage in a comparison of two is in some important way not temporally the same as the other (same or different period, different duration) (Blaauw 2012). At the finer scale, when basing a reconstruction on multiple species, one must be sure the remains of those species are contemporaries (Lundelius 1964; Semken et al. 2010;

Stegner 2015). Stratigraphically (or analytically) mixed assemblages can mute or alter the paleoenvironmental signal (e.g., Grayson 1981; Lundelius 1964; Redding 1978; Rhodes 1984).

The second reason we need robust chronological control of assemblages concerns the desired scale and resolution of one's paleoenvironmental reconstruction (Behrensmeyer et al. 2000, 2007a). This issue has been touched on earlier, but it warrants acknowledgement here that finer temporal resolution will tend to provide finer resolution in one's reconstruction. Related to the issue of temporal resolution is Avery's (2007:21) lament that "temporal scale in paleo samples is essentially unknowable in an ecologically useful sense but it can be argued that there is still information to be gleaned." Avery's concern is with the distinction between ecological time and paleoecological time (e.g., Peterson 1977; Roy et al. 1996; Turvey and Cooper 2009). The distinction is today signified in paleobiology with the term *time averaging* (Behrensmeyer 1982; Behrensmeyer et al. 2000, 2007a; see also Peterson 1977).

Time averaging does *not* refer to some duration of accumulation of faunal remains that produces a statistical average representation of multiple faunal communities. "It is incorrect to assume that the time-averaged faunal information represents a true 'average' of the ecology of the stratigraphic interval in question" (Behrensmeyer et al. 2007a:10). Rather the concept refers to the creation of an assemblage of faunal remains that represents temporally (or perhaps spatially) unrelated populations of organisms or communities (Behrensmeyer et al. 2000). The degree of time averaging – the amount of time represented by the fossil remains in a sample – can vary from essentially none to millions of years (see Figure 2.11), depending on taphonomic and depositional processes (taphonomic time averaging; Behrensmeyer and Hook [1992]) or analytical decisions concerning the aggregation of fossil samples (analytical time averaging; Behrensmeyer and Hook [1992]). Schematic models of time averaging are shown in Figures 3.1 and 3.2. In light of column B in Figure 3.1, Avery (1990:407) may be correct when she observes "each [paleozoological] sample in effect provides evidence of average [paleoenvironmental] conditions during the period of its accumulation and it therefore follows that short-term fluctuations will not be reflected." But column C in Figure 3.1 and also Figure 3.2 is what concerns those worried about time averaging – the muting or skewing of the paleobiological, paleoenvironmental, or paleoecological signal. As Roy et al. (1996:460) observed, there is a "primary trade-off in the fossil record: a uniquely long temporal perspective but with resolution that is coarse compared with samples taken by ecologists from modern environments." The non-equivalency of a biological community documented over the course of a season or year with a paleobiological assemblage accumulated over decades or centuries is what is meant to be highlighted by the concept of time averaging.

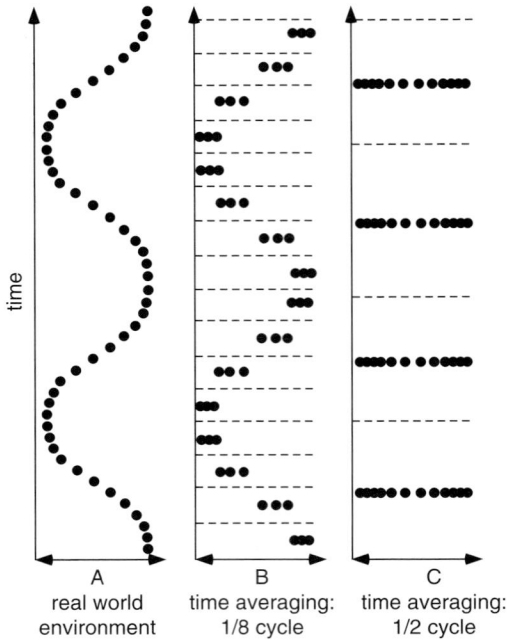

3.1. One model of time averaging (see also Figure 3.2). In all three columns, each dot represents an environmental indicator such as the presence or relative abundance of an animal species during one temporal increment such as a year or a decade. (A) the environment (e.g., climate) shifts causing the environmental indicator to shift over time. (B) time averaging resulting from stratigraphic lumping of three temporal increments per stratum, or 1/8 of a complete environmental cycle; note that a paleoenvironmental signal would still be present but with muted temporal resolution. (C) time averaging resulting from stratigraphic lumping of twelve temporal increments per stratum, or 1/2 of the complete environmental cycle; note that the paleoenvironmental signal is completely muted. Redrawn after Hopley and Maslin (2010:33).

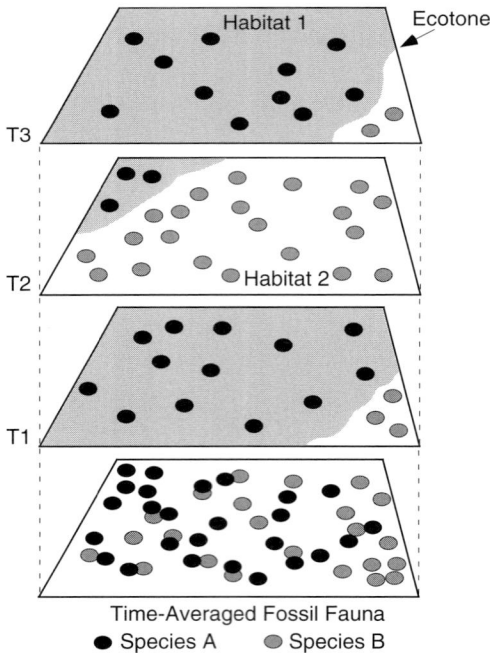

3.2. Another model of time averaging (see also Figure 3.1). Two distinct habitats, each with individuals (dots) of a distinct species restricted to that habitat, are separated by an ecotone. The ecotone is in one position at T3 (time 3), a different position at T2 (time 2), and another position in T1 (time 1); assume its location at any one time is a function of prevailing climate and its different locations signify different climates. If remains of each individual organism are deposited where the organism is shown to be living at T1, T2, or T3, the fossil assemblage in the bottom polygon will result. Study of that fossil assemblage will not reveal the shifting location of the ecotone nor suggest different climatic conditions existed during the period when the fossils were deposited. Redrawn after Behrensmeyer et al. (2007a:8).

North American paleozoologists have long recognized that collections of prehistoric animal remains are not necessarily equivalent to an ecological unit such as a community that has a limited spatio-temporal distribution because it was documented by a biologist over a season or two (Tedford 1970). This is reflected by the paleozoological concepts of *local fauna* and *faunule*. The former was defined by Hibbard (1958:3) as "an association of identifiable remains of animal life of the same age which have been collected from a restricted geographic area."

> A local fauna may represent one or more microenvironments of an area, with the relative abundance of the various species reflecting selective sorting, differential preservation, population density of an access-ible microhabitat, or the regional fauna of the area. As a rule, several microhabitats can be identified within a local fauna and the sum of these ecological associations gives a fairly accurate picture of regional ecology.
> (Semken 1966:167)

Faunules are the parts of a local fauna that can be stratigraphically and/or arch-aeologically distinguished (Graham 1981; Graham and Semken 1987; Walker 1982). Local faunas typically comprise multiple temporally (and perhaps spa-tially) unique communities, and each faunule represents a temporal duration of some length that may exceed the existence of an ecological-time community.

All assemblages of plant or animal remains are time averaged to some degree (Figure 2.11), but problems only arise when the duration of time averaging represents a span of time longer than the ecological or climatic event of interest (Roy et al. 1996). If an episode of environmental change spanned, say, a century, and the faunal assemblage under study accumulated during that 100 years plus 200 years before the episode and 200 years after the episode, that assemblage would be time averaged to a degree that prohibits recognition of the environ-mental event (Figure 3.1, column C). This does not mean that time averaging renders fossil assemblages analytically worthless, as implied in column B of Figure 3.1 (e.g., Behrensmeyer and Chapman 1993; Kidwell and Tomasovych 2013; Kowalewski 1996; Olszewski 1999). For one thing, time averaging mutes idiosyncratic (day to day weather) and seasonal fluctuation, things that may not be of interest. Attempts to analytically monitor time averaging have thus far focused on climatic events of multi-thousand-year durations (Hopley and Maslin 2010).

A time-averaged fauna could represent "an ecological *palimpsest*" (Su and Harrison 2007:282) (Figure 3.2). This means the assemblage(s) under study is(are) temporally coarse-grained (low resolution) with respect to the paleo-environmental variable(s) of interest. *Time perspectivism* (Holdaway and Wandsnider 2008) refers to aligning the temporal scale (duration) of the event or process of interest – the hypothesis being tested or research question answered, and the analytical techniques used – with the scale and temporal resolution of the assemblage(s) under investigation. Thus paleoecologists have astutely observed "to the extent that theories based on living communities do not explain the dynamics, structure, or patterns of change in ancient ones, we will need to modify, expand, or reject these theories and move toward a more comprehensive understanding of species associations in time and space" (Wing et al. 1992:10).

Paleontologists have come to recognize that whether or not time averaging poses an interpretive problem depends not just on the formational history of the assemblage(s) of remains under study (i.e., the amount of time averaging involved), but also on the scale and resolution of the environmental variable in the research question being asked. Alignment of the two is critical to valid analysis. Simulations mentioned earlier that used modern faunas with known habitat affiliations to indicate that time averaging does not always completely obscure or even significantly mute the paleoecological signal of a collection of faunal remains are important (Andrews 2006; Le Fur et al. 2011). But they do not encourage conceptualizing appropriate temporal scales and resolutions. They may in fact be misleading given that such simulations do not mean all time-averaged faunal assemblages will provide accurate paleoenvironmental signals of some desired scale and resolution (Figures 3.1 and 3.2).

Because fossil samples are often small (NISP per assemblage is low), some paleozoologists lump together assemblages of the same or nearly the same age to achieve sample sizes that are sufficiently large to foster confidence that analytical results are not a function of non-representative samples (e.g., Bobe and Eck 2001). Some research shows that such analytical time averaging may produce a rather different paleozoological signal than the original, non-lumped assemblages (Lyman 2003, 2010b). Deciding which of the two signals is the correct one requires independent evidence of paleoenvironmental change that correlates with one or the other of the two paleozoological signals. And this brings us back to the assumption that led us to this point – that the paleozoological record is of sufficient temporal resolution to align with those independent paleoenvironmental data.

ASSUMPTION 10: ECOTONE

It is assumed an *ecotone* (a transition area between two habitats or biomes; e.g., see Figure 2.2) or community edge is the best place to detect environmental change (Avery 1982; King and Graham 1981; Louthan et al. 2015; Rosvold et al. 2013; Semken 1980; Taylor 1965; Tchernov 1975; Woodcock 1992). This in turn seems to presume that at least some species are found in only one or a few kinds of habitat or biome (it is recognized some species are tightly adapted to habitat edges); outside of that area species are beyond their ecological tolerances (see Figure 2.4). A degree or two colder winters or warmer summers, a couple centimeters more (or less) annual precipitation, and either stress becomes too great and the species becomes locally extirpated or stress is reduced and the species increases in abundance, potentially leading to changes in a community's species composition that may be detected in the paleozoological record. The

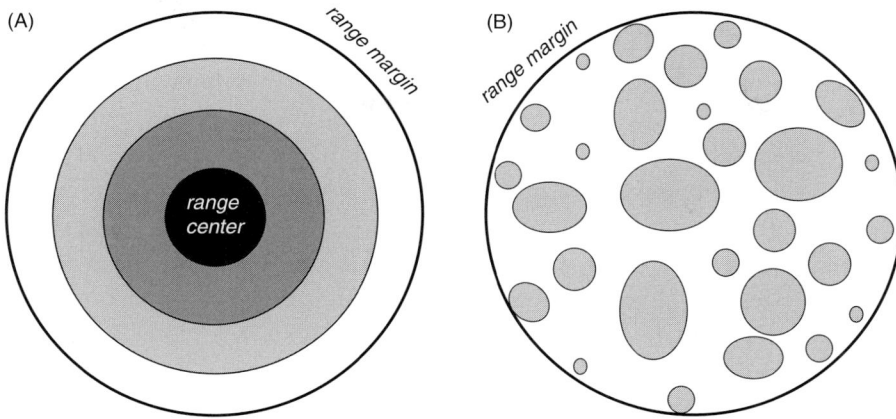

3.3. Models of the distribution of a species. (A) idealized distribution showing population density decline from range center to range margin; (B) individuals are discontinuously distributed across the range (size of oval within range indicates size and distribution of populations). After Guo (2014).

converse of this is equally important. If the paleoenvironment of the fossil site is within the "minimal stress zone" of the tolerance curve for a species whose remains are represented (Figure 2.4), a drop in winter temperatures of a couple degrees or the addition of a few centimeters of precipitation each year likely will not have a noticeable impact on the species of concern because it is well within its optimal tolerance range, in which case environmental changes may not be apparent in the faunal record.

A version of this assumption appears regularly in the biological literature and is referred to as the *central–marginal hypothesis* (e.g., Eckert et al. 2008; Guo 2014). Originally, and as implied by Figure 2.4, it was thought a species would display an abundance or frequency distribution such that few individuals occur near range boundaries and numerous individuals occur near the center of the species' range (e.g., Erasmus et al. 2002; Swihart et al. 2003) (Figure 3.3A). However, it is now clear this presumed geographic frequency distribution is idealized; reality tends to be more complex (Eckert et al. 2008; Guo 2014; Ries et al. 2004), with more or less isolated populations varying in size and density (Guo et al. 2005) (Figure 3.3B). Both the size and isolation of local populations may or may not increase as one moves from the center of the species' range to the range margin. Study of modern communities has shown that the sensitivity of a species to climatic change relates to its range size, niche breadth (specialized or generalized) (Thuiller et al. 2005), and the environmental gradients within the species' range (Guo et al. 2005; Holt and Keitt 2005). Thus, although the boundary of a species' range will suggest that species' niche edges and the limits of its ecological tolerances, the relationship may be less than perfect.

North American archaeologists may remember Rhoades's (1978) nearly forty-year-old scolding of archaeologists who simplistically borrowed the ecotone concept and the related notion of *edge effect*. In his view, where archaeologists had gone wrong was to conceive ecotones as transitions between communities or habitats that contained plants and animals representative of the adjacent communities and also organisms unique to the ecotone, and thus ecotones were thought to have more diverse taxa in greater frequencies than individual community centers (= edge effect) (after Odum 1971). Rhoades pointed out this essentialist, typological ontology was being replaced by a model of communities not as discrete entities but rather as more or less gradual continua from one to the other (e.g., Whittaker 1975). The replacement was sped along by mounting evidence for Gleason's (1926) individualistic hypothesis. This is not to say that one cannot sometimes detect distinct boundaries to at least plant communities (see Figure 2.3), but rather that the real world is more complex than one or the other simplistic model implies.

What does all of this have to do with reconstructing paleoenviroments on the basis of paleozoological remains? Assumption 10 seems to rest on one or both of the models represented in Figure 3.3. Both rest on the model of ecological tolerances in Figure 2.4 with the margins of those tolerances for a particular species corresponding to the margins of that species' geographic range, and the center of that species' range corresponding to the optimal tolerance zone. In other words the horizontal axis of Figure 2.4 represents not only gradients of environmental variables but also a corresponding geographic (latitude, longitude) gradient. Such a correspondence may, however, not exist. And that is the point of modern studies of the central–marginal hypothesis.

Assumption 10 is seldom mentioned in the paleozoological literature. It is likely infrequently mentioned because locations of prehistoric range edges (such as might have existed at the time the fossil assemblage under study was accumulated and deposited) are not always known. Whether a site that has produced remains of a particular species was near the edge of that species' range at the time the remains were deposited is also difficult to know. We say "difficult" rather than "impossible" because as our knowledge grows regarding the prehistoric distribution of particular species, so too does our knowledge of the locations of the boundaries of a species' range at particular times in the past. In North America, the development of the FAUNMAP database in 1994 (Graham and Lundelius 1994) and its evolution into the FAUNMAP II database (Graham 2005) and later the Miomap/Faunmap database is a prime example of our increasing knowledge. Alternatively, to approximate the prehistoric location of a range edge, the researcher can examine the modern distribution of a species that is of analytical interest and presume the modern edge of the range approximates the tolerance limits for that taxon. Or consider only more obvious range edges based on non-faunal data, such as those between terrestrial and aquatic habitats (e.g., lake shore, river bank). Whatever analytical path the researcher takes, if it involves assumption 10, the researcher should be cognizant of the central–marginal hypothesis and the key aspect of this hypothesis that there may be only a weak or no correspondence of the models in Figure 2.4 and Figure 3.3 for the species under consideration.

DISCUSSION

Any scientific research – by which we mean research centered on empirical phenomena and guided by a hunch, a hypothesis, or a theory (Lastrucci 1963) – rests on one or more assumptions about one or more aspects of the phenomena under study, such as how things interact under particular conditions. Sometimes the requisite assumption is implicit, but we prefer explicitness as the former can lead to misunderstanding. Further, unexpected or inconsistent analytical results suggest one or more of the requisite assumptions is inapplicable or inaccurate. And this is when we learn something about how that portion of the world under investigation works. A clear example in paleoenvironmental reconstruction is assumption 1 and its use revealing when and where niche conservatism (a good label for assumption 1) does not hold.

We have endeavored in this chapter to identify and describe the general and more or less universal analytical assumptions requisite to paleoenvironmental reconstruction based on paleozoological remains. Our focus has largely been on remains identified to taxon because those identifications often comprise the data upon which analysis and interpretation rest. Some of these and some other assumptions underpin analytical techniques that do not depend on taxonomic identifications, the so-called *taxon-free methods* described in Chapter 7, for instance. There are other assumptions, again often implicit, that are specific to a particular analytical technique, and we reserve discussion of these for chapters in which we describe those analytical techniques.

In the second edition of their informative *Reconstructing Quaternary Environments*, Lowe and Walker (1997:162) indicate that analysis of ancient biological remains for purposes of paleoenvironmental reconstruction involves "uniformitarian principles, namely that a knowledge of the factors that influence the abundance and distribution of contemporary organisms enables inferences to be made about environmental controls on plant and animal populations in the past." They add that "if the uniformitarian approach is to be employed [implying, strangely in our view, that it might not be] … [it must be assumed] that present plant and animal populations are in equilibrium with [the environmental parameters that govern their distributions and abundances]" and that "the ecological affinities of plants and animals have not changed through time" (Lowe and Walker 1997:163). The last of course concerns assumption 1 and niche conservatism, and the equilibrium part of their discussion seems to be concerned that a species – whether the modern analog or the ancient organism – was not in the process of responding to an environmental shift but had attained stasis in distribution, abundance, or both. We consider the last to be included in recognition of the differences between a fundamental and a realized niche.

In a more recent edited volume specifically concerned with paleoenvironmental reconstruction based on paleozoological remains (Bobe et al. 2007), most contributors do not mention the assumptions discussed here. In their introductory chapter, the editors (Behrensmeyer et al. 2007a) allude to what we discuss under assumptions 2, 6, 7, and 9, and worry particularly about sampling and taphonomic issues. In their concluding chapter the editors (Behrensmeyer et al. 2007b) indicate the volume's contributors hope for more actualistic research on modern faunal communities in order to derive general interpretive analogs and to avoid a transferred ecology that precludes the possibility of identifying non-analog faunas and environments. The volume is valuable and informative, and implicit adoption of many of the interpretive assumptions described here throughout its pages underscores the foundational nature of those assumptions: without them, there could be no paleoenvironmental reconstruction. It is with this in mind, and our twin goals of producing a useful text for students and an informative reminder to experienced researchers so that they do not become complacent, that we offer the discussion in this chapter. With this groundwork and that of ecological basics outlined in Chapter 2 out of the way, we can now turn to how to put this knowledge to work.

FOUR

BACKGROUND OF SELECT
PALEOZOOLOGICAL SAMPLES

The following chapters of this book (Chapters 5–10) summarize and illustrate the diverse suite of analytical techniques used to develop paleoenvironmental reconstructions from paleozoological assemblages. Though our discussion of those techniques draws upon a variety of assemblages from different times and places around the world, we illustrate our analyses using the same faunal assemblages as often as possible. We hope this commonality will allow the reader to focus on variability in the analytical techniques rather than on variability in the faunal assemblages.

The faunas we routinely turn to include the late Quaternary micromammals (rodents and assorted insectivores <0.15 kg adult body mass) and macromammals (mammals >0.75 kg adult body mass) from Boomplaas Cave in South Africa. These faunas are zooarchaeological in the sense that they were recovered from deposits that include abundant archaeological material, though as we outline below this does not mean that humans accumulated all of the faunal remains. We also consider the late Quaternary small mammals (rodents and lagomorphs) from Homestead Cave in Utah (western United States). The Homestead Cave faunas are paleontological; human occupation of the site was limited, there are very few artifacts, and there is no evidence to implicate people in the accumulation of the faunal remains. We selected these sites for several reasons. First, both provide stratified sequences that span long periods of time and encompass substantial environmental changes (based on associated non-faunal data). Second, they provide sufficiently large samples to reasonably illustrate how various analytical techniques work. There is also substantial variability in sample sizes between assemblages, providing us with an opportunity to illustrate how to contend with sampling issues. And lastly, both sets of faunas come from sites associated with a large body of published literature concerning the stratigraphy, chronology, paleoenvironments, and archaeology.

Because we frequently turn to these sites in the remainder of this book, we provide a brief discussion of each below. We have included in this discussion a synopsis of their relevant paleoenvironmental histories, emphasizing previous inferences derived from the faunas. These brief summaries are not meant to represent the definitive paleoenvironmental histories of each site or the respective regions in which they are found. Rather we hope that highlighting a few key patterns that will emerge in our forthcoming paleoenvironmental analyses will make it easier for our readers to follow and critically engage with those analyses. For those readers interested in delving into the environmental history in more detail, we recommend Chase and Meadows (2007) and Marean et al. (2014) for reviews of environmental archives relevant to Boomplaas Cave. Grayson's (2011) *The Great Basin: A Natural Prehistory* is the definitive source for paleoenvironmental records relevant to Homestead Cave.

BOOMPLAAS CAVE

Boomplaas Cave is a key late Quaternary archaeological and paleoenvironmental archive for southern Africa's Cape Floristic Region. This region comprises an area of ~88,000 km² along the southern and western-most portion of southern Africa, including the mountains of the Cape Fold Belt and the coastal lowlands. The Cape Floristic Region is best known for its spectacular floristic diversity, including the world's highest frequency of endemic plant species (Goldblatt and Manning 2002; Linder 2003), but to archaeologists it is also well known for its Middle and Later Stone Age archaeological sites that feature prominently in our understanding of modern human origins, with some of the best-known sites including the Klasies River Mouth caves, Blombos Cave, and the Pinnacle Point caves.

Boomplaas Cave is situated at an elevation of ~700 m above sea level within the cliffs of a limestone seam on the southern foothills of the Swartberg mountain range, approximately 60 m above the Cango Valley. The east–west trending Swartberg range forms the northern boundary of the intermontane basin known as the Klein Karoo, with the Outeniqua range marking its southern boundary 50 km to the south. The lowlands of the Klein Karoo, which sit in the rain-shadow of the Outeniquas, are a semi-desert; rainfall is higher in the mountainous uplands and Boomplaas Cave receives around 400 mm annual precipitation. Compared with much of the Klein Karoo, the Cango Valley is well-watered by streams draining from the flanks of the Swartberg and into the eastward-flowing Grobelaars River (at the foot of Boomplaas Cave) and the westward-flowing Matjes River (10 km west of Boomplaas Cave).

The vegetation in the immediate vicinity of Boomplaas Cave is part of a transitional shrubland whose component species vary in relation to temperature and rainfall gradients from the valley floor up the slopes of the Swartberg (see Vlok and Schutte-Vlok [2010] for a detailed summary). The transitional shrublands are dominated by single shrub species, though grasses and short-lived herbs flourish after fires. In the low-lying areas just south of Boomplaas Cave occurs a shrubby habitat known as renosterveld, characterized by the renosterbos (*Elytropappus rhinocerotis*) and a sparse understory of grasses. Along watercourses and ravines in the Cango Valley are more densely wooded habitats that include sweet thorn trees (*Acacia karroo*) and ironwood (*Olea* spp.) among others (Moffett and Deacon 1977). The transitional shrublands give way to fynbos habitats – hard-leaved evergreen shrublands typically dominated by restios, ericas, and proteas – as one moves up the slopes of the Swartberg. These include a grassy fynbos habitat known as waboomveld, indicated by the presence of *Protea nitida* (waboom or wagon tree) and relatively abundant grasses, just north of Boomplaas. The grasses that occur in the vicinity of Boomplaas include a mix of C_3 (cool-season) and C_4 (warm-season) species, reflecting the fact that rainfall is fairly evenly distributed through the year.

Excavations conducted by Hilary Deacon (University of Stellenbosch) from 1974 to 1979 uncovered a stratified sequence extending to 5 m in depth and spanning the past >65,000 years (H. J. Deacon 1979, 1995; H. J. Deacon and Brooker 1976; H. J. Deacon et al. 1984; see also J. Deacon 1984). Deacon (1979) organized the stratigraphy according to a hierarchical scheme of stratigraphic members, units, and sub-units. We use the coarser-scale members in our analyses primarily because these stratigraphic aggregates provide larger sample sizes. Our goal in these analyses is to illustrate the application of certain techniques, so finer-scale stratigraphic and temporal control – which might be important if our goal were to address particular paleoenvironmental questions – is not needed here.

The chronology of the Boomplaas Cave deposits is supported by radiocarbon dates (primarily on charcoal) for the middle to upper portions of the sequence and a combination of amino acid racemization (AAR) on ostrich eggshell (Miller et al. 1999) and U-series ages on speleothems (Vogel 2001) for the lower section. The Boomplaas Cave chronology is summarized in Table 4.1. The lowest dated member (OCH) is associated with a broad range of age estimates but it includes Middle Stone Age artifacts attributed to the Howieson's Poort industry, which has been dated elsewhere in southern Africa by optically stimulated luminescence to ~59 to 66 ka (Jacobs and Roberts 2017). The basal member (LOH) is estimated to date to 80 ka (H. J. Deacon 1979) but this is not supported by any radiometric age estimates.

TABLE 4.1 *The stratigraphy and chronology of Boomplaas Cave. Radiocarbon dates reported here are those obtained on charcoal. Age ranges represent calibrated Bayesian models from Sealy et al. (2016).*

Member	Age	Modeled age range (kcal BP)
DGL	1,630 ± 50 (^{14}C)	1.6 to 1.4
	1,700 ± 50 (^{14}C)	
	1,510 ± 75 (^{14}C)	
BLD	1,955 ± 65 (^{14}C)	2.3 to 1.6
BLA	6,400 ± 75 (^{14}C)	8.0 to 6.4
BRL	9,100 ± 135 (^{14}C)	12.3 to 10.1
	10,425 ± 125 (^{14}C)	
CL	12,060 ± 105 (^{14}C)	16.9 to 13.9
	12,480 ± 130 (^{14}C)	
	14,200 ± 240 (^{14}C)	
GWA	17,830 ± 180 (^{14}C)	22.5 to 20.6
LP	Undated	23.1 to 22.2
LPC	21,110 ± 420 (^{14}C)	25.8 to 25.1
	21,220 ± 195 (^{14}C)	
YOL	–	32.3 to 25.8
BP	32,400 ± 700 (^{14}C)	39.7 to 36.0
	33,920 ± 770 (^{14}C)	
OLP	37,400 ± 1370 (^{14}C)	42.9 to 40.3
	44,000 ± 4,000 (AAR)	
BOL	–	–
OCH	>49,000 (^{14}C)	–
	56,000 ± 6,000 or	
	65,000 ± 6,000 (AAR)	
	59,000 ± 2,000 (U-Series)	
	64,000 ± 2,000 (U-Series)	
	66,000 ± 7,000 (U-Series)	
LOH	–	–

Boomplaas was excavated using 3 mm mesh screens, though select 1×1 m excavation squares were sieved through 2 mm mesh screens to enhance recovery of the microfauna (Avery 1982). The recovered material has been reported in numerous publications spanning the past several decades. These include reports on the cultural remains (H. J. Deacon et al. 1976; H. J. Deacon et al. 1978; J. Deacon 1984), the chronology (Miller et al. 1999; Vogel 2001), fossil charcoal and pollen (H. J. Deacon et al. 1983; Scholtz 1986), micromammals (Avery 1982, 2004; Thackeray 1987), macromammals (Brink 1999; Driesch and Deacon 1985; Faith 2013a; Klein 1978, 1983), and isotope geochemistry of ungulate tooth enamel (Sealy et al. 2016).

Table 4.2 reports Avery's (1982) taxonomic abundances (minimum number of individuals = MNI) for the rodents and insectivores (shrews, elephant shrews,

TABLE 4.2 *Taxonomic abundances (MNI) for the Boomplaas Cave microfauna (after Avery 1982).*

Family	Taxon	DGL	BLD	BLA	BRL	CL	GWA	LP	LPC	YOL	BP	OLP	BOL	OCH	LOH
Chrysochloridae	*Chlorotalpa duthieae*	2	11	1	11	40	60	61	5	16	58	447	53	10	3
Macroscelididae	*Elephantulus edwardii*	8	12	6	46	8	1	1	0	1	6	11	5	1	1
Soricidae	*Myosorex varius*	44	69	41	100	138	505	463	56	87	638	6443	657	128	33
	Suncus varilla	2	7	10	22	19	32	25	2	7	64	890	33	7	2
	Crocidura cyanea	24	37	10	44	34	0	0	0	0	30	277	13	7	2
	Crocidura flavescens	91	164	67	126	93	29	47	3	10	49	170	51	15	3
Bathyergidae	*Cryptomys hottentotus*	128	224	79	200	279	106	89	3	19	155	560	100	49	5
Gliridae	*Graphiurus ocularis*	1	1	0	3	4	4	4	0	0	8	37	10	4	0
Nesomyidae	*Dendromus melanotis*	1	3	2	8	7	18	26	2	5	48	464	47	4	1
	Dendromus mesomelas	0	0	0	5	2	0	1	0	0	2	25	1	1	0
	Mystromys albicaudatus	32	89	39	60	29	5	6	1	1	9	102	15	10	3
	Steatomys krebsii	6	27	10	10	0	0	0	0	0	0	0	0	0	0
	Saccostomus campestris	10	20	3	6	0	0	0	0	0	0	0	0	0	0
Muridae	*Acomys subspinosus*	4	9	4	16	9	5	4	1	1	16	131	20	2	1
	Aethomys namaquensis	73	99	55	275	124	33	37	3	4	91	387	60	13	3
	Dasymys incomtus	2	3	1	5	4	2	2	0	1	10	15	6	2	1
	Mus minutoides	3	8	6	6	3	6	7	1	1	18	313	18	1	1
	Myomyscus verreauxii	9	9	5	53	15	3	8	1	1	14	81	18	2	1
	Rhabdomys pumilio	17	39	6	27	19	8	13	3	1	18	117	20	7	1
	Gerbilliscus afra	0	5	3	11	1	0	0	0	0	3	9	0	0	1
	Gerbillurus paeba	0	2	1	4	1	0	0	0	0	0	3	1	0	1
	Otomys laminatus	5	6	3	17	14	3	1	0	2	4	32	5	2	0
	Otomys saundersiae	53	103	50	175	189	777	761	38	97	709	3113	588	170	35
	Otomys irroratus	233	362	191	504	411	128	203	10	44	297	1583	379	84	22
	Otomys unisulcatus	9	26	11	28	7	98	80	6	17	17	29	14	10	0

and golden moles) from Boomplaas Cave. The sample includes more than 30,000 individuals distributed across twenty-five species. Based primarily on an assessment of the ecology of the prey species, most of which are nocturnal, Avery (1982) suggests that barn owls (*Tyto alba*) were the likely accumulators of the assemblage, an argument consistent with the presence of modern barn owl roosts in rockshelters adjacent to Boomplaas Cave. The micromammals are especially dense in deposits lacking archaeological remains (H. J. Deacon 1979), suggesting they were deposited when the cave was unoccupied by people.

The Boomplaas macromammal data are derived from specimen counts (typically referred to as number of identified specimens, or NISP) provided by Faith (2013a), reported here in Table 4.3. Note that Faith (2013a) did not examine the faunas from the uppermost pastoralist occupation (member DGL), which is dominated by sheep. The sample includes more than 6,400 specimens distributed across thirty-six non-overlapping taxa, though many of our ana-lyses focus specifically on the ungulates (>2,600 specimens distributed across twenty-one non-overlapping taxa). Given the highly fragmentary nature of the Boomplaas Cave material, which rendered most specimens unidentifiable to lower taxonomic groups, the vast majority of taxonomic identifications for ungulates are based on dental remains. Analysis of bone surface modifications of those specimens corresponding in size to the ungulate taxa (>5 kg) at Boomplaas Cave indicates a complex taphonomic history of bone accumu-lation (Faith 2013a). The mammals from the bottom of the sequence were accumulated primarily by carnivores – leopards (*Panthera pardus*) being a likely candidate – with large raptors, probably the Cape eagle owl (*Bubo capensis*), also introducing remains belonging to the smallest bovids (*Oreotragus oreotragus* and *Raphicerus* spp.). From members BOL to GWA, there are variable amounts of bone accumulation related to people, carnivores, and raptors, with the anthropogenic component related mainly to the largest ungulate species. And in the upper members (CL and above), people accumulated most of the faunal remains. This complex taphonomic history poses some challenges for interpreting the environmental implications of the Boomplaas macromammals, and we discuss how this might be dealt with in subsequent chapters.

Paleoenvironmental Summary

Our task of providing a summary of the environmental history is complicated by the fact that some of the most basic details concerning the Cape Floristic Region's paleoenvironments – were glacial phases of the Pleistocene wetter or drier than the present? – are actively debated (e.g., Chase and Meadows 2007; Chase et al. 2018; Faith 2013b; Marean et al. 2014). The debate is not due to a lack of data – the Cape Floristic Region has been a focus of paleoenviron-mental research for decades (e.g., J. Deacon and Lancaster 1988) – but instead reflects a combination of seemingly contradictory lines of evidence coupled with a good measure of not yet well-understood regional variation. With this in mind, we focus below on what has been inferred from the Boomplaas mammals.

From the base of the sequence to the Last Glacial Maximum, the large mammals are interpreted as indicating a transition from shrubland habitats – perhaps not unlike the contemporary vegetation – to open grassland, with the grasslands replaced by shrubland at the onset of the Holocene (Faith 2013a; Klein 1978, 1983). Isotopic analysis of the Last Glacial Maximum grazers indicates a dominance of C_3 grasses in the diet, implying an intensification of winter rainfall systems in the region (Sealy et al. 2016). The vegetation history inferred from the microfauna complements this scenario, though Avery (1982) documents other subtle changes superimposed on this general trend. Avery (1982) and Thackeray (1987) provide independent analyses of the microfauna indicating a general decline in temperatures from the base of the sequence to the Last Glacial Maximum, with the Holocene characterized by the warmest temperatures in the sequence.

An important point of contention concerns the precipitation history. Previous interpretations of the Boomplaas faunas are in complete opposition, with the Last Glacial Maximum interpreted as either the driest portion of the sequence (Avery 1982; H. J. Deacon et al. 1984; Thackeray 1987) or the wettest (Faith 2013a, 2013b). These contradictions are worth keeping in mind, if only because they demonstrate that faunal-based paleoenvironmental reconstructions are neither infallible nor unambiguous – far from it! As is the case with all paleoenvironmental archives, confidence in interpretation is enhanced whenever multiple lines of evidence are in agreement. There are paleoenvironmental records not far from Boomplaas Cave (~70 km west) that indicate greater moisture availability during the Last Glacial Maximum compared with the Holocene (Chase et al. 2018), though the implications of environmental archives from elsewhere in the Cape Floristic Region are less clear.

HOMESTEAD CAVE

Homestead Cave provides perhaps the most detailed late Quaternary mammal sequence for the Great Basin (Grayson 2006, 2011), the vast region of internal drainage in the arid western United States. Homestead Cave is located at the northwestern-most spur of the Lakeside Mountains just west of the Great Salt Lake in north-central Utah. This low-lying spur, known as Homestead Knoll, is a rocky promontory lacking active springs or perennial streams and receiving very little rainfall throughout the year (~225 mm). The cave is formed within a small limestone ridge and sits at an elevation of 1,406 m, approximately 100

TABLE 4.3 *Taxonomic abundances (NISP) for the Boomplaas Cave macromammals (after Faith 2013a).*

Family	Taxon	BLD	BLA	BRL	CL	GWA	LP	LPC	YOL	BP	OLP	BOL	OCH	LOH
Lagomorpha	*Lepus capensis*	17	2	46	9	0	1	1	2	1	2	7	4	3
	Bunolagus monticularis	98	21	94	41	1	0	1	4	16	26	6	11	0
	Leporidae indet.	13	2	18	18	0	3	1	0	8	6	1	2	0
Rodentia	*Hystrix africaeaustralis*	8	3	14	4	0	0	0	0	0	0	4	0	0
Primates	*Papio ursinus*	215	34	78	19	0	2	4	13	6	16	16	101	12
Carnivora	*Canis* cf. *mesomelas*	1	1	0	1	0	0	0	1	0	1	0	0	0
	Lycaon pictus	0	0	0	0	0	0	0	0	0	0	0	4	0
	Mellivora capensis	2	0	0	0	0	0	0	0	0	0	0	0	0
	Genetta sp.	1	0	0	0	0	0	0	0	0	0	0	1	0
	Herpestes ichneumon	2	0	0	0	0	0	0	0	0	0	0	0	0
	Herpestes pulverulentus	0	0	3	1	1	0	0	0	0	1	0	0	0
	Herpestes sp.	2	0	0	0	0	0	0	0	0	0	0	0	0
	Hyaenidae indet.	0	0	0	0	0	0	0	0	0	0	2	0	0
	Caracal/Leptailurus	4	0	1	2	1	0	0	0	1	1	0	1	1
	Felis silvestris	21	1	0	0	0	0	0	0	0	0	0	0	0
	Panthera pardus	2	1	0	0	0	0	0	1	1	4	2	14	2
	Felidae indet.	0	0	1	0	0	0	0	0	0	0	0	0	0
Hyracoidea	*Procavia capensis*	1377	126	258	293	32	20	13	36	44	101	100	187	95
Equidae	*Equus capensis*	0	0	0	15	1	2	1	0	0	0	0	0	0
	Equus zebra/quagga	14	4	53	419	28	24	10	20	12	4	2	3	0
Suidae	*Potamochoerus larvatus*	0	0	1	0	0	0	0	0	0	0	0	0	0
Bovidae	*Taurotragus oryx*	1	0	9	55	5	0	2	0	2	0	0	7	0
	Tragelaphus strepsiceros	0	0	11	0	0	0	0	0	0	0	0	0	0

Species													
Tragelaphini indet.	1	0	8	15	0	0	0	0	2	0	0	1	0
Hippotragus leucophaeus	3	0	4	11	1	2	0	0	0	0	0	2	0
Hippotragus equinus	0	0	1	5	0	0	0	0	0	0	0	0	0
Hippotragus sp.	0	0	11	15	1	0	0	1	2	0	0	1	0
Redunca fulvorufula	21	1	2	4	0	0	0	0	2	7	5	27	1
Redunca arundinum	0	1	1	0	0	0	0	0	0	0	1	2	2
Redunca sp.	35	2	5	4	1	0	0	0	0	4	2	28	2
Alcelaphus buselaphus	12	0	1	0	0	1	0	0	2	0	0	0	0
Connochaetes cf. *taurinus*	0	0	0	0	0	0	0	0	6	0	0	0	0
Connochaetes cf. *gnou*	0	0	0	0	0	1	5	0	6	1	0	0	0
Connochaetes / Alcelaphus	0	0	4	11	11	5	13	2	46	1	0	1	0
Damaliscus cf. *dorcas*	0	0	4	0	1	1	1	0	11	3	0	8	2
Alcelaphini indet.	0	0	1	0	3	1	0	0	23	2	0	0	0
Extinct caprin	0	0	75	233	8	4	4	0	3	0	2	1	0
Pelea capreolus	17	10	31	17	3	2	5	12	13	22	24	80	21
Antidorcas cf. *marsupialis*	0	0	0	4	1	1	1	0	0	1	1	10	4
Oreotragus oreotragus	65	19	162	17	0	2	7	7	7	16	11	32	2
Raphicerus melanotis	2	1	2	0	0	0	0	0	0	1	1	1	1
Raphicerus campestris	0	0	1	0	0	0	0	0	0	0	0	0	0
Raphicerus sp.	63	16	83	14	0	1	2	1	6	11	28	49	8
Oreotragus / Raphicerus	42	8	126	17	0	0	0	1	2	4	13	19	7
Syncerus antiquus	0	0	0	4	5	5	0	0	0	0	1	0	0
Syncerus caffer	0	0	2	11	2	0	0	0	0	1	0	2	0
Syncerus sp.	0	0	0	2	0	0	0	0	0	0	0	0	0

m above the valley floor. To the immediate west and northwest is the saline playa of Pleistocene Lake Bonneville, the pluvial lake that formerly covered much of western Utah until the Pleistocene came to an end. Although the playa is barren, the vegetation on the knoll itself is dominated by grasses and shrubs – the dominants being shadscale (*Atriplex confertifolia*) and horsebrushes (*Tetradymia* spp.) – with a few scattered junipers (*Juniperus osteosperma*). Greasewood (*Sarcobatus vermiculatus*) and big sagebrush (*Artemisia tridentata*) are common on the valley floor, along with invasive cheat grasses (*Bromus* spp.).

Excavations at Homestead Cave were directed by David Madsen (Utah Geological Survey) in 1993 (Madsen 2000). His team excavated a 1 × 1 m square down to bedrock – at a depth of ~2.7 m – providing a finely stratified sequence that begins ~13,000 years ago and continues into historic times. The stratigraphy is aggregated according to eighteen analytical units, from Stratum I at the base to Stratum XVIII at the top (Table 4.4). The original chronology reported by Madsen (2000) is provided by a series of twenty-one radiocarbon dates on various organic materials (e.g., fecal pellets, hackberry endocarps, charcoal), with an additional eighty radiocarbon dates obtained on kangaroo rat (*Dipodomys* spp.) femora more recently provided by Terry and Novak (2015). For the sake of simplicity, we report Madsen's (2000) chronology in Table 4.4.

Excavated deposits were passed through 1/4″ (6.4 mm), 1/8″ (3.2 mm), and 1/16″ (1.6 mm) mesh screens, from which organic and (rare) cultural remains were recovered. Madsen's (2000) monograph, which includes contributions from a variety of specialists, provides an excellent account of the excavated materials (see also Madsen et al. 2001). There are numerous other reports on Homestead Cave, including studies of the fecal pellets from woodrats (*Neotoma* spp.) (Smith and Betancourt 2003) and artiodactyls (Broughton et al. 2008), fishes (Broughton 2000; Broughton et al. 2000, 2006), mammals (Grayson 1998, 2000b; Grayson and Madsen 2000; Lyman and O'Brien 2005; Rowe and Terry 2014; Terry 2007, 2010a; Terry and Rowe 2015; Terry et al. 2011), and the chronology (Terry and Novak 2015).

The very limited evidence for human occupation of Homestead Cave, in contrast to sites elsewhere in the Bonneville Basin, is probably related to the lack of available water. But this did not detract from the suitability of the cave for owls. Roosting screech owls (*Megascops kennicottii*) and dense piles of owl pellets were observed in the cave when excavations began (Madsen 2000), and owl pellets in various states of decay were found throughout the sequence, with many of the recovered fossils having remains of pellets adhering to them (Grayson 2000a, 2000b). Like the micromammals from Boomplaas Cave, owls accumulated the vast majority of the Homestead faunal assemblage, which is dominated by rodents and lagomorphs. There are rare remains of large mammals, including artiodactyls and carnivores. These are represented

TABLE 4.4 *The stratigraphy and chronology of Homestead Cave (after Madsen 2000). Radiocarbon ages are calibrated (2σ range) using OxCal 4.3 (Bronk Ramsey 2009) and the IntCal13 calibration curve (Reimer et al. 2013).*

Stratum	¹⁴C age	Cal yrs BP
XVIII	–	
XVII	1,020 ± 40	799–1,051
XVI	1,200 ± 50	986–1,264
XV	–	
XIV	2,850 ± 50	2,848–3,143
XIII	3,480 ± 40	3,640–3,849
XII	3,400 ± 60	3,483–3,830
XI	–	
X	5,330 ± 65	5,946–6,278
IX	–	
VIII	–	
VII	6,160 ± 85	6,802–7,260
	6,185 ± 105	6,797–7,313
VI	7,120 ± 70	7,791–8,154
V	8,230 ± 69	9,022–9,406
IV	8,195 ± 85	8,996–9,425
III	–	
II	8,520 ± 80	9,320–9,682
	8,790 ± 80	9,561–10,154
	8,830 ± 240	9,241–10,564
I (upper 5 cm)	10,160 ± 85	11,396–12,127
	10,350 ± 80	11,836–12,527
I (general)	10,910 ± 60	12,696–12,942
I (lower 5 cm)	11,065 ± 105	12,729–13,096
	11,181 ± 85	12,811–13,213
	11,263 ± 83	12,975–13,303
	11,270 ± 135	13,796–14,892

primarily by small bones of the hands and feet (e.g., carpals, phalanges) and are thought to have been introduced by woodrats.

Our analyses of the Homestead Cave faunas make use of Grayson's (2000a) specimen counts (NISP) for rodents and lagomorphs (Table 4.5). Grayson's (2000a) data are based on identification of all mammals from the 1/4″ (6.4 mm) and 1/8″ (3.2 mm) sample fractions from fourteen of the eighteen stratigraphic units. Only the kangaroo rats (*Dipodomys* spp.) from Stratum X were identified so this stratum is not considered here. As is clear from Table 4.5, sample sizes are massive, with counts for individual assemblages ranging from 1,045 in Stratum XVIII – a solid figure by most paleozoological standards – to a whopping 28,525 in Stratum IV. These impressive samples are precisely why

TABLE 4.5 *Taxonomic abundances (NISP) for the Homestead Cave small mammals (after Grayson 2000a).*

Family	Taxon	I	II	III	IV	V	VI	VII	VIII	IX	XI	XII	XVI	XVII	XVIII
Sciuridae	*Ammospermophilus* sp.	0	0	0	0	0	0	0	3	8	0	0	0	0	0
	Ammospermophilus cf. *leucurus*	0	0	0	5	2	10	11	4	6	18	6	7	6	0
	Ammospermophilus leucurus	2	6	5	110	19	123	56	37	88	41	117	26	41	1
	Tamias sp.	2	0	0	0	0	0	0	0	0	0	0	0	0	0
	Tamias minimus	2	0	0	0	0	0	0	0	0	0	0	0	0	0
	Marmota cf. *flaviventris*	30	4	0	7	0	1	2	0	0	0	0	0	0	0
	Marmota flaviventris	13	4	0	8	0	3	0	1	1	0	0	0	0	0
	Urocitellus sp.	0	1	2	3	0	0	10	23	32	0	0	0	0	0
	Urocitellus cf. *mollis*	0	0	1	8	1	39	18	0	89	183	306	52	119	10
	Urocitellus mollis	5	4	4	38	17	148	76	54	227	205	523	231	556	35
Geomyidae	*Thomomys* sp.	107	404	238	2952	506	2492	1153	665	1573	520	1144	393	1008	48
	Thomomys bottae	0	30	18	2158	35	129	86	44	141	42	79	57	87	4
	Thomomys talpoides	2	0	0	0	0	0	0	0	0	0	0	0	0	0
Heteromyidae	*Chaetodipus formosus*	2	2	1	3	0	1	0	0	0	0	6	2	14	0
	Dipodomys sp.	310	1212	964	12629	2713	14016	9086	5286	14378	6343	15518	4048	9692	671
	Dipodomys microps	7	83	75	1033	245	1094	775	451	1075	467	1201	276	704	48
	Dipodomys ordii	43	34	17	50	7	63	7	5	24	10	34	11	22	1
	Microdipodops sp.	1	7	0	0	0	0	0	0	0	0	0	0	0	0
	Microdipodops megacephalus	6	10	0	0	0	0	0	0	0	0	0	0	0	0
	Perognathus longimembris	0	4	8	77	7	37	21	12	22	3	12	18	36	2
	Perognathus parvus	121	86	9	12	1	4	1	0	0	0	0	2	6	0

	1	2	3	4	5	6	7	8	9	10	11	12	13	14
Cricetidae														
Lemmiscus curtatus	0	9	3	0	1	2	0	3	0	2	2	8	121	552
Microtus sp.	0	1	2	1	4	4	1	2	16	3	53	44	197	247
Neotoma sp.	48	50	26	56	57	180	53	147	182	43	258	59	150	50
Neotoma cf. *cinerea*	0	7	0	0	0	0	4	3	3	0	196	250	1274	2310
Neotoma cinerea	0	2	0	0	0	1	1	1	2	1	46	56	234	267
Neotoma cf. *lepida*	118	1810	660	2322	1257	2454	807	1340	2873	786	4281	277	224	37
Neotoma lepida	11	394	90	287	225	522	178	287	572	144	883	68	56	4
Ondatra zibethicus	0	1	0	0	0	2	0	0	3	0	1	0	0	0
Onychomys sp.	0	5	0	0	1	3	1	1	0	0	0	0	4	7
Onychomys leucogaster	0	10	5	5	1	4	1	1	2	0	5	1	4	8
Peromyscus sp.	17	205	61	187	49	147	52	88	178	52	531	205	1124	1550
Pitimys sp.	0	0	0	0	0	0	0	0	0	0	0	0	0	1
Reithrodontomys sp.	0	8	0	13	0	6	1	2	9	5	36	4	39	4
Reithrodontomys cf. *megalotis*	0	0	0	0	0	0	0	0	0	0	4	0	0	0
Reithrodontomys megalotis	1	88	33	12	3	10	5	5	31	7	94	18	52	54
Leporidae														
Brachylagus idahoensis	0	0	0	0	1	0	1	1	4	1	4	3	32	192
Lepus sp.	14	355	138	642	420	618	407	422	806	202	680	91	577	2243
Lepus californicus	0	0	0	0	0	0	0	0	0	0	0	0	2	0
Lepus townsendii	0	0	0	0	0	0	0	0	0	0	0	0	2	18
Sylvilagus sp.	15	294	109	349	221	424	181	278	1443	295	2332	450	1832	2020
Sylvilagus cf. *audubonii*	0	6	2	6	2	5	1	0	8	2	13	6	13	2
Sylvilagus cf. *nuttallii*	1	1	1	1	3	3	2	4	7	5	13	1	20	28

the Homestead Cave faunas feature so prominently in the biogeographic histories of Great Basin mammals.

Paleoenvironmental Summary

The Great Basin has a spectacularly well-documented late Quaternary environmental history derived from geological evidence, plant macrofossil and pollen archives, and small mammal fossil assemblages (Grayson 2011). The Homestead Cave mammals have been used to inform on the nature of past climate change during the late Pleistocene and Holocene, as well as to understand the response of species to previously documented climatic changes during the middle Holocene (e.g., Grayson 1998, 2000a, 2000b; Lyman and O'Brien 2005). Consistent with other paleoenvironmental indicators – including faunal assemblages from elsewhere in the Bonneville Basin (Schmitt and Lupo 2012; Schmitt et al. 2002) – the Homestead mammals have been interpreted as indicating a late Pleistocene and early Holocene that was moister and cooler than what came afterwards. These conditions are suggested to have favored an expansion of sagebrush habitats with a prominent grass understory. A variety of sources indicate a middle Holocene that was warmer and drier than what came before or after, and this too has been inferred from the Homestead mammals. The mammals suggest that this phase of reduced moisture availability was associated with a decline of sagebrush and expansion of shadscale (*Atriplex confertifolia*), a shrub found in dry sediments that are highly saline. After the phase of middle Holocene aridity, environmental conditions broadly similar to the present prevailed.

SUMMARY

Boomplaas Cave and Homestead Cave are, in some important ways, ideal collections with which to illustrate the variety of analytical techniques described in subsequent chapters of this volume. They are well studied and well known, they produced large samples for each of several chronologically tightly controlled stratigraphically delimited assemblages, the collections represent temporal spans known to include major episodes of climatic variability, and the taphonomic histories of the assemblages of each are sufficiently well known as to not introduce insurmountable biases or skewing of paleoenvironmental signals.

Not all collections of ancient faunal remains provide such exemplary samples as Boomplaas Cave and Homestead Cave, so do not be misled into thinking all collections are of equal value. As should be clear from Chapter 3 and this chapter, not only do analyzing and interpreting all collections require certain analytical assumptions, some collections may simply not be amenable to some kinds of analysis for any of a plethora of reasons. We thus call upon a variety of collections to illustrate particular analytical techniques or to underscore certain points in subsequent pages. It is our hope that, as we indicated earlier, in frequently referring to the same collections the reader need not focus too much on the particulars of those collections but instead can focus on the techniques under discussion. With the background of this and preceding chapters in hand, it is now time to turn to the focus of the volume, the analytical techniques that have been used to manipulate faunal data in such a way as to reveal their paleoenvironmental implications.

FIVE

ENVIRONMENTAL RECONSTRUCTIONS BASED ON THE PRESENCE/ABSENCE OF TAXA

Whether analyzing a zooarchaeological assemblage excavated from a rockshelter in southwest France or a paleontological assemblage surface-collected from fluvial deposits in Alaska, there are certain procedures universally shared by faunal analysts. Over a period of days, weeks, months, or (often) longer, fossil remains are painstakingly identified to skeletal element and to the lowest taxon possible. Taxonomic identifications are usually made through direct comparison of the fossil remains with specimens of known taxonomic identity (see Chapter 3, assumption 6), although published morphological descriptions, illustrations, and the like may also be used. Once the identification phase is completed, a catalogue of those taxa present in the assemblage is compiled and some measure of abundance is provided for each taxon.

Although it can take years to learn the anatomical subtleties required to confidently identify fossil remains to taxon, amassing this technical skillset is largely a function of time and effort. What comes next can be far more challenging. Once taxonomic identifications are completed, faunal analysts are faced with "lists of identified specimens with which they must now do something" (Grayson 1984b:16). That "something" can be any number of things (e.g., Reitz and Wing 2008). One of those things is the subject of our discussion in this chapter: how do we get from a taxonomic list to a paleoenvironmental reconstruction? The generation and analysis of such lists has long been a staple of paleontology (Bate 1937; Howard 1930; Stock 1929) and zooarchaeology

(Grayson 1984b; Klein and Cruz-Uribe 1984; see Lyman 2015a for some history), so it comes as no surprise that the data contained within them, namely observations on the presence/absence and abundance of species, underpins many of the analytical tools for reconstructing paleoenvironments from faunal remains. Our aim in this chapter is to describe the analytical techniques that focus on the presence/absence of taxa in an assemblage of faunal remains. We discuss other techniques that consider other variables in subsequent chapters.

A BIT OF HISTORY

During the nineteenth century, paleontologists and naturalists of Western Europe grappled with an extraordinary and puzzling phenomenon. A burgeoning sample of Pleistocene fossil assemblages excavated from rockshelters, caves, and ancient fluvial systems documented the occurrence of arctic mammals, such as reindeer (*Rangifer tarandus*), musk-ox (*Ovibos moschatus*), and lemmings (*Lemmus* spp.), in apparent stratigraphic association with species whose modern relatives are found in temperate and even tropical environments, including hippopotamus (*Hippopotamus antiquus*), hyena (*Crocuta crocuta spelaea*), and lion (*Panthera leo spelaea*). Explaining these remarkable associations prompted decades of speculation and debate (see reviews in Grayson 1983a, 1984a; O'Connor 2007) on topics ranging from site formation processes and stratigraphic mixing (Anderson 1875; Dawkins 1874; Gieke 1881) to the migratory capabilities of the hippopotamus (Dawkins 1869, 1871; Gieke 1872; Lyell 1863), and especially relevant to our purposes, the nature of Pleistocene climates (Dawkins 1869; Lartet 1867, 1875).

The co-occurrence of reindeer, musk-ox, and hippopotamus was particularly inspiring for those interested in the climate of Pleistocene Europe. Recognizing the contrasting environments inhabited by contemporary representatives of these species, paleontologist Edouard Lartet (1867, 1875) proposed that this unusual association reflected a more equable climate, with cooler summers permitting the persistence of reindeer and musk-ox, and warmer winters allowing the survival of hippopotamus. Geologist W. Boyd Dawkins suggested a very different scenario. He refused to accept that climates suitable for the arctic group of mammals would also be suitable for hippos, noting "we have no reason to believe that the powers of resisting heat or cold possessed by these animals differed from those which they now possess" (Dawkins 1871:392). While he did not use the term niche conservatism (see Chapter 3, assumption 1), the assumption of unchanging ecological tolerances is precisely what led Dawkins to formulate his climatic inferences. He likened Pleistocene winters to those of northern Siberia, remarking "[t]he presence of the arctic group of mammals, the Reindeer, Musk-sheep [musk-ox today], and the like, implies that the climate under which they lived was severe" (Dawkins 1869:215). At the same time, however, the presence of hippopotamus must imply warm summer temperatures because "so far that we can judge by the habits of the living species, [it] could not have endured the low temperatures now necessary for the wellbeing of the Musk-sheep or reindeer" (1869:215). Echoing previous speculations of Charles Lyell (1863) concerning the extraordinary migratory capabilities of the hippopotamus, Dawkins (1869, 1871) reconciled the co-occurrence of warm-adapted and cold-adapted species by suggesting a migratory system in which northern mammals moved south in the winter and southern species moved north in the summer. According to Dawkins, and in direct opposition to Lartet (1867, 1875), this was a reflection of a Pleistocene climate characterized by greater seasonal extremes than found in the present.

While this nineteenth-century debate corresponds to a time when absolute chronologies were inaccessible and our understanding of past climate and environment (not to mention the migratory capabilities of hippos!) was in its infancy, Dawkins' reliance on the taxonomic identity of a fossil organism to infer paleoenvironmental conditions by way of explicit analogy with its living representatives would, as we showed in Chapter 3, not be out of place today (or in the earlier literature; see Lyell 1832). For example, in their study of the late Pleistocene fauna from Jaguar Cave in Idaho, Guilday and Adam (1967:29) reasoned that the presence of collared lemming (*Dicrostonyx* sp.), "an obligatory tundra form with a long evolutionary association with a boreal climate," is indicative of a tundra-like environment. Similarly, Klein et al. (1991:104) suggested "the presence of lechwe [*Kobus leche*], an antelope that is closely tied to shallowly inundated floodplains or swamps," in late Pleistocene deposits at Equus Cave in South Africa implies a more humid climate. Similar logic is even extended deeper in time. Price et al. (2009:44) noted the presence of an extinct koala (*Phascolarctos ?stirtoni*) in Pliocene deposits from Chinchilla in eastern Australia is indicative of woodland habitats, given that its living relatives are "mostly restricted to open forests and woodlands."

As we showed in Chapter 3, such paleoenvironmental interpretations are based on the assumptions of taxonomic uniformitarianism and niche conservatism when interpreting extant taxa. As we also showed in Chapter 3, when the fossil taxon is an extinct form, one assumes the ecological tolerances of that extinct taxon were the same as those of its nearest living relative. And recall also that, as we outlined in Chapter 2, stenotopic species (those with narrow ecological tolerances) are more informative of past environments than are eurytopic species (those with a broad range of tolerances).

WHAT ABOUT THE ABSENCE OF A TAXON?

Paleoenvironmental reconstructions tend to emphasize those species that are present in an assemblage rather than those that are absent. However, there are occasional cases where the absence of a taxon is interpreted as the absence of environments preferred by that taxon. For example, in his study of the Middle Pleistocene hyena den assemblage from Swartklip at False Bay in South Africa, Klein (1975:278) remarks "the absence of Vaal rhebuck (*Pelea capreolus*), klipspringer (*Oreotragus oreotragus*), rock hyrax (*Procavia capensis*), and baboon (*Papio ursinus*), all historic inhabitants of rocky, high-relief habitats in the South-Western Cape, implies an environment in which the modern Swartklip cliffs did not exist." The cliffs to which he refers are carved into the lithified remnants of a coastal dune succession (aeolianites) that extended seaward in the past, but which have been subject to ongoing erosion and truncation due to the Holocene marine transgression (Roberts et al. 2009). Klein's inference is reasonable considering that the missing taxa are known from the region historically (Skead 2011), are well documented in other late Quaternary assemblages in the region (Klein 1972, 1978; Klein and Cruz-Uribe 2000; Schweitzer and Wilson 1982), and could therefore be expected to have occurred in the Swartklip assemblage if suitable terrain were present (see Chapter 3, assumption 4). Thus, the absence of this set of taxa is both conspicuous and potentially informative of the past environment.

As we discussed in Chapter 3 under assumption 4, there is a fundamental problem with paleoenvironmental inferences derived from the absence of a taxon in a fossil assemblage. The absence of a taxon is not evidence of a taxon's absence in the past faunal community (e.g., Guthrie 1968b; Simpson 1937; White 1954). This is why paleozoologists typically refer to presence/absence data as asymmetrical, in the sense that the interpretive emphasis is placed on taxonomic presences rather than absences (Ervynck 1999; Grayson 1981; Lyman 2008a). Barring long-distance transport of skeletal remains or stratigraphic mixture, the presence of a taxon in a faunal assemblage provides reasonably secure evidence of its presence in the corresponding local faunal community, in which case it is also reasonable to use that taxon as a source of paleoenvironmental information. Not so for the absence of a taxon, which could be a reflection of unsuitable environments but also any number of taphonomic or sampling issues, in addition to ecological factors such as the presence of competitors or predators that might prohibit a species from occupying otherwise suitable environments.

FROM FAUNAL LIST TO PALEOENVIRONMENTAL RECONSTRUCTION

There are numerous ways one can go about making the inferential leap from faunal list to paleoenvironmental reconstruction. While there is no right or wrong way, a reasonable starting point for geologically recent faunas is to compile a list of those taxa one expects to observe in the contemporary or historic environment, drawing from modern wildlife censuses, historic accounts, or knowledge of the habitats found in the region today. This provides a base-line for identifying differences in taxonomic composition between fossil and modern communities that may be indicative of past environmental change. The next step is to evaluate those taxa found in a given assemblage, drawing from published observations of their modern ecological tolerances, and syn-thesize this information into a coherent paleoenvironmental interpretation.

Using the Boomplaas Cave ungulates as an example, our task is complicated by human impacts in historic times. Because of over-hunting and espe-cially landscape transformation in the colonial era, the present distribution of ungulates in the region is a poor reflection of their former historic distributions (Boshoff and Kerley 2001; Boshoff et al. 2001; Kerley et al. 2003). As noted in Chapter 2, modern data likely reflect the realized as opposed to the fundamental niche of many species. However, based on historic accounts (Skead 2011) and modern vegetation (Boshoff and Kerley 2001; Vlok and Schutte-Vlok 2010), those ungulates certainly or likely found in the vicinity of Boomplaas during the past several hundred years include Cape mountain zebra (*Equus zebra*), moun-tain reedbuck (*Redunca fulvorufula*), grey rhebuck (*Pelea capreolus*), klipspringer (*Oreotragus oreotragus*), steenbok (*Raphicerus campestris*), and grysbok (*R. melanotis*), in addition to a handful of others (Table 5.1). Examining the presence of ungu-late species across the Boomplaas Cave sequence (Table 5.1), we note that all of those taxa one can reasonably expect to have occurred in the vicinity of Boomplaas Cave are present, as are ten additional ungulate species (four extinct and six extant) whose presence is not anticipated by historic accounts or by con-temporary vegetation. Their occurrence is potentially indicative of past environ-ments distinct from the shrubland habitats found there today.

An initial assessment of the faunas from each assemblage reveals a not atyp-ical problem with historical records. Beginning with the top of the sequence, the late Holocene assemblage from member BLD (2.3 to 1.6 ka) documents many of the ungulates we expect to have occurred in the historic environ-ment, including Cape mountain zebra, mountain reedbuck, grey rhebuck, klipspringer, steenbok/grysbok, hartebeest, and eland. The recently extinct (as of ~1800 AD) blue antelope (*Hippotragus leucophaeus*), for which historic accounts are known only from the coastal lowlands ~200 km south-southwest of Boomplaas Cave (Skead 2011), represents the sole species not included on our historic list. Its presence could be indicative of environments unlike those today, but there are also many gaps in the historical record (Boshoff and Kerley 2001; Boshoff et al. 2001; Faith 2012a). Considering the handful of late Holocene assemblages documenting blue antelope outside the distri-bution inferred from historic accounts (Inskeep 1987; Klein and Cruz-Uribe 1987, 2000; Schweitzer and Wilson 1982), its presence in BLD at Boomplaas Cave may not be a reflection of different environments, but rather a reflection of limitations in the historical record.

TABLE 5.1 *The presence of ungulate species through time at Boomplaas Cave.*

Taxon		BLD	BLA	BRL	CL	GWA	LP	LPC	YOL	BP	OLP	BOL	OCH	LOH
Historical ungulates														
Cape mountain zebra	*Equus zebra*[a]	x	x	x	x	x	x	x	x	x	x	x	x	x
Bushpig	*Potamochoerus larvatus*			x										
Eland	*Taurotragus oryx*	x		x	x	x		x		x			x	x
Greater kudu	*Tragelaphus strepsiceros*			x										
Mountain reedbuck	*Redunca fulvorufula*	x	x	x	x					x	x	x	x	x
Grey rhebuck	*Pelea capreolus*	x	x	x	x	x		x	x	x	x	x	x	x
Hartebeest	*Alcelaphus buselaphus*	x		x			x			x				
Klipspringer	*Oreotragus oreotragus*	x	x	x	x		x	x	x	x	x	x	x	x
Steenbok/grysbok	*Raphicerus* spp.	x	x	x	x		x	x	x	x		x	x	x
Buffalo	*Syncerus caffer*			x	x	x					x		x	
Extinct and extralimital ungulates														
Cape zebra	*Equus capensis*				x	x	x	x						
Roan antelope	*Hippotragus equinus*			x	x									
Blue antelope	*Hippotragus leucophaeus*	x	x	x	x	x	x						x	
Southern reedbuck	*Redunca arundinum*			x								x	x	x
Black wildebeest	*Connochaetes* cf. *gnou*						x	x		x	x			
Blue wildebeest	*Connochaetes* cf. *taurinus*									x				
Bontebok	*Damaliscus* cf. *dorcas*			x	x	x	x			x	x	x	x	
Extinct caprin	*Caprini*			x	x	x	x			x		x	x	
Springbok	*Antidorcas* cf. *marsupialis*				x	x	x	x			x	x	x	x
Long-horn buffalo	*Syncerus antiquus*				x	x	x					x		

Note: [a] Specimens assigned to *Equus* cf. *zebra.*

We next examine the environmental preferences of the BLD ungulates and synthesize this information into a paleoenvironmental interpretation. We cannot emphasize enough that knowledge of the species involved – especially their environmental predilections – is crucial. In this case, our task is greatly facilitated by excellent zoological observations (Kingdon and Hoffman 2013; Skinner and Chimimba 2005). Synthesizing the paleoenvironmental implications of the recovered taxa often can be streamlined by considering groups of taxa with similar ecological requirements (e.g., Semken 1980), although there are certainly cases where individual taxa have particularly important implications (e.g., Guilday and Adam 1967). Briefly, the presence of mountain reedbuck, grey rhebuck, klipspringer, and grysbok is expected of shrubland vegetation in mountainous terrain, with Cape mountain zebra, hartebeest, and blue antelope indicative of at least some grasses. Eland is found in a wide variety of habitats (i.e., it is *eurybiomic*), limiting its value as a paleoenvironmental indicator, but its presence is not surprising. Together, this set of taxa from member BLD, all of which we can reasonably expect to have occurred in the vicinity of Boomplaas Cave in historic times, implies an environment similar to the present.

Ungulates in the Last Glacial Maximum assemblage from member LPC (25.8 to 25.1 ka) suggest a very different environment. While several of the taxa indicative of shrubby and mountainous terrain are present (grey rhebuck, klipspringer, and steenbok/grysbok), LPC also includes black wildebeest (*Connochaetes* cf. *gnou*) and springbok (*Antidorcas* cf. *marsupialis*), both of which are absent from the region in historic times, in addition to two extinct taxa, the Cape zebra (*Equus capensis*) and an unnamed caprin (Table 5.1). Black wildebeest and springbok are closely associated with open and grassy habitats today, and a similar environmental preference is also inferred for the extinct species (Faith 2014; Klein 1980). The implication is a Last Glacial Maximum environment that was grassier than the present, with a reduced shrubland component.

The preceding highlights the fact that providing a stratum-by-stratum assessment of all taxa found at Boomplaas Cave or anywhere else could become a massive and cumbersome undertaking. In some cases, providing such detail may be the objective of analysis. Particularly in the case of long stratified sequences, however, our main objective might (and often does) instead involve the identification and interpretation of temporal trends in the taxonomic compositions of faunas (often referred to as *faunal turnover*). How did the environment change through time? While this can be addressed through qualitative analysis or even visual inspection of the presence/absence list of taxa, making sense of the distribution of numerous taxa across many stratigraphic intervals can pose a substantial interpretive challenge. Fortunately this task can be greatly facilitated through quantitative procedures. We turn to such approaches next.

ORDINATION

Broadly speaking, *ordination* involves putting phenomena in order (P. Legendre and L. Legendre 2012). With respect to the analysis of biotic communities, ordination is typically used to arrange phenomena such as faunas (or faunal assemblages in the case of paleozoology) along one or more axes according to species composition (ter Braak 1987). One of the principal advantages of ordination in faunal analysis is that it allows us to reduce multiple variables, namely the many species in a faunal assemblage, into a smaller number of index variables that ideally account for a substantial amount of the variation in faunal composition (e.g., Semken and Graham 1996). These can be plotted in two or three dimensions, greatly facilitating interpretations of changes in faunal composition.

There are two distinct approaches to ordination. The first is known as constrained ordination or direct gradient analysis. In such analyses, the taxonomic composition of faunal samples is ordinated relative to one or more known environmental variables associated with each sample, with changes in species composition considered a response to the environmental gradient. Constrained ordinations are common in ecology, where it is possible to directly measure environmental variables associated with samples, but they are rarely applied in paleozoological contexts because the environmental data are unavailable (but see Semken and Graham 1996). For this reason, they are not considered here. The second approach, known as unconstrained ordination or indirect gradient analysis, is suitable for those cases where the objective is to infer environmental change from changes in taxonomic composition. Unconstrained ordination arranges faunal samples according to variation in taxonomic composition, and one examines this arrangement to assess whether it is revealing environmental information.

Correspondence Analysis

Correspondence analysis (CA) is perhaps the most widely used ordination method in paleozoological contexts (e.g., Alemseged 2003; de Ruiter et al. 2008; Dortch and Wright 2010; Faith 2013a; Matthews et al. 2011; Rector and Reed 2010; D. N. Reed 2007; K. E. Reed 2008; Semken and Graham 1996; Tryon and Faith 2016; Tryon et al. 2010). Following Greenacre and Vrba (1984:986), CA is "a technique for displaying the rows and columns of a data matrix … as points in corresponding low-dimensional space." In other words, when the relevant data matrix is a table of species occurrences or abundances across faunal samples (e.g., Table 5.1), CA ordinates both the taxa and samples (i.e., assemblages) in a small number of dimensions, often displayed in a bivariate scatterplot. The ability to simultaneously ordinate taxa and samples contributes substantially to the appeal of CA in paleozoological analysis, as it allows examination of the relationship between taxa and samples. In contrast to some other ordination techniques (e.g., principal components analysis), CA is well suited for ordination of assemblages and taxa where the taxa have a unimodal response to the environmental gradient (see the ecological tolerance curve in Figure 2.4).

We refer the reader elsewhere for details on the computations underlying CA (Greenacre and Vrba 1984; P. Legendre and L. Legendre 2012; Manley 2005), and provide here only the key points concerning interpretation of results. CA ordinates species occurrences or abundances – it is suitable for both (P. Legendre and L. Legendre 2012) – across faunal assemblages based on the contribution of each cell to the χ^2 statistic for the overall matrix, producing a number of axes equal to one less than the number of species or assemblages, whichever is fewer. These axes account for progressively less of the inertia, which roughly corresponds to the variation in faunal composition, meaning that typically only the first two or three axes are examined. Remaining axes are rarely considered, because they explain relatively little variation in the data. When displayed in two dimensions on a bivariate scatterplot, assemblages that plot close together are similar in taxonomic composition, while taxa plot nearest to those assemblages with which they are closely associated. It is important to note that taxa with very few occurrences in the matrix contribute disproportionately to the total inertia, meaning that the dominant pattern in the ordination may be driven by the presence or absence of a single rare taxon. Some suggest excluding infrequent taxa from the analysis, because the objective of CA is to highlight the primary axes of variation in faunal composition, not to emphasize the outliers (P. Legendre and L. Legendre 2012). Then again, it might also be the case that infrequent taxa have important paleoenvironmental implications, in which case they might be retained. We suggest that the decision to exclude rare taxa is a judgment call best informed by initial exploration of the data, such as we discussed above regarding the extinct blue antelope and the historic record.

When the objective is to identify environmental trends across faunal assemblages through time or space, a successful CA is one that has ordinated assemblages according to an underlying environmental gradient. CA does not tell us what that gradient might be, and it may be the case that the ordination instead reflects taphonomic (e.g., Semken and Graham 1996) or sampling issues (see below). Provided that assemblages are ordinated according to an environmental gradient, identification of the gradient is usually a matter of interpretation informed by an evaluation of the ecological preferences of the taxa in relation to their positions along the ordination axes. For example, if the CA ordinates taxa such that those with positive scores along the first axis inhabit warm climates and those with negative scores inhabit cool climates, it would be reasonable to infer that this axis is related to temperature. One could then proceed with interpreting the position of assemblages along the first axis in terms of temperature change (positive scores = warm; negative scores = cold). In most paleozoological applications of CA, typically only the first one or two axes provide useful environmental insight.

To illustrate CA we turn again to the Boomplaas Cave ungulates using data from Table 5.1. Those taxa occurring only once in the sequence are excluded because inspection of a CA with all taxa indicates two of them, bushpig (*Potamochoerus larvatus*) and greater kudu (*Tragelaphus strepsiceros*), contribute disproportionately to the first axis. We also excluded blue wildebeest (*Connochaetes* cf. *taurinus*). Its occurrence in member BP represents the only known record of this taxon in the Cape Floristic Region, suggesting we have sacrificed potentially important paleoenvironmental information by omitting it. But including it would skew results and render interpretation difficult. With the three taxa excluded, the first three axes of the CA account for 59.3% of the inertia (axis 1: 25.3%, axis 2: 18.0%, axis 3: 16.0%). These values are low compared with ordinations making use of abundance data, but they are typical of ordinations based on taxonomic presences.

Our interpretation of the CA begins with examination of the ordination along axis 1 (Figure 5.1A). Because the presence and absence of taxa in a faunal assemblage may be influenced by sample size (Grayson 1984b; Lyman 2008b) – an issue that we deal with in greater detail in the next chapter – it is prudent to first note that the ordination of assemblages along this axis is uncorrelated with (log-transformed) assemblage sample size (r = -0.283, p = 0.485). This means that variation in the ordination of assemblages, and the taxa associated with them, is not a sampling artifact, and may instead be reflecting paleoenvironmental change. We therefore may, with confidence that the ordering is not significantly influenced by sample sizes, turn our attention to the ordination of the taxa. With the exception of the southern reedbuck (*Redunca arundinum*), all taxa with positive scores on this axis are on our historic list (Table 5.1) and are indicative of shrubby mountainous terrain (mountain reedbuck, klipspringer, steenbok/grysbok, grey rhebuck, and Cape mountain zebra). Like the extinct blue antelope, southern reedbuck is yet another taxon for which historic records are vague (Skead 2011). Kingdon and Hoffman (2013:428) note its "most significant habitats in South Africa are valleys in which the grass cover is tall (or there is suitable herbaceous cover) and permanent water is available." The nearby Cango Valley would have provided such habitat whenever there was sufficiently dense vegetation cover along the banks of its rivers, in which case its association with the historic group is not incongruous.

Among those species with negative scores on axis 1, several are extinct (*Equus capensis*, *Syncerus antiquus*, *Hippotragus leucophaeus*, and an unnamed caprin) or unlikely to occur in shrubland habitats like those occurring near

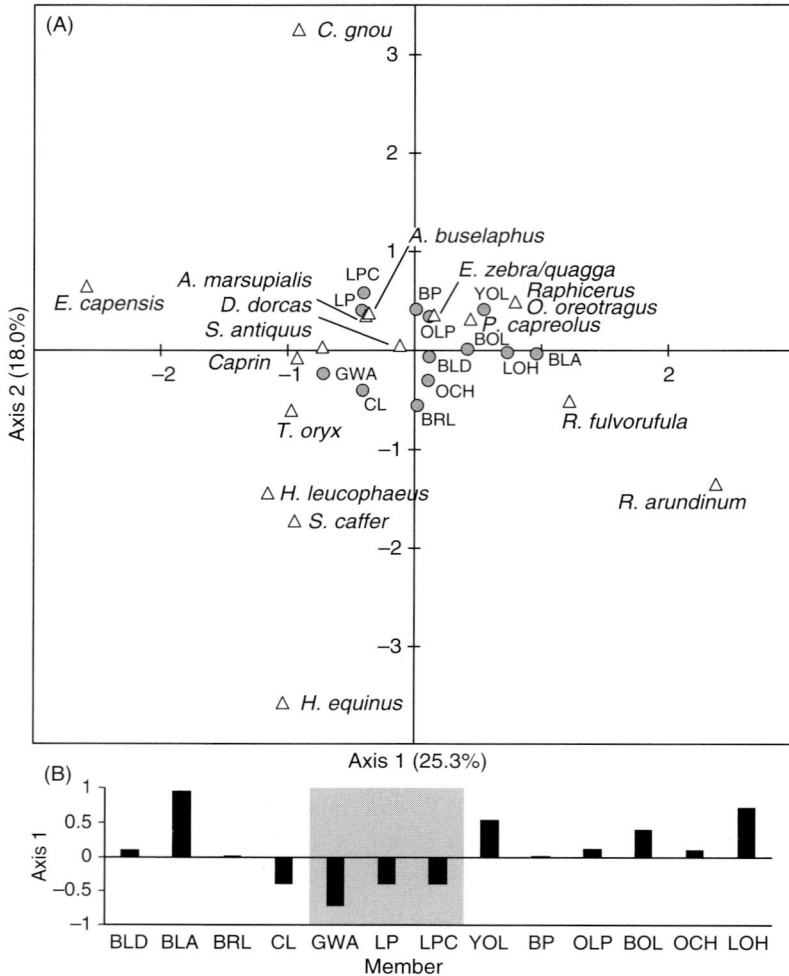

5.1. (A) Correspondence analysis of taxonomic presences for the Boomplaas Cave ungulates. Axis 1 (x-axis) accounts for 25.3% of the inertia; axis 2 (y-axis) accounts for 18.0% of the inertia. Data from Table 5.1. (B) Plot of axis 1 scores through the Boomplaas sequence, with assemblages arranged from youngest (left) to oldest (right). Members corresponding to the Last Glacial Maximum are indicated by grey shading.

Boomplaas Cave today, including springbok (*Antidorcas* cf. *marsupialis*), roan antelope (*Hippotragus equinus*), black wildebeest (*Connochaetes* cf. *gnou*), and bontebok/blesbok (*Damaliscus* cf. *dorcas*). These taxa prefer habitats that are relatively open and grassy, and the same is probably true of their extinct associates (Faith 2014; Klein 1980). The contrast in habitat preferences between those species with positive or negative values along axis 1 suggests this axis captures a contrast between habitats that are more shrubby (positive values) or more grassy (negative values). We infer that assemblages with positive scores on this axis, including all Holocene assemblages (BRL, BLA, and BLD) and

Pleistocene assemblages pre-dating the Last Glacial Maximum (YOL, BP, OLP, BOL, OCH, LOH), are characterized by shrubby habitats with similarities to the present. Assemblages with negative scores on axis 1, all of which date to the Last Glacial Maximum (LPC, LP, GWA) or Late-glacial (CL), have, we infer, a more open and grassy vegetation structure. Plotting axis 1 scores through time (Figure 5.1B), we observe a decline from the base of the sequence to the Last Glacial Maximum in member GWA, followed by a return to more positive values in the Holocene (i.e., shrubland to grassland to shrubland).

Because axis 1 accounts for only 25.3% of the inertia, there is substantial variation in faunal composition represented by the other axes, implying these are worthy of further exploration. Turning our attention to axis 2, we note there is a significant negative relationship between (log-transformed) assemblage sample size and the axis 2 scores for various assemblages ($r = -0.758$, $p = 0.003$). The implication is that the ordination of assemblages along this axis is likely telling us more about sample sizes than it is about paleoenvironments. We therefore exclude this axis from consideration. Because axis 3 explains only slightly less variation in faunal composition than axis 2, exploration of trends along this axis might be fruitful. However, our goals are merely to highlight the use of CA in faunal analysis, so we need only note here that the procedure for interpreting axis 3 would be identical to those illustrated above for axis 1.

In our example from Boomplaas Cave, CA facilitates synthesis and interpretation of a complex matrix of species occurrences across assemblages within a single site. The same approach can be used to examine taxonomic data across sites (Semken and Graham 1996), or to compare fossil assemblages with modern communities (Alemseged 2003; de Ruiter et al. 2008; Tryon et al. 2010). While this chapter is concerned with the taxa present in a fossil assemblage, CA is also used in paleoenvironmental reconstructions where taxonomic units are replaced by functional traits, such as diet and locomotor adaptations (Rector and Reed 2010; Reed 2008). Thus, black wildebeest and Cape mountain zebra become "terrestrial grazers." A detailed exploration of these "taxon-free" approaches is provided later in this volume (Chapter 7).

Detrended Correspondence Analysis

Two problems occasionally encountered in CA complicate interpretation (Hill and Gauch 1980). The first is known as the arch effect, an artifact of the CA algorithm – one that does not reflect the ecological structure of the data. In this case the second axis is a polynomial (arched) function of the first axis. In other words, the taxa and samples, as well as any underlying environmental gradient, take on a parabolic arch in a bivariate scatterplot of axis 1 and axis 2. Consider, for example, the abundances of five taxa along a hypothetical environmental gradient spanning nineteen sites (Figure 5.2A). The taxa respond unimodally to

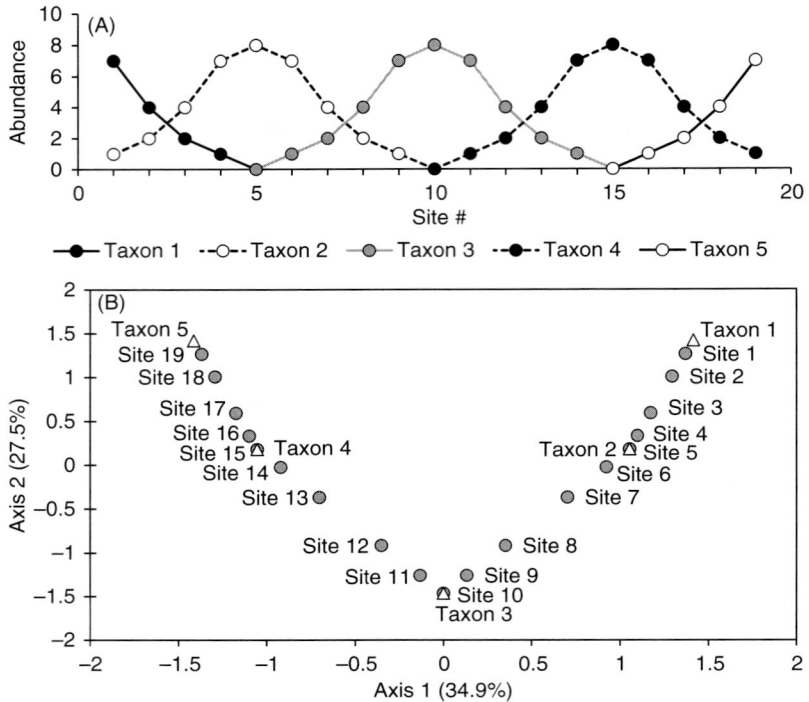

5.2. (A) Abundances of taxa along a hypothetical environmental gradient spanning nineteen sites. (B) The first (x-axis) and second (y-axis) axes of a correspondence analysis of the taxonomic abundances (circles = sites; triangles = taxa). Note how taxonomic abundances in A fluctuate inversely (one taxon increases as another decreases).

the environmental gradient, with Taxon 2 replacing Taxon 1, Taxon 3 replacing Taxon 2, and so on. A CA of these data generates a parabolic arch, in which there is a very clear polynomial relationship between the first and second axes (Figure 5.2B). Although the correct sequence of sites is preserved along axis 1 (i.e., their position tracks the hypothetical environmental gradient), if the goal of the CA is to reveal environmental gradients, one clearly does not want the second axis to be a function of the first.

The second problem is that CA does not preserve taxonomic distances. There is compression at the ends of the axes such that faunal samples (sites, assemblages, etc.) with equivalent taxonomic distances will plot closer together at the ends of the axis and farther apart in the middle. Turning back to our hypothetical environmental gradient, it is clear that taxonomic distance, however one may choose to measure it, between Site 2 and Site 3 is identical to the distance between Site 7 and Site 8 (Figure 5.2A). However, inspection of the CA indicates Sites 2 and 3, which appear at the positive end of axis 1, plot close together, whereas Sites 7 and 8, which appear toward the middle of axis 1, plot far apart. This gives the impression Sites 2 and 3 are more similar to each other than are Sites 7 and 8.

Detrended correspondence analysis (DCA) is a modified version of CA that fixes these problems through a two-step procedure. The first step is to eliminate the arch effect through a process referred to as detrending. This is accomplished by dividing the first axis into a number of segments of equal length (twenty-six segments is standard). Within each segment, the ordination scores for the second axis (and higher axes) are adjusted by centering them to a mean value of zero, effectively straightening out the arch. The second step (rescaling) is to rescale the axes such that equal distances in taxonomic composition translate to equal distances along the axes. Although there is debate about the appropriateness of these modifications (Peet et al. 1988; Wartenburg et al. 1987) – even its proponents describe DCA as an "ad-hoc, brute-force" solution (Peet et al. 1988:926) – it remains the case that DCA can facilitate a more straightforward interpretation of the data (see examples in Hill and Gauch 1980).

To illustrate an application of DCA, we turn to the small mammals from Homestead Cave. We begin with a CA, in this case using abundance data (for purposes of illustration) with taxa aggregated by genus and with genera occurring only once excluded. A bivariate scatterplot of axis 1 (80.6% of inertia) versus axis 2 (12.5% of inertia) results in a pronounced arch (Figure 5.3A), which renders the environmental significance of axis 2 (if any) difficult to decipher. When the same data are examined using DCA (Figure 5.3B), the arch is eliminated. The ranking of assemblages along axis 2 is inversely correlated with that for axis 1 ($r_s = -0.840$, $p < 0.001$), implying the two axes are capturing a similar environmental gradient. Both axes are still useful for interpretation, however, because axis 1 seems to be more sensitive to changes in taxonomic composition at the base of the sequence (Stratum I to IV), whereas axis 2 is reflecting changes in the upper portions of the sequence. For the sake of clarity, we intentionally omitted the taxon scores from Figure 5.3, but if these were plotted alongside the assemblage scores, we could then proceed with a paleoenvironmental evaluation similar to that provided above for Boomplaas Cave.

Other Ordination Techniques

CA and DCA are only two of many unconstrained ordination methods, with other common methods including principal components analysis (PCA), principal coordinates analysis (PCoA), and non-metric multidimensional scaling (NMDS). All of these approaches allow for ordination of samples according to their taxonomic composition, but they are less commonly used to interpret paleoenvironments from faunal remains for various reasons. The drawback to PCA, which is suitable for abundance data (expressed as proportions or percentages), is that it assumes taxa respond linearly to the environmental gradient. This assumption may be valid over short gradients (e.g., from Sites 1–5

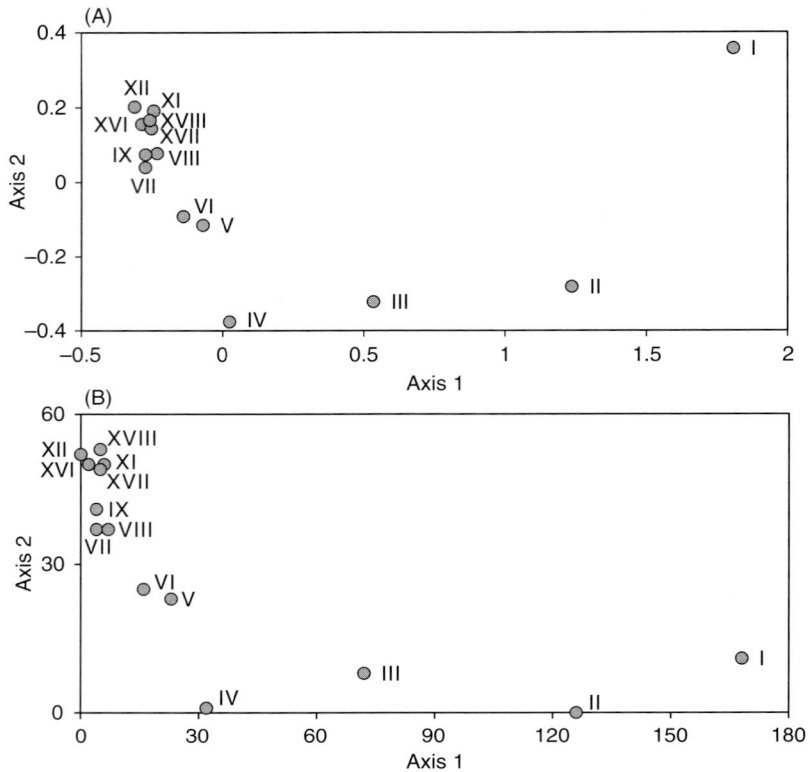

5.3. (A) Correspondence analysis of Homestead Cave small mammal assemblages (Roman numerals) aggregated by genus. For the sake of clarity, ordination of taxa is not shown. Note the pronounced arch effect. (B) The same data subjected to a detrended correspondence analysis.

in Figure 5.2), but we expect a non-linear, unimodal response over longer ones (e.g., from Sites 1–19 in Figure 5.2). Both PCoA and NMDS allow samples to be ordinated according to any number of taxonomic similarity metrics, but they do not provide simultaneous ordination of taxa and samples. Ordination of samples alone, as in Figure 5.3, tells us little about why they are positioned where they are. However, it is possible to evaluate the relationship of taxa to the samples by calculating the correlation between the taxon vector (e.g., the abundance of a taxon across samples) and the sample scores along the ordination axes (e.g., the axis 1 score across samples). The correlation coefficients can then be rescaled and plotted as vectors in conjunction with the sample ordinations (L. Legendre and Gallagher 2001). If this were done, one could proceed with a paleoenvironmental interpretation similar to that provided by our CA of Boomplaas Cave. Because the interpretive steps remain the same, and the key methodological differences relate only to the algorithms used to generate the ordination, we refer the reader elsewhere for detailed overviews of other ordination methods and their potential applications in paleozoological analysis (Hammer and Harper 2006; Shi 1993).

ONE SPECIES AT A TIME (USUALLY)

To this point, we have introduced several analytical techniques that can be used to decipher the paleoenvironmental significance of the presence of taxa in an assemblage. The techniques we have discussed have concerned all or most of the taxa present in a prehistoric fauna (at least as represented by the assemblage[s] at hand). Other techniques that we now turn to usually consider only one or perhaps two or three taxa at a time. These techniques focus on paleoclimatic inferences as opposed to paleohabitats, though it should be obvious they can be used for the latter. As will become clear once again, in either instance, knowledge of the ecological tolerances and preferences of the studied taxon must be very well known.

Indicator Taxa

There are several techniques for reconstructing past environments and climates dependent only on the taxonomic identities of ancient faunal remains. A couple of these typically consider only one species at a time. One technique uses what are sometimes referred to as *indicator taxa* or *indicator species* (e.g., Behrensmeyer et al. 2007a; Churcher and Wilson 1990; Dalquest 1965; Kingston 2007; Woodcock 1992). A single species that has well-known, relatively narrow ecological tolerances is the analytical focus. Because indicator taxa are stenotopic, the occurrence of one in an assemblage is taken to signify a particular environment or climate existed in the site area when the species' remains were accumulated and deposited. Only the presence of the indicator taxon is considered, not its abundance. Species associated with the indicator may or may not also be consulted.

A classic example in North America is John Guilday and Paul Parmalee's (1972) synopsis of Quaternary records of voles of the genus *Phenacomys*. They describe the genus as including four species, three of which are presently confined to the Pacific Coast of North America; the fourth species, the heather vole (*Phenacomys intermedius*), is more widespread but largely restricted to boreal forest habitats. Guilday and Parmalee (1972:171) suggest "this vole is an excellent boreal indicator in small vertebrate samples of suspected Pleistocene age from south of its present range." They interpret late Quaternary *extralimital* (outside the modern range) records of this species, of which they document several, in just this way – as indicative of the presence of boreal forests in areas that today do not include such habitats.

If the ecological tolerances of a particular species of interest were poorly known or the species seemed to occur in a relatively wide range of environments (was eurytopic), a common analytical procedure is to use the geographic range of the species as a coarse, low-resolution signal of paleoenvironment (e.g., Gidley and Gazin 1938; Lundelius 1967, 1983; Moine et al. 2002; Rymer 1978; Slaughter 1967). For instance (from a northern hemisphere perspective), a taxon today found to the north of the deposit containing its remains is thought to indicate a cooler, perhaps moister, environment in the past than at present; a taxon today found to the south of the deposit is interpreted to indicate a warmer, perhaps drier environment than at present. This is still an important method (Birks et al. 2010). Early examples again derive from the work of John Guilday (e.g., Guilday et al. 1964, 1969). For instance, Guilday (1971) reported that the late Pleistocene occurrence of yellow-cheeked vole (*Microtus xanthognathus*) remains in Pennsylvania, Virginia, West Virginia, Kentucky, Illinois, and Missouri some 2,200 km south and southeast of the species' modern distribution indicated the absence of dense (boreal) forest and the "presence of taiga or Hudsonian Zone conditions [Figure 2.1] in the sites from which remains of this animal have been recovered" (Guilday 1971:246). Many of the sites producing the remains presently occur in deciduous forests.

Indicator taxa, then, are stenotopic, having a narrow range of ecological tolerances, whether for particular habitats or climatic variables. It is indeed the narrow ecological tolerances that allow a species' presence to alone serve as a proxy indicator of environmental conditions, based of course on the critical assumption that a species' ecological tolerances have not evolved (see Chapter 3). It should be clear that analytical use of indicator species depends on their presence in an assemblage, something that does not seem to occur very often based on the few published instances when a particular vertebrate species has been used as an indicator. Nevertheless, when their remains are found, the mere presence of a particular stenotopic species is sufficient to reveal details of local paleoenvironmental conditions (e.g., Hafner 1993; Mead 1987; Mead and Grady 1996; Mead et al. 1992), making for a minimal-effort sort of analysis (beyond the usual drudgery of taxonomic identification of remains). Not surprisingly, more labor-intensive analytical protocols focusing on one or several individual species are also available.

Climatograph

An intriguing technique that implicitly rests equally on the notions of stenotopic indicator species and the belief that the biogeographic range of a species is indicative of that species' ecological tolerances involves the construction of what is known as a *climatograph*. Originally described in an unpublished Master's degree thesis (Dulian 1975), the technique was first used in a publication by Russell Graham (1984). In the few published studies of which we are aware, a climatograph is constructed for a single species, or at least one species at a time (Lyman 1986; Wood and Barnosky 1994). To construct a climatograph, the modern biogeographic range of the species of interest must be known and shown on a map. Then, the modern values of the climatic variable(s) of interest at a series of geographic points along the species' range edge are determined. In published examples, typically only a handful of points are used (~10–50), though with GIS software it is now feasible to obtain hundreds or even thousands of points. In particular, the WorldClim database of global climate data (Fick and Hijmans 2017; Hijmans et al. 2005; see www.worldclim.org) provides access to nineteen climatic variables across the globe. The implicit assumption is that the positions of the range edge are a function of climatic variables rather than, say, an impassible geographic barrier such as an uncrossable river or a population of organisms that precludes the species of interest from occupying an area (e.g., predators or competitors). That is, it must be assumed that the range of the species under study is a function of the environmental variables of interest rather than some other (biotic or abiotic) variable; in short, the geographic range must reflect the fundamental niche rather than the realized niche of the species. And provided all of this, the reasoning behind examining the climate along the range edge is that this is thought to approximate the climatic tolerance limits beyond which a species cannot exist (i.e., the boundary between the zone of physiological stress and the zone of intolerance in Figure 2.4).

Graham (1984) drew a climatograph that plotted the mean annual precipitation and mean annual high temperatures at multiple locations along the border of the modern range of the eastern chipmunk (*Tamias striatus*). This revealed a kind of climatic envelope for the chipmunk when plotted on a bivariate graph of the two climatic variables. The modern values of the same climatic parameters were determined for the several sites that had produced late Pleistocene remains of the eastern chipmunk, and then graphed against those values of the modern climatic envelope. Differences between the two were interpreted to reflect environmental change in the geographic area producing the fossils. Graham (1984) found that in order for eastern chipmunks to occupy the sites producing their remains today, temperatures would have to be about 7–8°C cooler, and annual precipitation would have to be about 120 mm greater.

An example of constructing a climatograph will clarify the analytical protocol (after Lyman 1986). First, this technique can be applied to any species represented in a fossil assemblage. It is most revealing, however, when applied to a species that today does not occur in the site area. A species that occurs in the site area today would only suggest environmental conditions like those today existed at the time the fossils were accumulated and deposited. Early Holocene fossils of Columbian ground squirrel (*Urocitellus columbianus*) recovered from the Lind Coulee archaeological site (45GR97) in eastern Washington State occur in an extralimital location (Figure 5.4). The fossil location seems today to be warmer and drier than the climatic conditions currently occupied by

TABLE 5.2 *Modern climate data (°C temperature, mm precipitation) for Columbian ground squirrel (Urocitellus columbianus).*

Point	Temperature			Precipitation		
	January	July	Annual	January	July	Annual
A	−1	21	7	152	100	800
B	−1	21	10	100	100	350
C	1.5	21	10	76	100	300
D	−1	18.5	7	76	100	300
E	−4	21	7	100	100	250
F	−6.5	21	4.5	100	100	250
G	−6.5	15.5	1.5	100	100	800
H	−4	18.5	4.5	100	127	500
I	−9.5	15.5	4.5	100	127	500
J	−1.5	15	1★	100★	100★	500
Lind Coulee	−1.5	18	10	45	0	250

Notes: See Figure 5.5 for map and point locations. ★ = estimate.

the species. To determine if this is in fact so, geographic points (ten, in this case) at the edge of the species' current range are plotted (A − J), and the mean annual temperature and precipitation, and the modern mean July and January temperature and precipitation at each of those points and the site location are recorded based on any one or more of several sources (e.g., Steinhauser 1979; Visher 1954) (Table 5.2).

The climatograph is constructed on a bivariate plot of the two climatic variables, with each geographic point plotted and the climatic envelope defined by connecting each of the geographic points at the edge of the geographic range (Figure 5.5). We have been rather liberal in connecting the geographic points such that a large, smooth rather than a small, irregular envelope results. In light of similarities in and differences between the modern climate of the site and the modern climatic envelope for the species, and assuming niche conservatism, we conclude that the Columbian ground squirrel remains at the Lind Coulee site imply that early Holocene climates were about the same temperature as today (because the site falls within the range of temperatures delimited by the climatic envelopes), but they were moister both in the winter and particularly in the summer (because the site falls below the range of precipitation values delimited by the climatic envelopes).

Barnosky (1998) modified the climatograph technique to produce what he referred to as a "bivariate climate space" (BCS, hereafter). In an early iteration of the analysis, Wood and Barnosky (1994) plotted ~50 points along the range edge of two species that are today largely, but not completely, allopatric. Remains of these two taxa occur in the same series of several strata

5.4. Modern distribution (cross hatched) of Columbian ground squirrel (*Urocitellus columbianus*) in northwestern North America (after Hall 1981). Lettered points are locations where climatic data were recorded (see Table 5.2). X is the location of the Lind Coulee site where 8,700 ¹⁴C yr BP remains of Columbian ground squirrel were recovered. Redrawn from Lyman (1986).

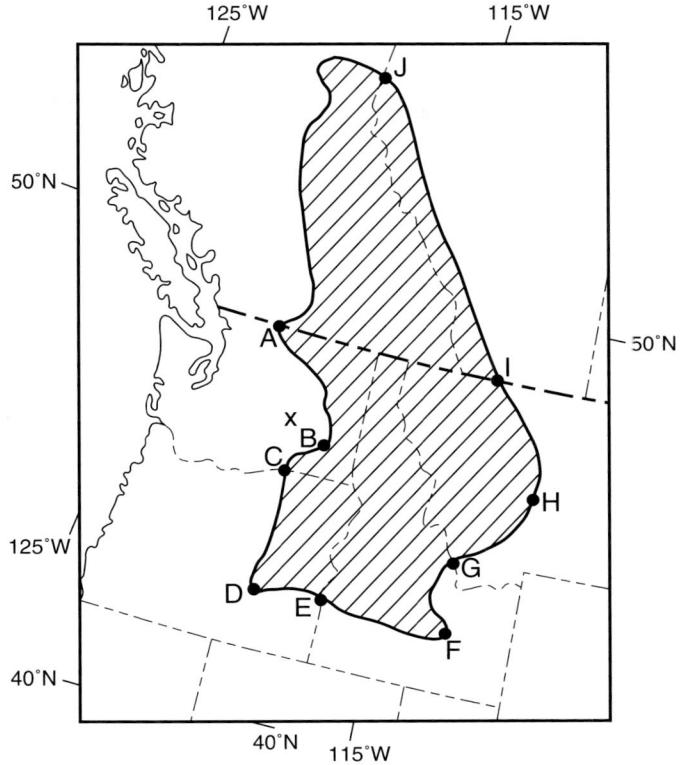

deposited between 487,000 and 365,000 BP at Porcupine Cave in Colorado. The two species are sagebrush vole (*Lemmiscus curtatus*) and an extinct lemming (*Mictomys* cf. *metoni*). The BCSs are defined by a plot of monthly mean precipitation and monthly mean temperature values at the ~50 locations on the border of each species' range (Figure 5.6). Wood and Barnosky's (1994; see also Barnosky 1998) analytical innovation was to include (i) mean monthly climate values and (ii) climatographs for two species on the same graph, making visual interpretation relatively straightforward (see also Shabel et al. 2004). Recall that *Mictomys metoni* is extinct. Wood and Barnosky (1994; Barnosky 1998) provide solid reasons to use the modern geographic range of the extant *Mictomys borealis* for the extinct *M. meltoni*. In short, the latter is thought to be a lineal descendant of the former, the two would be considered the same species were one to adopt an evolutionary species concept, *Mictomys* has always had boreal biogeographic affinities, and the two species have very similar dental morphologies, suggesting minimal differences in their ecological tolerances. The modern range of *M. borealis* is used to construct the BCS graph; this is a strongly warranted use of the NLR or nearest living relative approach (see Chapter 3, assumption 2).

Turning to Figure 5.6, we assume niche conservatism for both plotted species. The two species are today largely but not quite completely allopatric.

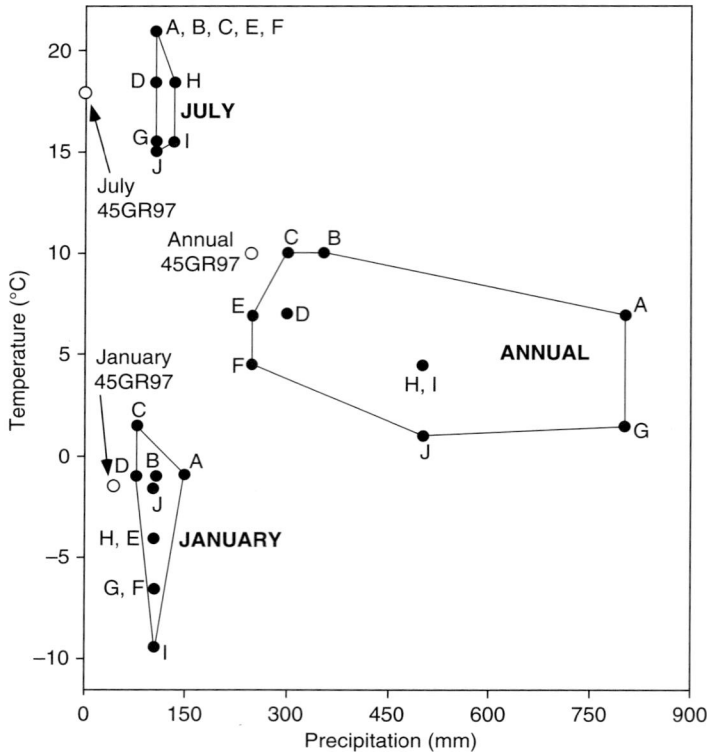

5.5. Climatograph for Columbian ground squirrel (*Urocitellus columbianus*) derived from data in Table 5.2. The lettered points are the geographic locations where climatic data were recorded (see Figure 5.4). Polygons on the left summarize July mean precipitation plotted against July mean temperature, and January mean precipitation plotted against January mean temperature. The polygon on the right represents the annual mean temperature and precipitation for all geographic points. Modern July, January, and annual mean climate of the Line Coulee (45GR97) site (open circles) are labeled. Corrected and redrawn from Lyman (1986).

M. borealis occupies northern latitudes and prefers moist climates whereas *L. curtatus* occupies more southern latitudes and prefers warmer, more xeric climates. The climatic envelopes for the two species overlap slightly (Figure 5.6A, B, C). Modern seasonal temperature extremes at Porcupine Cave would not have to be reduced (warmer winters, cooler summers) for the two species to be sympatric there (Figure 5.6D). The BCS graphs suggest for the two species to be sympatric; January effective precipitation (that which actually influences vegetation, not that which merely falls and quickly evaporates) must have been greater. One way to increase winter effective precipitation is to reduce effective solar insolation during the winter months (which in turn reduces evaporation). Wood and Barnosky (1994; Barnosky 1998) go to some lengths discussing shifts in solar insolation necessary to influence effective precipitation sufficiently, but their arguments need not concern us here. It suffices to note that their interpretations of flux in solar insolation align with independent evidence for such. Close study of the ecology of modern representatives of both species shows that low precipitation and high summer temperatures prevent modern bog lemmings from occupying areas south of their modern range. Less solar insolation in January and ~15 mm greater effective precipitation would increase winter snow cover and bog lemmings could occupy local habitats around Porcupine Cave.

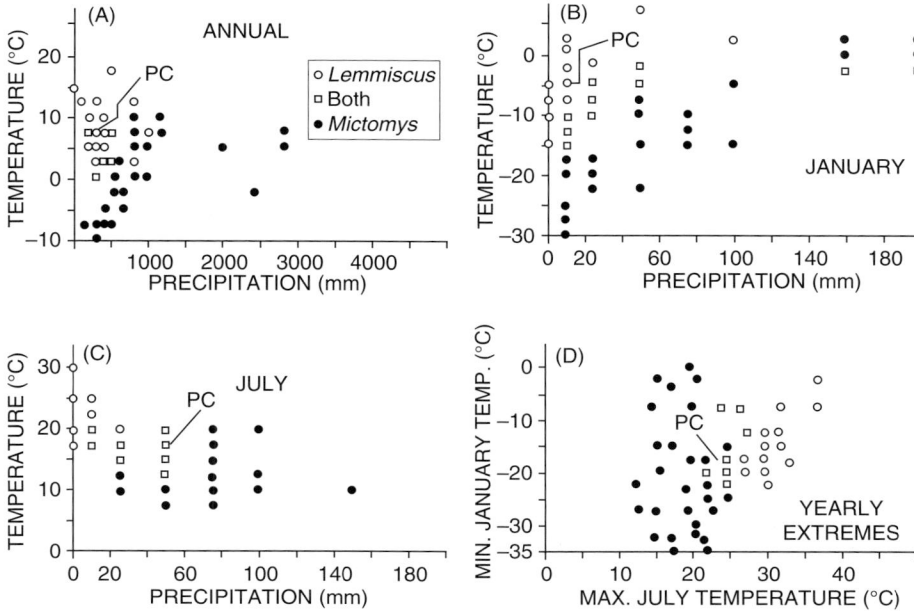

5.6. Bioclimatic range graphs for *Lemmiscus curtatus* and *Mictomys borealis* (after Wood and Barnosky 1994; Barnosky 1998). The two species are largely but not completely allopatric today. Each dot represents a distinct geographic location at the edge of the species' modern range; some dots represent multiple locations. The point labeled "PC" represents the modern climate of Porcupine Cave, the fossil location. (A) Annual bioclimatic range; (B) January bioclimatic range; (C) July bioclimatic range; (D) temperature extremes in January and July.

It should come as no surprise that when multiple indicator taxa are available, they often are used simultaneously (e.g., Lyman 2014a). Among the few climatograph applications there are none in which more than a single taxon per assemblage is considered, but the bivariate climate space (BCS) approach includes just such a plot, as exemplified by Wood and Barnosky's (1994) use of the technique (Figure 5.6). There are two other analytical techniques that simultaneously consider multiple, if not all, species in an assemblage.

MULTIPLE TAXA

Likely as a result of the assumption that the geographic range of a species was significantly influenced by environmental factors, in the middle of the twentieth century paleozoologists developed several techniques that simultaneously considered particular details of the ranges of multiple taxa. The techniques of course rest on the key assumption that the ecological tolerances of extant species represented in an assemblage of ancient animal remains have not changed in the time between the accumulation and deposition of those remains and today (Chapter 3, assumption 1). And as we note at the end of this section, there are other assumptions involved as well.

Area of Sympatry

What became known as the area of sympatry (AOS) technique emerged from analyses of North American Quaternary paleozoologist Claude W. Hibbard's fossil collections at the University of Michigan (Semken 1988). The extent to which Hibbard influenced the development of the technique is unclear from the published record (see Lyman [2016] for additional details). Whatever the case, by the middle 1950s Hibbard was aware that some of the prehistoric faunas he was studying included species that were today allopatric but which seemed on the basis of stratigraphic associations to have had sympatric distributions in the past (e.g., Hibbard 1955, 1960). The analytical utility of modern geographic ranges of multiple taxa for paleoenvironmental reconstruction was likely on his mind, and his collaborators and students were likely also thinking along these lines.

Beginning in the 1950s, several researchers associated with Hibbard's University of Michigan lab constructed maps representing what became known as the *area of sympatry* indicated by superimposed modern range maps of the multiple species represented in the ancient faunas they were studying (Etheridge 1958; Guilday et al. 1964; Schultz 1967, 1969; Smith 1954; Stephens 1960). Semken (1966) was the first to use the term *sympatry* when he determined the AOS for eleven extant species of mammals in an Illinoian assemblage from Kansas. He also determined the AOS for ten and the AOS for nine of the species represented in the assemblage, noting that as the number of species included decreased, the AOS increased in size. All of the early researchers inferred that the climate characterizing the modern AOS – the area where all (or most) of the species in the fossil assemblage were found today – represented the climate that existed at the fossil site at the time the remains were accumulated and deposited. For example, the AOSs Semken mapped indicated a cooler and moister climate than the fossil-producing site presently experienced.

The AOS technique has been used fairly regularly since the 1960s (see Lyman [2016] for references). It involves constructing a map showing the geographic location and shape of the AOS, and the location of the site that produced the remains of animals that were stratigraphically associated and whose overlap in modern species-specific ranges defines the AOS. The modern climate of the AOS is typically (but not necessarily) derived from isopleth maps of climatic variables such as average coldest temperature in mid-winter and average warmest temperature in mid-summer, though GIS software can now be used to evaluate numerous climatic variables using regional or global climate layers. The modern climate of the AOS is inferred to closely approximate the climate of the site of deposition at the time the assemblage of faunal remains was accumulated and deposited.

Although the AOS technique seems more rigorous than the simple "high latitude range signifies a cool climate whereas a low latitude range signifies a warm climate" sort of interpretive algorithm based on a species' modern range, the analytical protocol does include some potential pitfalls (Graham and Semken 1987). Finding the geographic area where the species represented in a prehistoric assemblage of remains today co-occur (the AOS) is the first step (after taxonomic identification and stratigraphic and chronometric analyses), and it involves consultation of range maps for all (or as many as possible, see below) species in the assemblage. The shape and location of the AOS depends on the set of distribution maps used. For example, for North American mammals, the two major sources of maps are Burt and Grossenheider (1964) and Hall (1981; Hall and Kelson 1959). Differences (some major, some minor) between range maps for particular species exist in these two sources, and thus an AOS for an assemblage will vary with the set of range maps used.

Whatever set of maps is used, one must assume those maps are accurate reflections of non-anthropogenically influenced distributions of all included species. This may not always be so, as noted in our discussion of the historic distribution of the Boomplaas Cave ungulates. And one must also assume that the geographic ranges are robust reflections of each included species' fundamental niche. Finally, more than one AOS may be delimited for any given assemblage depending on which taxa are included, because sometimes not all taxa in an assemblage are today sympatric with one another (i.e., non-analog associations). In this case, isopleth maps showing the AOS for the maximum number of species and each of several successively smaller numbers of included species can be drawn (e.g., Rhodes 1984; Rhodes and Semken 1986). We do have to worry about the possibility that non-analog associations (see Chapter 2) are reflecting environments without a modern analog (e.g., Williams and Jackson 2007). Therefore, when an AOS excludes an allopatric species associated with the species defining the AOS, an inaccurate climatic interpretation based on modern geographic ranges and climates may result.

The next analytical step is to superimpose the AOS over isopleth maps of climatic variables, such as maximum annual (summer) temperature, minimum annual (winter) temperature, annual precipitation, number of frost-free days per annum, and the like. The climatic implications of the AOS will depend on the maps of climatic variables used as these too can vary from source to source. Ultimately, the modern climate of the AOS is taken to represent the climate of the prehistoric fauna. The mapped variables are (i) the geographic AOS of the taxa represented in an assemblage and (ii) the location of the site that produced the assemblage. Figure 5.7A illustrates schematically how an AOS for three taxa is determined.

An intriguing real example of the technique involved determination of three AOSs for three temporally distinct faunas at one site. Vertebrate paleozoologist

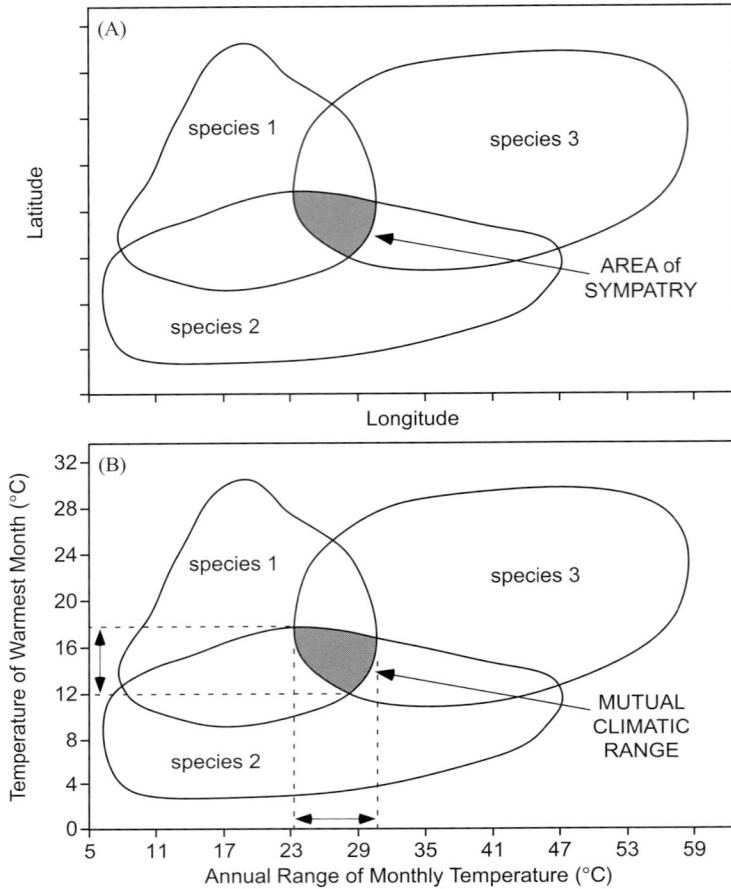

5.7. (A) A schematic illustration of how an area of sympatry (AOS) is determined for a fossil fauna that includes three species. Typically an AOS is determined based on all species in an assemblage (usually more than three), in which case the AOS would include a smaller area. The modern climate of the AOS is inferred to represent the ancient climate at the time the fossil fauna was deposited. (B) A fictional mutual climatic range (MCR) for three species. Temperature ranges for each species are indicated; the MCR is the shaded area. Dashed lines indicate the annual range of monthly temperature and range of the warmest month that would be inferred for a fossil assemblage that includes these three species.

Holmes Semken (1980) identified the mammal remains at the Cherokee Site in northwestern Iowa. Those remains were distributed across three temporally distinct strata, as follows: Cherokee I – 6,350 ^{14}C yrs BP, Cherokee II – 7,300 ^{14}C yrs BP, and Cherokee III – 8,400 ^{14}C yrs BP. Sample sizes for each stratum were not large, but multiple taxa were present in each. Rhodes and Semken (1986) revised the AOSs originally determined by Semken (1980) by omitting Chiroptera and Carnivora and by using different modern distribution maps. They used only insectivores, lagomorphs, and rodents to construct an AOS for each assemblage; the faunas are summarized in Table 5.3, and the three AOSs are shown in Figure 5.8. Rhodes and Semken (1986) indicate the temporal placement and sequence of the three AOSs reflects increasing aridity when superimposed on isopleth maps of climatic variables. They infer this because the AOSs in the younger samples correspond to areas characterized today by drier climates than those in the older samples. And other independent data suggest increasing aridity at this time as well.

TABLE 5.3 *Small mammal faunas from the Cherokee archaeological site, Iowa (Semken 1980). Frequencies are MNI. Taxonomic names have been updated, following Wilson and Reeder (2005).*

Taxon	Cherokee I	Cherokee II	Cherokee III
Masked shrew *Sorex cinereus*	1	2	1
Pigmy shrew *Microsorex hoyi*	1		
Short-tail shrew *Blarina brevicauda*	1	2	2
Eastern mole *Scalopus aquaticus*		1	
Rabbit *Sylvilagus* sp.		1	
Eastern chipmunk *Tamias striatus*			1
Franklin's ground squirrel *Spermophilus franklinii*	1		
Red squirrel *Tamiasciurus hudsonicus*		3	
Plains pocket gopher *Geomys bursarius*	1	1	1
Hispid pocket mouse *Perognathus hispidus*	1	1	
Meadow vole *Microtus pennsylvanicus*	2	7	3
Prairie vole *Microtus (Pedomys) ochrogaster*	2	4	1
White-footed mouse *Peromyscus leucopus*	3	11	3
Deer mouse *Peromyscus maniculatus*		4	1
Western harvest mouse *Reithrodontomys megalotis*		3	1
Northern grasshopper mouse *Onychomys leucogaster*	1		
Southern bog lemming *Synaptomys cooperi*		2	
Meadow jumping mouse *Zapus hudsonius*		2	1
Beaver *Castor canadensis*			1
Σ MNI	14	44	16
Richness	10	14	11

The Cherokee AOSs are informative for our purposes here because they also display variability. The Cherokee III AOS includes all eleven taxa in the faunal assemblage; the Cherokee III fauna represents what is sometimes referred to as a "harmonious" or "analog" fauna because all represented taxa are sympatric in at least one area. The Cherokee II fauna is "disharmonius" or "non-analog" (Graham 2005) because only thirteen of the fourteen mammals are found in sympatry today. Moreover, one species (*Perognathus hispidus*) is missing from the two eastern AOSs, and one species (*Tamiasciurus hudsonicus*) is missing from the southwestern AOS (Figure 5.8). The Cherokee I fauna is also non-analog as only nine of ten taxa occur sympatrically today. The northern Cherokee I AOS lacks *Perognathus hispidus* and the southern Cherokee I AOS lacks *Microsorex hoyi*.

It should be clear that if the AOS includes the location of the site from which the fossil assemblage was collected, the inference would be that the climate at the time the fossils were accumulated and deposited was much like the modern climate. It should also be clear that the more taxa included in

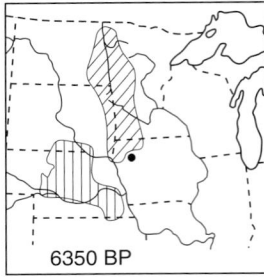

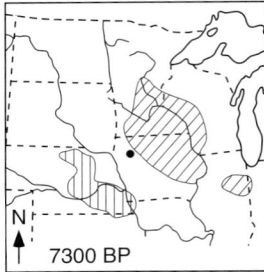

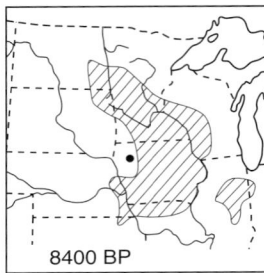

5.8. Areas of sympatry (AOS) (cross hatched) for three mammalian faunas of different ages from the Cherokee Site (dot) in northwestern Iowa state, USA. Each of the two 8,400 BP AOSs contain all eleven insectivore, lagomorph, and rodent (ILR) taxa represented in the prehistoric assemblage. Each of the three 7,300 BP AOSs contain thirteen of fourteen ILR taxa; a squirrel species occurs only in the two diagonally (eastern) cross-hatched areas, and a pocket mouse occurs only in the vertically (western) cross-hatched area. Each of the two 6,350 BP AOSs contains nine of ten ILR taxa; a species of pygmy shrew occurs only in the diagonally (northeastern) cross-hatched AOS, and a pocket mouse occurs only in the vertically (southwestern) cross-hatched AOS. Redrawn after Rhodes and Semken (1986).

constructing an AOS, not only will the AOS quite likely become smaller but the more probable that one or more species represented in the assemblage will not be sympatric with the others. This could raise some objections, but recall that an AOS is based on what is known of modern distributions of species, and there have in many areas been major changes in those distributions over the past couple of centuries. Further, an AOS is meant to be a measure of paleoenvironments and change therein is interpreted at an ordinal scale. That this is so should be clear from the fact that a geographic area, even a relatively small one of, say, 100 km², can display climatic variability across space, particularly in topographically heterogeneous environments. Rhodes and Semken (1986) chose to use the subjectively identified geographic center of an AOS as indicative of its overall climate, which likely becomes less valid as the AOS increases in size. Nevertheless, researchers who study ancient insect remains developed a technique specifically to circumvent the potential that not all species would be sympatric today. We turn to that technique next, and note here that there is no reason to limit application of the technique to insects.

Mutual Climatic Range (Coexistence Approach)

Early efforts to reveal paleoclimates by paleoecologists studying ancient beetle remains involved attempts to determine the modern area of co-occurrence of all species in an assemblage of prehistoric remains (e.g., Coope 1959). This technique was referred to as the *range overlap method* (Coope 1959; Lowe and Walker 1997). As with the AOS technique, the range overlap method involved determination of the area where the ranges of species represented in an assemblage today overlapped. Then, data such as mean annual temperature and mean annual precipitation were derived from weather stations (global climate databases such as WorldClim could be used today) within the area of overlapping ranges and the ancient climate was inferred to approximate that within the area of range overlap (Lowe and Walker 1997). Because coleopteran species were understood to be sensitive to temperature, researchers believed their analytical technique had to contend with such things as latitudinal variability in insolation and species that were today allopatric relative to the majority of species represented in an assemblage (Atkinson et al. 1986). This group of researchers came up with a different solution for contending with allopatric species than those working in North America with the AOS technique.

The *mutual climatic range* (MCR) technique was derived from palynological techniques (Atkinson et al. 1987) and was originally developed and named in the 1980s. It involves plotting climatic envelopes within which each of the species represented in an assemblage of remains is found today (Figure 5.7B). It requires detailed knowledge of climatic variables across a species range (Atkinson et al. 1987; Elias 1997, 2001). Climatic variables typically used in the MCR technique are the annual range of temperature on the x-axis and the mean temperature of the warmest month on the y-axis, as these have been found to be the ones often influencing the biogeography of Coleoptera (Atkinson et al. 1986, 1987), though any variables of interest could be used. The temperature range defined by the overlap of the included species' climatic envelopes is the MCR (Figure 5.7B). The climate indicated by the MCR is inferred to represent the prehistoric climate of the site at the time the beetle remains were accumulated and deposited. Because the MCR technique plots the overlapping climatic ranges of multiple species, the included species need not be geographically sympatric but instead can be allopatric. This was viewed as an advantage relative to the geographic range overlap method (Atkinson et al. 1986, 1987).

The MCR technique has been applied most often to insects (e.g., Alfimov and Berman 2009; Elias et al. 1999; and papers associated with Elias and Whitehouse 2014). A similar set of techniques used by paleobotanists is known by various names readily subsumed under the label *coexistence approach* (Utescher et al. 2014). In short, the name says it all. Analytically one determines the climatic parameters under which all (or as many as possible) of the species in an assemblage coexist today, and then concludes the ancient climate in which the paleo-flora lived was the same. Not all paleobotanists think it is a robust technique, and some of the critical literature regarding the coexistence approach identifies problems with any such technique (e.g., Grimm and Potts 2016), including the MCR. These problems include such things as biogeographic distributions of species not reflecting climatic factors but instead competition or some other variable, lack of niche conservatism or violation of the assumption of "physiological uniformitarianism" (e.g., Tiffney 2008), assemblages of taxa that do not represent communities (Grimm and Potts 2016), and what paleozoologists often think of as time-averaged assemblages (e.g., Behrensmeyer et al. 2007a).

We are unaware of applications of the MCR or coexistence approaches to mammal fossils (but see Polly and Eronen [2011] for an example of the application to mammals of an MCR-like technique). It should be easy to apply the MCR technique to any group of organisms using a protocol like that associated with climatographs to determine a climatic envelope. Unlike the AOS technique, the MCR technique graphs modern climate ranges occupied by species rather than geographic ranges occupied today by species (compare Figures 5.7A and 5.7B). Both techniques map or graph multiple species simultaneously, and both ignore taxonomic abundances. Efforts have been made to incorporate taxonomic abundances (e.g., Huppert and Solow 2004) and ubiquity data (e.g., Bray et al. 2006) into modified versions of the MCR technique. But considering only species' presences, several paleozoologists recently developed an analytical technique that is something of a hybrid of the AOS and MCR techniques.

The UTM-MCR Technique

Blain et al. (2009:57) suggest determination of "the overlap of the current distribution of all taxa occurring in a [fossil] locality or [stratigraphic] level permits us to calculate potential palaeoclimatic conditions (= mutual climatic range method = MCR)." Their analytical protocol involves determination of the particular 10 ×10 km UTM (Universal Transverse Mercator) grid squares where all species in a prehistoric assemblage currently live. Once that is done, they determine the modern climatic parameters of each of the occupied UTM squares (Blain et al. 2009). Fernández-García and López-García (2013:231) explain that the "aim of this method is to find the present geographical region that exhibits the same species assemblage as that documented in a given stratigraphical level through the intersection obtained from the overlap of current species distribution maps. The current climatic conditions of the intersected area are used to infer past conditions." In other words, the climate of those UTM squares where various species represented in a fossil assemblage are currently sympatric is compared with the climate at the location of the fossil site to determine how (if at all) the climate changed in the past (as in the area of sympatry technique). This technique has been applied to herptofaunas (e.g., Blain et al. 2009, 2010, 2015) and mammalian faunas (e.g., Fernández-García et al. 2016; López-García et al. 2010, 2015b) (see Lyman [2016] for additional references). This MCR technique of Blain et al. (2009) and López-García et al. (2010) shares features with the AOS technique described above – producing a geographic map of the modern distribution of all species in a fossil assemblage (an area of sympatry), consulting maps of climatic variables – except they use 10 × 10 km UTM grid squares as the mapping units because the species-specific distribution maps they consult provide distribution data "based on a 10 × 10 grid network" (López-Garcia et al. 2010:458). In turn, because of this, "climatic data are resolved to 10 × 10 km squares" (López-Garcia et al. 2010:458). To keep the original MCR and this new MCR analytical protocol distinct, Lyman (2016) suggested Blain et al.'s (2009) analytical protocol be referred to as the UTM-MCR technique. While this protocol is based on 10 × 10 km UTM squares, analysis of geographic ranges and climatic variables using GIS software could allow for higher or lower resolution analyses.

A point that should be underscored here is that any technique that uses the range of a taxon to construct a climatic envelope for one or more species must, as noted above, assume a tight relationship between each species' range and its physiological tolerances to climatic variables. In other words, one must assume each species' range is a robust reflection of its fundamental niche. Further, a climatograph, an AOS, and an MCR, or any other analytical result that is based on the geographic distribution of a species must rest on the assumption that the maps are accurate reflections of a species' fundamental niche. Because many range maps involve interpolations of multiple geographic points, and choice of points depends on the time period from which biogeographic data are derived, different researchers will often produce different range maps for a species. The same goes for maps of climatic variables.

Discussion

The techniques discussed in the three immediately preceding sections are somewhat intuitive and commonsensical. This most decidedly does not mean they are foolproof, as we have tried to make clear. What it does mean, we think, is that each of them is readily grasped conceptually. This is so because each technique, in a relatively straightforward manner, takes advantage of the fact that each species has a geographic distribution that is more or less known with some degree of *accuracy* and precision, and that distribution likely reflects, at least to some degree, the climatic tolerances of the species. While these facts in a way explain how and why each technique works analytically, they also provide a roadmap to potential analytical pitfalls.

The area of sympatry technique involves establishing the geographic location and shape of the modern area of co-occurrence of multiple species in a prehistoric assemblage. The mutual climate range technique involves determination of the climatic co-occurrence (not necessarily geographic sympatry) of the species in a prehistoric assemblage. The UTM-MCR adopts elements of both the AOS and the MCR techniques. All three assume that all included species were at least approximately members of the same paleocommunity. And all three focus on taxonomic presences, not taxonomic absences and not taxonomic abundances.

SUMMARY

In this chapter we have discussed research techniques that focus analytically on the presence, and to a lesser degree the absence, of taxa in an assemblage. As we have shown, these techniques might consider one, two, three, or any number of taxa up to all of those represented in an assemblage. Because of this the techniques we have discussed tend to be fairly commonsensical. And some researchers likely will consider it a good thing that the techniques described do not require a great deal of statistical sophistication. What is required is good knowledge of the ecological tolerances and preferences of the species considered.

In the next chapter, we turn to techniques that analyze taxonomic abundances. In that chapter we also consider further some of the drawbacks of examining only the presence of taxa. We reserve discussion of that topic for the next chapter because, to highlight the drawbacks, taxonomic presences must be compared and contrasted with taxonomic abundances.

SIX

ENVIRONMENTAL RECONSTRUCTION BASED ON TAXONOMIC ABUNDANCES

In Chapter 5 we described several straightforward techniques that focus analytical attention on the presence of one or more taxa in an assemblage. Just as the presence and absence of taxa will vary through time or space, it is also common for the taxa in any given set of assemblages to vary to greater or lesser degrees in the frequency of their remains, suggesting abundances of those taxa on the landscape may have varied as well. Ecologists have long observed not only that different species of animals occur in different habitats or environments and have different ecological tolerances, but that species also tend to vary in abundance across landscapes, often in concert with one or more environmental variables. Paleozoologists are aware of this correspondence and they recognize that taxonomic abundance data are often informative regarding ancient environments, sometimes even more informative than taxonomic presences.

It is easy to demonstrate that the abundances of taxa in a faunal assemblage, usually quantified according to the number of identified specimens (NISP) or minimum number of individuals (MNI), can provide paleoenvironmental insight of finer resolution than taxonomic presences alone. Consider the case of the chisel-toothed kangaroo rat (*Dipodomys microps*) from Homestead Cave. The xeric-adapted chisel-toothed kangaroo rat is closely associated with desert valleys dominated by the shrub *Atriplex confertifolia* (shadscale). It is uniquely adapted to feeding on shadscale, using its chisel-shaped lower incisors to shave off the hypersaline outer surface of the leaves, exposing and then consuming

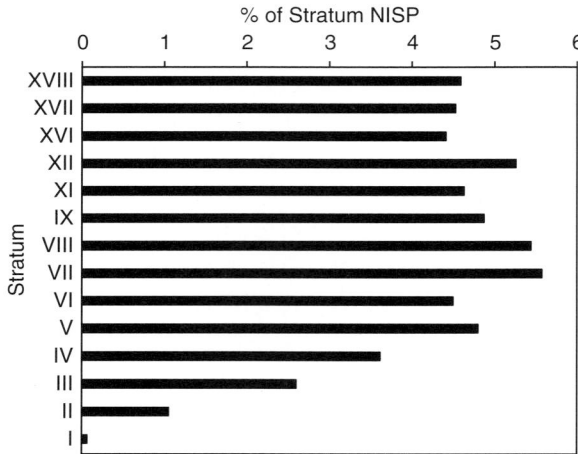

6.1. Relative (% of all NISP in a stratum) abundances of chisel-toothed kangaroo rats (*Dipodomys microps*) across the Homestead Cave stratigraphic sequence. See Table 4.5 for absolute abundance data.

the more palatable inner portions (Kenagy 1973). This close association with desert shrubland means the chisel-toothed kangaroo rat is an environmentally informative stenotopic taxon.

The chisel-toothed kangaroo rat is, however, one of many taxa to occur in all levels of the Homestead Cave sequence (see Table 4.5), meaning that its presences reveal nothing about the dramatic environmental changes that occurred since the terminal Pleistocene. But when we consider its relative abundance (% of all NISP in a stratum) through time, a striking pattern emerges (Figure 6.1). From initial rarity in the terminal Pleistocene (Strata I, II, and III), the chisel-toothed kangaroo rat increases steadily in abundance until the middle Holocene (Strata VI–IX). This parallels the trend documented for kangaroo rat specimens that could not be firmly identified to species (*Dipodomys* sp.), which account for the majority of identified specimens from Stratum IV and above. It follows that the dominant taxon since the middle Holocene at Homestead Cave is likely the chisel-toothed kangaroo rat. This is one of several lines of evidence that led Grayson (2000a, 2000b) to infer progressive aridification from the terminal Pleistocene through the middle Holocene, insight that could not be obtained from a consideration of chisel-toothed kangaroo rat presences alone.

To the extent that taxonomic abundances in fossil assemblages reliably track the abundances of species on past landscapes (that our *measured variable* accurately reflects our *target variable*), abundance data potentially hold greater potential for paleoenvironmental inference than taxonomic presences. The reason is that taxonomic presences represent a *nominal scale* of measurement, one that provides no indication of ordering or magnitude, whereas abundance data may provide just such information (Grayson 1984b; Lyman 2008b). Notice we said "may provide" rather than "does provide." But one might protest that ten of species A means it is twice as abundant as the five of species B. That is, it might be expected that taxonomic abundances provide a *ratio scale* of measurement, one with a meaningful zero value (e.g., an NISP of 0 for taxon A implies it is absent, not only from the assemblage, but from the ancient landscape as well). Were this so, it would be possible to quantify the magnitude of difference between any two taxa (e.g., taxon A is twice as abundant as taxon B). However, both Grayson (1984b) and Lyman (2008b) demonstrate that taxonomic abundance measures typically used in paleozoological contexts (NISP and MNI) behave in a manner that is *ordinal scale* at best, a scale of measurement that preserves ranks but not magnitude when it comes to the measured variable (it is a taphonomic issue whether the measured variable is also a robust estimate of the target variable; this must be assumed [see Chapter 3]). Thus, we can often infer that taxon A is more abundant than taxon B, but not that it is twice as abundant as taxon B. In the remainder of this chapter, we assume taxonomic abundance data are in fact at least ordinal scale, and note that many quantitative analyses of taxonomic abundances treat the data as if they are ratio scale. This is usually acceptable provided that interpretations of those analyses are presented in ordinal-scale terms (Lyman 2008b).

In practice, the basic process by which taxonomic abundance data are used to reconstruct past environments closely parallels that for taxonomic presences. They share the same critical assumptions, for instance (Chapter 3). The key difference is that when synthesizing the environmental associations of various taxa into a paleoenvironmental interpretation, abundances are used to assess the importance of the ecological signal of a taxon, with greater weight given to more abundant taxa and less weight to the rare taxa. While rarely made explicit, the rationale underpinning paleoenvironmental reconstructions derived from abundance data is that the more abundant taxa are those that represent the overall local (meso-scale) environment; these are the taxa at or closest to the peak of their unimodal response to an environmental gradient (see Figure 2.4). In contrast, there are many ecological reasons why a taxon may be rare in a faunal community (Brown 1984; Hubbell 2001; Magurran and Henderson 2003), but at least some will be rare because they are at the limits of their ecological tolerances, or because they are associated with micro-environments that are a minor (micro-scale) component of the broader landscape. Thus, if taxa associated with grasslands dominate a fossil assemblage that includes both grassland and woodland taxa, one would infer that grasslands were the dominant vegetation type on the ancient landscape, with woodlands being less abundant. This does not imply that the most abundant taxon is always the most important or significant environmental indicator. An abundant but eurybiomic taxon reveals little about paleoenvironments, while a rare *stenobiomic* taxon may provide essential information (recall our discussion of indicator species in Chapter 5).

At this point it should be clear that taxonomic abundance data might provide more resolution to our paleoenvironmental reconstructions than taxonomic presence data alone. Before turning to analytical techniques that can be applied to such quantitative data, it is important to first delve more deeply into this issue (as promised in Chapter 5). Notice we said taxonomic abundance data "might provide more resolution." The fact that it is a possibility rather than a given has led to numerous statements on the pros and cons of taxonomic abundance data, particularly relative to the pros and cons of taxonomic presence data. We next turn our attention to those statements.

TAXONOMIC PRESENCES OR ABUNDANCES?

As Grayson (1981) noted several decades ago, faunal analysts tend to favor either taxonomic presences or abundances for paleoenvironmental interpretation. Whereas taxonomic abundances may hold greater potential because such data can provide an ordinal-scale measurement, it also assumes that taxonomic abundances in a fossil sample (the measured variable) are representative of those in the ancient faunal community (the target variable) (see Chapter 3, assumption 7). To some, this assumption is sufficiently tenuous to warrant a focus on taxonomic presences alone (e.g., Andrews 1996; Behrensmeyer et al. 2007a; Nesbit Evans et al. 1981). Grayson (1981:34), for one, argued that "analyses which depend only upon the taxa recorded as present within a fauna are to be preferred," although he later abandoned this viewpoint when the assemblages he was comparing were exceptionally large (Grayson 2000a, 2000b; Grayson et al. 2001).

Researchers have suggested many reasons why the abundances of taxa in a fossil collection will not be accurate measures of those taxa on the ancient landscape. Terry (2009) implied that measuring taxonomic abundances in a fossil collection is so difficult (not to mention contentious among analysts as to which quantitative unit to use – NISP, MNI, something else [e.g., Domínguez-Rodrigo 2012; Lyman 2018a]) that it is ill advised. Behrensmeyer et al. (2000:124) note that because taphonomic process can skew taxonomic abundances in sometimes significant ways, "vertebrate paleontologists regard both relative abundances and rank-ordering of species as suspect (i.e., guilty of bias unless proven otherwise)." According to the guilty-unless-proven-otherwise viewpoint, nominal-scale presence/absence data may be the best that we can hope for.

The votes explicitly and implicitly cast against using taxonomic abundance data for paleoenvironmental reconstruction would seem to outnumber those cast in favor of such usage. Interestingly, the votes fall the other way when it comes to study of zooarchaeological collections to assess such things as human diet. In those cases potential taphonomic problems are acknowledged and diet summarily inferred based on taxonomic abundances. We are aware of few empirical studies of how taphonomic processes might or might not bias the

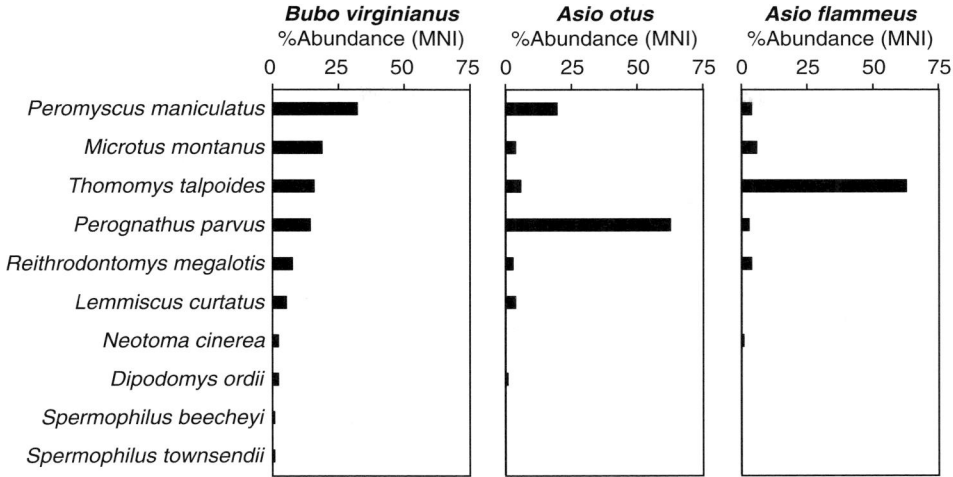

6.2. Relative abundances of small mammals (% of MNI in assemblage) recovered from pellets of three owl species in central Oregon. Data from Maser et al. (1970).

paleoenvironmental signal of taxonomic abundances (discounting the growing number of fidelity studies; see below). An early example of just such a study shows that processes of fossil accumulation can skew taxonomic abundances (Grayson 1981).

Just as it was easy to demonstrate that abundances of taxa can provide enhanced paleoenvironmental insight, it is also easy to demonstrate how such data can lead one astray. Here we turn to Grayson's (1981) example concerning the predation patterns of owls, which draws from observations on the mammals recovered from owl pellets regurgitated by great horned owl (*Bubo virginianus*), short-eared owl (*Asio flammeus*), and long-eared owl (*A. otus*) in central Oregon (Maser et al. 1970). Despite coming from habitats that are broadly similar, the taxonomic abundances of select species recovered from pellets of the three owl species are very different (Figure 6.2). Had these assemblages come from three different sites or from three strata within a site, paleoenvironmental interpretations derived from the abundance data would be misleading. However, rather than sampling different environments through space or time, the three faunal assemblages are sampling the same environment through the taphonomic filters of three different agents of bone accumulation.

Does the possibility that fossil taxonomic abundances track factors other than the paleoenvironment mean we should consider only taxonomic presences? While the answer is yes among some faunal analysts, others remain comfortable interpreting abundance data in paleoenvironmental terms. Detailed taphonomic analysis provides one means of enhancing confidence in the paleoenvironmental interpretation, whether taxonomic presences or taxonomic abundances are the analytical focus. While taphonomic analysis will never reveal whether a fossil assemblage is truly representative of species presences or abundances on the ancient landscape (Lyman 1994, 2008b), it does allow for an assessment of whether or not changes in taxonomic composition between assemblages are influenced by changing taphonomic circumstances (e.g., Alemseged 2003; Bobe and Behrensmeyer 2004; Bobe et al. 2002; de Ruiter et al. 2008; Faith 2013a; Grayson et al. 2001; Terry 2007). If changes in species composition through a faunal sequence are unrelated to taphonomic change, as measured by any number of taphonomic proxies (e.g., skeletal part frequencies, fragmentation, number and kind of bone surface modifications), then we can be reasonably confident that inferred temporal trends are meaningful.

The example described by Grayson (1981) is not unique. Andrews (1990) presented a similar study in which he compared the modern faunas generated by different species of avian predator that had exploited the same area. Taxonomic presences and abundances varied by predator, just as they had in Grayson's (1981) example. In a clever later study, Andrews (2006) manipulated nineteen modern mammalian faunas from a variety of habitats in various ways to determine how depleted one fauna had to be, and how mixed faunas had to be, in order to skew or mute the environmental (habitat) signal (see also Le Fur et al. 2011). Although important, these studies are of the "what if?" sort, meaning that they illustrate potential outcomes, in this case, that a faunal paleoenviromental signal can in fact be distorted. What they do not do is tell us when those outcomes might apply to an actual temporal series of faunal assemblages that we hope reflects paleoenvironmental conditions.

A growing number of so-called *fidelity studies* illustrate the other side of the same coin. As originally defined by Behrensmeyer et al. (2000:120), fidelity studies assess "the quantitative faithfulness of the [fossil] record of morphs, age classes, species richness, species abundance, trophic structure, etc. to the original biological signals." For instance, does the frequency of faunal remains created by, say, a particular avian predator species accurately reflect the abundances of represented taxa on the landscape exploited by the predator? Such studies have become more frequent over the past twenty years or so (e.g., Hadly 1999; Kidwell 2001; Lyman 2012d; Lyman and Lyman 2003; Miller 2011; Miller et al. 2014; Terry 2008, 2010a, 2010b; Western and Behrensmeyer 2009). By and large these studies tend to confirm that, indeed, accumulations of faunal remains, such as those under an avian predator's roost, can retain an accurate paleoenvironmental signal. But like those studies by Grayson (1981), Andrews (1990, 2006), and others noted above, the possibility of ecological fidelity does not mean a particular fossil assemblage necessarily faithfully reflects the paleoenvironment. We describe ways to gain (or abandon) confidence in thinking assemblages retain ecological fidelity shortly.

Is the Analysis of Taxonomic Presences a Safer Alternative?

As we have seen, awareness of the potential taphonomic pitfalls of abundance data plays a role in driving some analysts to focus their efforts on taxonomic presences. Is this a safer alternative than focusing on taxonomic abundances? We believe that analyses of taxonomic presences can be prone to pitfalls just as pernicious as those that can plague abundance data. We begin by noting that focusing on the presence of only one or a few indicator species can produce inaccurate insight to paleoenvironments. This is so because, as we saw in Chapter 3, one must assume the ecological tolerances and habitat preferences of species have not evolved, and we in fact know and can demonstrate in some cases that this assumption is not always true. Thus, some analysts argue use of one or a few indicator species is ill-advised (e.g., Nesbit Evans et al. 1981). Instead, they argue, one should consider the presences of all taxa in an assemblage (Nesbit Evans et al. 1981; Van Couvering 1980).

If all taxa present in an assemblage are to be considered simultaneously, a pertinent fact is that species abundance distributions in living (Magurran 1988, 2004; Magurran and Henderson 2003; McGill et al. 2007; Preston 1948) and fossil communities (Grayson 1984b; Lyman 2008b) are characterized by a small number of very abundant taxa and a large number of rare taxa (they are structurally uneven). An example is provided in Figure 6.3, which illustrates the distribution of taxonomic abundances from Stratum IV at Homestead Cave. Kangaroo rats (*Dipodomys* spp.) are most abundant, accounting for a little more than 48% of the identified specimens, whereas most taxa individually account for <1% of the assemblage.

Let us assume, for the sake of argument, that the taxonomic abundances from Stratum IV at Homestead Cave accurately reflect taxonomic abundances on the ancient landscape. The dominance of kangaroo rats (48.5% of small mammal specimens), most of which likely belong to chisel-toothed kangaroo rats (95% of *Dipodomys* specimens identified to species from Stratum IV belong to this taxon), is consistent with xeric environments. Also present in Stratum IV, however, are six taxa that Grayson (2000a, 2000b) associates with environments that are more mesic than the contemporary environment at Homestead Cave. These are pygmy rabbit (*Brachylagus idahoensis*), sagebrush vole (*Lemmiscus curtatus*), yellow-bellied marmot (*Marmota flaviventris*), bushy-tailed woodrat (*Neotoma cinerea*), Great Basin pocket mouse (*Perognathus parvus*), and western harvest mouse (*Reithrodontomys megalotis*) (Figure 6.3). These taxa are rare, together accounting for only 1.6% of the small mammal assemblage. Rather than indicating a mesic environment, their overall rarity likely reflects a micro-environment or rare habitat that is incapable of supporting high abundances of these taxa (e.g., Graham and Semken 1987). Following the typical protocol for interpreting abundance data, we attach less significance

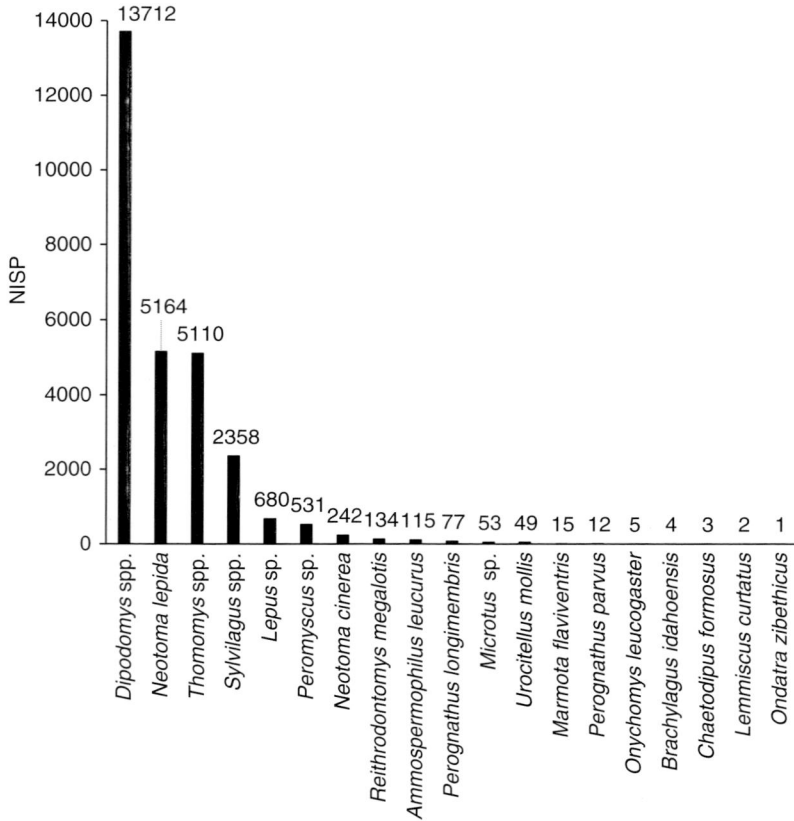

6.3. Species abundance distribution (NISP) for small mammals from Stratum IV at Homestead Cave.

to the micro-scale paleoenvironmental implications of these rare taxa (rare mesic micro-environment patches) than we do to the dominant kangaroo rats (meso-scale, abundant, large xeric environment patches).

When sampling from species distributions such as the one in Figure 6.3, it is easy to recover specimens belonging to the more abundant taxa (kangaroo rats), but it requires larger and larger sample sizes to detect the increasingly rare taxa (e.g., pygmy rabbits). As noted above, at least some of the rare taxa, which may be at the limits of their ecological tolerances or associated with small microenvironmental patches on the landscape, are likely to have ecological associations seemingly at odds with the dominant taxa. The implication is that well-sampled fossil assemblages are more likely to detect those rare taxa with eco-logical implications distinct from those of the dominant taxa. In perhaps clearer language, large samples are necessary to detect the presence of rare taxa. When taxonomic abundances are the focus of paleoenvironmental reconstruction, this is a minor problem because the dominant taxa remain dominant regardless of how many specimens or species are recovered. What is lost when interpreting a

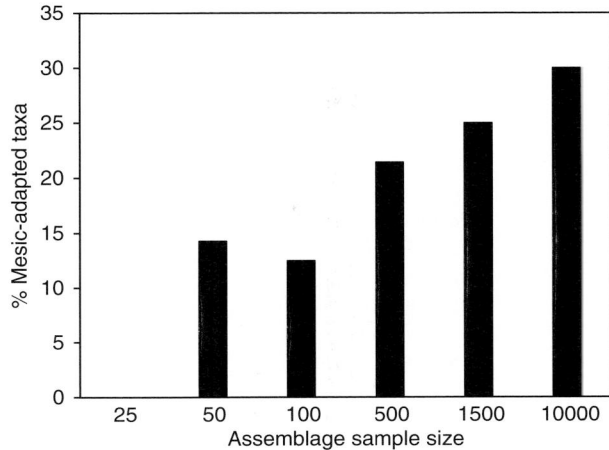

6.4. Percentage of taxa associated with mesic environments for six random sub-samples of Homestead Cave Stratum IV. Data from Table 6.1.

poorly sampled assemblage is the finer nuances provided by a greater number of species. And there is another issue to keep in mind when considering whether to interpret taxonomic presences or abundances. When abundances are discounted and species presences become the focus of analysis, there is scope for paleoenvironmental interpretations to be influenced by sampling because rare taxa are given the same interpretive weight as abundant taxa. To provide an example, we turn once again to Stratum IV from Homestead Cave.

Stratum IV includes an impressive 28,525 identified small mammal specimens (= NISP). It is this massive sample that allowed the detection of rare taxa that prefer mesic conditions (and prompted Grayson to abandon his earlier suggestion that taxonomic presence data were to be preferred over taxonomic abundance data). But what if fewer specimens had been recovered? Would smaller samples have revealed the ancient presence of mesic micro-environments? To answer these questions, we randomly sampled Stratum IV at sample sizes of 25, 50, 100, 500, and 10,000 specimens (Table 6.1). Not surprisingly, the dominant taxon (kangaroo rat) is present in all simulated small assemblages. This is not so for the six rare taxa indicative of mesic environments. At a sample size of 25, none are recovered. As sample size increases, however, so too does the number of mesic-adapted taxa, with all six occurring in the assemblage of 10,000 specimens. Importantly, not only are the rarely represented mesic-adapted taxa more likely to occur as sample size increases, but they also account for an increasing proportion of the total number of taxa recovered (Figure 6.4). This is so because rare taxa account for only a small number of the taxa in a poorly sampled assemblage that primarily includes the handful of dominant taxa. As sample size increases, an increasing number of rare taxa are recovered relative to the few dominant taxa. Thus, the increasing proportion of mesic-adapted taxa (Figure 6.4) at higher sample sizes is simply a reflection of the enhanced recovery of rare taxa.

TABLE 6.1 *Taxonomic abundances (NISP) of small mammals from Homestead Cave Stratum IV and randomly sub-sampled assemblages at different sample sizes.*

Taxon	IV All	Sub-samples 25	50	100	500	1,500	10,000
Ammospermophilus leucurus	115	0	0	0	1	6	44
Brachylagus idahoensis	4	0	0	0	0	1	2
Chaetodipus formosus	3	0	0	0	0	0	2
Dipodomys	12,629	11	16	50	199	665	4,454
Dipodomys microps	1,033	0	3	3	27	57	395
Dipodomys ordii	50	0	0	0	1	1	17
Lemmiscus curtatus	2	0	0	0	0	0	1
Lepus sp.	680	0	1	2	22	35	237
Marmota flaviventris	15	0	0	0	1	0	4
Microtus	53	0	0	1	0	2	12
Neotoma	258	1	1	4	4	12	108
Neotoma cinerea	242	0	1	0	1	7	89
Neotoma lepida	5,164	7	8	17	93	272	1,779
Onychomys leucogaster	5	0	0	0	0	3	2
Ondatra zibethicus	1	0	0	0	0	0	0
Perognathus longimembris	77	0	0	0	3	5	23
Perognathus parvus	12	0	0	0	0	2	3
Peromyscus	531	0	1	1	8	31	192
Reithrodontomys megalotis	134	0	0	1	3	6	63
Spermophilus	49	0	0	0	4	3	21
Sylvilagus cf. *audubonii*	13	0	0	0	0	1	2
Sylvilagus cf. *nuttallii*	13	0	0	0	0	0	5
Sylvilagus sp.	2,332	5	8	6	46	120	779
Thomomys bottae	5,110	1	11	15	87	271	1,766

Note: Bold indicates taxa associated with relatively mesic environments.

Another critical point must here be made explicit. If our simulated assemblages (Figure 6.4) represented a fossil sequence, it would be very easy to infer from a consideration of species presences that there are important changes in moisture availability between samples. Some have none or only a few mesic-adapted taxa, while others have more (Table 6.1). However, all that has actually happened is a change in sampling intensity, not a change in the environment. The same interpretive mistake would not result from a consideration of taxonomic abundances. Xeric-adapted kangaroo rats are dominant irrespective of sample size, and their abundances do not fluctuate in any meaningful way over the simulated samples ($\chi^2 = 5.290$, p = 0.382). Likewise, the six mesic-adapted taxa are absent or rare throughout all assemblages, with no statistically significant changes in their abundances ($\chi^2 = 3.690$, p = 0.595).

Our analysis demonstrates that if abundance data are discounted and equal weight is given to all species in an assemblage, then paleoenvironmental interpretations may be influenced by sampling effort. Through a perhaps surprising twist related to the relationship between species abundance distributions and sampling intensity, the better sampled assemblages are the ones most likely, when the focus is on taxonomic presences, to lead one astray because they provide more taxa at the limits of their ecological tolerances. Another way of thinking of this issue is to return to the point we raised at the beginning of this chapter concerning the presence of chisel-toothed kangaroo rats at Homestead Cave. Recall that because this species occurs throughout the sequence, its presence alone tells us nothing about past environmental *changes* (its consistent presence indicates the consistent presence of at least some arid habitat). But if the sample sizes at Homestead Cave were considerably smaller, we might expect chisel-toothed kangaroo rats to be absent from the lowest levels (because it is rare and hard to sample) and present in the upper levels of the sequence (because it is more abundant and easy to sample). In such a case, one might infer that its presence in the upper levels signifies a more xeric environment than in the lower levels from which its remains would likely be absent. This inference would be consistent with other lines of evidence from the site, but the pattern is only evident because smaller sample sizes mean that the species was not sampled when it was rare. The point of this thought experiment is that the presence of a taxon in a fossil sample is closely related to its abundance, and in turn to sampling effort.

Does this mean that abundance data should be always preferred? Of course not. Our objective is only to point out that there are interpretive pitfalls inherent to both approaches; only through awareness of those pitfalls is it possible to formulate sound, or at least adequately cautious, paleoenvironmental inferences. The presence of a (rare) taxon in a fossil assemblage is but one of numerous variables of paleozoological interest known to be influenced by sampling artifacts (Bulinski 2007; Faith and Gordon 2007; Grayson 1984b; Lyman 2008b). Fortunately, the identification of sample-size effects is a straightforward process; simply examine the relationship between assemblage sample size and the variable of interest (Lyman and Ames 2004, 2007). Turning back to our simulated assemblages from Homestead Cave Stratum IV, if the proportion of mesic-adapted taxa were the focus of paleoenvironmental inference, it is readily apparent that this variable is influenced by sampling artifacts; there is a strong correlation between assemblage NISP (log-transformed) and the proportion of mesic-adapted taxa ($r = 0.932$, $p = 0.007$). It follows that this variable is not tracking paleoenvironmental change but rather is tracking variability in sample size. Similar sampling phenomena have been observed in the fossil record. For instance, Faith (2011c:224) noted that the proportion of mammalian taxa that are grazers in various Pleistocene South African faunas is correlated with assemblage sample size – larger assemblages include a greater proportion of grazer taxa (similar to Figure 6.4) – meaning that this variable is a problematic indicator of past environments. We will encounter sample size issues again in discussions that follow.

How to Gain (or Lose) Confidence

Above we indicated that taphonomic concerns caused some analysts to deem that taxonomic abundance data provide skewed to completely inaccurate reflections of paleoenvironments. We believe the same can be said of the influence of taphonomic processes on study of taxonomic presences. We also indicated above that detailed taphonomic analysis might enhance our confidence that a particular assemblage or set of assemblages retained a robust paleoenvironmental signal. Even with detailed taphonomic analysis, however, we can never know the precise nature of the relationship between the measured variable (taxonomic presences and abundances in a fossil assemblage) and the target variable (taxa present and their abundances on an ancient landscape). What we can do to enhance our confidence in the validity of a fossil assemblage's paleoenvironmental implications is to determine if taxonomic abundances (or presences) covary with taphonomic data. For instance, given arguments that greater fragmentation reduces identifiability (Cannon 2013, and references therein), one could determine whether the degree of fragmentation of a taxon increases as the abundance of that taxon decreases. If the two variables are significantly correlated, then the researcher would be ill-advised to interpret shifts in the taxon's abundance as reflecting environmental change. In an early example, Higgs (1967:159) considered potential biases related to human prey choice, if in an unsophisticated way by today's standards, when with respect to fossil faunas in Libya, he noted that "changes [in taxonomic abundances] are not coincident with cultural changes … [making] it positively unlikely that hunting custom played any appreciable part in the large bovine fluctuations [in abundance]."

Any number of taphonomic variables might be measured and a correlation sought between one or more of those variables and taxonomic abundances (e.g., Faith 2013a). Alternatively, if studied deposits in a site allow, the paleozoologist could compare stratigraphically bounded assemblages associated with artifacts and assemblages with no (or few) associated artifacts (e.g., Grayson 1991b; Klein and Cruz-Uribe 2000). Comparisons between stratigraphically bounded assemblages could be of taxonomic presences, abundances, body size distributions, skeletal part frequencies, or any one or more taphonomic variables. Similarities suggest taphonomic agents or processes had little influence on the faunas, whereas differences would suggest those agents or processes had some influence. And of course, the paleoenvironmental implications of the fauna, however detected, should be compared with the implications of independent paleoenvironmental data (Chapter 3).

ABUNDANCES OF A FEW TAXA

In the search for paleoenvironmental trends through a temporal sequence of shifting (or perhaps static) taxonomic abundances, it is not unusual to examine the abundances of select (usually stenotopic) taxa or all taxa simultaneously through time. At the outset, one typically hopes to be able to identify the principal changes in faunal composition by eye. There are two ways one might initially investigate the paleoenvironmental significance of taxonomic abundances. Both concern how abundance data are presented for interpretation. The first, listing frequencies of each taxon in a table (e.g., Tables 4.2, 4.3, and 4.5), must precede the second, graphing those frequencies. The paleozoologist does not necessarily have to produce a graph of taxonomic abundances, though it does sometimes facilitate interpretation of the tabled data when the table includes numerous taxa and assemblages. Analysts have long produced tables and various forms of graph (e.g., Bate 1937; Cartmill 1967; Fowler and Parmalee 1959; Harris 1963; Howard 1930; Sparks 1961; Stock 1929). Abundances can be in the form of absolute or relative (percent) frequencies of NISP, MNI, biomass, or some other variable, for each taxon, and included taxa can be at any one or more levels in the taxonomic hierarchy. Relative abundances are thought to be preferred over absolute abundances because the former will correct for variability in sample size between assemblages. As we will see below, the problem of closed arrays may counterbalance this benefit.

Tables listing taxonomic abundances and graphs (of whatever style) of such abundances are, in many ways, the bread-and-butter of many published analyses. We were therefore a bit baffled while writing this chapter to discover that discussions on how to analyze and interpret taxonomic abundances for purposes of paleoenvironmental reconstruction are quite rare. We are aware of several relatively recent brief discussions of analytical protocols (e.g., Andrews 1996; Graham and Semken 1987; Reed 2013; Reed et al. 2013; Terry 2009; Yalden 2001), yet very little is said in any of them about how one might contend analytically with taxonomic abundance data. Not even tables or graphs of taxonomic abundances are mentioned. Following the template of Chapter 5, we begin discussion here with relatively simple techniques that focus on one or a few taxa in assemblages before turning to techniques that simultaneously consider many or all taxa.

Abundances of Indicator Taxa

An abundance of data is typically better than the alternative, but as indicated in the preceding section, large datasets with many taxa or assemblages can make it difficult to intuitively grasp their paleoenvironmental implications. As with analyzing taxonomic presences, one way to manage a plethora of data is to focus on the taxonomic abundances of indicator taxa, particularly those species that are stenotopic. An excellent example is provided by Guilday's (1969) analysis of small mammal remains recovered from the Wasden Site-Owl Cave, in southeastern Idaho. Eighteen taxa variably distributed across numerous strata were represented by a total MNI of nearly 6,000. Excavations were in twenty-six, mostly 10-cm-thick levels. Radiocarbon dates and geomorphology suggest the deposit accumulated beginning sometime during the late Pleistocene and spans the entire Holocene. Guilday (1969:48) wrote that "some degree of climatic change is indicated by the gopher–rabbit–ground squirrel ratio up and down the [stratigraphic] column." He presented absolute taxonomic abundance data for all taxa in table form, absolute abundance data for (northern pocket) gopher (*Thomomys talpoides*), rabbit (*Sylvilagus* spp. [including what is now *Brachylagus idahoensis*]), and ground squirrel (*Spermophilus* spp.) in both table and graphic form (Figure 6.5), and relative taxonomic abundance data for the three indicator taxa in graphic form (Figure 6.6). He interpreted the taxonomic abundance data as indicative of increasing sagebrush and decreasing grass over time, that is, increasing aridity, particularly stratigraphically above the depth of 100–110 cm.

Guilday (1969) noted that graphing relative (percent) abundances of species could be misleading. Because relative abundances of taxa in an assemblage must always sum to 100 percent, an increase in one (or more) taxa will cause a corresponding decrease in one (or more) taxa. We cannot improve on Guilday's (1969:48) own words: relative abundance graphs possess, he said, "an intrinsic weakness. As the percent of one item [read *taxon*] goes up, the percent of another perforce goes down relatively, whether it does in actual numbers or not." This is undoubtedly why he presented both absolute and relative taxonomic abundance data. Percentage or relative abundances represent what are known as "closed arrays"; they must sum to 100 percent, and this places various statistical and interpretive constraints on analyses (Jackson 1997). A year before Guilday, paleontologist Eitan Tchernov (1968) noted that although studying relative (percent) abundances of taxa should reveal evidence for reconstructing past environments, all taxa in a fauna should be consulted simultaneously precisely because of the closed array problem:

> It is impossible to determine climatic fluctuations on the basis of the relative [frequency] distribution of any two animals; when one diminishes in number the other does not necessarily multiply at its expense. One must take into account the entire groups of animals whose representatives have successfully adapted themselves to an appreciable portion of biotopes. The relationship between two biotopes is not always correlated, and only the interrelation between all biotopes represents a picture closer to the truth.
>
> (Tchernov 1968:139)

Subsequent to Guilday's (1969) analysis, archaeologist B. Robert Butler (1969:59) plotted "the ratio of pocket gophers to rabbits through time" on the basis of the fact that the two taxa had different food habits and occupied different plant associations. In short, pocket gophers (*Thomomys* sp.) prefer mesic grasslands whereas the rabbit species (*Sylvilagus* sp., *Brachylagus idahoensis*) prefer more xeric brushy habitats (note that *B. idahoensis* was previously considered

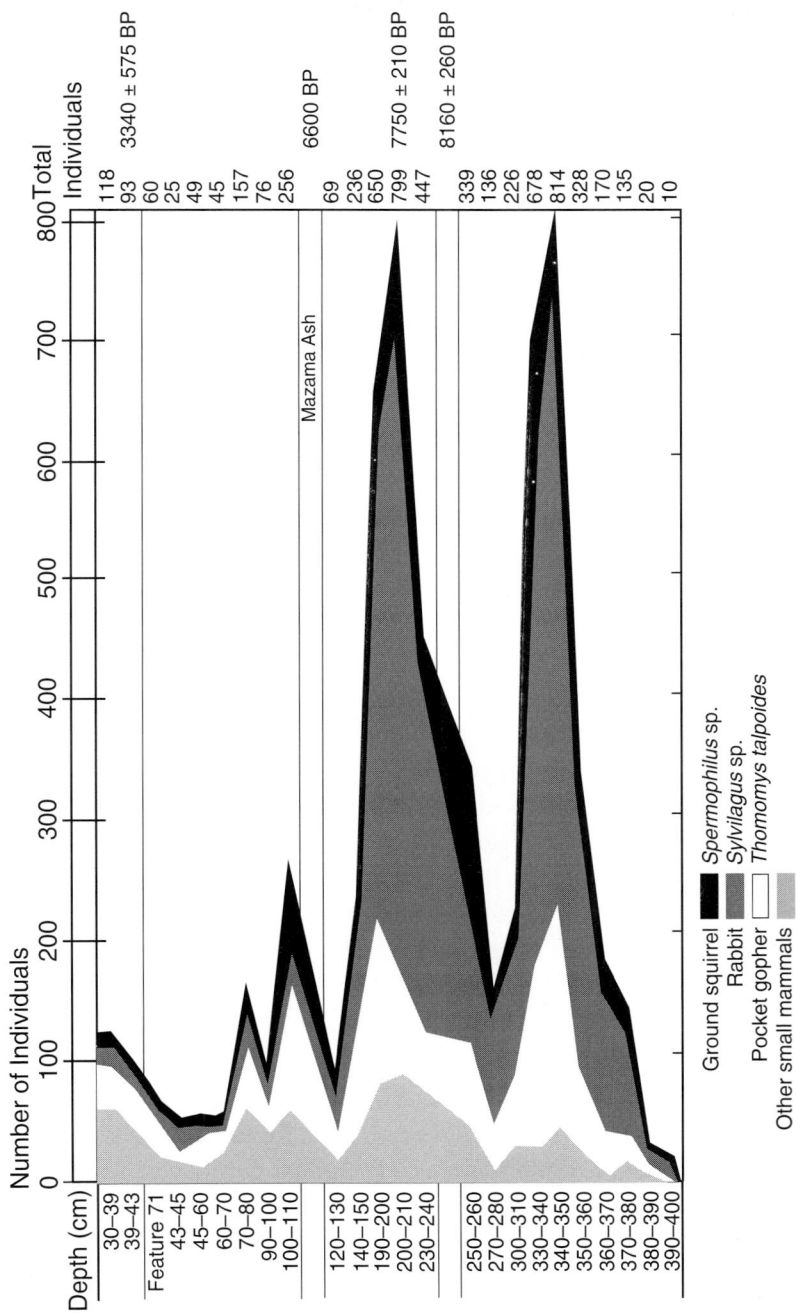

6.5. Absolute abundances (MNI) of small mammal taxa across the Wasden Site–Owl Cave stratigraphic sequence. After Guilday (1969). Compare with Figure 6.6.

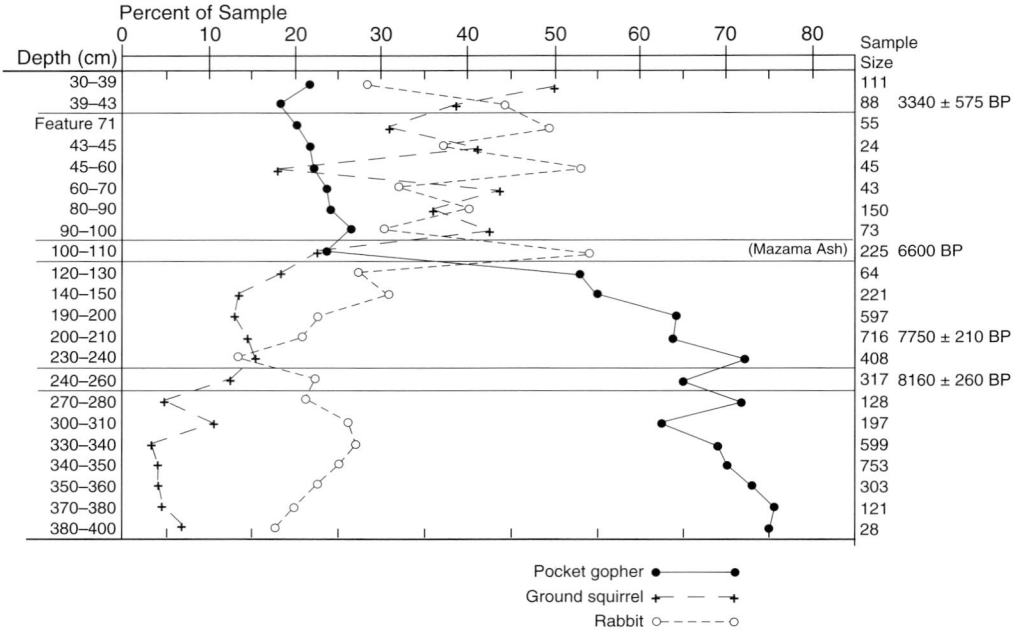

6.6. Relative abundances (%MNI) of small mammal taxa across the Wasden Site-Owl Cave stratigraphic sequence. After Guilday (1969). Compare with Figure 6.5.

a mesic-adapted taxon in our discussion of Homestead Cave; this relates to the issue of relative tolerances discussed in Chapter 2 [see Figure 2.7]). Butler (1969) used "ratios" of taxa that had been calculated by a colleague. We describe the formula in the following subsection. Here it suffices to note that Butler graphed the ratios (Figure 6.7), noting that there was a marked shift from pocket gophers outnumbering rabbits prior to about 6,600 BP, to rabbits becoming the dominant species after that time. He interpreted this shift to represent (i) a shift from abundant grass and forbs, perhaps with some sagebrush (*Artemisia* sp.), to reduced grass and forbs with increased sagebrush, and (ii) greater aridity or less effective precipitation late in time relative to earlier. He continued to publish the graph (or a variant of it) and his interpretation for the next decade (Butler 1972a, 1972b, 1976, 1978). This is, in part, what makes it a classic example of analysis of abundances of indicator taxa. The basic approach has subsequently been used by others, sometimes with a graph (e.g., Fernandez-Jalvo et al. 1998; Grayson 1976, 1977), sometimes without (e.g., Driver 2001; Lyman 2014b).

Abundance Indices

In the text and on the figure he published, Butler (1969) indicates the formula used to determine the ratio of gophers to rabbits was:

$$[(f_1 - f_2)/(f_1 + f_2)] 10 = \text{ratio of abundance}$$

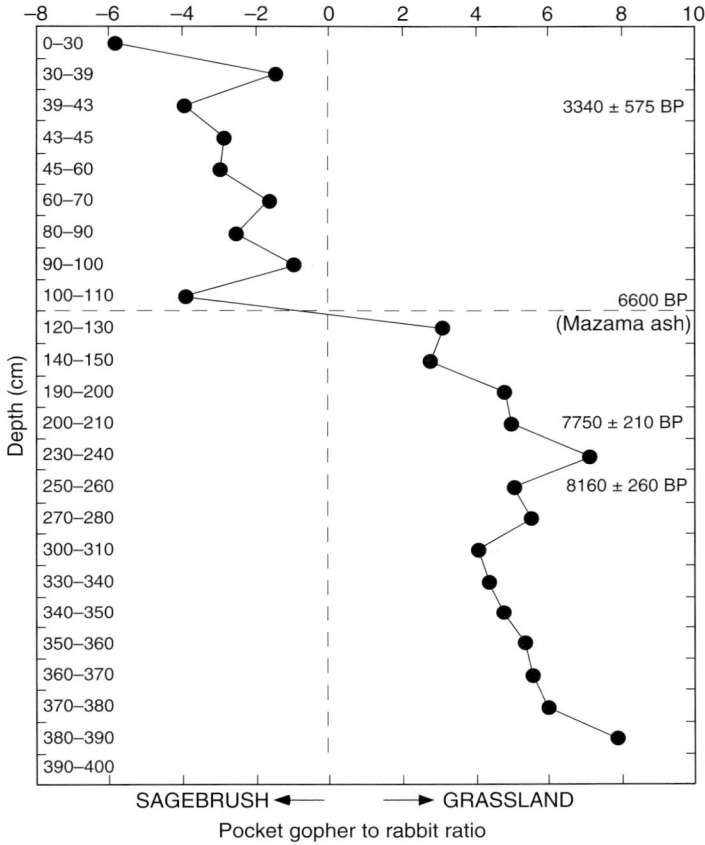

6.7. Ratio abundances (MNI) of pocket gophers and rabbits across the Wasden Site-Owl Cave stratigraphic sequence. After Butler (1969). Compare with Figures 6.5 and 6.6.

where f_1 was the abundance of gophers and f_2 was the abundance of rabbits.[1] Apparently independently, several researchers working in western North America calculated similar ratios of other pairs of taxa during the 1970s and 1980s (Driver and Woiderski 2008). Collectively these ratios have become known as "abundance indices" (e.g., Codding et al. 2010a), though none of them includes multiplying the result of the division by 10 (an arbitrary and unnecessary procedure), thus index values range from −1 to 1. Although originally developed largely for the purpose of monitoring human influences on

---

[1] This is in fact not the formula Butler's colleague used; using the formula Butler describes does not produce the values plotted in Butler's graph, though the general plot of values is quite similar to Butler's original. In his discussion Butler (1969:59) quotes an unpublished presentation by the individual who designed the formula and calculated the ratios. We used the formula Butler presented to generate the values plotted in Figure 6.7. It does not matter that our graph differs slightly from Butler's as the trend in relative abundances of the two species is the same in Butler's and our analyses.

faunas, the indices are readily adapted to paleoenvironmental reconstruction. One of these is particularly exemplary in this regard.

The Lagomorph Index or Leporid Index (LI) has been used in western North America to track environmental change (e.g., Quirt-Booth and Cruz-Uribe 1997; Sanchez 1996; Shaffer and Schick 1995). The LI allows such tracking because it employs two taxa of leporids that differ in preferred habitats. Jackrabbits or hares of the genus *Lepus* occupy open areas and use escape speed to avoid predators; rabbits or cottontails of the genus *Sylvilagus* (and *Brachylagus*) occupy open areas but prefer relatively denser vegetation cover in order to hide from predators. Habitats occupied by *Lepus* tend to be more arid than those occupied by *Sylvilagus*. Both genera are members of the taxonomic family Leporidae and the order Lagomorpha. In North America, Lagomorpha includes the families Leporidae and Ochotonidae; members of the latter (pikas) are rare in North American archaeological sites. We thus prefer the name Leporid Index for the abundance index, unless remains of ochotonids are included in calculations.

The LI has been given several different names and calculated several different ways (Driver and Woiderski 2008). The key variable analysts typically hope to monitor is the abundance of *Sylvilagus* (rabbit) relative to *Lepus* (hare) abundances. Thus, keeping the math as simple as possible, one formula that has been used to calculate the LI is:

(*Sylvilagus* abundance)/(*Sylvilagus* abundance + *Lepus* abundance) = LI

Index values >0.5 indicate *Sylvilagus* is more abundant than *Lepus*; index values <0.5 indicate *Sylvilagus* is less abundant than *Lepus*. One can list the index values in a table, plot them in a graph, or both. Jacob Fisher's (2012) recent analysis of the LI values for the Five Finger Ridge archaeological site in southwestern Utah provides an example. Fisher found that shifts in the LI correspond well with independent paleobotanical data. *Lepus* remains became less abundant and *Sylvilagus* remains became more abundant just as might be predicted given paleobotanical evidence for increasing woodlands and a more closed vegetation structure.

An abundance index could be designed for any two (or several) taxa whose abundances relative to one another might provide a sensitive paleoenvironmental signal. For instance, in one of the earliest examples of such, Bate (1937) calculated the relative (percentage) abundances of *Dama mesopotamica* and *Gazella gazella* and plotted the values through time, suggesting the first indicated woodlands (mesic) and the second indicated steppe (xeric) habitats. Her conclusions were later questioned (Tchernov 1968; see also the review and references in Garrard 1982), but her method was subsequently used by others (e.g., Higgs 1967; Tchernov 1968) and advocated in a later publication (Ziegler 1973), despite the recognized problem of closed arrays. Vrba (1974, 1975, 1980) developed strong warranting arguments that the percent abundance of Alcelaphini (the tribe of Bovidae that includes wildebeest and allies) plus Antilopini (the tribe of Bovidae that includes gazelles and allies) – both archetypal open habitat lineages – among all Bovidae in Africa was an excellent indicator of the openness of habitats. Based on a correspondence analysis of census data (taxonomic abundances) from modern bovid communities spanning a broad range of habitats (published later by Greenacre and Vrba 1984), she observed that bovids of these tribes tend to be abundant (>60% of bovids) in environments with a low proportion of bush and tree cover, but rare (<30% of bovids) in environments with a high proportion of cover (Vrba 1980); this is frequently known as the alcelaphin plus antilopin criterion (AAC). Despite the potential problem of being a closed array, the solid uniformitarianist foundation (Greenacre and Vrba 1984) has prompted many paleozoologists working on late Cenozoic African faunas to examine the relative abundances of these tribes – sometimes also including other indicator species – to develop paleoenvironmental inferences (e.g., Bobe 2006; Bobe and Behrensmeyer 2004; Garrett et al. 2015; Kappelman 1984; Potts 1988; Tryon et al. 2010). Shipman and Harris (1988; see also Plummer et al. 2009) later expanded on the AAC to consider a few other bovid tribes indicative of other habitat types, but the basic premise – linking their relative abundances to particular habitats – remains the same.

To be explicit, we note that one does not need to have species-level data to construct a ratio of taxonomic abundances. Vrba's AAC, for instance, relies on tribal-level data. In their analysis of the early Pleistocene faunas from Olduvai Gorge, Fernandez-Jalvo et al. (1998) used the ratio of Gerbillinae (indicator of open habitats) to Murinae (indicator of closed habitats), calculated as the abundance (MNI in this case) of Gerbillinae divided by the abundance of Murinae (Gerbillinae/Murinae index). This ratio had been used previously by others (Dauphin et al. 1994; Denys et al. 1996; D. N. Reed 2007), and thus like Butler's index of gopher and rabbit abundances and like the leporid index, it is a classic one. Fernández-Jalvo and colleagues (1998) observed a decline in the relative abundance of murines and an increase in the relative abundance of gerbils though time at Olduvai Gorge, which they interpreted as indicating a shift toward more open and arid environments.

Ratios of abundances need not be restricted to taxonomic categories; they can be used in a taxon-free manner, in which ecological categories are substituted for taxonomic identities (see Chapter 7). Grayson (1976) identified the avifaunal remains recovered from an archaeological site on Lower Klamath Lake in northern California, many of which represented waterfowl. Although Grayson was able to identify many of the remains to genus or species, he noted that the bird taxa tended to fall into two major behavioral groups – non-diving taxa and diving taxa. The latter habitually dive to forage and escape from danger; the former are sometimes referred to as dabblers that feed by swimming along the surface with their bill cutting through the water or tipping their tail into the air and their head underwater to feed on a shallow bottom. Grayson (1976) calculated the ratio of non-divers to divers (using MNI values) across five temporal periods within the Holocene under the assumption that if divers outnumbered non-divers the water was deeper than if the opposite was the case. Over time, an increase in water depth would be accompanied by increasing numbers of divers and decreased abundances of non-divers. He found that divers decreased and non-divers increased over time, suggesting decreased lake depth that corresponded with greater aridity implied by independent paleoecological data.

One important thing to keep in mind when calculating index values is that although those values eliminate the influence of differences in absolute sample size such as are apparent in Figure 6.5, they are not necessarily free of closed array problems (Jackson 1997). This is so because the index values are proportions; they are exactly the same as percentages except that proportion values range between 0 and 1 whereas percentage values range between 0 and 100. The abundance indices (or ratios of abundances of one taxon to another) are measures of relative abundances.

Discussion

To this point, we have discussed several simple techniques for deciphering the paleoenvironmental significance of a few (usually stenotopic) taxa in assemblages. We think of these as analysis "by hand" because the researcher compiles the tables and draws the graphs (assisted these days by computer graphing programs) that are subsequently interpreted. The simplicity comes from the fact that the number of variables that one must simultaneously keep track of is equivalent to the number of taxa included; two taxa means two variables that fluctuate, three taxa means three variables, and so on. But as the number of taxa increases, the difficulty of simultaneously keeping all taxa in mind increases. Modern computer technology provides what we think of as "mechanical" analysis. Many software programs allow simultaneous consideration of all taxa and simplification of multivariate phenomena into a few "axes" the meaning of which is left to the researcher. Before turning to some of these mechanical analytical techniques we offer a little more history to contextualize analyses of abundances of all taxa.

HISTORY OF ANALYZING TAXONOMIC ABUNDANCES

Analysis and interpretation of taxonomic abundances of complete faunas was, it seems, unusual in the middle twentieth century. De Graaf (1961) devoted some space to estimating the "relative density" of several species of micromammals from South African cave sites on the basis of MNI values, but when it came to interpreting the paleoenvironmental significance of shifts in those abundances, he was extremely cautious because he lacked sufficiently detailed information on the ecological tolerances of many of the species. Among European paleontologists, Björn Kurtén (1952), at the suggestion of another individual, determined the abundances (MNI) of numerous mammal species from various paleontological sites in a small area of China, and concluded on the basis of various evidence that two communities were represented. He then categorized each species represented as (a) *brachydont*, browser, forest preferring; (b) intermediate; or (c) *hypsodont*, grazer, plains preferring. Frequencies of these summed taxa per category indicated to Kurtén the two communities differed in terms of paleoenvironment. He did not evaluate the paleoenvironmental implications of each species or their abundances.

Many Old World researchers suggested all taxa be consulted, but focused heavily (if not exclusively) on the most abundant and/or most stenotopic species rather than analyze or interpret the relative abundances of all taxa across a series of assemblages (e.g., Cartmill 1967; Garrard 1982; Levinson 1985). Paleo-Africanist Richard Klein sometimes focused on taxonomic presences (e.g., Klein 1972), other times on the relative frequencies of species over time (e.g., Klein 1976b, 1983) but interpreted relative abundances of general categories of species such as grazers and browsers, and still other times used both presences and abundances (Klein 1980).

In North America, Claude Hibbard, a vertebrate paleontologist famous for (among other things) his introduction of screening techniques that enhanced recovery of microfauna (Hibbard 1949), never (so far as we can determine) considered taxonomic abundances (e.g., Hibbard 1955, 1963). Nor did George Gaylord Simpson, although Simpson, like Hibbard, did reconstruct past environments based on the presence of taxa (Simpson 1947) and used quantitative measures of the similarity of faunas based on species present (Simpson 1936, 1943b). Vertebrate paleontologist J. Arnold Shotwell, well known for proposing a (what is now known to be flawed [Grayson 1978]) technique for distinguishing local (proximal) taxa of a prehistoric fauna from taxa originally in distant (distal) faunal communities based in part on NISP values per species (Shotwell 1955), focused more on species presences than taxonomic abundances when it came to deciphering the paleoenvironmental significance of faunas (e.g., Shotwell 1958, 1963). Other North American researchers followed suit (e.g., Dalquest 1965; Guthrie 1968b; Semken 1966; Slaughter 1966). Even Ernest Lundelius, Jr., who studied Quaternary fossils in both Australia (e.g., Lundelius 1960, 1983) and North America (e.g., Lundelius 1979), interpreted taxonomic presences rather than abundances.

But as implied earlier in this chapter, taxonomic abundances were not ignored by everyone in the mid-twentieth century (e.g., Butler 1969; Flannery 1967; Guilday et al. 1964; Guthrie 1968a; Harris 1963). In fact, this time period seems to be when greater analytical and interpretive focus on taxonomic abundances emerged, at least in zooarchaeology (Lyman 2015a, 2018a, 2018b, 2018c). Early programmatic statements produced by paleontologists on how to use animal remains to reconstruct past environments did not mention taxonomic abundances (e.g., Case 1921, 1936; George 1958; Ladd 1959). Recall, as noted above, Kurtén (1952) indicates someone else suggested a study of taxonomic abundances might prove revealing. It is in the 1970s and 1980s that detailed considerations appear, largely in the zooarchaeology literature, on the validity of particular quantitative units (NISP, MNI) and how values for each might be determined (see discussion and references in Lyman 2008b). Paleontologists had long determined taxonomic abundances using MNI (e.g., Howard 1930; Kurtén 1952; Stock 1929), so the pertinent literature in that discipline was not about which quantitative units to use (see Badgley [1986] and Holtzman [1979] for notable exceptions). Rather, much of it was about sampling and sample size influences (e.g., Tipper 1979; Voorhies 1970; Wolff 1975), and more recent literature continues along that avenue (e.g., Bulinski 2007; Davis and Pyenson 2007; Moore et al. 2007). Why, then, did early paleozoologists seemingly not often consider the abundances of all represented taxa and why have they only in the past couple of decades come to do so?

In her early study of abundances of all or most taxa in a series of assemblages, Margaret Avery (1982) provides what we think is the critical variable for answering the question just posed. She pointed out that when there was an abundance of data, it was "difficult to isolate by simple graphic methods alone the underlying patterns which [are] assumed to exist. For this reason it was considered appropriate to employ some form of multivariate statistical analysis as an aid to interpretation of the evidence" (Avery 1982:229). Avery chose factor analysis "because of its data-reduction capabilities" (p. 229), but even then she did not include the rarely represented taxa. The reason to mention Avery's analysis, however, concerns the fact that she specifically states the necessity of a multivariate statistical analytical technique originates in the abundance of data requiring study, and she identifies the computer technology she used (see also Cruz-Uribe 1983; Klein 1976b). The latter is what we find important. It seems to us that the beginnings of intensive analysis of taxonomic abundances of most or all taxa for purposes of paleoenvironmental reconstruction coincides temporally with the emergence of the necessary computer technology. This is certainly the case with the emergence of modern paleobiology from paleontology (Sepkoski 2012).

Thus we have John Guilday and colleagues (1964) graphing in some detail the abundances of species within each of several taxonomic families, but interpreting the shifts in abundance in broad habitat terms such as "the boreal element" or "Hudsonian life zone." He does precisely the same thing fourteen years later (Guilday et al. 1978), and so far as we can ascertain, never used computer technology to assist his analyses (Guilday passed away in 1982). A contemporary of Guilday's, Holmes Semken, a student of Hibbard's, initially focused on taxonomic presences (e.g., Semken 1966), but eventually used computer-facilitated correspondence analysis to make sense of taxonomic abundances (Semken and Graham 1996). But even today, not everyone finds it necessary to perform Avery's (1982) recommended multivariate statistical analysis so as to reduce complex data to a couple of index values, vectors, or axes (e.g., Georgina 2001; Lyman 2014b). It is beyond our scope here to explore the timing, magnitude, and nature of the influence of emerging computing technology on paleoenvironmental reconstruction based on faunal remains, but we think it would be an interesting study. The important point here is analysis of abundances of all represented taxa in collections did not become more commonplace (and certainly not ubiquitous) until the necessary computer technology was available. In the next section, we discuss techniques for considering abundances of all taxa that can be undertaken with that technology.

ABUNDANCES OF ALL TAXA IN ASSEMBLAGES

Examples described to this point in this chapter are relatively straightforward because they consider one or a few taxa. Such techniques ignore the other taxa typically found in assemblages. Thus things are, for better or worse, not always so simple as the examples described to this point might imply. For instance, turning to the Boomplaas Cave ungulates (Table 4.3, Figure 6.7), and considering all (or most) of those taxa, one might be struck by the rise and decline of zebras (*Equus zebra/quagga*) and the extinct caprin, coupled with an approximately inverse pattern for steenbok/grysbok (*Raphicerus* spp.), klipspringer (*Oreotragus oreotragus*), and mountain reedbuck (*Redunca fulvorufula*). There are other trends that follow a somewhat different trajectory, such as the long-term decline of grey rhebuck (*Pelea capreolus*), while some of the rarer taxa (e.g., *Taurotragus oryx*) do not seem to follow any obvious trends at all. As was the case with our initial investigation of the presence/absence data for Boomplaas Cave (Chapter 5), it is clear that synthesizing this information may not be a straightforward process. Once again, ordination can facilitate the process, so we turn to it first. We subsequently consider a couple of other techniques that focus attention on all taxa in an assemblage.

Ordination of Taxonomic Abundances

In Chapter 5 we discussed a handful of ordination techniques that can be used to translate multiple variables – namely the presences or abundances of multiple taxa across assemblages – into a smaller number of index variables

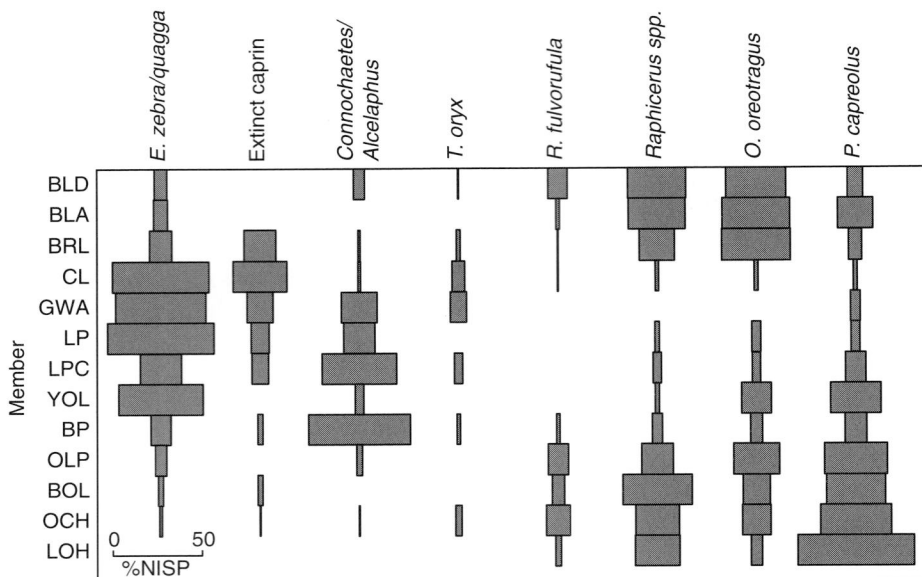

6.8. Relative abundances (%NISP) of some ungulate taxa across the Boomplaas Cave strati-graphic sequence.

(or axes) that summarize the variation in faunal composition. The ordination of assemblages can be scrutinized to identify changes that are (one hopes) indicative of changing environments. Some early examples of ordinations conducted on taxonomic abundance data used principal components analysis (e.g., Cruz-Uribe 1983; Klein 1976b) or the closely related factor analysis (e.g., Avery 1982; Deacon et al. 1984; Thackeray 1987), but we prefer correspondence analysis (CA) because it is suitable for taxa that have a unimodal response to an environmental gradient and because it simultaneously ordinates taxa and assemblages, facilitating interpretation (Chapter 5).

Complementing our previous CA on presence/absence data from Boomplaas Cave (Chapter 5), we now provide a parallel analysis using abundance data. Unlike the previous analysis, we have included the indeterminate taxonomic category *Connochaetes/Alcelaphus* (wildebeest and hartebeest) in order to take into account the many specimens that could not be confidently assigned to either genus (Figure 6.8). To avoid repetition, we focus on the principal similarities and differences between ordinations, and how interpretations differ when abundances are taken into account. As is typical for abundance data, the first two axes account for the majority of variation in faunal composition (axis 1: 44.5%; axis 2: 27.4%) (Figure 6.9). Unlike the case of the CA on taxonomic presences (see Figure 5.1), the ordination of assemblages along both axes is unrelated to (log-transformed) sample size (axis 1: $r = -0.124$, $p = 0.685$; axis 2: $r = 0.395$, $p = 0.181$), meaning that both may reflect paleoenvironmental conditions independent of sampling artifacts.

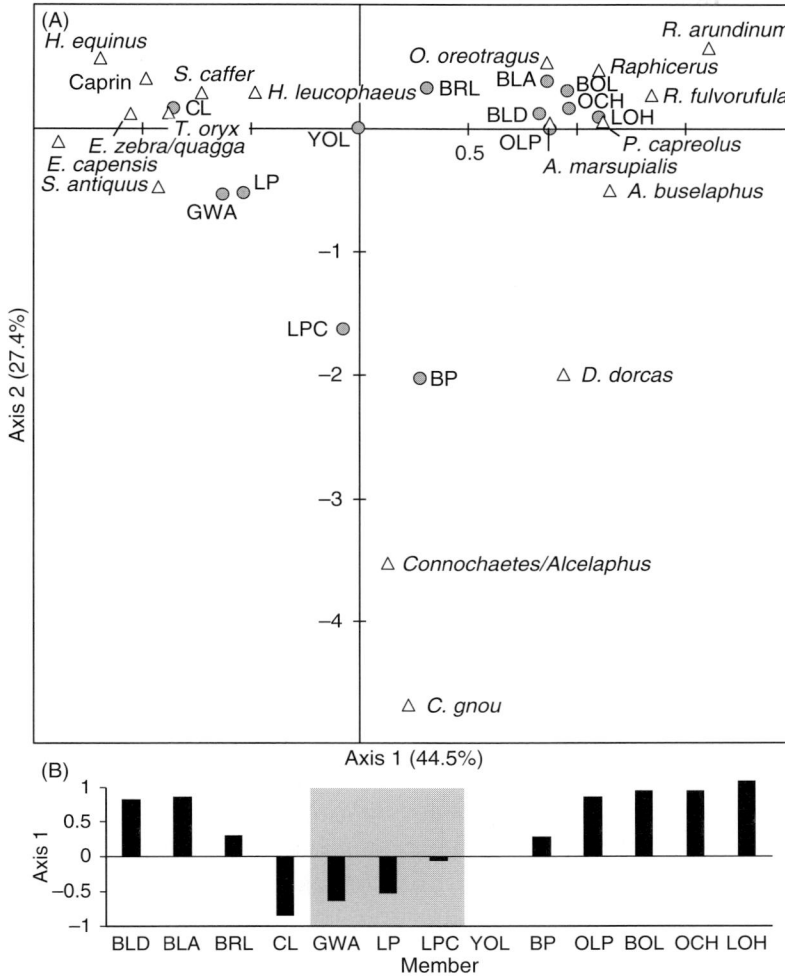

6.9. (A) Correspondence analysis of taxonomic abundances for the Boomplaas Cave ungulates. Axis 1 (x-axis): 44.5% of inertia; axis 2 (y-axis): 27.4% of inertia. (B) Plot of axis 1 scores through the Boomplaas sequence, with assemblages arranged from youngest (left) to oldest (right). Members corresponding to the Last Glacial Maximum are indicated by grey shading.

The arrangement of taxa along axis 1 (45.2%) is similar in both the taxonomic presences and taxonomic abundances ordinations (r_s = 0.776, p < 0.001), with a few important differences (compare Figure 5.1 and Figure 6.9). Zebras (*Equus zebra/quagga*) now group with the set of grassland taxa associated with Last Glacial Maximum and Late-glacial assemblages (LP, GWA, and CL), reflecting their numerical dominance at this time (Figure 6.8). Likewise, hartebeest (*Alcelaphus buselaphus*) and springbok (*Antidorcas* cf. *marsupialis*) are now grouped with the set of shrubland taxa associated with the basal and Holocene assemblages, reflecting the fact that springbok are most abundant in members OCH and LOH and hartebeest in BLD.

The arrangement of assemblages along axis 1 is also similar to that provided by taxonomic presences (r_s = 0.758, p = 0.003), broadly capturing a contrast between Last Glacial Maximum and Late-glacial assemblages versus all others. The most prominent difference is that member YOL now occupies an intermediate position along axis 1. YOL includes only taxa expected of the historic environment, and in the ordination of taxonomic presences it is associated with the set of taxa suggestive of shrubby vegetation (see Figure 5.1). However, the most abundant taxon in YOL is *Equus zebra/quagga*, suggesting an important grassland component that is missing from the contemporary environment (assuming of course that its abundance in this assemblage reflects its abundance on the past landscape), and this is consistent with its intermediate position in the ordination of abundance data (Figure 6.9).

Whereas axis 2 is strongly influenced by sample size in the CA of taxonomic presences, this is not so for the CA of taxonomic abundances. In this case, axis 2 primarily captures a contrast between assemblages with abundant (negative values) or rare (positive values) specimens assigned to the alcelaphin (wildebeest and allies) antelope taxa *C. gnou*, *Damaliscus* cf. *dorcas*, and indeterminate *Connochaetes/Alcelaphus*. The close association of the latter with *C. gnou* implies that most of these specimens of uncertain taxonomic identity probably belong to *C. gnou*. Elevated abundances of these taxa, which are indicative of grassland habitats, in members BP and LPC render these assemblages distinct from all others at Boomplaas Cave, a feature not readily apparent in the first two axes of the CA on taxonomic presences. The implication is an important contrast between the grassland faunas of members BP and LPC compared with those from LP, GWA, and CL; the former set of assemblages have more alcelaphin antelopes and the latter more zebras. Noting that zebras are capable of subsisting on large amounts of low-quality forage, including dry grasses, Faith (2013a) suggests that this trend may indicate a decrease in moisture availability following the Last Glacial Maximum, a pattern now known to be consistent with other regional paleoenvironmental archives (Chase et al. 2018).

We previously noted for the CA on taxonomic presences from Boomplaas Cave (Figure 5.1) that axis 2 is strongly influenced by assemblage sample size, such that those assemblages with negative axis 2 scores are those with the largest samples. In light of our previous discussions, it is now clear why this is the case. Those assemblages with negative values along axis 2 are associated with roan antelope (*Hippotragus equinus*), buffalo (*Syncerus caffer*), and southern reedbuck (*Redunca arundinum*) (Figure 5.1), taxa that are consistently rare throughout the Boomplaas Cave sequence (see Table 4.3), accounting for only a small percentage of the entire ungulate assemblage (roan antelope: 0.28%; buffalo: 0.84%; southern reedbuck: 0.33%). With this rarity in mind, it is not surprising that those assemblages lacking some or all of these taxa, and with corresponding positive values along CA axis 2 (Figure 5.1), are those with smaller sample sizes (Table 4.3). In contrast, sample size is far less important in our CA of taxonomic abundances (Figure 6.9), for which the position of assemblages along axes 1 and 2 is unrelated to assemblage sample size. Because taxonomic abundances are the focus of this ordination, the rare taxa and, by extension, sampling intensity have less influence on the outcome.

The results we have just described raise the question again: should paleozoologists focus their efforts on taxonomic presences or abundances? As we hope to have made clear earlier in this chapter, there is no universal answer; there are pitfalls inherent to both methods. One would hope that in most cases, the researcher will analyze both taxonomic presences and abundances, and the results will provide complementary paleoenvironmental information. In the case of Boomplaas Cave, for example, ordinations of taxonomic presences and abundances both indicate an expansion of open and grassy vegetation during the Last Glacial Maximum and Late-glacial (see Figures 5.1 and 6.8). But when the signals are discordant, the faunal analyst must explain why. Are the abundance data misleading because of taphonomic processes, or are the taxonomic presences misleading because of sampling issues? The answer ultimately allows an evaluation of whether taxonomic abundances or presences are the appropriate target of analysis.

Other Techniques to Examine Abundances of All Taxa

If there are, say, fewer than eight to ten taxa in a temporal series of assemblages, it may be a simple matter to graph relative frequencies of those taxa across temporal periods and interpret fluctuations in all represented taxa. In these cases, the graph of fluctuating taxonomic abundances is interpreted (e.g., Guilday et al. 1978; Semken 1983). If more than a dozen or so taxa are represented, then correspondence analysis or some other ordination technique can facilitate interpretation. But besides these approaches, there are a few other straightforward techniques that can also be used to explore taxonomic abundances, and we briefly highlight those here.

In a report on the ungulates from the Middle and Upper Paleolithic deposits at Grotte XVI in southwest France, Grayson and Delpech (2003) conducted a contingency table analysis to identify changes in taxonomic abundances through time. They used chi-square tests to determine whether those abundances differed significantly from one stratum to the next, and consulted adjusted residuals (ARs) to identify which taxa were driving observed differences. ARs represent standard normal deviates (absolute values greater than 1.96 are statistically significant at $\alpha = 0.05$), with a positive AR indicating an increase in abundance and a negative value indicating a decrease in abundance. Grayson and Delpech (2003) observed that reindeer (*Rangifer tarandus*) increased significantly in abundance across all adjacent pairs of stratigraphic units from the base of the sequence upwards. This trend was paralleled by significant declines in other taxa such as red deer (*Cervus elaphus*) and roe deer (*Capreolus capreolus*). In a previous study of the Grotte XVI ungulates, Grayson et al. (2001) showed that increase in the relative abundance of reindeer tracked decreasing summer temperatures, as reconstructed from a pollen core elsewhere in France. Although Grayson et al. (2001) and Grayson and Delpech (2003) framed their interpretation of the reindeer temporal increase in terms of a *response* to previously documented climate change, in other circumstances they could have used knowledge of the ecological tolerances of reindeer relative to other ungulates at Grotte XVI to infer a change in climate across stratigraphic intervals.

Others have implemented similar contingency table analyses to identify taxonomic shifts for the purpose of paleoenvironmental reconstruction (e.g., Klein 1976b; Rowan et al. 2015) and in zooarchaeological analyses of subsistence change (e.g., Cannon 2003; Fisher and Valentine 2013; Wolverton 2005). Here we turn to the Homestead Cave faunas to illustrate the approach, focusing our attention (for the sake of brevity) on Stratum I through IV with taxa aggregated at the genus level (see Table 6.2). We first note that chi-square tests indicate highly significant differences across adjacent stratigraphic intervals (Stratum I versus Stratum II: $\chi^2 = 2099.034$, p < 0.001; Stratum II versus Stratum III: $\chi^2 = 686.178$, p < 0.001; Stratum III versus Stratum IV: $\chi^2 = 949.975$, p < 0.001). This provides a strong indication that there are statistically meaningful differences in species composition through these levels. The ARs in Table 6.2 indicate those taxa whose changing abundances are the source of these significant differences, with the sign of the AR indicating the direction of change (+ = increase; − = decrease) relative to the underlying stratigraphic interval. For example, increase in the relative abundance of kangaroo rats (*Dipodomys*) from Stratum I to II is indicated by an AR of 30.75. The gist of the analysis is to use the ARs to identify taxa that change significantly in abundance, which in turn can be used to guide paleoenvironmental interpretation. For example, the significant decline in specimens attributed to *Lemmiscus* – all of which belong to the sage vole (*Lemmiscus curtatus*), a taxon associated with sagebrush habitats – from Stratum I to IV suggests a loss of sagebrush habitats. The parallel increase in kangaroo rats is harder to interpret when specimens are aggregated at the genus level; we have lost resolution by combining two taxa (*D. microps* and *D. ordii*) with different ecological tolerances. However, it is also clear from inspection of Figure 6.1 (see also Table 4.5) that as *Dipodomys* increases from Stratum I to IV, so too does the proportion of specimens assigned to chisel-toothed kangaroo rat (*D. microps*). Given what we have already said about this taxon, the obvious implication is that the loss of sagebrush habitats was associated with an expansion of xeric shadscale habitats.

While our brief interpretation of trends in two genera (*Lemmiscus* and *Dipodomys*) at Homestead Cave seems fairly straightforward, further examination of the ARs in Table 6.2 reveals an interpretive problem perhaps not

TABLE 6.2 *Taxonomic abundances (NISP) and adjusted residuals (ARs) derived from a contingency table analysis of the Homestead Cave mammals aggregated at the genus level (Stratum I through IV).*

Taxon	Stratum I	Stratum II		Stratum III		Stratum IV	
	NISP	NISP	AR	NISP	AR	NISP	AR
Ammospermophilus	2	6	1.81	5	1.39	115	1.91
Tamias	4	0	-1.75	0	–	0	–
Marmota	43	8	**-4.00**	0	-1.71	15	1.23
Spermophilus	5	5	0.42	7	**2.46**	49	-0.86
Thomomys	109	434	**17.45**	256	**6.27**	5,110	**12.28**
Chaetodipus	2	2	0.27	1	0.25	3	-1.10
Dipodomys	360	1,329	**30.75**	1,056	**21.75**	13,712	**11.73**
Microdipodops	7	17	**2.72**	0	**-2.50**	0	–
Perognathus	121	90	-0.21	17	**-2.57**	89	**-2.45**
Lemmiscus	552	121	**-13.55**	8	**-5.33**	2	**-7.76**
Microtus	247	197	0.43	44	**-3.05**	53	**-12.36**
Neotoma	2,668	1,938	**-2.06**	710	-0.07	5,664	**-6.07**
Ondatra	0	0	–	0	–	1	0.32
Onychomys	15	8	-0.83	1	-1.07	5	-0.64
Peromyscus	1,550	1,124	-1.51	205	**-10.05**	531	**-17.76**
Reithrodontomys	58	91	**4.38**	22	-1.78	134	**-2.14**
Brachylagus	192	32	**-8.84**	3	**-2.45**	4	**-3.09**
Lepus	2,261	581	**-26.87**	91	**-8.05**	680	**-2.55**
Sylvilagus	2,050	1,865	**6.08**	457	**-8.82**	2,358	**-13.59**

Notes: The sign of the ARs (+ = increase; – = decrease) pertains to the change in abundance relative to the underlying stratum. Significant ARs in bold.

unexpected whenever samples sizes are large, there are numerous taxa to contend with, and there are substantial changes in taxonomic composition. The problem is that most of the taxa display significant changes through time, so synthesizing all of this information into a coherent environmental interpretation is not exactly straightforward. To make life a little easier one might focus attention on a restricted subset of indicator taxa, or alternatively use an ordination technique to tease out the primary taxonomic changes.

Sometimes our goal is not to identify how various taxa changed in abundance through time (or space), but instead to quantify the magnitude of change in a temporal (or spatial) sequence of assemblages. Is the degree of change stable or are there intervals of gradual change punctuated by dramatic change? If the latter, then when (or where) are the most prominent changes taking place? These sorts of questions are often addressed in paleozoological contexts through analysis of faunal similarity and distance metrics (e.g., Bobe and Behrensmeyer 2004; Bobe et al. 2002; de Ruiter et al. 2008; Faith 2013a; Faith and O'Connell 2011; Frost 2007; Lyman 2008b, 2014b). Such analyses make use of an array of metrics – some suitable for taxonomic presences and others for taxonomic abundances – that quantify the similarity (or distance) between any two faunal assemblages (for discussion of such indices see Hammer and Harper 2006; Lyman 2008b; Magurran 1988, 2004). Similarity and distance metrics are used in a variety of paleozoological analyses, including certain ordination techniques (e.g., Jones 2015; Terry 2010b) and cluster analysis (e.g., Frost 2007; Reed 2008). But when used to address questions of the sort posed above, one typically calculates (and plots) the metric across adjacent stratigraphic intervals and inspects the index values for temporal trends or phases of pronounced change. Published examples of such analyses typically make use of metrics designed for abundance data (e.g., Bobe and Behrensmeyer 2004; Bobe et al. 2002; de Ruiter et al. 2008; Faith 2013a; Faith and O'Connell 2011; Frost 2007; Lyman 2014b), though one could also use those designed for taxonomic presences, with the caveat that these metrics are often sensitive to sample size (Lyman 2008b). And to be clear, analysis of similarity/distance metrics can only indicate the degree to which taxonomic composition has changed. Determining what the change is (and its paleoenvironmental implications) will require examination of additional lines of evidence.

Here we provide such an analysis for the Homestead Cave mammals. In this case we calculated chord distance across successive pairs of stratigraphic units with taxa aggregated at the genus level. Chord distance is a distance metric used for abundance data, with values ranging from 0 for assemblages with identical abundances to $\sqrt{2}$ for assemblages with no taxa in common. It is calculated as

$$CD_{jk} = \sqrt{2 - 2 \frac{\sum x_{ji} x_{ki}}{\sqrt{\sum x_{ji}^2 \sum x_{ki}^2}}}$$

where x_{ji} is the abundance of taxon i in assemblage j and x_{ki} is the abundance of taxon i in assemblage k. Chord distance or the closely related squared-chord distance (= CD_{jk}^2) is often used to quantify faunal change in the paleozoological literature (e.g., Bobe et al. 2002; Faith 2013a; Frost 2007) and it is also a preferred metric for analysis of pollen abundance data (e.g., Gavin et al. 2003). Figure 6.10 plots chord distances across the Homestead sequence. Perhaps not surprisingly, changes in taxonomic composition are greatest in the basal assemblages corresponding to the end of the Pleistocene and the beginning of the Holocene (Stratum I through IV). In contrast, the magnitude of faunal change is relatively muted in overlying Holocene assemblages. This does not mean that things are not changing in the upper layers – Grayson (2000a, 2000b) highlights many important shifts in taxonomic composition during the middle Holocene – but that the amount of change is less prominent than what came before. So to the extent that taxonomic changes are signaling paleoenvironmental changes, the implication is that paleoenvironmental changes across the Pleistocene–Holocene transition were far more pronounced than those that occurred during the Holocene.

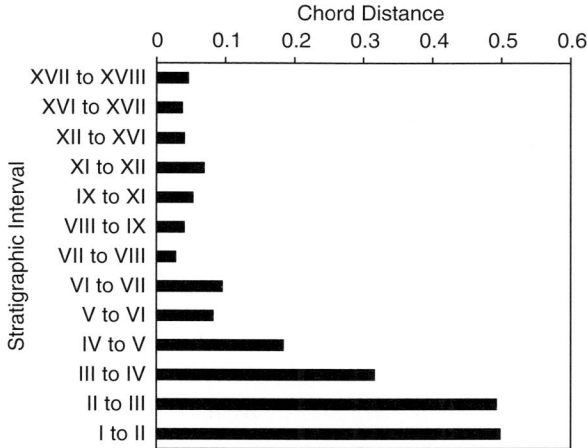

6.10. Chord distances calculated for adjacent pairs of stratigraphic units across the Homestead Cave sequence. Taxa aggregated at the genus level.

Others have used distance metrics to examine the extent to which taxonomic changes are related to taphonomic changes. Bobe et al. (2002) and de Ruiter et al. (2008) calculated one set of chord distances for taxonomic abundances and another for skeletal element abundances, reasoning that the latter should be sensitive to changing taphonomic circumstances. By comparing the taxonomic and taphonomic chord distances either graphically or using correlation coefficients, one can evaluate "the extent to which taphonomic changes may be responsible for apparent changes in taxonomic abundances" (Bobe et al. 2002:483). Though skeletal part data were the taphonomic variable of choice in these studies, any number of taphonomic variables could be used. For instance, Faith (2013a) provided a similar analysis for the Boomplaas Cave ungulates, except that he arrayed taxonomic chord distances against those calculated for abundances of different kinds of bone surface modification (e.g., carnivore tooth-marks, stone-tool butchery marks, gastric etching). In all of the studies noted here, the authors found that changes in taxonomic composition varied independent of taphonomic change. The interpretation was that some mechanism other than taphonomic processes – such as climate-driven environmental change – was responsible for changes in faunal composition.

SUMMARY

In this chapter we have described several approaches to paleoenvironmental reconstruction in paleozoology that utilize taxonomic abundances. While all analytical assumptions are important to keep in mind (Chapter 3), it is worth emphasizing that these approaches (and those on taxonomic presences [Chapter 5]) rely heavily – more so than some other methods discussed in this book – on the uniformitarian assumption that the ecological preferences of taxa observed today can be extended to the past. When taxonomic presences are the focus of analysis, the occurrence in a fossil assemblage of a species whose modern relatives inhabit a particular environment is interpreted as evidence of that environment in the past. The reasoning is identical when taxonomic abundances are considered, except that the ecological implications of a particular taxon are weighted according to that taxon's abundance.

While analyses of taxonomic presences are thought to be more conservative than analyses of taxonomic abundances, both approaches have distinct advantages and disadvantages. Analyses focusing on taxonomic presences are immune to taphonomic processes that may alter relative abundances of taxa, but they are vulnerable to sample-size effects. In contrast, taxonomic abundances have potential to provide greater paleoenvironmental resolution, but the outcome can be heavily skewed by taphonomic biases. Ideally, both presences and abundances should provide the same paleoenvironmental signal. When they do not, it is up to the analyst to understand why a discrepancy has occurred, with likely reasons including taphonomic or sampling issues. We have provided examples of how both factors can influence paleoenvironmental interpretations.

At this point, it should be increasingly clear that there are many ways of approaching a species list in order to generate paleoenvironmental inferences, a theme that will continue in subsequent chapters. While our focus in this and the preceding chapter has been on approaches based on the taxonomic identity of individual species, there are many cases where it may be desirable to generate paleoenvironmental reconstructions derived from other lines of faunal evidence, perhaps because of concerns regarding the assumption of taxonomic uniformitarianism (e.g., for assemblages with many extinct taxa) or simply to provide a different perspective on the data. In the next chapter, we examine those approaches that consider the ecological traits of the species in a community.

SEVEN

TAXON-FREE TECHNIQUES

We begin this chapter with what may seem to be an unusual question. What does Australia's western grey kangaroo (*Macropus fuliginosus*) have in common with East Africa's Thomson's gazelle (*Eudorcas thomsonii*)? If we approach this question from a taxonomic perspective, then the answer is not very much. Both are mammals, but the former a marsupial and the latter a placental, meaning that they have not shared a common ancestor since the Jurassic (Bininda-Emonds et al. 2007; Luo et al. 2010). From an ecological perspective, however, numerous similarities are apparent. Both species are terrestrial with cursorial adaptations; they are *mixed feeders* that consume grasses, forbs, and shrubs; and they are similar in body mass (~20 kg). Parallels between these and other parameters (e.g., lifespan, number of offspring, sociality) imply similarities not apparent from a strictly phylogenetic comparison of the two species.

The replacement of taxonomic with ecological information in our comparison of two distantly related mammals exemplifies the subject of this chapter: *taxon-free* approaches to paleoenvironmental reconstruction. Unlike the types of analyses presented in preceding chapters, taxon-free approaches are those in which species and communities are characterized according to ecological or morphological variables, such as those related to diet, habitat preferences, locomotor behavior, or body mass. Taxonomic diversity is sometimes considered a taxon-free approach in the sense that a faunal assemblage is characterized by one or several diversity metrics (Andrews and Hixson 2014; Damuth 1992), but because it poses a unique set of analytical problems, we

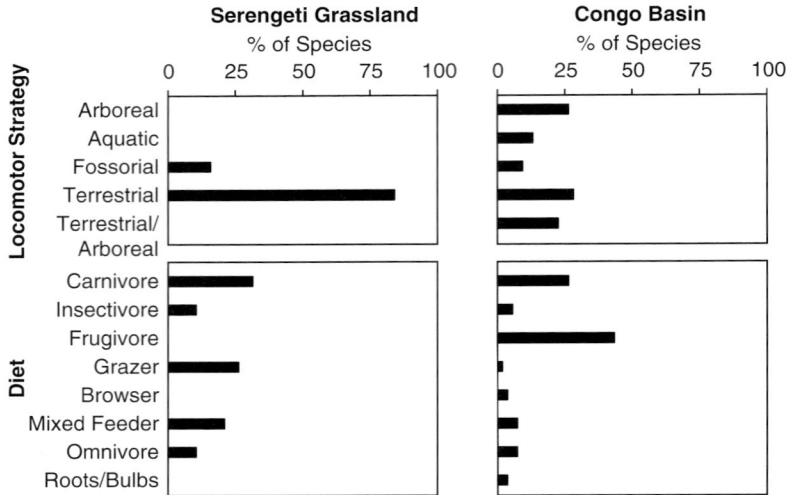

7.1. The distribution of large mammalian taxa (>500 g) across locomotor and dietary classes in the Congo Basin (right) and the Serengeti Plains (left). Data from K. E. Reed (1998).

consider this topic separately in the next chapter. Our aim here is to focus on taxon-free approaches to understanding the ecology of individual taxa (*paleoautecology*) and the environmental implications of the assemblages in which they are found (*paleosynecology*).

TAXON FREE: WHAT IT IS AND WHAT IT IS NOT

The basis of taxon-free analysis involves the description of a taxon or community not according to taxonomic identity, but instead according to one or several ecological descriptors. For example, Thomson's gazelle can be characterized by its locomotor regime and diet as a terrestrial mixed feeder. Extending such characterizations to the community-level, an herbivore community can be characterized according to the distribution of species (or specimens/individuals) across dietary or locomotor classes (Figure 7.1). There is an enormous array of variables to choose from, but when paleoenvironmental reconstruction is the goal, the focus is on those that are known or can be inferred to track environmental gradients, and these are most frequently those related to diet, body size, habitat preference, and locomotor habitats. In Figure 7.1, for example, one might reasonably infer that the greater percentage of arboreal species and frugivorous species in the Congo Basin signals a more forested environment compared with the community from the Serengeti plains.

By characterizing a taxon or fossil assemblage according to one or a few variables, a tremendous amount of information is sacrificed. So why bother with a taxon-free approach? First, the aim is not to provide an exhaustive description of a single taxon or community, but rather to provide a characterization that can be meaningfully compared with others (Andrews et al. 1979; Damuth 1992). In doing so it becomes possible to compare communities through time and space regardless of the species involved (hence the label "taxon free"). For example, although it is difficult to compare ecological patterns in a community of rainforest mammals from the Amazon with those from the Congo on a taxonomic level, we can readily compare them according to the trophic or locomotor adaptations of the species that are present. This opens the door to identifying regional, continental, or global relationships between environmental gradients and the ecological structure of modern communities, which in turn provides a framework for generating inferences from the fossil record (e.g., Andrews et al. 1979; Eronen et al. 2010b; Hernández-Fernández and Peláez-Campomanes 2003, 2005; Lintulaakso and Kovarovic 2016; Louys et al. 2011a, 2015; Mendoza et al. 2005; Nesbit Evans et al. 1981; Rector and Reed 2010; Reed 1998). The second advantage to some taxon-free approaches concerns their potential to circumvent problems related to the assumption of taxonomic uniformitarianism (Chapter 3, assumption 1). In the case of extinct species, taxon-free methods can allow for more robust ecological inferences than provided by analogy with their nearest living relatives (assumption 2 in Chapter 3 is unnecessary) (e.g., Spencer 1997; Sponheimer et al. 1999; Stynder 2009). As we noted in Chapter 3, even when dealing with extant species, the assumption of taxonomic uniformitarianism can be problematic, as there is always the possibility that ecological preferences varied in the not-too-distant past (e.g., Faith 2011a; Mead and Spaulding 1995; Owen et al. 2000).

Before we proceed with a more detailed description of the history and application of taxon-free methods, it is worth clarifying what taxon-free approaches are not meant to do. First, the objective is not to provide a paleoenvironmental reconstruction that is somehow independent of taxonomic information. Despite what the name suggests, all taxon-free approaches rely to varying degrees on taxonomy (Andrews and Hixson 2014; Reed 1998), in part because fossils must be identified to some taxonomic level before they can be characterized on an ecological basis. John Damuth (1992:184) puts it best: "it is the result (i.e., the characterization of the community) that is independent of explicit knowledge about the taxa and their relationships, not necessarily the inferential pathways to that characterization." Second, while taxon-free approaches can in some cases provide a stronger ecological foundation than those relying on taxonomic uniformitarianism (Andrews and Hixson 2014), particularly in the case of the deeper-time fossil record (e.g., Andrews et al. 1979; Nesbit Evans et al. 1981; Reed 1998), this does not mean the main purpose of these approaches is to circumvent the general assumption of uniformitarianism.

As indicated in Chapter 3, paleoenvironmental reconstruction based on animal remains depends on the general uniformitarian assumption that the relationship between a biological variable (e.g., particular species, tooth morphology, body size) and an environmental variable (e.g., biome, mean winter temperature, habitat type) observed today also held in the past (Cleland 2001, 2011). Expressing this assumption in the specific terms of one of its myriad forms produces, among others, the taxonomic uniformitarian assumption (assumption 1 in Chapter 3, that a species today has, and has always had, certain environmental tolerances). The important point here is that the extent to which taxon-free approaches are independent of taxonomic uniformitarianism varies. For instance, one can characterize an ancient Thompson's gazelle represented by fossils as a mixed feeder based on direct observations of this extant species' diet in modern ecosystems and the uniformitarian assumption that this species had the same diet in the past that it is observed to have today. This is not exactly the taxonomic uniformitarian assumption but a different, diet-specific version thereof, and one that undergirds many taxon-free analyses (e.g., Assefa et al. 2008; Matthews et al. 2011; Rector and Reed 2010). But one could also provide a dietary characterization of the ancient Thompson's gazelle based, for instance, on its mandibular morphology (see below), an inference that does not rely on taxonomic uniformitarianism.

We demonstrate in this chapter that it is possible to make paleoenvironmental inferences based on analyses of fossils that depend to varying degrees on the taxonomic identity of those fossils. The degree of dependence is dictated by the nature of the relationship between the taxonomic identity of the fossil and the environmental property of interest. If the environmental property of interest is in deciphering which habitat type was once present, say closed forest, open woodland, or grassland, then we might list the species preferring each and see which of them are present in the fossil collection; this would require the taxonomic uniformitarian assumption. Or we could count up how many hypsodont molariform teeth and how many brachydont teeth there were, irrespective of species represented by those teeth. If hypsodont teeth greatly outnumber brachydont teeth, we would likely conclude grassland was abundant, and we could do so without referring to the species represented by any of the teeth. But the inference requires the uniformitarian assumption that hyposodont teeth indicate grassland whereas brachydont teeth represent brushy or forested habitats.

A BRIEF HISTORY

Paleoecologists distinguish between paleoautecology (the ecology of fossil taxa) and paleosynecology (the ecology of fossil communities). These concepts are not mutually exclusive, because the former is often used to inform the latter, but they do provide a reasonable framework for exploring the history and practice of taxon-free approaches to paleoenvironmental reconstruction.

The earliest examples of taxon-free approaches to reconstructing the ecology of individual taxa (paleoautecology) are as old as paleontology itself. French naturalist Georges Cuvier set the stage for what is now called *ecological morphology* (Wainwright and Reilly 1994) by observing that an animal's form was related to the ecological context of its function. For example, in his description of the giant Pleistocene ground sloth *Megatherium* of South America, Cuvier (1796) notes that "[t]he great thickness of the sides of the lower jaw, which even surpasses that of the elephant, seems to indicate that the large animal we are examining did not content itself with leaves, but – like the elephant and the rhinoceros – broke and crushed the branches themselves" (translation from Rudwick 1997:30). This inference is taxon-free because it is based not on the phylogenetic identity of *Megatherium* (i.e., its phylogenetic relationship to modern sloths, anteaters, and armadillos), but instead upon anatomical similarities with more distantly related species. Cuvier's reasoning would hardly be out of place today, with the only distinction being that contemporary studies typically provide a more detailed morphometric analysis of the fossils and modern skeletons being compared. There is now a vast literature exploring how morphological features of skulls, teeth, and limbs can be used to understand aspects of diet, locomotor habits, or habitat preferences that can then be linked to past environments (e.g., Barr 2015, 2017; DeGusta and Vrba 2005a, 2005b; Faith et al. 2012; Janis 1995; Kappelman 1991; Kappelman et al. 1997; Kovarovic and Andrews 2007; Palmqvist et al. 2003; Plummer and Bishop 1994; Plummer et al. 2008, 2015; Reed 1997, 1998, 2008; Van Valkenburgh 1987, 1988).

Turning to fossil communities (paleosynecology), the earliest taxon-free approaches can be traced to the paleobotanical literature. In their studies of the global distribution of dicot plants, botanists Irving Bailey and Edmund Sinnott (1915:831–832) noted "certain interesting correlations between structural characters [of the leaves] and climate." They observed that among woody plants today, those species found in tropical and subtropical climates tend to have leaves with continuous margins, whereas those from temperature regions tend to have discontinuous (e.g., toothed, serrated, lobed) margins (see also Bailey and Sinnott 1916). While Bailey and Sinnott were not the first to document relationships between leaf form and climate (e.g., von Humboldt 1850), they were the first to extend their observations to the fossil record, providing an analysis of fossil leaves from Cretaceous deposits in the Arctic that implied a tropical climate very different from today. Although the term taxon-free was not in use at the time, Bailey and Sinnott (1915:833) noted that by substituting taxonomic with ecological (phenotypic) information, their method "rests upon a physiological and ecological basis rather than the usual phylogenetic one." Their taxon-free approach provided the foundation for what is now a staple in paleobotanical analysis (Peppe et al. 2011).

AUTECOLOGICAL APPROACHES

Before we can begin making community-level paleoenvironmental inferences within a taxon-free framework, we must first provide ecological characterizations of the individual taxa, and perhaps even individual specimens, in a fossil assemblage. It is possible to do so by way of taxonomic uniformitarianism (Chapter 3, assumption 1), provided we are comfortable with that assumption, but in some cases it might be preferable to infer ecology directly from the fossil material rather than the taxa represented by those fossils. This might, for example, include analyses involving extinct species with unknown ecological preferences or focusing on subsets of a fossil assemblage (e.g., postcranial elements of many families) that cannot be reliably identified to lower, ecologically informative taxonomic levels. We focus here on the various approaches geared towards paleoautecological inference, although we note that all of them can also be used to make inferences at the community level when applied to entire fossil assemblages rather than to individual taxa.

Ecomorphology

Dutch zoologist Cornelis van der Klaauw (1948:27) coined the term ecological morphology, usually shortened to *ecomorphology* today, to describe "the connection between the shape of the animal in its entirety and its surroundings." Broadly, ecomorphology is concerned with understanding the relationships between an animal's morphology and the ecological context of its function (i.e., what that morphology does). How does dental or cranial morphology relate to an animal's diet? How does the morphology of the limb bones relate to the habitat through which an animal moves? Answering these questions can provide insight into a range of ecological and evolutionary questions (Wainwright and Reilly 1994), but for our purposes it also provides a robust framework for inferring ecology from the morphology of fossil specimens.

The central premise or assumption of ecomorphology is that there are meaningful relationships between functional morphology and the environment, with the former representing adaptations to selective environmental pressures. While not all morphological traits necessarily represent solutions to environmental problems (Gould and Lewontin 1979), this premise is supported by convergent evolution of ecomorphological traits between distantly related groups of species (Andrews and Hixson 2014) and by analyses of taxa that demonstrate morphological convergence independent of phylogenetic relationships (e.g., Barr 2014; Louys et al. 2013; Pérez-Barbería and Gordon 2001; Raia et al. 2010). Because ecologically relevant functional morphologies are the outcome of natural selection, this means that the morphologies we observe in a living or fossil taxon are telling us about the ecological context of its evolutionary history (Losos and Miles 1994), and not directly about the ecology of that taxon in life. It is therefore sensible to view ecomorphology as telling us about the *potential* ecology of an animal as opposed to its *actual* ecology (Reed et al. 2013).

In many ways this renders ecomorphology a complement of taxonomic uniformitarianism; both involve assumptions about the relationship between a biological variable and an environmental or ecological variable. Ecomorphology, however, makes inferences about the ecology of a fossil taxon based on an assumption about its evolutionary past (i.e., selective [ecological] pressures faced by its phylogenetic predecessors built the morphology), whereas taxonomic uniformitarianism makes inferences about a fossil taxon based on the present (i.e., the ecological tolerances of the fossil's phylogenetic successors are the same as those of the fossil). In both cases, there is a temporal gap between the fossil specimen(s) under study and the ecological variable(s) inferred. One temporal gap is between the selective ecological conditions that built (selected for) the fossil animal, and the fossil animal. The other temporal gap is that between the fossil animal and the fossil animal's living, observable descendants.

When paleoenvironmental reconstruction is the objective of ecomorphological analysis, most studies follow a broadly similar approach, which we outline here. The first step involves selecting a target taxonomic group, usually at the level of order or family. The ideal taxonomic group of course depends on what is present in the fossil collections of interest, but also on the ecological diversity of that group. For example, there is a disproportionately large body of literature concerned with the ecomorphology of the Bovidae (antelopes and buffaloes), especially in Africa (e.g., Barr 2014, 2015, 2017; DeGusta and Vrba 2003, 2005a, 2005b; Kappelman 1988, 1991; Kappelman et al. 1997; Kovarovic and Andrews 2007; Plummer et al. 2008; Scott and Barr 2014; Spencer 1995, 1997), and there are several reasons why this is so. First, the bovids are a dominant component of most African mammal assemblages from the Pliocene onwards, meaning there are a lot of fossils to work with. Second, living bovids are both taxonomically and ecologically diverse, occurring in a broad range of habitats and feeding on a wide range of plants, making them a particularly useful paleoenvironmental indicator, especially compared with other families that might be commonly represented in a fossil assemblage but have considerably less ecological diversity. And third, the exceptional diversity means postcranial remains are notoriously difficult to identify to lower taxonomic levels (horns and teeth are most useful for identification). This means that in the absence of ecomorphological analysis, large portions of African fossil assemblages are of limited use for paleoenvironmental reconstruction.

The second step involves selecting a skeletal element that has a clear functional relationship to the ecological variable of interest and that is reasonably abundant in the fossil record. Most ecomorphological studies aimed at paleoenvironmental reconstruction are concerned with relationships between diet and anatomy of the dentition, cranium, or mandible (e.g., Figueirido et al. 2009; Janis 1988; Janis and Ehrhardt 1988; Kay 1975; Mendoza et al. 2002; Spencer 1995; Van Valkenburgh 1988), and between locomotor regime or habitat preference and the anatomy of the appendicular elements (e.g., Barr 2014; Curran 2012; DeGusta and Vrba 2003, 2005a, 2005b; Kappelman 1988; Plummer et al. 2008; Scott et al. 1999; Van Valkenburgh 1987). Continuing with bovids, for example, there are pronounced differences in femoral anatomy that are strongly related to habitat structure (Kappelman 1988). Those species inhabiting structurally open habitats (e.g., grasslands) typically rely on speed to outdistance cursorial predators, and this is associated with anatomical (skeletal) features that enhance joint stability in the anteroposterior plane, such as a femoral head that is cylindrical in shape. Conversely, those species inhabiting structurally closed settings (e.g., forests) must navigate a more complex three-dimensional substrate of trees, bushes, branches, roots, etc., which restrict high-speed cursorial locomotion. This is associated with a femoral head that is more spherical, providing a more mobile hip joint. While these and other morphological features mean that ecomorphological analysis of the bovid femur can provide useful paleoenvironmental information (Kappelman 1991; Kappelman et al. 1997; Scott et al. 1999), it is also sensible to worry about taphonomy. Measurable femora or portions thereof, particularly the epiphyses, are not particularly common in the fossil record. Those elements that also play a role in locomotion but are more frequently preserved intact (e.g., the astragalus) have broader potential for paleoenvironmental reconstruction (e.g., Barr 2015; DeGusta and Vrba 2003; Plummer et al. 2008).

The third step involves selection of *dimensions* of the skeletal element in question, with the aim of capturing aspects of a bone's shape that are ecologically relevant. This requires a good deal of knowledge about functional anatomy, as the relationships between morphology and ecology are obviously strengthened when the dimensions capture a morphological feature of known functional relevance. Traditionally, these involve a series of linear dimensions measured with standard osteometric tools (Figure 7.2), although it is now possible to quantify complex aspects of shape using laser scanners (Barr 2014, 2015) or three-dimensional point digitizers (Curran 2012). The traditional approaches are faster and cheaper, although the latter may ultimately prove more useful because they allow for the measurement of complex shapes otherwise impossible to capture with measurements of linear dimensions.

With the relevant taxonomic group, skeletal element(s), and dimensions in mind, the faunal analyst proceeds with measuring a large sample of modern species, in addition to relevant fossil specimens. The modern sample should be representative of the fossil group's ecological diversity in order to capture

7.2. An example of linear dimensions measured by Barr (2015) to study the ecomorphology of the bovid astragalus. Measurements are as follows: (A) width at the level of flange on lateral surface; (B) maximum medial length; (C) radius of circle fit to the margin of the proximal trochlea; (D) functional length of the astragalus; (E) radius of circle fit to the margin of the distal trochlea. See Barr (2014) for a discussion of their functional significance.

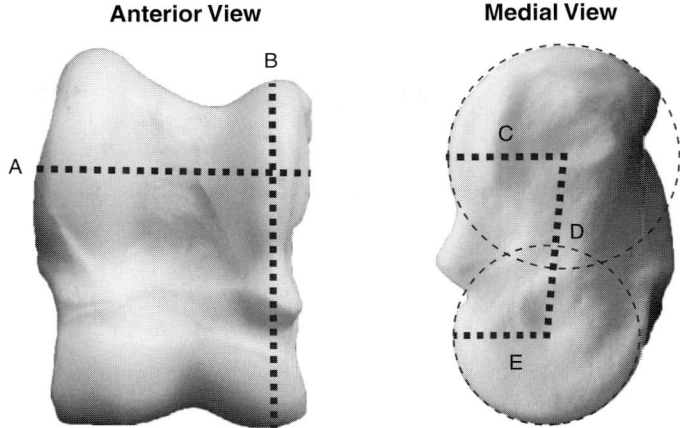

Anterior View

Medial View

the range of variation likely to be encountered in the fossil record. Next, the modern species are usually classified into discrete ecological categories. For example, the habitat preferences of living Bovidae are often classified according to four broad categories (open, light cover, heavy cover, or forest) that span a continuum of canopy cover from open to closed condition (e.g., Barr 2015; Kappelman et al. 1997; Plummer et al. 2008). In her analysis of Cervidae, Sabrina Curran (2012) used a very similar scheme, but also considered the substrate directly encountered by the hoof (e.g., dry, wet, mountainous) to explore the ecomorphology of the distal phalanges. Obviously the classifications will depend on the taxonomic group in question. Instead of focusing on habitat type, for example, Van Valkenburgh (1987) classified modern carnivores according to their locomotor behavior: arboreal (dwells in the trees), scansorial (capable of climbing), terrestrial (dwells on the ground; rarely or never climbs), semi-fossorial (regularly digs for food and shelter). Ultimately, what we want is a scheme that strikes a balance between being sufficiently narrow to be ecologically informative while remaining sufficiently broad to be consistently detected in the fossil's anatomy.

Prior to quantitative analysis, *measurements* for each modern specimen are usually corrected for body size. Particularly when the species span a wide range of body mass, size will often account for most morphological variation in the sample, obscuring the subtler relationships between the shape of a skeletal element and the corresponding ecology. When dealing with linear dimensions, there are several ways to correct for body size, including calculating ratios between measurements (e.g., Plummer et al. 2008), calculating the residuals of the relationship between the measurements and body size (e.g., Kovarovic and Andrews 2007), dividing each measurement by an independent measure of body mass (e.g., Spencer 1995), or by dividing each measurement by the geometric mean of all measurements (e.g., Barr 2015). The latter is typically recommended in morphometric analysis of linear dimensions (Jungers et al. 1995).

We are now ready to proceed with data analysis. Probably the most common approach used in ecomorphological analyses aimed at paleoenvironmental reconstruction is discriminant function analysis (DFA). DFA is a multivariate technique designed to classify individuals (specimens) into predefined categories based on an observed set of measured variables (for important discussions and examples see Barr 2015; Barr and Scott 2014; Kovarovic and Andrews 2007; Kovarovic et al. 2011; Scott and Barr 2014). In the context of our discussion here, it is used to examine how well morphology, reflected by numerous measurements, discriminates between modern species assigned to previously determined ecological categories (e.g., dietary classes, habitat preferences, loco-motor regime). If the DFA indicates that morphology reliably distinguishes between ecological categories, we can then predict the likely ecological cat-egory of a fossil specimen, allowing paleoenvironmental inferences based on individual taxa or entire assemblages. For example, Faith et al. (2012) show that metatarsal anatomy of African bovids accurately predicts habitat membership (open, light cover, heavy cover, forest) in 283 out of 321 modern specimens belonging to 37 species (see also Plummer and Bishop 1994). The discriminant functions were then used to predict the habitat classification of two metatarsals belonging to the extinct East African antelope *Damaliscus hypsodon*, which were assigned to the "open" habitat classification with high probabilities (98.6% and 88.2%), an ecological characterization consistent with its phylogenetic relationships (it belongs to a tribe of open-habitat species). Barr (2015) used a similar approach to make community-level inferences. He showed that astragalus morphology also effectively identifies the habitat categories of modern African bovids, but instead of using DFA to predict habitat preferences of a single taxon, he examined 234 fossil astragali of uncertain taxonomic affiliation (below the family level) spanning the past 3.4 to 1.9 Ma from the Shungura Formation in Ethiopia. Predicted habitats of the fossil sample show clear and significant changes through the sequence (Figure 7.3), suggesting variation in the propor-tion of open (light cover) and closed (heavy cover) habitats through time.

Paleodietary Reconstruction

The foods that an animal eats are perhaps its most direct link to its environ-ment. And because those foods must be present on the landscape on which that animal lives, knowing the diet of a fossil organism can play a key role in paleoenvironmental reconstruction. Ecomorphology provides one means of reconstructing dietary ecology, but it can only tell us about the foods that an animal was adapted to eat and not directly about the foods that it did eat. While we can expect close correspondence between the two in many cases,

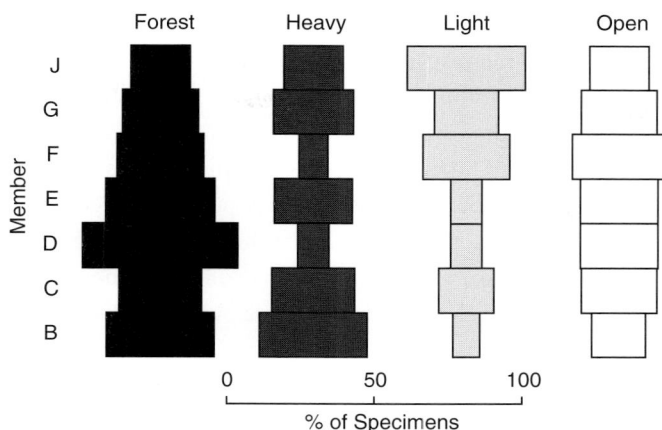

7.3. Temporal variation in the proportion of bovid astragali assigned to various habitat categories in the Shungura Formation. Data from Barr (2015).

observations of living organisms tell us that morphology does not always predict diet (see examples in Calandra and Merceron 2016; Gailer et al. 2016; Ungar 2015). What is needed in our paleoenvironment toolkit are direct paleodietary proxies, and we describe several of the more important approaches here.

Biogeochemical proxies are perhaps the best-known and most widely celebrated paleodietary proxies. They also encompass a family of methods that are sufficiently broad to warrant multiple standalone volumes and major reviews (e.g., Ambrose and Katzenberg 2000; Ambrose and Krigbaum 2003; Lee-Thorp 2008; Lee-Thorp and Sponheimer 2006, 2013; Pate 1994; Sandford 1993; Schoeninger 1995), so we limit ourselves to a brief synopsis here. The basic premise of the biogeochemical proxies, namely trace elements and stable isotopes, is that the chemical composition of the food and water that an animal consumes is recorded in its tissues (e.g., Ambrose and Norr 1993; Burton and Price 1990; Burton et al. 1999; Cerling and Harris 1999; DeNiro and Epstein 1978, 1981; Elias et al. 1982; Lee-Thorp and van der Merwe 1987; Lee-Thorp et al. 1989; Schoeninger and DeNiro 1984), including the hard tissues (e.g., bone, tooth enamel, dentine, and cementum) preserved in the fossil record. Provided that diagenesis has not altered the original chemical signature (e.g., Hedges 2002; Koch et al. 1997; Nelson et al. 1986; Price et al. 1992; Sponheimer and Lee-Thorp 2006; Wang and Cerling 1994), this means we can recover a direct paleodietary proxy from the biogeochemistry of fossil bones and teeth.

Trace elements, especially strontium (Sr) and barium (Ba), have potential to distinguish between trophic levels (e.g., carnivores versus herbivores), between grazers and browsers, or between marine and terrestrial food sources, although a detailed understanding of how they vary in modern foodwebs is needed before they can be broadly applied in paleodietary reconstruction (Lee-Thorp and Sponheimer 2013). In contrast, the stable isotopes of carbon (C), oxygen (O), and nitrogen (N) are particularly well understood and have inspired a vast literature on paleodietary reconstruction. The ratio of ^{13}C to ^{12}C is widely used to identify the consumption of plants that follow the C_3 photosynthetic pathway, including most dicotyledons (e.g., trees, shrubs, herbs) and temperate grasses, and those that follow the C_4 pathway, namely tropical grasses and sedges. When measured in modern or fossil herbivores, it is possible in certain ecosystems (i.e., those with C_4 and C_3 plants) to discriminate between grazers, browsers, and mixed feeders (e.g., Ambrose and DeNiro 1986; Cerling et al. 2003; Koch et al. 1998; Sponheimer et al. 2003), and to infer vegetation structure and how it has changed through time (e.g., Cerling et al. 1993, 1997; Prideaux et al. 2007; Quade et al. 1992; Sealy et al. 2016; Sponheimer and Lee-Thorpe 2003). For example, the carbon isotope composition of teeth assigned to *Damaliscus hypsodon*, the extinct open-habitat antelope noted above, indicates it was a grazer that consumed almost exclusively C_4 grasses (Faith et al. 2015), another inference consistent with its phylogenetic relationships (Marean 1992). The stable isotope ratios of oxygen ($^{18}O/^{16}O$), which are primarily related to their behavior in water, are sensitive to the influence of climate on hydrology (e.g., Ayliffe and Chivas 1990; Blumenthal et al. 2017; Cormie et al. 1994; Fricke et al. 1998; Levin et al. 2006; Longinelli 1984), and can be used to discriminate between animals that obtain water from free-standing sources or from the foods they eat (e.g., Harris and Cerling 2002; Kohn 1996; Sponheimer and Lee-Thorp 2001). Though not related to paleodietary reconstruction, it is also possible to infer ancient moisture availability (aridity) by comparing the oxygen isotope signature of animals that obtain water primarily from leafy vegetation with those that regularly drink surface water (e.g., Blumenthal et al. 2017; Levin et al. 2006; Secord et al. 2012; Yann et al. 2013). This is based on the premise that the oxygen isotope signature of plant leaves is influenced by evaporation whereas that of surface water often (but not always) reflects that of meteoric rainfall, and increased (decreased) aridity translates to more (less) evaporation of plant leaves relative to surface water (Blumenthal et al. 2017; Levin et al. 2006; see Faith 2018 for a discussion of complicating factors). And lastly, the stable isotope ratios of nitrogen ($^{15}N/^{14}N$) are often used to detect trophic levels and to discriminate between consumption of marine versus terrestrial foods (DeNiro and Epstein 1981; Minagawa and Wada 1984; Schoeninger and DeNiro 1984), although the deeper-time suitability of nitrogen isotopes is limited by preservation of bone collagen.

The second family of paleodietary proxies is based on tooth wear, and includes analysis of *dental microwear* and *dental mesowear*. Dental microwear analysis involves the examination of microscopic pits and scratches on the surface of an animal's tooth as they relate to the foods that it consumes (for important reviews see Calandra and Merceron 2016; Teaford 1991, 1994, 2006; Teaford et al. 2013; Ungar 2015; Ungar et al. 2008). The basic premise is that the varied physical properties of different food types, as well as any adhering grit, translate to

Grazer: *Macropus giganteus*   Browser: *Wallabia bicolor*

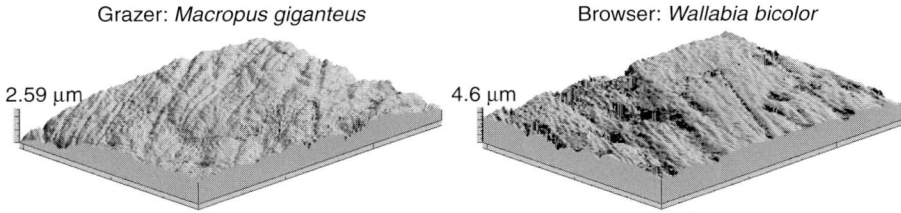

7.4. Three-dimensional photosimulations of dental microwear surfaces of a typical grazer, the eastern gray kangaroo (*Macropus giganteus*) (left), and a typical browser, the swamp wallaby (*Wallabia bicolor*) (right). Images courtesy of Larisa DeSantis (see also DeSantis 2016).

different wear patterns. These relationships have been studied across a tremendous diversity of modern and fossil vertebrate taxa, including primates (e.g., Grine and Kay 1988; Kay 1987; Merceron et al. 2005b; R. S. Scott et al. 2006, 2012; Shapiro et al. 2016; Teaford and Walker 1984), artiodactyls (e.g., Merceron and Ungar 2005; Merceron et al. 2004, 2005a; Schubert et al. 2006; Scott 2012a; Solounias and Moelleken 1992a, 1992b; Ungar et al. 2007), perissodactyls (e.g., Hayek et al. 1991; MacFadden et al. 1999; Rivals et al. 2015; Semprebon et al. 2016a), proboscideans (e.g., Rivals et al. 2012; Semprebon et al. 2016b; Zhang et al. 2017), rodents and shrews (e.g., Burgman et al. 2016; Caporale and Ungar 2016; Hopley et al. 2006; Withnell and Ungar 2014), carnivorans (e.g., Anyonge 1996; DeSantis and Haupt 2014; DeSantis et al. 2012a; Donohue et al. 2013; Goillot et al. 2009; Van Valkenburgh et al. 1990), and xenarthans (e.g., Green 2009; Green and Kalthoff 2015; Green and Resar 2012; Haupt et al. 2013). With respect to modern and fossil ungulates, dental microwear is particularly useful for distinguishing browsers (dicot feeders) from grazers (monocot feeders), with the former typically characterized by enamel surfaces dominated by pits and the latter by linear scratches (e.g., Merceron and Ungar 2005; Merceron et al. 2005a; Scott 2012a; Solounias and Moelleken 1992a; Ungar et al. 2007) (Figure 7.4). Experimental studies indicate microwear patterns have a very fast turnover rate, potentially on the order of one or two weeks (Teaford and Oyen 1989). This means microwear documents only the most recently consumed foods, a phenomenon Grine (1986) termed the "Last Supper" effect. For those species with diets that vary seasonally, the time of death will dictate the observed wear pattern.

Early microwear studies utilized binocular light microscopy to produce qualitative descriptions of wear patterns (Baker et al. 1959; Butler 1952; Mills 1955). In the late 1970s, scanning electron microscopes (SEMs), which provide enhanced resolution and depth of field, became the new standard in microwear analysis (e.g., Rensberger 1978; Ryan 1979; Walker et al. 1978), and this was followed by research aiming to quantify the pits and scratches on teeth, including their frequencies, lengths and widths, and orientations (e.g., Gordon 1988; Grine 1986; Teaford and Walker 1984). While this promised to provide finer-scale ecological information, the time and cost associated with SEM analysis and, most

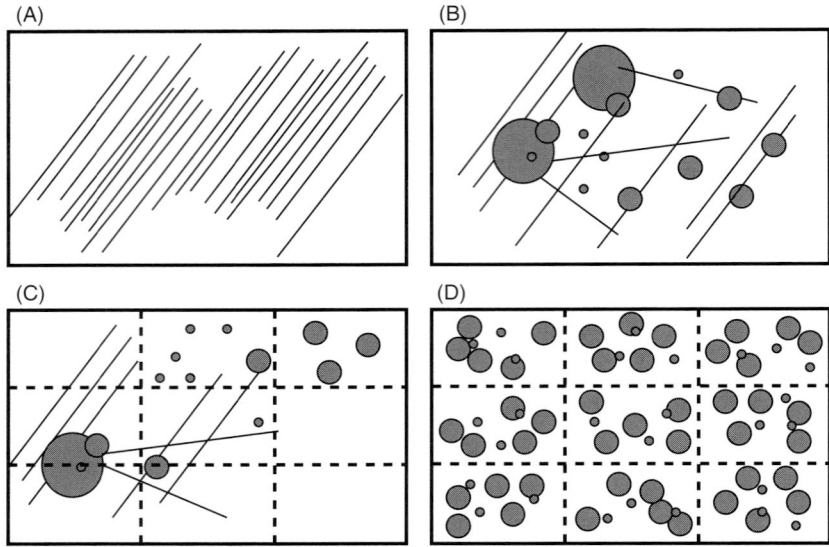

7.5. Schematic of dental microwear surfaces (lines = scratches and circles = pits) demonstrating (A) anisotropic texture: surface relief with similar orientations (e.g., multiple parallel scratches); (B) complex texture: surface relief at a high multiplicity of scales (e.g., pits and scratches of variable size overlying each other); (C) heterogeneous texture: surface relief that varies in texture across the tooth surface; and (D) homogeneous texture: surface relief that does not vary in texture across the tooth surface. Redrawn from Scott et al. (2006).

importantly, lack of replicability and high observer error (Grine et al. 2002) demanded development of alternative methods. Today, the state-of-the-art approach is called *dental microwear texture analysis* (DMTA) (Scott et al. 2005, 2006; Ungar et al. 2003). Briefly, a very small portion of the enamel surface, typically on the order of 0.01 mm² (= 100 × 100 μm), is scanned using a white-light confocal microscope, creating a three-dimensional point cloud of the tooth surface (Figure 7.4). It is then possible to analyze the "texture" of the tooth surface by using computer software to quantify variables related to its complexity, anisotropy, and homogeneity (Figure 7.5). This provides a framework for comparing wear patterns between species of known diets and for inferring the diets of fossil specimens. For example, in her comprehensive study of modern African Bovidae, Jessica Scott (2012a) showed that grazers are characterized by highly anisotropic surfaces (i.e., numerous parallel striations), whereas browsers are characterized by highly complex surfaces (i.e., greater surface relief due to pits and scratches of multiple sizes overlying each other) (Figure 7.6). These modern baselines can then be used to interpret the wear patterns documented in the fossil record. For instance, a small subset of Scott's (2012b) DMTA data from Pliocene deposits at Laetoli in Tanzania (Figure 7.6) includes teeth that show microwear signals consistent with modern-day grazers (*Parmularius*), mixed feeders (*Simatherium*), and browsers (*Antidorcas*).

7.6. A plot of mean dental microwear complexity versus anisotropy for modern African bovids (circles) and fossil taxa (crosses) from Laetoli in Tanzania. Points represent mean values for a given taxon (data from Scott 2012a, 2012b). Obligate grazer = >90% monocots; variable grazer = 60–90% monocots; mixed feeder = 30–70% monocots; browser = >70% dicots. Complexity is measured as area-scale fractal complexity, which is the change in surface roughness across scales. Anisotropy is measured as the differences in lengths of depth profiles sampled across a surface at different orientations.

A closely related but analytically less demanding approach to paleodietary reconstruction is provided by dental mesowear analysis. Developed by Mikael Fortelius and Nikos Solounias (2000), mesowear analysis is based on the idea that different diets are associated with different macroscopic wear patterns on the occlusal surface of herbivore molars. In species that feed primarily on dicots (i.e., browsers), most of the wear is related to *attrition*, which is the result of tooth-on-tooth contact that occurs when tooth surfaces slide past each other, sharpening the teeth and promoting tall cusps with pointed tips. However, when grasses are consumed, the mastication of grit and siliceous phytoliths – there is substantial debate about the relative importance of the two (e.g., Baker et al. 1959; Damuth and Janis 2011; Jardine et al. 2012; Kubo and Yamada 2014; Lucas et al. 2013; Mendoza and Palmqvist 2008; Rabenold and Pearson 2014; Williams and Kay 2001) – translates to abrasive wear. *Abrasion* is caused by food-on-tooth contact, which obliterates the attritional signal and results in shorter and either rounded or blunted cusps. In contrast to dental microwear, these meso-scale wear patterns provide an ecological signal that is thought to be averaged over a temporal scale of months to years (Davis and Pineda-Munoz 2016; Fortelius and Solounias 2000; Rivals et al. 2007a).

In the original scheme devised by Fortelius and Solounias (2000), the buccal apices of the maxillary second molar cusps (paracone and metacone) are scored according to cusp shape and occlusal relief (Figure 7.7). Cusp shape is scored as sharp, round, or blunt, depending on the degree of facet development, with preference given to the sharper cusp in the rare case of specimens where the two cusps differ in shape. Following their criteria, "[a] sharp cusp has (practically) no rounded area between the mesial and distal phase I facets, a rounded cusp has a distinctly rounded tip without planar facet wear but retains facets on the lower slopes, while a blunt cusp lacks distinct facets altogether"

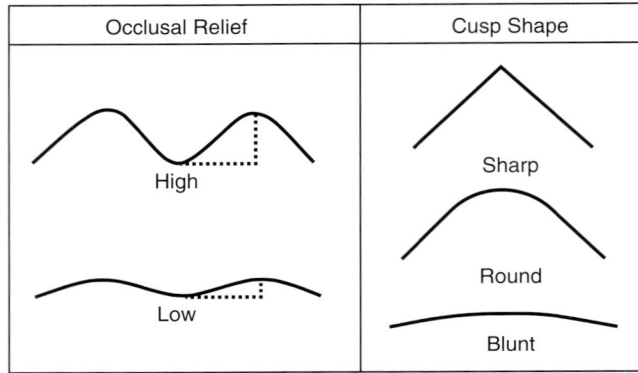

7.7. Schematic of mesowear variables according to the system devised by Fortelius and Solounias (2000).

(Fortelius and Solounias 2000:9). Occlusal relief is scored as either high or low, depending on the height of paracone and metacone above the valley between them (Figure 7.7). For bovids, cervids, and equids, they considered relief to be high when that height is at least 10% of the occlusal length of the entire tooth; in all but borderline cases, this can be gauged by eye without relying on direct measurements. Fortelius and Solounias (2000) demonstrated that when enough teeth are scored for a given taxon (~20), the mesowear signatures reliably distinguish between browsers (mostly high and sharp), grazers (mostly high/low and round/blunt), and mixed feeders (mostly high and sharp/round), provided that specimens in very early or late stages of wear are excluded.

Since its development, the application of mesowear analysis has been prolific. Mesowear patterns have been explored across the tooth row (e.g., Franz-Odendaal and Kaiser 2003; Kaiser and Fortelius 2003; Kaiser and Solounias 2003; Louys et al. 2011b), for various taxonomic groups (e.g., Butler et al. 2014; Croft and Weinstein 2008; Fraser and Theodor 2010; Mihlbachler and Solounias 2006; Taylor et al. 2013; Ulbricht et al. 2015), and novel systems for scoring and quantifying meso-scale wear patterns have been proposed (e.g., Mihlbachler and Solounias 2006; Mihlbachler et al. 2011; Rivals and Semprebon 2006; Rivals et al. 2007b; Solounias et al. 2014). The method is now widely used across diverse temporal and geographic contexts to reconstruct the dietary ecology of extinct species, identify dietary shifts, and characterize the diets of fossil communities (e.g., Butler et al. 2014; Faith 2011a; Faith et al. 2011; Kubo et al. 2015; Pérez-Crespo et al. 2016; Rivals and Lister 2016; Rivals et al. 2015; Stynder 2009; Yamada et al. 2016). For example, Butler et al. (2014) studied wear patterns across modern Australian marsupials using an ordinal-scale scoring system spanning a gradient from attrition-dominated to abrasion-dominated wear patterns: 0 = high occlusal relief and sharp cusps, 1 = high occlusal relief and round cusps, 2 = low occlusal relief and sharp cusps, 3 = low occlusal relief and round cusps, 4 = low occlusal relief and blunt cusps. For the maxillary

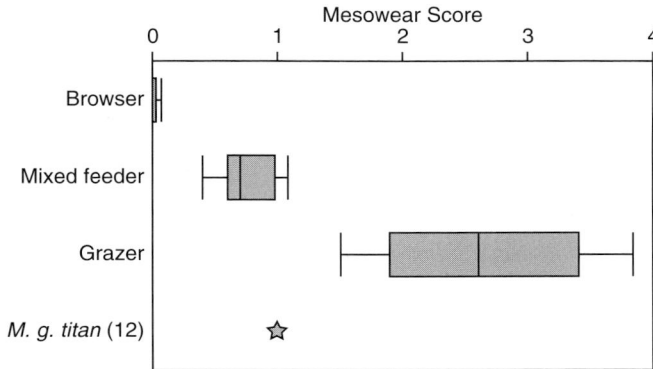

7.8. Box plots illustrating the univariate mesowear scores for maxillary third molars of modern Australian marsupials across dietary categories (mean for eight species in each category) compared with fossil *Macropus giganteus titan* (mean of twelve specimens). Data from Butler et al. (2014).

third molar, the average mesowear score readily distinguishes between species that are grazers, mixed feeders, or browsers. Turning to the fossil record, they showed that the mesowear signature of the giant Australian Pleistocene kangaroo *Macropus giganteus titan* implies a mixed feeding diet (Figure 7.8), unlike the grazing diet typical of larger species of *Macropus* today (Helgen et al. 2006). The paleoenvironmental implication of the mixed feeding mesowear signal is that the habitat in which this kangaroo lived comprised a mixture of grass and brushy species.

SYNECOLOGICAL APPROACHES

While it is possible to develop initial environmental inferences from a taxon-free characterization of a single species (e.g., a grazer implies presence of grasses), it is obvious that resolution is enhanced when the entire fossil community or a substantial portion thereof is considered. As others have done before us (Andrews and Humphrey 1999), we treat the term "community" as equivalent to "fossil assemblage." This might include all taxa in an assemblage or a particular subset of that assemblage (e.g., the rodent community, the ungulate community, the large mammal community).

Paleoecologists have developed a handful of taxon-free techniques for interpreting the environmental implications of fossil communities, but before we turn to these we think it is important to briefly describe an example of a fairly typical analysis. The example is provided again by Semken's (1980) analysis of the Cherokee archaeological site micromammals; recall that we discussed his area of sympatry analysis of the Cherokee faunas in Chapter 5 (see Table 5.3 and Figure 5.8). Semken also assigned many of the mammal taxa to one of three habitats (meadow, woodland, steppe) and we list his assignments in Table 7.1; taxonomic abundance data for each of the three Cherokee faunas are given in

TABLE 7.1 *Preferred habitats of Cherokee archaeological site stenotopic micromammals (after Semken 1980).*

Meadow (moist grassland)	Woodland	Steppe (xeric)
Sorex cinereus	*Blarina brevicauda*	*Spermophilus franklinii*
Microsorex hoyi	*Scalopus aquaticus*	*Geomys bursarius*
Microtus pennsylvanicus	*Sylvilagus* sp.	*Perognathus hispidus*
Synaptomys cooperi	*Tamias striatus*	*Microtus (Pedomys) ochrogaster*
Zapus hudsonicus	*Tamiasciurus hudsonicus*	*Peromyscus maniculatus*
	Peromyscus leucopus	*Reithrodontomys megalotis*
	Lasiurus cinereus	*Onychomys leucogaster*
	Eptesicus fuscus	
	Mephitis mephitis	
	Castor canadensis	

Note: See Table 5.3 for taxonomic abundances and common names of most taxa across the three temporally distinct assemblages.

Table 5.3. Semken (1980) summed absolute (MNI) abundance data for all taxa within a habitat category in each of the faunas, and calculated relative abundances of each habitat type's summed taxa per temporally distinct fauna to produce what we think of as a habitat index value (Cherokee I: meadow = 28.5, woodland = 28.5, steppe = 42.7; Cherokee II: meadow = 27.8, woodland = 44.6, steppe = 27.6; Cherokee III: meadow = 31.4, woodland = 43.9, steppe = 25.2); these represent taxon-free characterizations of each assemblage because they are based not on taxonomic identity, but rather on ecological characterizations of each taxon. Semken (1980) interpreted the fluctuations in habitat indices to reflect shifts in vegetation driven by changes in climate. It is not necessary to summarize his detailed paleohabitat and paleoclimatic inferences; it suffices to note that he perceives in the habitat indices a gradual loss of woodland and increased aridification through the sequence.

As we noted, Semken's (1980) approach is a fairly typical sort of analytical protocol and interpretation. Semken exemplifies many others who have performed similar analyses focused on explicit assessment of temporal shifts in either presences or abundances of taxa according to their habitat associations or, in other cases, dietary preferences (e.g., Bobe and Behrensmeyer 2004; Cuenca-Bescós et al. 2009; Faith 2013a; García-Alix et al. 2008, 2009; Gunnell 1994; Guthrie 1968a; Price and Sobbe 2005; Rector and Verrelli 2010; Sesé Benito 1994). Most of these studies depend on the assumption that the ecology of species in the present accurately reflects their paleoecology (Chapter 3, assumption 1), but as we discussed above it is possible to derive paleoecological inferences using ecomorphology or any number of paleodietary proxies. And like Semken (1980), the paleoenvironmental interpretation is often based on a somewhat subjective interpretation of the presence or abundances of taxa across ecological categories.

But as we turn to next, there are techniques that allow fossil communities to be more directly related to habitat types found in the present.

Habitat Metrics

There are a handful of taxon-free techniques that quantitatively relate the habitat associations of the taxa in a faunal assemblage to a particular habitat (e.g., Andrews 1990, 2006; Andrews et al. 1979; Emery and Thornton 2008; Hernández-Fernández 2001; Hernández-Fernández and Peláez-Campomanes 2003; Matthews et al. 2005; Nesbit Evans et al. 1981). Following Lyman (2017), we refer to these approaches as *habitat metrics*. Their appeal is that the outcome of analysis can potentially include a quantitatively supported assignment of a fossil fauna to a habitat type, such as evergreen tropical deciduous woodland, boreal forest, or tundra. Here we examine how these techniques work, and highlight their potential and pitfalls in paleoenvironmental reconstruction.

Ecological Diversity Analysis

While the earliest taxon-free characterizations of fossil communities are to be found in the paleobotanical literature (Bailey and Sinnott 1915, 1916), it took decades for vertebrate paleoecologists to develop comparable methods. The history begins with the work of ecologist Theodore Fleming (1973). At the time, many of his contemporaries were concerned with documenting and explaining latitudinal gradients in the taxonomic diversity of vertebrate communities (e.g., Karr 1971; MacArthur 1964; Simpson 1964). Fleming (1973:558) was interested not only in latitudinal gradients in taxonomic diversity (for more on this see Chapter 8), but especially in "changes in the *kinds* of species inhabiting different regions." He measured the kinds of species present in a community by their ecological diversity, which he defined as the distribution of species across body size, trophic, and locomotor adaptations; it should now be clear that these characterizations are taxon-free.

Fleming (1973) was not the first to provide a taxon-free description of a mammal community (e.g., Harrison 1962; Valverde 1964), but his work did catch the attention of paleoecologist Peter Andrews, who developed what is now called *ecological diversity analysis*. In their landmark study, Andrews et al. (1979) used Fleming's (1973) framework to characterize the ecological diversity of twenty-three modern African mammal communities (excluding bats) and several Miocene and Pleistocene fossil faunas from East Africa. These were characterized by the distribution of species across taxonomic orders, but also according to the same taxon-free variables used by Fleming (1973): body size, locomotor adaptation, and feeding habits. The rationale for doing so is that "[b]y establishing patterns of community structure for a series of modern mammal communities, and then calculating corresponding patterns for fossil communities, it is possible to compare the communities directly even though species composition of the latter is completely different" (Andrews et al. 1979:179–180). Andrews et al. (1979) showed that contemporary patterns of ecological diversity are remarkably similar across similar habitats, and later studies using identical or similar taxon-free characterizations showed the same to be true around the world (e.g., Andrews 1992, 1996; Andrews and Humphrey 1999; Kovarovic et al. 2002; Lintulaakso and Kovarovic 2016; Louys et al. 2011a; Mendoza et al. 2005; Reed 1998).

The convergence of mammalian community structure across similar habitats irrespective of their taxonomic composition means that ecological diversity analysis can be used to make inferences about the habitat(s) sampled by a fossil assemblage. Here we briefly describe one example provided by Kovarovic et al. (2002), who used ecological diversity analysis to explore the 2.66 Ma mammalian fauna surface-collected from the Upper Ndolanya Beds in Laetoli, Tanzania. Following Andrews et al. (1979), they characterized the forty-four mammalian taxa in the Ndolanya fauna according to body mass, diet, and locomotion; the categories they used are indicated in Figure 7.9. Body masses for extant taxa were taken from contemporary observations, whereas those for extinct taxa were inferred from regressions relating tooth size to body mass in extant organisms. Their assignments of fossil taxa to locomotor categories are based on "morphological adaptations of the limb bones" (Kovarovic et al. 2002:397), although very similar conclusions could be obtained from taxonomic uniformitarianism in this particular case. And lastly, inferences concerning the diet were derived from dental morphology and wear.

With ecological characterizations provided for each taxon, the next step is to calculate the proportion (or percentage) of taxa across ecological categories, producing what Andrews (1992) calls *ecological diversity spectra*. Kovarovic et al. (2002) compared the Ndolanya diversity spectra with those across forty-four modern communities averaged over fifteen broad habitat types found in Africa, Asia, and Central America (habitat descriptions provided in Andrews and Humphrey 1999). Using a combination of ordination (principal components analysis) and qualitative comparisons of ecological diversity spectra, which we show in Figure 7.9, they reached two conclusions relevant to our discussion here. First, the fossil assemblage includes fewer small-bodied taxa (<1 kg) than in any modern community, which Kovarovic et al. (2002) attributed to taphonomic bias. This highlights a critical assumption of the comparative framework used in ecological diversity analysis; in addition to other assumptions (Chapter 3), it is assumed that the ecological diversity of the taxa recovered in the fossil sample is directly comparable to living communities. It is obviously unreasonable to require that a fossil sample include *all* species that constituted an ancient community, but we do have to worry about biases that systematically alter the distribution of taxa across ecological categories. These can

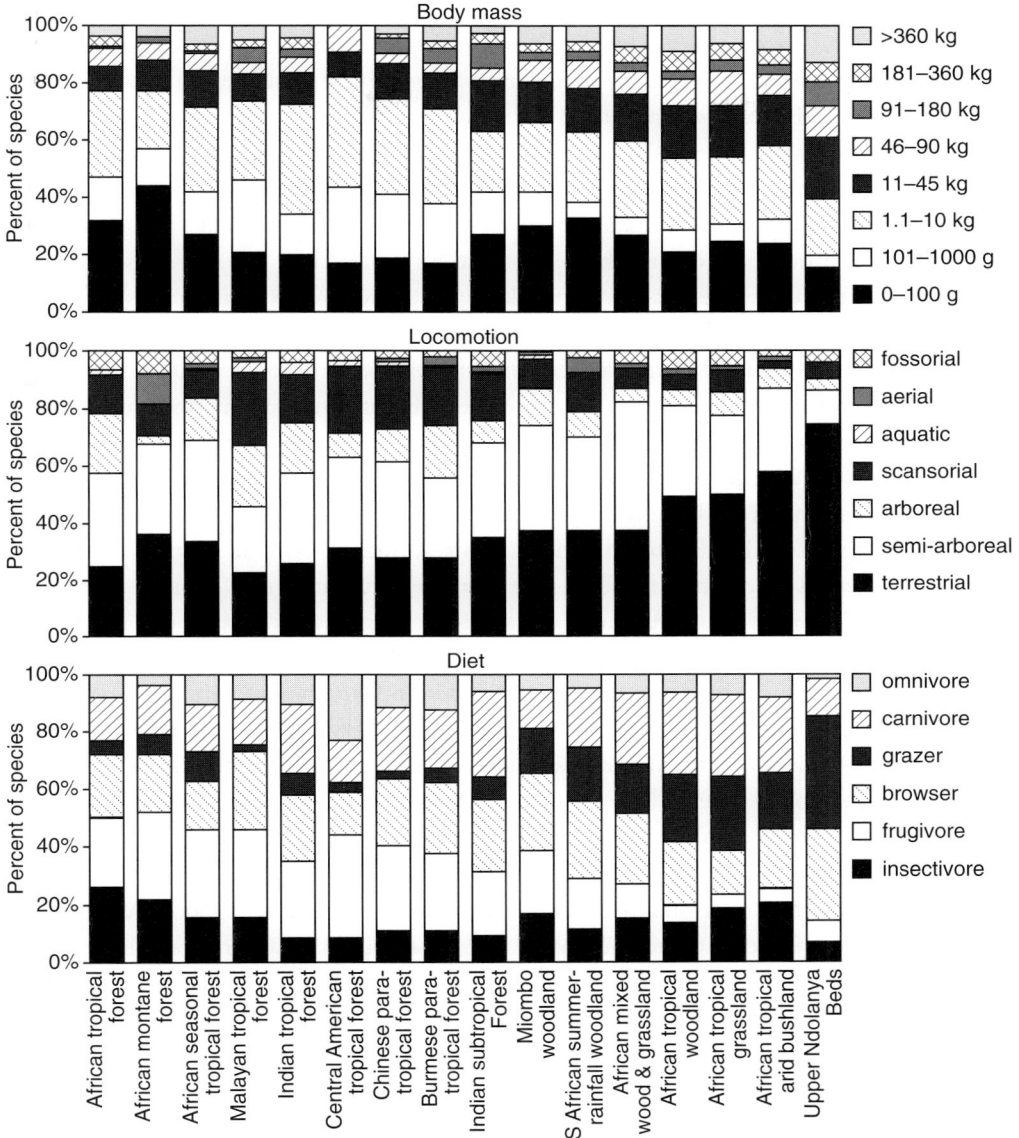

7.9. The distribution of species across body size classes (top), locomotor regimes (middle), and dietary classes (bottom) across modern habitats relative to the Upper Ndolanya Beds (Laetoli). Redrawn from Kovarovic et al. (2002).

include sampling problems (see Chapter 6) or any number of taphonomic processes that might cause, for example, large-bodied species to be over-represented or arboreal species to be under-represented. In this particular example, the absence of small-bodied taxa in the Ndolanya fauna not only precludes an ecologically informative comparison of body mass distributions (Figure 7.9), but also can influence the representation of taxa across dietary or locomotor categories. For the sake of argument, it is not unreasonable to suggest that the missing small-bodied taxa might disproportionately belong to the insectivorous or semi-arboreal categories, as is the case of many shrews (Soricidae), a taxonomic group that is poorly represented in the Ndolanya sample (Kovarovic et al. 2002). If this were true, then the rarity of small-bodied species will also translate to a rarity of taxa with these ecological traits. To get around this problem, Kovarovic et al. (2002) provided a parallel analysis that excluded those taxa smaller than 1 kg from the modern and fossil communities. Another option might be to focus on a particular taxonomic subset (e.g., rodents, ungulates) that is particularly well sampled and (one hopes) representative of that portion of the paleocommunity.

The second relevant conclusion reached by Kovarovic et al. (2002) is that the Ndolanya diversity spectrum is most similar to modern communities from African tropical bushlands, with or without small-bodied taxa included in the analysis. We can see in Figure 7.9, for example, that the dominance of terrestrial taxa coupled with high frequencies of grazers and browsers is certainly atypical for forest habitats, and much more similar to African tropical bushlands. Of course, one might question the extent to which the Ndolanya fauna is all that different from those of African tropical grasslands and woodlands, which are quite similar to each other and to the Ndolanya fauna. These similarities imply that, at least for the habitat types and ecological variables considered here, ecological diversity analysis may be useful for distinguishing forests from non-forests, but lacks resolution to confidently discriminate between finer-scale habitat types. This is not necessarily a limitation of ecological diversity analysis in general, however, as other studies using slightly different sets of ecological variables have shown that it is possible to reliably discriminate between grasslands, woodlands, and shrublands in Africa (Reed 1998, 2008). We can also see that, in addition to the rarity of small-bodied taxa, there are other ways in which the Ndolanya fauna is unlike modern communities; it has too many terrestrial and grazer taxa coupled with a lack of carnivores. This could reflect any of several possible mechanisms: systematic taphonomic bias, a paleocommunity without modern analogs, or a lack of relevant analogs in the modern sample. We do not attempt to disentangle these possibilities here, but emphasize that these are relevant concerns for any comparative analysis of this sort.

The analytical approach of Kovarovic et al. (2002) is fairly typical of ecological diversity analysis. Other studies differ in their ecological variables, modern communities, or quantitative procedures (e.g., Andrews 1996; Assefa et al. 2008; Kay and Madden 1997; Louys et al. 2015; Mendoza et al. 2005; Rector and Reed 2010; Reed 1997, 1998, 2008), but the basic approach remains the same: a list of taxonomic presences is converted to a taxon-free characterization based on ecological traits, and this is compared with modern communities from a range of habitat types to infer the likely habitat of the fossil sample. It should be clear that the comparison is based on the uniformitarian assumption that modern communities with a particular combination of values for certain ecological variables (e.g., body mass, diet, locomotion) are effective analogs for reconstructing ancient environments. There is only indirect dependence on the taxonomic uniformitarian assumption (Chapter 3, assumption 1).

Before we move on, it is important to briefly note the range of quantitative tools that have been used in ecological diversity analysis. Kaye Reed (1998) used DFA to show that the distribution of taxa across fourteen ecological categories pertaining to diet and locomotion discriminates between modern African mammal communities from a range of habitat types (forests, closed woodland, bushland, open woodland, shrubland, and grassland). She then used the discriminant functions to predict the habitats represented by the Pliocene faunas from Makapansgat in South Africa (see also Mendoza et al. 2005), providing quantitatively supported habitat reconstructions; this approach could eliminate some of the ambiguity related to qualitative comparisons (Figure 7.9). In a later study, Reed (2008) used correspondence analysis (see Chapter 5) to visualize the relationships between the ecological adaptations of fossil and modern communities from various habitats (see also Assefa et al. 2008; Matthews et al. 2011; Rector and Reed 2010). This is arguably a more effective means of synthesizing multiple variables than the three stacked bar graphs shown in Figure 7.9 because the relevant information is presented in a single plot. But it is also worth noting that even relatively simple techniques, such as bivariate scatterplots (e.g., percentage of terrestrial species versus percentage of frugivorous species) can be very useful in discriminating between communities from different habitats (Andrews 2006; Assefa et al. 2008; Reed 1997, 1998, 2008).

Taxonomic Habitat Index

Building on the ecological diversity analysis of Andrews et al. (1979), Nesbit Evans et al. (1981) developed the *taxonomic habitat index* (THI), an approach that has since been implemented across numerous geographic and temporal contexts (e.g., Andrews 1990, 1996; Bañuls-Cardona et al. 2012, 2014; Emery and Thornton 2008, 2012, 2014; Fernández-Jalvo et al. 1998; Nel and Henshilwood 2016; Reed 2007). The aim of the THI is to combine the habitat preferences of each taxon in a faunal assemblage into a taxon-free index describing the habitats preferred by those taxa. The first step is to establish a classificatory scheme that broadly describes the habitats occupied by the taxa in a faunal assemblage. For example, in the context of African faunas, Andrews (2006) recognized five habitat types defined by vegetation cover: forest, woodland, bushland, grassland, and semi-desert. He defined these as:

- forest – closed multiple canopy, tall tree, evergreen to semi-deciduous, herbaceous ground vegetation;
- woodland – open to closed single canopy, tall tree, deciduous, herbaceous and grass ground vegetation;

- bushland – open to closed single canopy, low to tall shrubs, deciduous, mainly grass ground vegetation;
- grassland – trees and shrubs less than 20% canopy cover or absent; and
- semi-desert – widely spaced shrub canopy, low deciduous, mainly annual grass and herb ground vegetation.

It should be obvious that the habitat classification determines the resolution provided by the analysis. If the habitat types are defined broadly (e.g., open habitats versus closed habitats), then the analysis will provide only coarse-grained resolution. If the types are defined narrowly (e.g., Mediterranean lowland mixed forest), then the analysis will require so many different types to encompass the possible range of habitats in which a taxon occurs that the next step becomes impractical.

Once habitat types are established, taxa are assigned scores summing to 1 across each habitat, with values weighted according to the importance of a habitat to the taxon in question. For example, as described by Andrews (2006:575), "a species living mainly in woodland, but partly also in bushland and occasionally in forest and grassland, might have scores of forest 0.1, woodland 0.5, bushland 0.3, grassland 0.1, semi-desert 0." Reed (2007:227) refers to these weights as *niche models*, noting that "[t]heoretically, a niche model should be based on the probability of [observing] a particular habitat … for a given taxon." In practice, the development of niche models is subjective and prone to inter-analyst disagreement. How do we decide whether a species occurs in woodland 50% of the time versus 60% of the time? Andrews, one of the method's initial developers (Nesbit Evans et al. 1981), readily admits that this step is the "most controversial" aspect of the THI, but he is correct in noting that this is "a general problem not specific to the present method" (Andrews 2006:575).

To some extent, creation of the niche model merely formalizes what is done intuitively (and sometimes implicitly) when the faunal analyst makes an assessment of the environmental implications in traditional paleoenvironmental reconstructions (those approaches discussed in Chapters 5 and 6), albeit at a level of resolution that may be relatively coarse, depending on the habitat types that are involved. In their rendition of the technique, Emery and Thornton (2008:162) explicitly state that assigning what they refer to as "habitat fidelity values" to taxa "is largely subjective and the assignations are broad and generalized." But they also consulted with three wildlife ecologists who work in the geographic area that produced the faunas Emery and Thornton were studying, to ensure their assessments of each taxon's habitat preferences were not significantly far off the mark. A more objective approach is provided by Campell et al. (2011, 2012), who plotted the occurrence of modern South African rodents from museum collections on maps showing the distribution of different vegetation biomes (e.g., desert, grassland, forest, thicket). They generated niche models by calculating the percentage of individual specimens or localities in which a taxon is present (both were consulted and shown to provide similar results) for each biome (i.e., habitat type). Campell et al. (2011, 2012) recognized that the sampling strategies employed by the various zoologists who had generated the museum collections may lead to enhanced recovery of a species in some habitat types and poor recovery in others (i.e., collector bias), but their approach is certainly an improvement over a best-guess sort of niche model, and it has the added benefit of being replicable.

Niche models can be generated for extinct taxa and for taxa identified above the species level. Consider the extinct Cape zebra (*Equus capensis*) known from Boomplaas Cave and elsewhere in southern Africa, whose use of various habitat types is unknown. Following the protocol outlined by Nesbit Evans et al. (1981), the taxon can be accommodated in the THI by taking the average niche model for living representatives of *Equus*, which in Africa would represent an average of wild ass (*E. africanus*), Grevy's zebra (*E. grevyi*), plains zebra (*E. quagga*), and mountain zebra (*E. zebra*). The exact same approach can be used to handle the case of an assemblage that has produced specimens identified only to the genus *Equus*, with none assigned to species. In the case of an extinct genus, one might instead use an average niche model for living representatives of the same tribe or family. The assumption is that the average habitat preferences of the extant representatives are suitable analogs for the extinct taxon (see assumption 2 in Chapter 3 on using nearest living relative taxa of extinct taxa for an ecological analog).

The next step in implementation of the THI is to calculate the average niche model for all species in an assemblage, producing what some call a *habitat spectrum* (Van Couvering 1980) and Emery and Thornton (2008) refer to as *habitat representation indices*. Following Andrews' (2006) scheme, a faunal community sampling a grassland habitat might be characterized by the following habitat spectrum: forest 0.05, woodland 0.15, bushland 0.30, grassland 0.40, and semi-desert 0.10. In most applications of the THI, all taxa are weighted equally in generating the habitat spectrum, meaning that the analysis relies solely on taxonomic presences, although there are examples where the habitat spectrum is weighted by taxonomic abundances (Emery and Thornton 2008; Reed 2007). Like ecological diversity analysis, because habitat spectra are based on taxon-free niche models rather than individual taxa, it is possible to compare fossil communities from any place or time, regardless of whether or not they share the same species.

Once habitat spectra are compiled for various fossil assemblages, they can be compared with each other or with habitat spectra from modern communities. Here we provide an example using micromammal remains collected by Denné Reed (2007) from nine owl roosts associated with both barn owl (*Tyto alba*) and

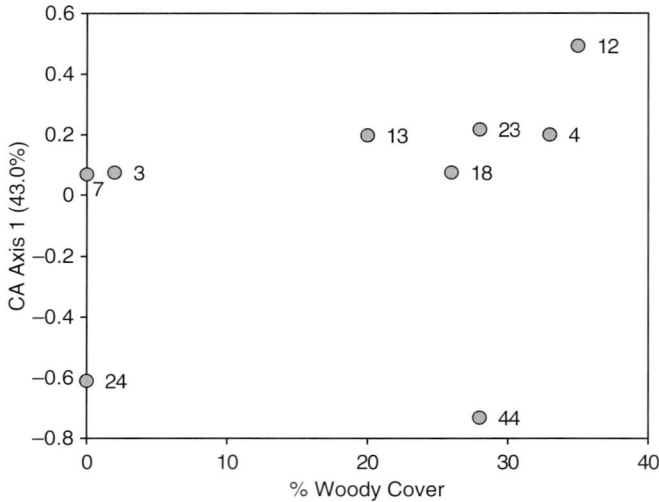

7.10. The relationship between woody cover and the first axis of a CA of taxonomic presences from Reed's (2007) modern Serengeti raptor roosts. The number adjacent to each symbol indicates the number assigned to each roost (see also Figure 7.11).

spotted eagle owl (*Bubo africanus*) in the Serengeti of northern Tanzania. These roosts encompass a range of environments, from grasslands to dry woodlands to mesic woodlands. For a 1.5 km radius around each roost, which corresponds to the foraging radius of the barn owl, Reed calculated the percentage of woody cover on the landscape using satellite imagery, with values ranging from 0% to 35%. He shows that traditional approaches to paleoenvironmental inference based on taxonomic abundance data strongly reflect variation in the degree of woody cover between samples. The same is also true of the taxonomic presences. Figure 7.10 illustrates the relationship between the primary axis of a correspondence analysis of taxonomic presences (taxa occurring only once excluded) in each of nine roosts and the proportion of woody cover. There is a weakly significant relationship between axis 1 (43.0% of inertia) and the proportion of woody cover ($r_s = 0.641$, $p = 0.068$), a trend that becomes highly significant when the outlier (roost 44) is excluded ($r_s = 0.928$, $p = 0.002$). Reed (2007) noted the anomalous absence of arboreal taxa from roost 44, despite the fact that the roost is located near a dense riverine forest. Arboreal taxa are especially abundant in the accumulations of spotted eagle owls, which roost in trees and opportunistically prey upon tree-dwelling rodents, whereas they are less common in assemblages generated by barn owls, which include roost 44. Thus, the absence of arboreal taxa from roost 44 is, Reed suggested, likely an artifact of predator bias. With this in mind, however, both taxonomic abundances (Reed 2007) and presences (Figure 7.10) from the Serengeti owl roosts faithfully track habitat change across the landscape.

Compared with other analyses based on taxonomic abundances, Reed (2007) found that the THI analysis provides a less robust environmental signature. His niche models, which build on previous work (Fernández-Jalvo

TABLE 7.2 *Niche models for Serengeti rodents based on Fernández-Jalvo et al. (1998) and Reed (2007).*

Taxa	Forest	Woodland	Bushland	Grassland	Semi-arid
Arvicanthis	0.00	0.00	0.25	0.75	0.00
Aethomys	0.18	0.25	0.40	0.18	0.00
Mastomys	0.00	0.33	0.33	0.33	0.00
Mus	0.35	0.19	0.26	0.20	0.00
Oenomys	0.50	0.50	0.00	0.00	0.00
Pelomys	0.00	0.00	0.50	0.50	0.00
Thallomys	0.00	0.50	0.50	0.00	0.00
Grammomys	0.40	0.35	0.20	0.00	0.05
Zelotomys	0.00	0.00	0.20	0.70	0.10
Gerbillus	0.00	0.00	0.20	0.20	0.60
Tatera	0.00	0.00	0.40	0.60	0.00
Steatomys	0.20	0.20	0.20	0.20	0.20
Dendromus	0.05	0.27	0.40	0.28	0.00
Saccostomus	0.00	0.33	0.33	0.33	0.00
Otomys	0.00	0.25	0.50	0.25	0.00
Xerus	0.00	0.33	0.66	0.00	0.00
Heterocephalus	0.00	0.20	0.30	0.40	0.10
Acomys	0.00	0.17	0.17	0.17	0.50
Dasysmys	0.00	0.00	0.20	0.80	0.00
Lemniscomys	0.00	0.10	0.10	0.50	0.30
Praomys	0.80	0.20	0.00	0.00	0.00

et al. 1998), are reported here in Table 7.2 and the habitat spectra for taxonomic presences illustrated in Figure 7.11. Based on inspection of the habitat spectra, differences between roosts are not especially clear. When the roosts are ranked according to the proportion of woody cover, there are no significant relationships between that ranking and the value for each habitat type across habitat spectra (forest: r_s = -0.122, p = 0.751; woodland: r_s = 0.548, p = 0.130; bushland: r_s = 0.336, p = 0.375; grassland: r_s = -0.312, p = 0.407; semi-arid: r_s = -0.321, p = 0.392). If roost 44 is excluded, there is a significant positive correlation only for the woodland category (r_s = 0.747, p = 0.040), which increases with the percentage of woody vegetation cover. This is an encouraging result, but one would have hoped for a parallel trend in the bushland category, as bushland vegetation is included in Reed's calculation of woody cover. One would have also hoped for a significant inverse trend for the grassland category, as grassland must decline as woody cover increases because this is a closed array (values must sum to 100 percent). In this example, then, the THI lacks the precision needed to capture the environmental variation represented in the modern faunal assemblages associated with the roosts.

There is one primary reason why Reed (2007) found that the THI is outperformed by more traditional analyses of taxonomic data. Even if we accept that the niche models for individual taxa are reasonable, meaning, for

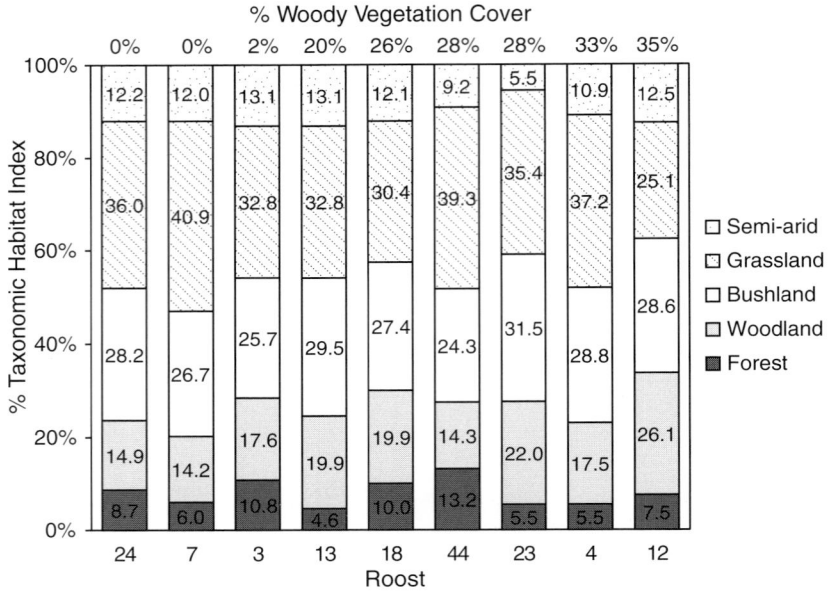

7.11. Habitat spectra for Serengeti raptor roosts across a gradient from open to wooded habitats. Redrawn from Reed (2007).

example, that the proper score for *Dendromus* in the forest category is 0.27 (Table 7.2) and not 0.25 or 0.30, the process of transforming a taxon into a niche model and combining these models into a habitat spectrum involves a loss of information. Rather than providing an analysis of twenty-one rodent taxa (Table 7.2), whose presences and abundances may vary substantially between assemblages, the THI collapses these variables (taxa) into the number of habitat types recognized in the model, which in this example is five. This substitution of taxonomic with ecological information is a necessary step in taxon-free analysis, but it also translates to lower resolution, which in turn means less power to detect subtle environmental differences between assemblages. Also lost is the detailed ecological information provided by individual taxa. For example, the chisel-toothed kangaroo rat is no longer a xeric-adapted taxon associated with desert valleys dominated by shadscale (see discussion in Chapter 6), but instead might become the following: forest 0, woodland 0, bushland 0.2, grassland 0, and desert 0.8 (ignoring for the moment the subjectivity in assigning such values).

There are cases, however, where the loss of resolution might be an acceptable tradeoff. As is true of all taxon-free methods, when comparing faunal assemblages that have few or no species in common, perhaps because they are from different regions or time periods, condensing the taxonomic data into a taxon-free habitat spectrum greatly facilitates comparison. Likewise, the THI is well-suited for assemblages of primarily extinct taxa. The THI was initially

TABLE 7.3 *Climatic zones and corresponding vegetation biomes used in Hernández-Fernández and Peláez-Campomanes' (2003) bioclimatic model.*

Zone	Climate	Zonobiome (main vegetation type)
I	Equatorial	Evergreen tropical rainforest
II	Tropical with summer rains	Tropical deciduous forest
II/III	Transitional tropical semi-arid	Savanna
III	Subtropical arid	Subtropical desert
IV	Winter rain and summer drought	Sclerophyllous woodland-shrubland
V	Warm-temperate	Temperate evergreen forest
VI	Typical temperate	Nemoral broadleaf-deciduous forest
VII	Arid-temperate	Steppe to cold desert
VIII	Cold-temperate (boreal)	Boreal coniferous forest (taiga)
IX	Arctic	Tundra

developed to facilitate paleoenvironmental reconstruction of Miocene faunas from East Africa, most of which are dominated by extinct genera (Nesbit Evans et al. 1981). Given that the THI readily accommodates niche models for extinct taxa, the method provides a useful means of providing relatively coarse habitat reconstructions, especially when compared with modern analogs. Yet again, any such comparison and subsequent interpretation requires a uniformitarian assumption.

The Bioclimatic Model

Building on foundations established by the THI, Hernández-Fernández (2001; Hernández-Fernández and Peláez-Campomanes 2003) developed a parallel habitat metric known as the *bioclimatic model*. This technique classifies faunal communities according to one of ten climatic and vegetation zones (Table 7.3) based on Walter (1970); these are analogous to the habitat types used in the THI. For each species found in modern faunal communities, a *climate restriction index* (CRI) is calculated based on the occurrence of a taxon across those zones. The CRI is equivalent to the niche model in the THI, except that in the former, the scores are not weighted according to the importance of a zone to a taxon. Thus, a taxon occurring only in zones XIII (Boreal coniferous forest) and IX (Tundra) receives a score of 0.5 for each, regardless of whether it prefers one environment over another. From here, the average CRI for those taxa in a faunal community is calculated, providing a bioclimatic spectrum (= habitat spectrum) composed of ten bioclimatic components (= the scores for each of the ten climatic zones). As in the THI, the bioclimatic spectra of fossil faunas can be compared with each other or with modern communities.

Bioclimatic analysis is effectively a variant of the THI, but one of the more noteworthy differences in its published applications concerns an additional quantitative step used in interpretation of fossil data. While many applications of the THI rely on qualitative comparisons of fossil communities to modern ones, developers of bioclimatic analysis employ DFA to predict the likely bio-climatic zone sampled by a fossil community. DFA of bioclimatic spectra from modern communities predicts, with high accuracy (90–98%), the bioclimatic zone in which a modern community is found (Hernández-Fernández 2001), with similar results obtained from an expanded dataset used to validate the model (Hernández-Fernández and Peláez-Campomanes 2003). It follows that, keeping in mind the core analytical assumptions (Chapter 3), discriminant ana-lysis may do a reasonable job of assigning a fossil assemblage to a bioclimatic zone. This means that it is possible to translate a faunal list to a quantitatively supported paleoenvironmental interpretation such as tundra or boreal forest.

To illustrate the approach, we provide here an analysis of the Boomplaas cave micromammals (rodents only). Following the protocol outlined by Hernández-Fernández (2001), we generated climate restriction indices for the rodent taxa observed at Boomplaas Cave (Table 7.4A), which were then used to create bioclimatic spectra for each assemblage (Table 7.4B). We ran a DFA on the bioclimatic spectra calculated for rodents from the fifty modern communities reported in Hernández-Fernández and Peláez-Campomanes (2003) to predict the likely bioclimatic zone(s) represented across the Boomplaas Cave sequence. All of the Boomplaas Cave assemblages are assigned with high probabilities (≥0.96) to zone IV: winter rain and summer drought. Our analysis reveals the two main weaknesses of this type of analysis. The first, also recognized by Hernández-Fernández and Peláez-Campomanes (2005), is that fossil communi-ties may sample ecotones that do not neatly fit within any of the ten bioclimatic zones. Boomplaas Cave is situated in an area that is today transitional between Walter's (1970) zones IV and V, receiving rainfall throughout the year. Thus, the assignment of all Boomplaas Cave assemblages to zone IV is probably off the mark, at least for the upper levels that sample climates that are likely similar to the present. Second, the qualitative zones provide only coarse-resolution paleo-environmental information, with all of earth's myriad environments distilled to only ten different types (Table 7.3). The assignment of all Boomplaas Cave levels to a single zone tells us nothing about the paleoenvironmental changes that occurred through time; it masks the dramatic shifts in faunal composition that Avery (1982) documented using more traditional analyses. This gross simplifica-tion highlights the coarse-grained nature of these qualitative descriptions; they are limited by the broad bioclimatic zones recognized in the analysis (Table 7.3). Do not misunderstand; we are not saying these descriptions are valueless but rather their coarse resolution might be misleading.

Ecometrics

The second family of taxon-free synecological techniques is based on the analysis of measurable, ecologically relevant morphological (anatomical) traits

TABLE 7.4 *Example of the bioclimatic model technique outlined by Hernández-Fernández (2001) and Hernández-Fernández and Peláez-Campomanes (2003).*

Taxon	Climate zone					
	I	II	II/III	III	IV	V

(A) Climate restriction indices (CRI) for Boomplaas Cave rodent species.

Taxon	I	II	II/III	III	IV	V
C. hottentotus	0	0.25	0	0.25	0.25	0.25
G. ocularis	0	0	0	0.33	0.33	0.33
D. melanotis	0	0.25	0.25	0	0.25	0.25
D. mesomelas	0	0.33	0	0	0.33	0.33
M. albicaudatus	0	0.50	0	0	0.50	0
S. krebsi	0	0.50	0	0	0.50	0
S. campestris	0	0.25	0.25	0.25	0	0.25
A. subspinosus	0	0	0	0	1	0
A. namaquensis	0	0.20	0.20	0.20	0.20	0.20
D. incomtus	0.25	0.25	0	0	0.25	0.25
M. minutoides	0.20	0.20	0	0.20	0.20	0.20
M. verreauxi	0	0	0	0	1	0
R. pumilo	0	0.20	0.20	0.20	0.20	0.20
G. paeba	0	0	0.33	0.33	0.33	0
T. afra	0	0	0	0.50	0.50	0
O. laminatus	0	0	0	0	0.50	0.50
O. saundersae	0	0.33	0	0	0.33	0.33
O. irroratus	0	0.33	0	0	0.33	0.33
O. unisulcatus	0	0	0	0.50	0.50	0

Stratum	Climate zone					
	I	II	II/III	III	IV	V

(B) Bioclimatic spectra for rodent assemblages across the Boomplaas Cave sequence. These represent the average CRI of all species in an assemblage.

Stratum	I	II	II/III	III	IV	V
DGL	2.81	20.42	5.63	12.08	39.69	19.38
BLD	2.50	18.15	6.85	15.37	39.91	17.22
BLA	2.65	19.22	7.26	14.31	40.29	16.28
BRL	2.37	18.95	6.49	14.56	39.56	18.07
CL	2.65	16.77	5.78	14.80	41.28	18.73
GWA	3.21	17.98	4.64	12.02	41.79	20.36
LP	3.00	19.00	4.33	11.22	41.22	21.22
LPC	1.82	20.61	5.91	12.27	43.33	16.06
YOL	3.46	19.36	5.00	10.39	42.44	19.36
BP	2.81	17.81	4.06	13.65	41.77	19.90
OLP	2.65	16.77	5.78	14.80	41.28	18.73
BOL	2.81	17.81	6.15	12.60	40.73	19.90
OCH	3.00	19.00	4.33	11.22	41.22	21.22
LOH	3.46	19.36	7.56	12.95	41.15	15.51

across the taxa in a community. Among vertebrates, these could include body mass, limb proportions, dental hypsodonty, or any other taxon-free trait that influences how an organism interacts with its environment. Eronen et al. (2010a) coined the term *ecometrics* to describe taxon-free trait analysis, and this term is increasingly used today (e.g., Barr 2017; Evans 2013; Fortelius et al. 2016; Lawing et al. 2012; Meloro and Kovarovic 2013; Polly et al. 2011; Sukselainen et al. 2015). However, although the term is new, ecometric analysis has a long history in paleoenvironmental reconstruction, with Bailey and Sinnott's (1915, 1916) early study of leaf shape in relation to climate representing a typical example of this family of techniques. The basic premise is simple; if a trait can be shown to vary across environmental gradients in modern ecosystems, then it can be used as a paleoenvironmental proxy in the fossil record. Recall the general uniformitarian assumption discussed in Chapter 3.

In this section we describe two ecometric variables: body size and dental hypsodonty. We do not wish to imply that ecometrics is limited to the study of these traits alone; the array of possibilities is highlighted by ecometric studies on traits including relative tail length and vertebral shape in snakes (Lawing et al. 2012), the calcaneal gear ratio of carnivorans (Polly 2010; Polly and Sarwar 2014), and the shape of bovid metatarsals and astragali (Barr 2017). We focus on body size and dental hypsodonty here because they are widely used in paleoenvironmental reconstruction and because they provide an opportunity to illustrate two very different analytical approaches.

Body Size Distributions

The body size of a taxon is perhaps one of the most important predictors of its ecology, life history, and physiology (e.g., J. H. Brown et al. 2004; Damuth 1981; Damuth and MacFadden 1990; Owen-Smith 1988; Peters 1983; Smith and Lyons 2011; West et al. 1997; White et al. 2007). It is not surprising then that the body size distribution of modern and ancient vertebrate communities is often explored in relation to the environment (e.g., Andrews 1996; Andrews et al. 1979; Fleagle 1978; Fleming 1973; García Yelo et al. 2014; Holling 1992; Lambert 2006; Lambert and Holling 1998; Legendre 1986, 1987, 1989; Lintulaakso and Kovarovic 2016; Lyman 2013; Travouillon and Legendre 2009). Body size is often an important variable used in ecological diversity analyses of the sort outlined above, but as we discuss here, it is sometimes analyzed on its own to generate paleoenvironmental inferences.

Cenogram analysis is easily the most popular technique used for the paleoenvironmental study of body size distributions (e.g., Costeur 2004; Costeur and Legendre 2008; Croft 2001; Deng 2009; Flynn 2003; García Yelo et al. 2014; Gingerich 1989; Gunnell 1994, 1997; Hernández-Fernández et al. 2006; Kay et al. 2012; Legendre 1986, 1987, 1989; Lyman 2013; Montuire and Marcolini 2002; Storer 2003; Tsubamoto et al. 2005). A cenogram is a plot of the logarithm of the body mass of each taxon in a community or local fauna (y-axis), with taxa ranked along the x-axis in order of descending body mass (Figure 7.12). They were first used by ecologist José Valverde (1964) to visualize predator–prey relationships across body size gradients in a modern mammalian

7.12. Cenograms for mammal communities associated with Kalahari thornveld vegetation in South Africa (data from Legendre 1989) and moist tropical forest near Cristóbal in Panama (data from Fleming 1973).

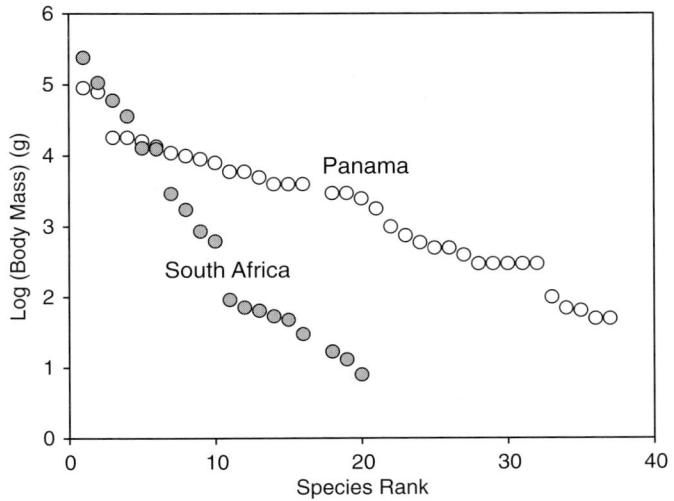

community from southern Spain. Serge Legendre (1986, 1987, 1989) later used this graphical technique to explore how body size – a variable that is directly accessible from the fossil record (Damuth and MacFadden 1990) – differs across modern communities (carnivores and bats excluded) from varied environments, which he then used to interpret cenograms of fossil assemblages. Legendre made three observations based on qualitative inspection of the modern communities. First, cenograms for communities from open environments have a gap in the medium-sized species (~500 to 8,000 g), which translates to a break in slope between the larger and smaller species, whereas those from closed environments (forests) include many species in this size range and are therefore continuous. Second, the slope of large species (>8,000 g) is steeper in arid environments, but more level in humid environments. And third, the slope for small species (<500 g) is steeper in cool environments than in warm environments.

Legendre modeled his observations as shown in Figure 7.13. In his view, an absence of gaps in the distribution of body sizes indicates a closed habitat (top panels in Figure 7.13); if gaps were present in the medium-size range (500 to 8,000 g, according to Legendre), then open habitats are implied (bottom panels in Figure 7.13). Finally, the more large-bodied species (>8,000 g, according to Legendre), the more mesic the environment (left panels in Figure 7.13) and the fewer large-bodied species, the more xeric the environment (right panels in Figure 7.13). The analytical protocol is, then, to construct a cenogram for the species represented in a fossil assemblage, estimate the mean adult body mass of each species, and qualitatively compare the fossil cenogram with the interpretive models, perhaps assisted by cenograms for a variety of modern communities to determine a good match for the fossil cenogram. Use the best match as the interpretive analog for the ancient faunal community.

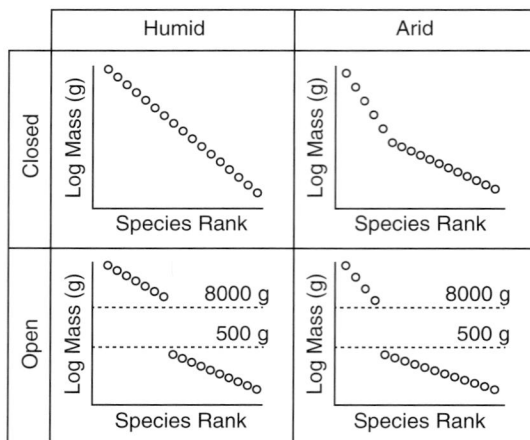

7.13. Legendre's (1986, 1987, 1989) interpretive algorithm for cenograms.

The example in Figure 7.12 provides cenograms for two modern mammal communities, the first associated with semi-arid Kalahari thornveld vegetation (a grassy shrubland) in South Africa (Legendre's [1989] Transvaal community 3), and the second with a moist tropical forest near Cristóbal in Panama (from Fleming 1973). The Kalahari thornveld community shows a gap from ~90 to 600 g (= 1.95 to 2.78 in log units) that is consistent with its relatively open habitat; this is less than the suggested 500 to 8,000 g gap (= 2.70 to 3.90 in log units), but contrasts with the more continuous distribution observed for the moist tropical forest. In addition, the cenogram for the Kalahari thornveld community shows a relatively steeper slope for large mammals, which correctly indicates the more arid conditions relative to the tropical forest. There is also a faint suggestion of a steeper slope for the smaller species, which also correctly indicates the cooler climate that characterizes the Kalahari community.

While our example and Legendre's cenogram analyses are qualitative in nature, Philip Gingerich (1989) provided the first quantitative cenogram analysis. In a reexamination of the modern communities originally compiled by Legendre, he measured the size of the gap above 500 g and used linear regression to calculate the slopes for small and large size classes. Gingerich showed that the gap is small in forest environments and large in open habitats, and that the slope of the regression for species ranging in mass from 500 g to 250 kg is steeper (more negative) in arid environments than in humid environments; he did not examine these parameters across temperature gradients. In the wake of his analysis, it is now not uncommon for paleoecologists to quantify these and numerous (10+) other variables that describe a cenogram's shape (e.g., Gómez Cano et al. 2006; Hernández-Fernández et al. 2006; Rodríguez 1999; Storer 2003; Travouillon and Legendre 2009).

Although Gingerich's (1989) study suggested that quantitative analysis of cenograms could be used to make paleoenvironmental inferences, subsequent work casts doubt on how exactly the body mass distributions illustrated in cenograms are related to climate. In an influential critique, Rodríguez (1999) analyzed sixteen cenogram variables across ninety-two modern mammal communities from around the globe. His analysis confirmed Legendre's observations concerning the gap in body size distributions and habitat structure, but failed to support the other generalizations. However, Rodríguez (1999) did find numerous other weak to moderate correlations between various cenogram parameters and some climate variables, especially evapotranspiration and temperature (but not precipitation). Similar mixed results between cenograms and various climate parameters have been documented in continental-scale analyses (Croft 2001; Kay et al. 2012; Travouillon and Legendre 2009), but unlike Rodríguez (1999) these studies indicate that precipitation plays an important role in the distribution of species across size classes at these smaller spatial scales. The implication of these contradictory findings is that cenograms capture an aspect of community structure that is somehow related to the environment, but that the relationship is complex and variable across different parts of the world and at different spatial scales. Understanding this complexity requires that we examine why the patterns captured in cenograms are related to the environment in the first place.

Since the introduction of cenograms as a paleoenvironmental tool, there has been a focus on documenting the empirical relationships between the structure of the cenogram and various climatic and environmental parameters (e.g., Croft 2001; García Yelo et al. 2014; Gingerich 1989; Hernández-Fernández et al. 2006; Hostetler 1997; Travouillon and Legendre 2009). Far less attention has been directed at understanding *why* such relationships exist, exemplifying a point we made in Chapter 3 that correlations between an environmental variable and a biological one are often used to interpret faunal assemblages despite no understanding of why the correlation exists. Concerning the preponderance of medium-sized (500 to 8,000 g) species in closed habitats (forests), one possibility is that the spatial patchiness of foods in arboreal settings may impose selective constraints that favor a narrow range of body sizes for arboreal taxa (Kay and Madden 1997). Too small, and the energetic costs of moving between patchily distributed resources separated by gaps in the forest canopy are prohibitive. But too large, and the structural limits of the supporting branches are exceeded. This hypothesis could explain why there tend to be more medium-sized species in closed environments, resulting in a fairly continuous size distribution without a clear gap, although it does not readily explain why taxa of this size are sometimes completely absent from open habitats (see also Hostetler 1997).

Turning to the slope of the line(s) defined by the distribution of body sizes, one issue is paramount; the slope calculated in cenogram analysis tracks taxonomic richness (the number of species). As Gingerich (1989) noted, when there are more species present in a particular body size interval, the slope tends to be shallower than if fewer species are present. This can be seen in our cenograms for the Kalahari thornveld and moist tropical forest communities (Figure 7.12). The steeper slope for large mammals above the gap in the former community is due to a relative lack of larger-bodied species. As we discuss in the next chapter, richness is often correlated with precipitation or other climate variables, and there are good reasons why this is so. However, these relationships are complex and variable for different taxonomic groups in different parts of the world. Thus, richness explains why the slope in cenogram analysis is related to climate, and why analysis of modern communities from different regions and different spatial scales have produced inconsistent and sometimes contradictory results.

The relationship between slope and richness poses an important, but all too often ignored, analytical problem for the analysis of cenograms in the fossil record. It is well known that the number of species recovered in a faunal assemblage is strongly influenced by sample size, with larger samples typically providing more species (Grayson 1984b; Lyman 2008b). Because the number of species in turn influences the slope in cenogram analysis, it follows that sampling effects will have a substantial influence on the outcome. To illustrate this point we turn again to the Boomplaas Cave sequence and provide a cenogram analysis of the early Holocene mammals from member BRL (Figure 7.14A). By many standards, the BRL assemblage can be considered reasonably well sampled, including forty-eight mammalian species (bats excluded). Following Gingerich's (1989) protocol, we calculate the slopes for small mammals (up to 500 g) and large mammals (500 g to 250 kg) (Figure 7.14B). Once again, these represent variables of paleoenvironmental interest thought to be related to temperature and aridity, respectively. To illustrate the influence of sample size on richness and our results, we ask: how would these values change if fewer specimens had been recovered? To answer this question, we generated ten random sub-samples of the BRL assemblage with sample sizes increasing from 100 to 1,000 specimens by increments of 100 (we treat the MNI counts [Table 4.2] for microfauna as if they are equivalent to NISP in this analysis; NISP data for the microfauna are unavailable). For each of these random assemblages, we calculated slopes and the number of taxa for the small and large mammal size classes. As can be seen in Figure 7.14B, as sample size increases, so too does the number of species, and this is associated with slopes that become increasingly shallow (more positive); the relationships between slope and both sample size (small mammals: $r_s = 0.863$, $p = 0.001$; large mammals: $r_s = 0.915$, $p < 0.001$) and richness (small mammals: $r_s = 0.963$, $p < 0.001$; large mammals: $r_s = 0.975$, $p < 0.001$) are strong and highly significant. If we were to uncritically treat our simulated assemblages as a fossil sequence, it would be easy to infer from a cenogram analysis that there were changes in temperature (small mammal

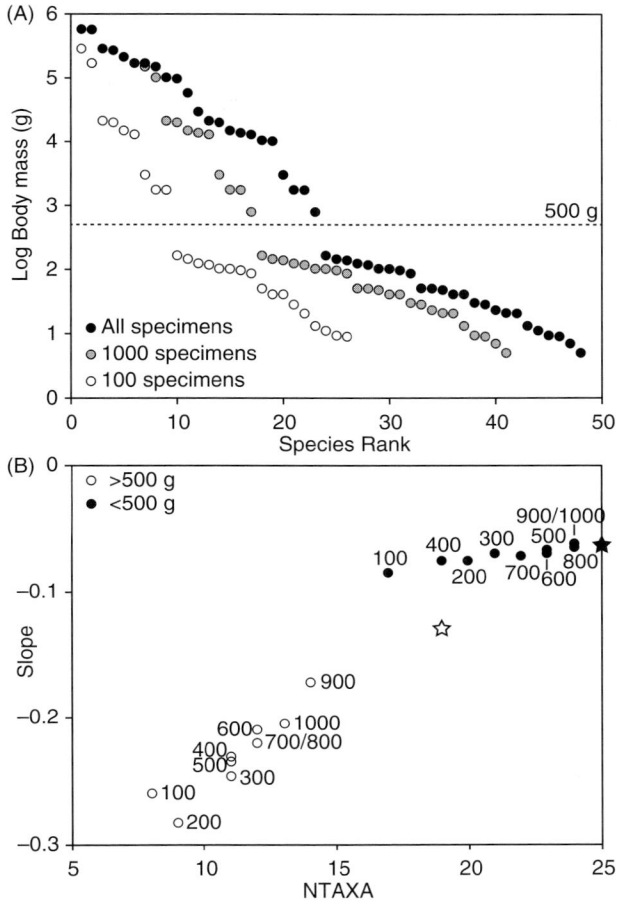

7.14. (A) Cenograms for Boomplaas Cave mammals (bats excluded) from member BRL and random sub-samples of BRL at sample sizes of 1,000 and 100 specimens. Note that the recovery of fewer species at lower sample size translates to relatively steeper slopes. (B) The relationship between richness (NTAXA) and slope for large (>500 g) and small (<500 g) mammals calculated for random sub-samples of Boomplaas Cave member BRL. Sub-sample size indicated for each point. Stars indicate slopes and richness for the BRL assemblage with all specimens included.

slope) and aridity (large mammal slope) through time. Following Legendre's (1986, 1987, 1989) generalizations, those assemblages with smaller samples would be interpreted as consistent with relatively cooler and drier conditions, whereas those with larger samples would be seen as suggesting warmer and wetter conditions. However, the only real difference in this case pertains to sample size, not a change in the environment.

The implications of our analysis are two-fold. First, the cenograms of fossil samples can only be meaningfully compared with each other if we control for differential sampling intensity between assemblages. If not, the observed patterns are likely to tell us more about sample size than any paleoenvironmental variable of interest. Although sample size effects can be sorted out with a bit of analytical gymnastics (e.g., standardizing assemblages to equivalent sample size using randomization procedures, such as *rarefaction*; see Chapter 8), we are not convinced it is worth the effort. We have just shown that the cenogram slopes track richness, and one can also quantify the gap between smaller and larger species by tallying the number of species in the size range (e.g., 500 to 8,000 g) that defines the gap (e.g., Rodríguez 1999; Travouillon and Legendre 2009). Thus, the ecologically relevant aspects of cenogram analysis can instead be monitored by an analysis of richness across size classes. Richness is also sensitive to sample size, but as we discuss in the next chapter, there are a handful of robust techniques for getting around sample size effects.

The second implication of our analysis is that cenograms based on fossil assemblages can only be meaningfully compared with those based on modern communities when all species in the fossil community have been sampled; this is because the absence of species from a fossil sample will alter cenogram parameters in a way that has nothing to do with the environment. As others have noted before us (Alroy 2000; George 2012), obtaining such comprehensive fossil samples is extremely difficult, resulting in interpretive problems. For example, we just characterized the assemblage from Boomplaas member BRL as reasonably well sampled, but we cannot be confident that *all* species in the paleocommunity are represented in the fossil sample. George (2012) made precisely this point when he examined several cenograms based on fossil mammals. Noting that Travouillon and Legendre (2009) indicated cenograms should be compared qualitatively and not quantitatively, George (2012) observed that depending on scale issues, particularly how high and how wide a cenogram is drawn, comparisons of a fossil cenogram with a modern cenogram can be misleading. And he noted many fossil assemblages contain a non-representative sample of the small-bodied taxa, a non-representative sample of the large-bodied taxa, or non-representative samples of both. This means of course that a cenogram based on a fossil assemblage likely will be taxonomically incomplete. For these reasons, and despite Lyman's (2013) earlier interpretation of cenograms (although in fairness to Lyman, he did use a sample of 2,486 NISP), we are skeptical of the use of cenograms as anything other than a graphical technique for displaying patterns in body mass distributions (and perhaps sampling issues).

Some researchers continue to work with cenograms, increasing the number of modern communities studied to see how they fit Legendre's original interpretive algorithm (Figure 7.13) and acknowledging criticisms, particularly that concerned with sampling issues (e.g., Costeur and Legendre 2008). Others have suggested alternatives, such as interpreting simple univariate summary statistics describing body mass distributions (mean, standard deviation, kurtosis, and skewness) (Alroy 2000). The problem with that alternative is that we currently lack detailed understanding of how these variables vary across modern environmental gradients (though they are certainly worth exploring). Our concern with cenograms should not be interpreted as implying body size is an unimportant ecological trait or paleoenvironmental indicator. It most certainly is (e.g., Smith and Lyons 2011). What is needed, however, are more robust approaches for the analysis of body size trends in the fossil record. Our brief study of the Boomplaas fauna implies the analysis of richness across size categories, provided sampling issues are properly dealt with (see Chapter 8), could be a more useful method of capturing the environmentally relevant variables that one might hope to recover through cenogram analysis.

Dental Hypsodonty

As we noted in our discussion of paleodietary reconstruction, the foods that an animal eats are one of its most direct links to its environment. By extension, it is reasonable to infer that any ecometric trait that tracks the dietary ecology of the species in a community has potential to inform on the environment. Among mammalian herbivores, one such trait is dental hypsodonty, or tooth crown height. The hypsodonty of a tooth determines the amount of wear it can sustain before it loses function, with elevated hypsodonty an adaptation that prolongs its functional lifespan (e.g., Damuth and Janis 2011; Fortelius 1985; Janis 1988; Janis and Fortelius 1988). Broadly speaking, and keeping in mind there are exceptions (e.g., Gailer et al. 2016), those taxa with high-crowned (hypsodont) teeth consume primarily grasses and/or inhabit open environments, whereas those taxa with low-crowned (brachydont) teeth consume more dicots and/or inhabit closed environments (Figure 7.15). Hypsodonty is usually measured as a dimensionless *hypsodonty index*, typically the *unworn* height of the tooth crown relative to its length or width (Damuth and Janis 2011). We emphasize a tooth must be unworn (or only lightly worn) to provide an ecologically meaningful hypsodonty index, because crown height of course declines as a function of age and wear (Gifford-Gonzalez 1991b; Klein and Cruz-Uribe 1983a; Lubinski 2000; Steele 2005).

Because grazers are typically more common in grasslands and browsers more common in forests, we can expect the average hypsodonty of the taxa in a community should track vegetation structure, as well as the climate parameters that influence the vegetation (e.g., temperature, precipitation). Emphasizing the apparent association between hypsodont taxa and more open and arid environments in the present, Fortelius et al. (2002) reasoned the average hypsodonty of the mammalian herbivores in a fossil community should track precipitation through time (greater average hypsodonty = increased aridity). They developed a coarse-grained system for scoring the hypsodonty of a fossil taxon: brachydont = 1, *mesodont* = 2, hypsodont = 3. These assignments are based on a hypsodonty index of crown height relative to length calculated for second molars (upper or lower), with brachydont teeth providing an

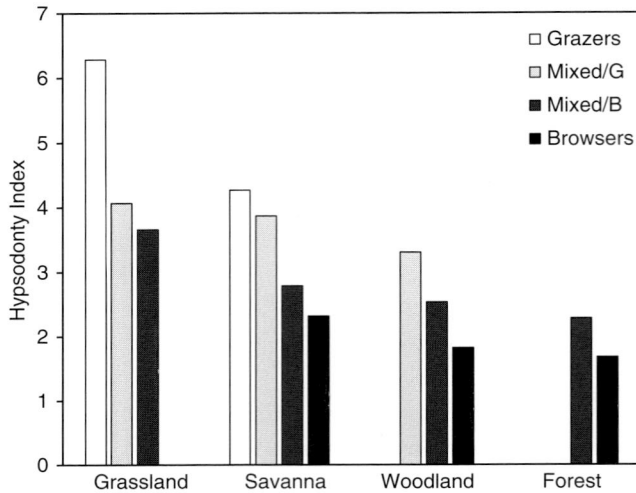

7.15. The average hypsodonty index of mandibular third molars for 133 extant ungulates across habitat and dietary classes. Following Damuth and Janis (2011), dietary classes are defined as follows: Grazers: ≥90% grass in diet; Mixed/Grazer: 50–89% grass in diet; Mixed/Browser: 11–49% grass in diet; Browser: ≤10% grass in diet. Habitat types are defined as follows: Grassland: open grassland lacking woody vegetation cover; Savanna: open habitat with scattered trees and bushes; Woodland; mostly continuous tree canopy but with grass understory; Forest: closed forest with no clearings. Redrawn from Damuth and Janis (2011).

index <0.8, mesodont teeth providing an index of 0.8 to 1.2, and hypsodont teeth >1.2. We refer to these values as *hypsodonty scores*. While the ordinal-scale hypsodonty score translates to reduced resolution compared with the hypsodonty index (a ratio-scale variable), the benefit is that the assignment of an extinct taxon to one of these categories can be reasonably evaluated without relying on unworn teeth to generate a true hypsodonty index. And for extant taxa, we do not even need fossil teeth to generate a reasonable assignment (e.g., *Equus* will always be assigned a score of 3). Using this scoring system, Fortelius et al. (2002) calculated the mean hypsodonty score of the large herbivores documented across Eurasian fossil localities spanning the past 20 million years. They documented spatial and temporal changes in the mean hypsodonty score, which they interpreted as indicating precipitation changes that were in broad agreement with other paleoclimate proxies. Studies implementing similar methods soon followed (e.g., Eronen 2006; Eronen and Rook 2004; Eronen et al. 2009; Fortelius et al. 2003, 2006; Liu et al. 2009), but only more recently have these relationships been rigorously evaluated in the present.

Using the same hypsodonty scoring system developed by Fortelius et al. (2002), Eronen et al. (2010b) showed that for modern communities across the globe (excluding Australia), the average hypsodonty of large herbivores (orders Artiodactyla, Perissodactyla, and Primates) is inversely correlated with mean annual precipitation (MAP; linear regression: $r^2 = 0.58$). In a related study, Liu et al. (2012) introduced a second ecometric variable known as longitudinal lophedness. This represents the number of longitudinal lophs (cutting edges) on the second molar (Jernvall et al. 1996). Liu et al. (2012) found the combination of mean hypsodonty score and mean lophedness of modern large herbivore communities (orders Artiodactyla, Perissodactyla, Primates, Proboscidea)

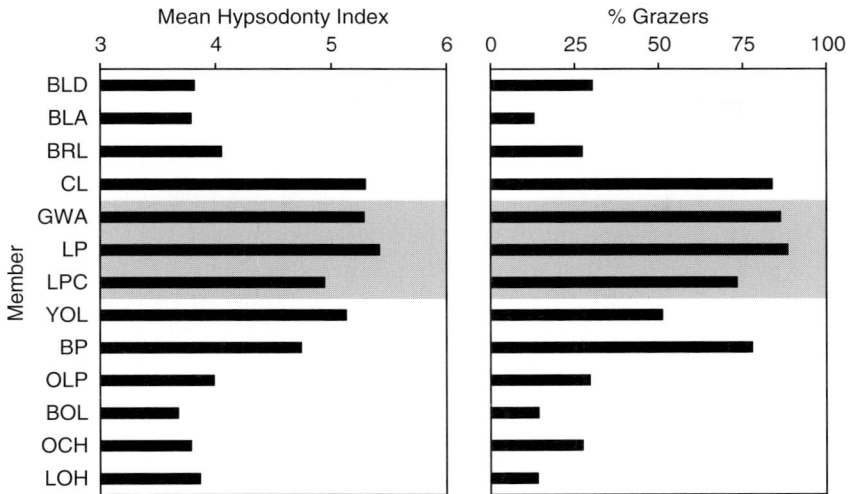

7.16. Changes in mean hypsodonty index of ungulate taxa (weighted by taxonomic abundances) and the relative abundance of ungulate grazers across the Boomplaas Cave sequence. Members corresponding to the Last Glacial Maximum are indicated by grey shading.

predicts mean annual temperature (MAT; $r^2 = 0.67$), MAP ($r^2 = 0.63$), and net *primary productivity* ($r^2 = 0.73$). They also provided regression equations that can be used to predict these parameters in fossil samples, potentially allowing for numerical paleoenvironmental reconstructions. We explore such reconstruction in Chapter 9. It is sufficient here to note the work of Eronen et al. (2010b) and Liu et al. (2012) provides crucial empirical support for a growing body of literature that relies on the hypsodonty of a community as an index of precipitation change in the past (e.g., Eronen et al. 2010c, 2015; Faith et al. 2013; Fortelius et al. 2016; Kaya et al. 2016; Meloro and Kovarovic 2013; Sukselainen et al. 2015).

Here we turn to the Boomplaas Cave ungulates to illustrate one way in which dental hypsodonty can be used to derive paleoenvironmental inferences. We use the hypsodonty indices reported by Janis (1988) and Damuth and Janis (2011) for the lower third molar of modern ungulates. In the case of extinct taxa, we use the average hypsodonty index of living African congeners. And for those taxa identified to genus or higher, we take the average of those species found at the site. For each assemblage, we calculate the mean hypsodonty index of all taxa, with values weighted according to taxonomic abundances (NISP counts) (Figure 7.16). We begin by noting there are highly significant differences in hypsodonty index across the sequence (Kruskal–Wallis test: $p < 0.001$). These differences relate to a prominent increase in the hypsodonty index from the base of the sequence to the Last Glacial Maximum, followed by a decline into the Holocene (Figure 7.16). Consistent with evidence linking elevated hypsodonty to grazing diets (Damuth and Janis 2011), this pattern matches the changing abundances of ungulate specimens that belong to grazers ($r_s = 0.874, p < 0.001$).

Together, the taxon-free trends in hypsodonty and diet illustrated in Figure 7.16 closely track the rise and decline of grassland species documented in our previous analysis of taxonomic presences (Chapter 5) and abundances (Chapter 6). Because taxa with hypsodont teeth are typically associated with more open environments (Figure 7.15), it is reasonable to interpret the shifts in mean hypsodonty as reflecting an expansion of open habitats (i.e., grassland vegetation) during the Last Glacial Maximum and Late-glacial. And based on the global relationships between hypsodonty and climate (Damuth and Janis 2011; Eronen et al. 2010b; Liu et al. 2012), it might also be reasonable to infer greater aridity at this time. However, as we shall see in Chapter 9, the extent to which global climate–hypsodonty relationships apply in this particular context is uncertain. Briefly, this uncertainty stems in part from the fact that relationships between animals and climate are mediated by relationship between vegetation and climate (see also Chapter 2). So the inference that increased hypsodonty signals increased aridity requires that more open and grassy habitats signal increased aridity, the latter being a topic of some debate in this part of southern Africa (Chase and Meadows 2007; Faith 2013b).

SUMMARY

Taxon-free methods of paleoenvironmental reconstruction encompass a diverse family of techniques that characterize individual organisms or entire communities by their ecological traits as opposed to their taxonomic identity. When applied to individual taxa, taxon-free analysis allows us to make coarse ecological inferences, especially those related to diet, locomotor behavior, or habitat preferences, without relying on the taxonomic uniformitarian assumption (Chapter 3, assumption 1); this is especially useful in the deeper-time record where the assumption becomes increasingly tenuous, but it can be applied in any case where a more conservative stance is favored. When communities are characterized by taxon-free traits, it becomes possible to compare communities from any time or place regardless of whether they share the same species. In doing so, we lose a good deal of information pertaining to the ecology of the species in a community, and this can translate to lower resolution. In many cases, however, this is a reasonable tradeoff; taxon-free techniques provide a framework for comparing fossil assemblages with modern animal communities from anywhere in the world. In turn, this allows us to relate fossil assemblages to modern habitats (habitat metrics) and interpret changes in taxon-free traits (ecometrics) in terms of modern environmental gradients.

EIGHT

ENVIRONMENTAL INFERENCES BASED ON TAXONOMIC DIVERSITY

As early as the late 1700s, European explorers and naturalists became aware of the fact that the diversity of plants and animals they encountered increased from high latitudes to the tropics (Hawkins 2001; Lomolino et al. 2010). German naturalist Alexander von Humboldt, who travelled widely in Latin America from 1799 to 1804, commented on the "variety and magnitude in vegetable forms" found in the tropics compared with the far north, where "only those organic forms are capable of full development, which have the property of resisting any considerable abstraction of heat … or can sustain a protracted interruption of their vital functions" (von Humboldt 1850:215). He summarized his observations and proposed a causal mechanism (temperature) with the following: "Those who are capable of surveying nature with a comprehensive glance, and abstract their attention from local phenomena, cannot fail to observe that organic development and abundance of vitality gradually increase from the poles toward the equator, in proportion to the increase of animating heat" (von Humboldt 1850:217). The pattern von Humboldt was describing is today one of the best-known ecological patterns: the latitudinal gradient in species richness, in which the number of species increases from the poles to the equator (e.g., Gaston 2007; Willig et al. 2003).

The latitudinal gradient is documented across taxonomic levels (e.g., species, genera, families) spanning major taxonomic groups (e.g., microbes, plants, and animals) in both terrestrial and aquatic environments, and even extends into the deep-time fossil record (Lomolino et al. 2010; Rosenzweig 1995; Willig et al.

2003). Although it is one of ecology's oldest known patterns (Hawkins 2001), the mechanisms underlying the latitudinal gradient in taxonomic richness are widely debated; Lomolino et al. (2010) cite thirty explanatory hypotheses that have been proposed since the late 1700s. Whatever the mechanism may be, one cannot help but notice, as von Humboldt did, that the gradient in richness corresponds to gradients across environmental variables, including primary productivity, temperature, rainfall, and seasonality. Although these factors on their own are probably insufficient explanations for the gradient (Lomolino et al. 2010), the obvious and widely documented correlations between richness and the environment have played an important role in prompting paleozoologists to use taxonomic diversity as an index of paleoenvironmental change (e.g., Andrews et al. 1979; Avery 1982). The analysis of taxonomic diversity as a potential paleoenvironmental indicator is now commonplace (e.g., Blois et al. 2010; Bobe and Eck 2001; Cruz-Uribe 1988; Faith 2013b; Grayson 1984b, 2000b; Lyman 2014b; Matthews et al. 2011; Nel and Henshilwood 2016). In some cases, it offers potential to provide insight into environmental variables not readily apparent from the taxonomic identity of the species in an assemblage (e.g., Faith 2013b; Faith et al. 2017), but it is also one of the more difficult variables to measure and interpret in a paleozoological context (Grayson 1984b; Grayson and Delpech 1998; Grayson et al. 2001; Lyman 2008b). In order to understand the promise and pitfalls of diversity as a paleoenvironmental indicator, paleozoologists need to know what it is, how it is measured, and what it means. Our aim in this chapter is to bring these issues to light.

WHAT IS DIVERSITY?

The concept of diversity, while perhaps intuitively familiar, is deceivingly complex. Magurran (1988:1) likens diversity to an optical illusion, because "[t]he more it is looked at, the less clearly defined it appears to be and viewing it from different angles can lead to different perceptions of what is involved." Diversity has been defined in numerous ways, and there are countless means of measuring various aspects of it (e.g., Hill 1973; Hurlbert 1971; Magurran 1988, 2004; Peet 1974; Whittaker et al. 2001). Many authors have used the terms richness and diversity interchangeably (see Spellerberg and Fedor 2003), potentially leading to substantial confusion. The resulting uncertainty regarding what any given researcher who uses the term might actually mean has led some to suggest the term be abandoned, and even to deny the very existence of the concept (Hurlbert 1971). This viewpoint has gained little traction, but it does highlight the need to be careful with terminology (e.g., Spellerberg and Fedor 2003). We use the term *diversity* as a generic label for any single property of a faunal community (or fossil assemblage) or any combination of multiple properties. Three specific properties or variables are typically subsumed within what we think of as diversity. *Taxonomic richness* refers to the number of taxa present; *taxonomic evenness* (also called *taxonomic equitability*) refers to the abundances of individuals within each taxon; *taxonomic heterogeneity* is a combined measure of both taxonomic richness and taxonomic evenness. We will consider each of these in some detail momentarily. It is important to first discuss a couple of significant aspects of the general concept of *taxonomic diversity*.

Taxonomic diversity is always specific to a given spatial scale, which can range from minute (e.g., a 1 × 1 meter quadrat) to massive (e.g., North America). Ecologists and paleoecologists (e.g., Sepkoski 1988; Whittaker et al. 2001) divide the continuum of spatial scales over which diversity can be measured into *alpha diversity*, *beta diversity*, and *gamma diversity*. These terms are almost always used to describe richness and this likely contributes to some disagreement over their definition (Whittaker et al. 2001). Alpha diversity refers to diversity at the local scale, usually within a single habitat. In paleozoological contexts, this hopefully equates to the diversity documented in a single assemblage (see Shotwell 1955, 1958). Beta diversity refers to the change in diversity between communities, also referred to as *turnover*, usually across adjacent habitats on a landscape scale (e.g., between a mountain slope and adjacent valley). The fossil equivalent is the comparison of diversity between nearby assemblages that are contemporaneous. Because beta diversity is concerned with turnover through space, it summarizes a different phenomenon than alpha diversity. Gamma diversity has been variably defined as the species diversity of large geographic areas (e.g., biomes, continents), equivalent to alpha diversity writ large (e.g., Lomolino et al. 2010), or as the change in diversity across such areas, equivalent to beta diversity at a larger spatial scale (e.g., Hammer and Harper 2006; Sepkoski 1988). This brief discussion underscores the fact that the analyst must be explicitly clear about the variable she seeks to measure, especially when a term that has a range of meanings is used to label that variable. When examining taxonomic diversity as an indicator of environmental change in the past, paleozoologists are typically concerned with faunal turnover through time at a single site (i.e., alpha diversity through time).

HOW IS DIVERSITY QUANTIFIED?

Both ecologists and paleoecologists describe the array of diversity metrics as bewildering (Hammer and Harper 2006; Magurran 1988). To paraphrase Hammer and Harper (2006:186), there is literally a diversity of diversity indices. The abundance of diversity indices is due in part to the fact that species diversity can include either or both of two variables, richness and evenness, which in turn gives rise to metrics that emphasize one or the other to various degrees (Peet 1974). High richness is indicative of elevated diversity (and also heterogeneity), as is high evenness, which occurs when all species are equally abundant

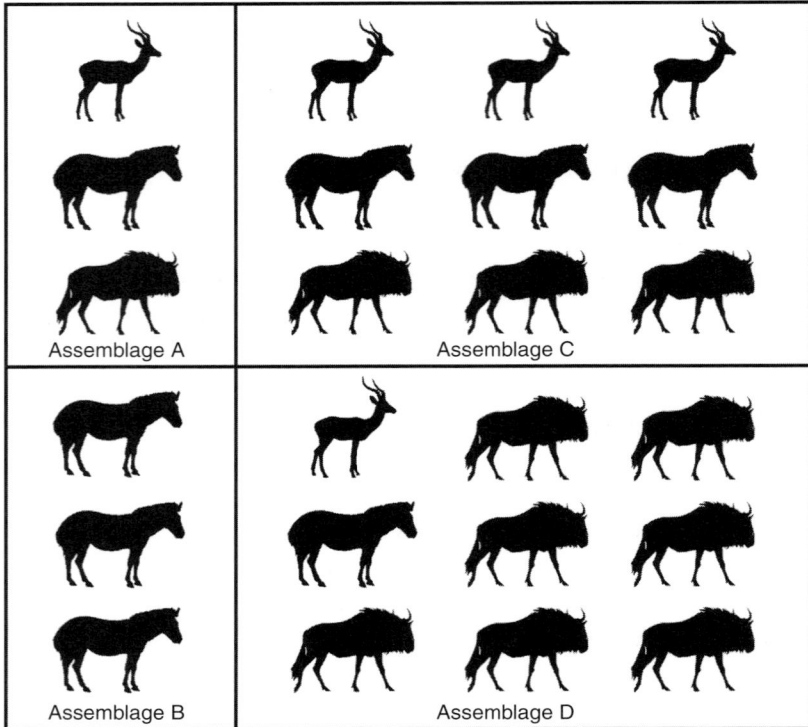

8.1. Four hypothetical faunal assemblages of varied richness and evenness.

(or nearly so). Borrowing from Magurran's (1988) example, these two concepts are illustrated here across four hypothetical faunal assemblages (Figure 8.1). Assemblages A and B have identical sample sizes of three individuals, but A has three ungulate species compared with only one in B. Thus, assemblage A is more diverse (and more heterogeneous) than assemblage B because it is the richer of the two assemblages. Turning to assemblages C and D, both include nine individuals distributed across three ungulate species, meaning that richness is identical. However, all species are equally abundant in assemblage C (it has an even distribution), whereas assemblage D is dominated by a single taxon (it has an uneven distribution). Thus, assemblage C is more diverse (and more heterogeneous) than assemblage D because of greater evenness. In what was perhaps an attempt to clarify the significance of the combination of richness and evenness, ecologist Eric Pianka (1988:343) pointed out that species heterogeneity "is high when it is difficult to predict the species of a randomly chosen individual organism and low when an accurate prediction can be made."

In this section we examine the measurement of taxonomic diversity. We do not aim to provide an exhaustive survey of all diversity metrics – see Magurran (1988, 2004) for an introduction to some of the many indices available – but instead focus on those that are frequently used in paleozoological contexts.

Taxonomic Richness

In ecology, the most straightforward and widely used diversity metric is species richness, which simply refers to the number of species in the sample (Gaston 1996). The same variable is routinely examined in paleozoological contexts, although it often goes by the more inclusive (from a taxonomic perspective) name of taxonomic richness or the number of taxa (NTAXA) (Grayson 1984b; Lyman 2008b). The difference between ecological and paleozoological terminology stems from the fact that, unlike observations of living organisms in ecological studies, it is not always possible to identify fossil remains to species. Published taxonomic lists often include taxa identified at multiple taxonomic levels, often ranging from species to family, such that an analysis of richness at the species level alone may not be the ideal approach.

There are several ways one can go about tallying NTAXA in faunal assemblages for which specimens are identified at multiple taxonomic levels. One of the more common approaches is to calculate all non-overlapping taxa in a manner that avoids double-counting the same taxon (e.g., Faith 2013b; Grayson 1991a; Grayson et al. 2001). Even in the case of faunal lists where all identifications are at the genus level, this is equivalent to counting the *minimum number of species* present because all specimens of a genus must belong to at least one species. Following Lyman (2008b), we use white-tailed deer (*Odocoileus virginianus*) and mule deer (*Odocoileus hemionus*), species that overlap biogeographically in parts of western North America, as an example of this approach (Table 8.1). In an assemblage including remains of *Odocoileus* sp., *O. virginianus*, and *O. hemionus* (assemblage A), NTAXA is two, no different from an assemblage including remains attributed only to *O. virginianus* and *O. hemionus* (assemblage B). For assemblage A, the specimens attributed to *Odocoileus* sp. are not included in the tally because they must belong to one of the two species that have already been counted. Likewise, in an assemblage including remains of *Odocoileus* sp. and either *O. virginianus* (assemblage C) or *O. hemionus* (assemblage D), NTAXA is only one. The specimens attributed to *Odocoileus* sp. are not counted as a second taxon because there is no definitive evidence to suggest a second species of deer is present, even if some of those specimens *might* belong to the second species. And lastly, for the assemblage including remains only identified to *Odocoileus* sp. (assemblage E), NTAXA is also one. We cannot be sure which of the two deer species is represented by these specimens, and it is even possible that both are present, but we can be certain that at least one distinct taxon is present.

Other approaches to counting NTAXA one might consider include restricting the analysis to a single taxonomic level, such as tallying only those specimens identified to species or counting higher, more inclusive, taxonomic levels (e.g., genera). A limitation to such alternative approaches includes a loss

TABLE 8.1 *The number of non-overlapping taxa for five hypothetical assemblages (A–E) of* Odocoileus *spp.*

Taxon	A	B	C	D	E
Odocoileus sp.	x		x	x	x
Odocoileus hemionus	x	x	x		
Odocoileus virginianus	x	x		x	
ΣNTAXA	2	2	1	1	1

of information. For example, turning back to the assemblages in Table 8.1, if we consider specimens identified to species only, then assemblage E provides an NTAXA of 0, even though there has to be at least one species of deer present. Likewise, if we consider genera only, then all five assemblages provide an NTAXA of 1, despite the fact that there is variation in the number of species documented in each assemblage. This might seem to warrant a tally of all non-overlapping taxa, but as Lyman (2008b) points out, the drawback to this approach is that differences between assemblages could be influenced by taphonomic issues such as differential fragmentation. For a fossil to be identified to taxon it must retain diagnostic anatomical features, but if that specimen is sufficiently fragmented then it may be identifiable, for example, to genus but not to species. If both species of deer were in fact present in assemblage E (Table 8.1), but fragmented to the point where only genus-level identification is possible, then the difference in NTAXA between assemblages E and A or B, for which two taxa are identified, could be due to taphonomic rather than ecological processes. Similar taphonomic problems could arise from differential skeletal element representation, particularly in the case of taxa where some elements can be identified to species (e.g., teeth) but others only to genus or family (e.g., postcranial elements). That is, intertaxonomic variability in the taphonomic destruction of skeletal parts identifiable to species versus those only identifiable to genus or family could be a concern. These taphonomic issues have much wider implications for the measurement of taxonomic diversity, and we address them in greater detail later in this chapter.

Whatever approach is used to tally NTAXA, it is well documented by ecologists (Gaston 1996; Gotelli and Colwell 2001; Magurran 1988, 2004) and paleozoologists (Grayson 1984b; Hammer and Harper 2006; Jamniczky et al. 2003, 2008; Lyman 2008b; Wolff 1975) that richness is dependent on sample size. Both Grayson (1984b) and Lyman (2008b) provide detailed discussions of these issues, so we do not belabor the point; smaller assemblages tend to have fewer taxa and larger assemblages tend to have more taxa. To illustrate how richness changes as a function of sampling effort, we turn to the small mammal assemblage from Homestead Cave Stratum IV (taxa aggregated at

8.2. The relationship between sample size (NISP) and various metrics of taxonomic richness (NTAXA, Margalef's index, and Menhinick's index) for random sub-samples of the small mammal assemblage from Homestead Cave Stratum IV (5,000 sub-samples for each NISP value from 1 to 500), with taxa aggregated at the genus level. Solid line indicates the mean and dashed lines indicate 95% confidence limits.

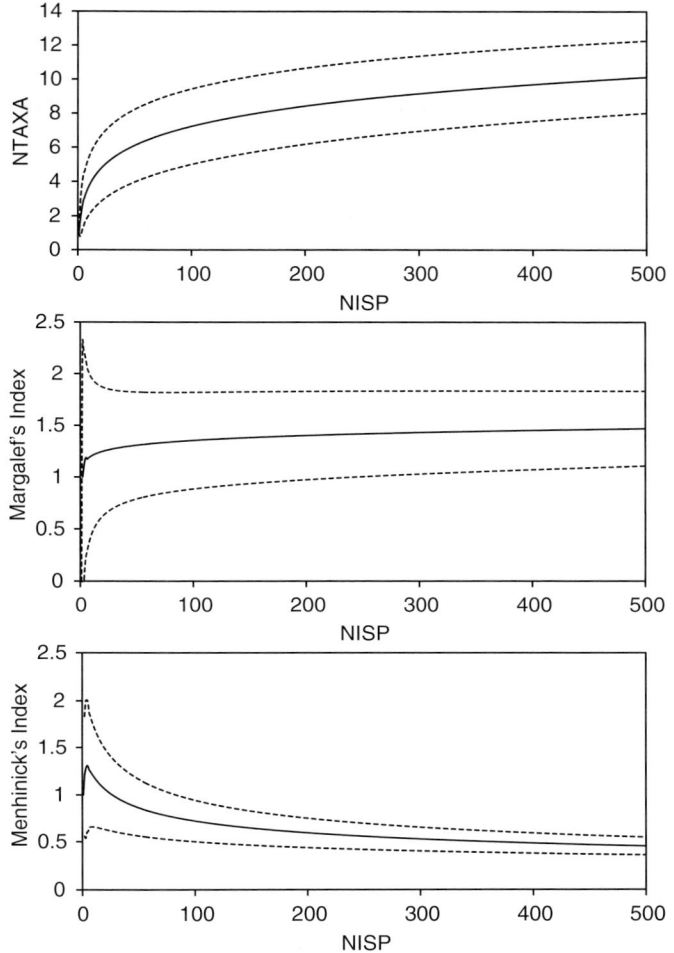

the genus level). We drew between 1 and 500 specimens from Stratum IV and tallied NTAXA (and several other diversity metrics – see below) for each of those sub-samples (i.e., NTAXA when 1 specimen is drawn, NTAXA when 2 specimens are drawn, and so on until we had NTAXA when 500 specimens are drawn). We repeated this process 5,000 times to generate the mean and 95% confidence limits for NTAXA at varying sample sizes. The uppermost panel in Figure 8.2 illustrates how those NTAXA values (with 95% confidence limits) vary as a function of NISP. It shows that as sample size increases, the number of taxa also increases, albeit at a decreasing rate. The prevalence of these relationships in the fossil record (Grayson 1984b; Lyman 2008b) implies that unless sample sizes are identical, a straightforward comparison of NTAXA between assemblages is more likely to tell us about differences in sampling than it is to reveal ecologically meaningful differences in richness.

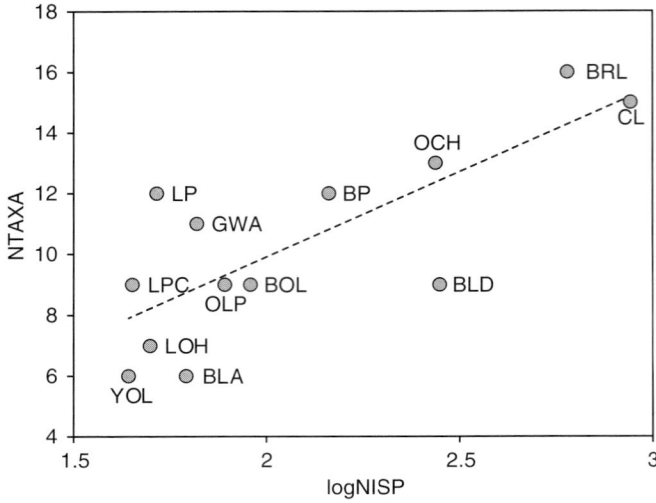

8.3. The relationship between sample size (logNISP) and NTAXA for the Boomplaas Cave ungulates in each stratum.

When comparing richness across multiple assemblages, it is easy to demonstrate that sample size has influenced NTAXA. Simply plot the relationship between the two variables and determine whether there is a correlation between them. Figure 8.3 illustrates the relationship between NISP (log-transformed) per stratum and NTAXA (all non-overlapping taxa) per stratum for the Boomplaas Cave ungulates (Table 8.2). Consistent with our expectations, there is a strong correlation between logNISP and NTAXA ($r = 0.776$, $p = 0.002$), with the coefficient of determination ($r^2 = 0.602$) implying that more than half the variance in NTAXA can be explained by sample size differences. Thus, a direct comparison of NTAXA across assemblages at Boomplaas Cave tells us little about changes in richness of the sampled faunal communities. As one of us has already noted, however, this does not mean "we need to stop, throw up our hands in dismay, and curse the day" (Lyman 2008b:181). There are several means of getting around sampling issues, and we turn to them below.

There are two simple indices of species richness that attempt to control for sample-size effects (Magurran 1988, 2004), both of which occasionally are used in zooarchaeological contexts (e.g., Baxter 2001; Codding et al. 2010b; Jones et al. 2016; Kaufman 1998). Margalef's index is calculated as:

$$\frac{NTAXA - 1}{\ln(n)}$$

where n is sample size. A similar index is provided by Menhinick's index:

$$\frac{NTAXA}{\sqrt{n}}$$

Use of these indices may allow for meaningful comparisons of richness between assemblages, but neither provides a universal solution to sample size

TABLE 8.2 *Sample size (NISP) and various indices of richness for the Boomplaas Cave ungulates.*

Member	NISP	NTAXA	Margalef index	Menhinick index	Residual	Rarefied NTAXA
BLD	281 (233)	9 (8)	1.42	0.54	-3.43	6.48
BLA	62 (54)	6 (5)	1.21	0.76	-2.72	5.00
BRL	601 (463)	16 (13)	2.34	0.65	1.72	8.52
CL	875 (841)	15 (11)	2.07	0.51	-0.18	7.04
GWA	66 (63)	11 (10)	2.39	1.35	2.11	8.82
LP	52 (51)	12 (10)	2.78	1.66	3.67	9.46
LPC	45 (45)	9 (8)	2.10	1.34	1.06	7.95
YOL	44 (43)	6 (6)	1.32	0.90	-1.88	6.00
BP	145 (118)	12 (10)	2.21	1.00	1.20	8.43
OLP	78 (72)	9 (9)	1.84	1.02	-0.28	7.95
BOL	91 (78)	9 (8)	1.77	0.94	-0.68	6.71
OCH	274 (254)	13 (12)	2.14	0.79	0.63	7.86
LOH	50 (43)	7 (6)	1.53	0.99	-1.22	6.00

Notes: NTAXA represents non-overlapping taxa. Values in parentheses represent NISP and NTAXA values for taxa aggregated at the genus level and used in the rarefaction analysis; the NISP values are somewhat smaller because they do not include specimens identified to bovid tribe only.

influences because both make assumptions about the underlying relationship between sample size and NTAXA. Margalef's index assumes that the number of taxa increases as a function of the natural logarithm of sample size, whereas Menhinick's index assumes that it increases as a function of the square root of sample size (Figure 8.4). If richness increases in a manner other than the one assumed by the index that is used, the index values can vary systematically as a function of sample size. At Boomplaas Cave, for example, there is a significant relationship between sample size (logNISP) and Menhinick's index (r = -0.715, p = 0.006), but not for Margalef's index (r = 0.195, p = 0.524; Table 8.2). Thus, the latter set of index values may tell us something about how richness changed through time, but the former tells us more about changing sample sizes. Also worth keeping in mind is that these indices are only appropriate for cases where the local species pool is incompletely sampled. Once all species are recovered, as sample size increases richness remains unchanged, leading to a decline in index values. This problem with these particular indices has been observed for well-sampled fossil assemblages (Avery 1982).

Turning back to the random sub-samples from Homestead Cave (Figure 8.2 lower two panels), we see that both indices vary systematically as a function of sample size. Margalef's index increases as samples become larger, while Menhinick's index declines. This indicates that as sampling effort increases, taxa are added at a greater rate than assumed by Margalef's index but at a lesser rate than assumed by Menhinick's index (Figure 8.4). Because of this potential

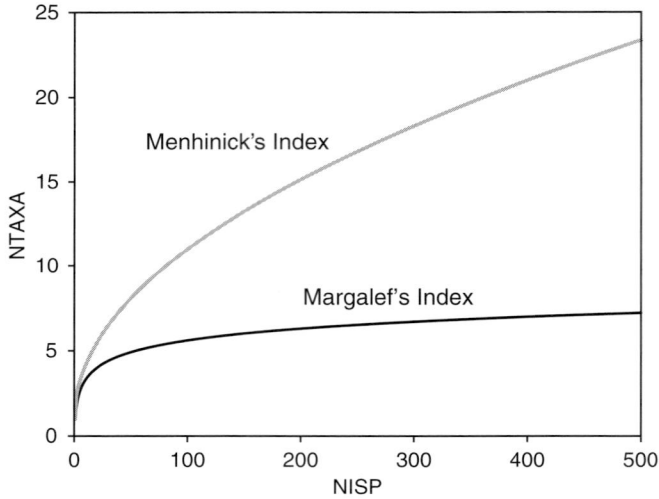

8.4. The general relationship between sample size (NISP = n) and NTAXA assumed by Margalef's richness index (a logarithmic function: NTAXA = ln(n)) and Menhinick's richness index (a power function: NTAXA = $n^{0.5}$).

for systematic variation, we recommend alternative approaches for overcoming the effects of sample size on taxonomic richness. But if Menhinick's index and Margalef's index are to be used, it is important to demonstrate their observed values are uncorrelated with sample size. Because of the critically important relationship between sample size and measures of taxonomic diversity, we devote some of the following discussion to three sets of techniques used by paleozoologists to contend with sample size issues.

Sampling to Redundancy

Simpson (1937:68) was among the first, if not *the* first, paleozoologist to worry explicitly about what he referred to as "sample adequacy." He suggested a means to empirically evaluate whether or not any particular sample was representative: "Probably the best criterion of the adequacy of a collection as a sample of the preserved fossils is that of repetition. When collecting begins to pile up mainly or only duplicates, it probably has achieved sampling adequacy for the local deposit, but as long as many species remain very rare in collections, it probably has not" (Simpson 1937:68). He was largely concerned with basing conclusions on a sample that was representative of what was in a stratum ("the local deposit"), and that what was in the stratum was representative of the fauna on the landscape. The former concerns sample size and the latter concerns taphonomy. We discuss taphonomic issues later in this chapter.

A less direct way to assess sample adequacy with respect to taxonomic richness was described by Wolff (1975; see also Gamble 1978; Wells 1978). He relied on the trophic pyramid (see Figure 2.9) and argued that small herbivores should be more abundant than small carnivores and that large herbivores should be less abundant than small herbivores but more abundant

than large carnivores. He plotted abundances of faunal remains in these categories to determine whether his fossil assemblages met the prediction, presuming that when his samples met the prediction, they were representative
of the prehistoric community because the fossil collection approximated the
normal structure of the trophic pyramid. Wolff (1975) also plotted the cumulative volume of sediment excavated and frequency of faunal remains on the
horizontal axis against the cumulative taxonomic richness of his excavations
on the vertical axis. He argued that when the taxonomic richness curve lev-
eled off, he had accumulated a sample that was representative of what was
in the stratum sampled. This is a somewhat more direct way to assess sample
adequacy than using the trophic pyramid, but there is an even more direct
technique to do so.

Following Wolff's second technique (1975), Lyman and Ames (2004, 2007)
suggested constructing a cumulative curve with sample size measured as NISP
increasing on the horizontal axis and the variable of interest such as taxonomic richness, evenness, heterogeneity, or other variable on the vertical axis
(e.g., Travouillon et al. 2007). Once the curve levels off and remains stable over
successively added samples (producing a cumulatively larger total sample), the
analyst can argue that based on the notion of *sampling to redundancy* (Simpson's
repetitiveness), the total sample is representative in terms of the variable of
interest. Other paleozoologists have independently come to a similar conclusion (Jamniczky et al. 2003, 2008).

An example of a sample-to-redundancy curve with respect to taxonomic
richness is shown in Figure 8.5A. The fauna is of late prehistoric–early historic period mammals, recovered from a site in the Portland Basin of the
northwestern United States, near Portland, Oregon (Lyman and Ames 2004).
The sample increments represent what was recovered during each of several
sequent annual six- to eight-week field seasons and the variable of interest
(on the y-axis) is taxonomic richness. The cumulative sample size measured
as ΣNISP grows from 519 to 6,420, and richness increases from 16 to 25. The
maximum of 25 species is attained when the cumulative sample size is 4,079
NISP, and remains stable over the two-step increase to 6,420 NISP, suggesting
researchers had sampled to redundancy and major increases in sample size
would not increase richness significantly.

In general, if using the sample-to-redundancy approach, the analyst should
consider how the sample from the first field season of excavation (or any temporal unit) compares with the summed samples of the first season and second
season, and with the summed samples of the first, second, and third seasons,
and so on. If that is not possible, or if preferred, the analyst could compare
multiple independent samples of different sizes in terms of the target variable to determine whether a pattern implicating sample size influences on the
target variable emerges. Or the analyst could generate random sub-samples at

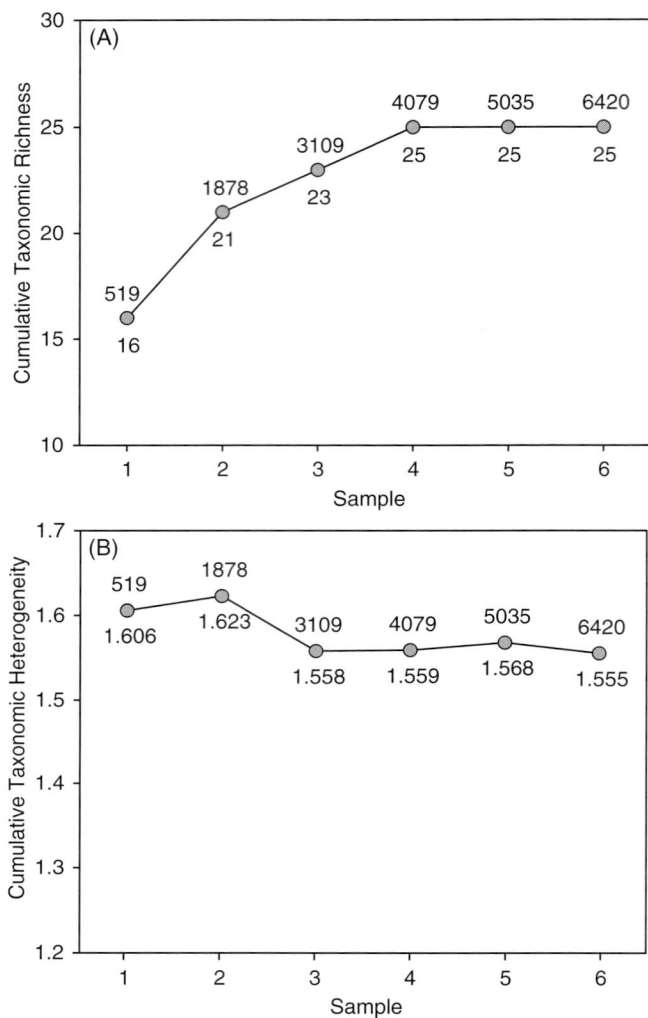

8.5. Sampling-to-redundancy curves showing (A) the cumulative taxonomic richness and (B) cumulative taxonomic heterogeneity across six annual field seasons; number above each point is cumulative NISP, number below each point is cumulative richness/heterogeneity. After Lyman and Ames (2004).

varying sample sizes (as in Figure 8.2) to determine whether the target variable has stabilized.

A point that should be obvious (but made here in case it is not) is that the sampling-to-redundancy approach concerns whether a single assemblage provides a representative measure of taxonomic richness (or other target variable, see below). If the analyst wanted to compare NTAXA as measured in several independent assemblages, a sampling-to-redundancy sort of analysis would have to be performed on each to determine which assemblages were adequate and which were not. The latter could then be deleted from further analysis. Alternatively, assemblages might be summed if they were adjacent in time and space, a technique used by some paleozoologists (see select chapters in Bobe et al. 2007). Such analytical aggregates of assemblages may eliminate sample size influences, but may simultaneously result in a loss of resolution from time and space averaging (see Chapter 3). Fortunately, there are other analytical techniques that contend with variability in the size of fossil samples.

Rarefaction

Rarefaction, which is commonly used by paleozoologists (e.g., Blois et al. 2010; Faith et al. 2017; Lyman 2014b) and ecologists (e.g., Colwell et al. 2012; Gotelli and Colwell 2001; Gotelli and Graves 1996), provides an alternative means of comparing richness across assemblages of different sizes. Developed by Sanders (1968) and revised by others (Hurlbert 1971; Tipper 1979), the aim of rarefaction is to estimate the expected NTAXA that would have been recovered had fewer specimens been collected. In doing so it provides a means of comparing NTAXA across assemblages of different sizes by rarefying larger samples down to a sample size equivalent to the smallest assemblage under consideration. There are different algorithms that can be used to conduct rarefaction analysis – the classic approach is provided by Hurlbert (1971) – and it is also possible to estimate NTAXA for any given sample size through randomized sub-sampling (Blois et al. 2010), as we demonstrated in Figure 8.2.

Here we provide a rarefaction analysis for ungulates aggregated at the genus level for Boomplaas Cave (Table 8.2). *Connochaetes* and *Alcelaphus* are combined because relatively few specimens could be securely assigned to either genus (and for purposes of illustration here). The assemblages are rarefied to a standardized sample size of forty-three specimens, corresponding to the size of the two smallest assemblages from members YOL and LOH. The importance of controlling for sample size is readily apparent from comparison of raw NTAXA to rarefied NTAXA (Table 8.2). For example, the Last Glacial Maximum assemblage from member LP emerges as the most taxonomically rich assemblage when its relatively small sample size is taken into account with rarefaction. In contrast, the assemblage from member CL is one of the richest in the sequence, but it is also the largest. Rarefaction analysis shows its richness to be intermediate relative to other assemblages.

When the expected NTAXA obtained from rarefaction analysis is plotted across all possible sample sizes, the outcome is a *rarefaction curve*. Figure 8.6 provides rarefaction curves for ungulates from Boomplaas Cave members BLD and BRL. For this taxonomic group, BRL has more specimens than BLD (BRL: 463; BLD: 233) as well as more ungulate genera (BRL: 13; BLD: 8). However, because there are almost twice as many identified specimens in BRL compared with BLD, it is unclear whether the presence of five more genera in BRL is related to its larger sample size or because it is sampling a taxonomically richer ungulate fauna. The rarefaction curves indicate that sample size cannot account for the difference in richness (Figure 8.6). At any given sample size except for an NISP of one (where both assemblages provide only one

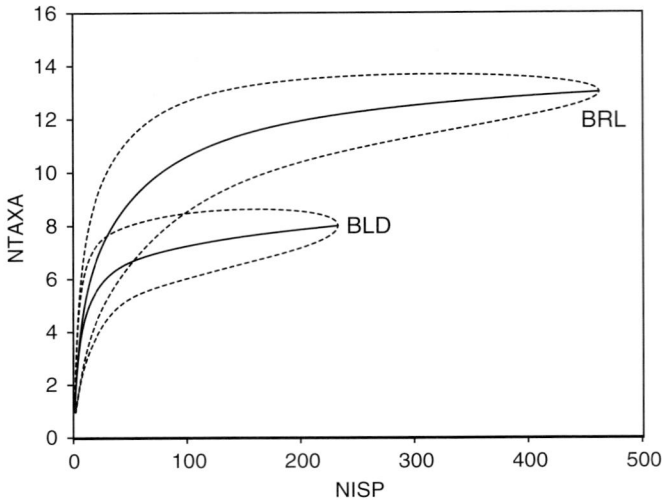

8.6. Rarefaction curves for Boomplaas Cave ungulates from members BLD and BRL. Dashed lines represent 95% confidence limits on the estimate of NTAXA for any given NISP.

taxon), BRL is predicted to provide more taxa than BLD. At a sample size of 233 specimens, which corresponds to the sample size of BLD, rarefaction analysis predicts the recovery of 12.2 (95% confidence limits: 10.7 to 13.6) genera for BRL, significantly larger than the 8 genera observed in BLD.

One might visually compare rarefaction curves derived for assemblages of disparate sizes. Rarefaction curves for each of four strata-specific samples from the Marmes site in eastern Washington State in the northwestern United States are superimposed in Figure 8.7 (after Lyman 2014b). The three assemblages from the A6–16 stratum, Harrison stratum, and U-I stratum all date to the latest Pleistocene and earliest Holocene; the Marmes stratum dates to the early Holocene. The 95% confidence intervals of the first three overlap with one another; the Marmes stratum assemblage hardly overlaps at all with the first three, indicating that for any given total NISP, it produces considerably fewer taxa than those other, more ancient three assemblages regardless of sample size.

Regression Analysis

Regression analysis provides a third approach to comparing NTAXA across assemblages of varying sample sizes (e.g., Cannon 2004; Faith 2008, 2011c, 2013b; Grayson 1984b, 1998; Grayson and Delpech 1998; Lyman 2012a; Marean et al. 1994). There are two ways in which it has been used to explore variation in richness across assemblages. The first involves residual analysis (Cruz-Uribe 1988; Faith 2013b; Grayson 1984b; Marean et al. 1994). In this approach, the relationship between sample size and NTAXA across multiple assemblages is first modeled using (usually) linear regression. Because the relationship between NISP and NTAXA tends to be curvilinear, it is first linearized through log-transformation of NISP (as in Figure 8.3) or of both NISP and NTAXA; the

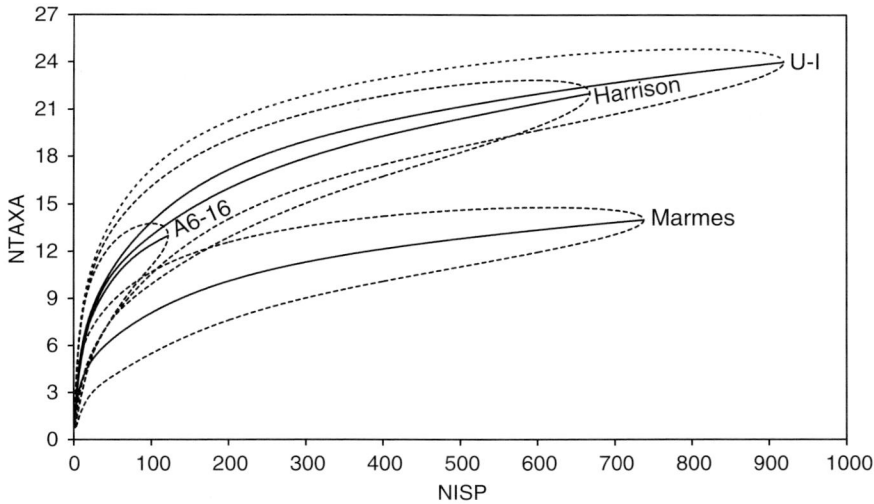

8.7. Rarefaction curves for four assemblages of mammal remains from the Marmes site, south-eastern Washington State (USA). The solid line is the mean value and dotted line the 95% confidence interval based on 1,000 iterations. Note the similarity of the curves for the A6–16, Harrison, and U-I strata, all dating to the Pleistocene–Holocene transition, and how they differ from the curve for the Marmes stratum that dates to the early Holocene. The implication is that more species were deposited per NISP during the Pleistocene–Holocene transition than during the early Holocene. After Lyman (2014b).

proper relationship (semi-log or log-log) between NISP and NTAXA can be determined via residual analysis, by ensuring that there are no systematic trends in the residuals as a function of sample size (Grayson 1984b). Once the proper regression is determined, the residuals – the difference between the observed NTAXA and that predicted by the regression – can be calculated and used as a proxy for richness independent of sample size. Turning again to Boomplaas Cave, those assemblages falling above the regression line include more species than expected for their sample sizes, while those falling below the line include fewer species (Figure 8.3). The residuals for all Boomplaas Cave assemblages (Table 8.2) are correlated with rarefied NTAXA ($r = 0.918$, $p < 0.001$), indicating that both approaches to removing the effects of sample size on NTAXA provide comparable results in this context.

The second means by which regression analysis has been used to compare NTAXA between assemblages is by fitting two or more regression lines to the data, and determining whether there are differences between those regressions (Cannon 2004; Faith 2008, 2011c; Grayson 1998; Grayson and Delpech 1998; Lyman 2012a). The classic example is provided by Grayson (1998), who observed that there are two distinct relationships between sample size and NTAXA for the Homestead Cave small mammals. This is illustrated in Figure 8.8A, which shows the relationship between NISP (log-transformed) and NTAXA, with NTAXA tallied at the genus level. For the three basal (terminal Pleistocene

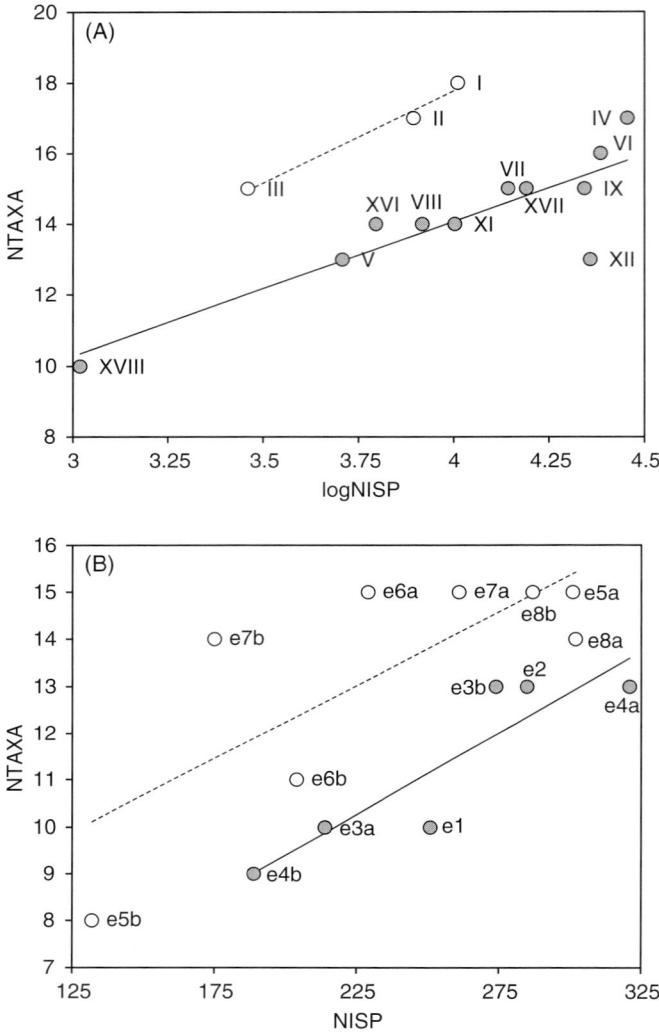

8.8. (A) The two different relationships between sample size (logNISP) and NTAXA (aggregated at genus level) at Homestead Cave. Late Pleistocene– early Holocene assemblages are represented by open circles and dashed (least-squares regression) line; middle and late Holocene assemblages are represented by filled circles and solid (least-squares) line. (B) The relationship between NISP and NTAXA per stratum (alpha–numeric label) at Samwell Cave, California. Late Pleistocene–early Holocene assemblages are represented by open circles and dashed (least-squares regression) line; middle and late Holocene assemblages are represented by filled circles and solid (least-squares regression) line. Data from Blois et al. (2010).

and earliest Holocene) assemblages (Strata I–III), the relationship is strong, although not statistically significant ($r = 0.956$, $p = 0.184$), but for the middle to upper levels (Stratum IV and above), all dating to the Holocene, there is a highly significant relationship ($r = 0.877$, $p < 0.001$). It is clear that for any given sample size, those assemblages from the base of the Homestead Cave sequence provide more small mammal genera for any given sample size than those that accumulated afterwards.

The regression analysis just described for the Homestead Cave faunas is hardly unique. Lyman (2012a) found the same relationship between sample size and richness for the same two temporal periods (late Pleistocene and Holocene) among the mammal remains recovered from Samwell Cave, California (Blois et al. 2010) (Figure 8.8B). In this case as well, comparison of the intercepts of the least-squares linear regression lines shows that more taxa had been input to the paleozoological record per NISP during one time than during the other, revealing details within samples of similar size that might have otherwise gone undetected. Those details have paleoenvironmental implications concerning the significance of greater taxonomic richness, a topic we come to later in this chapter.

If there were more assemblages constituting the high richness relationship in Strata I–III at Homestead Cave (Figure 8.8), it would be possible to statistically compare regression coefficients to determine whether the rate at which taxa accumulate as a function of sample size differs significantly between the lower and middle to upper levels (e.g., Cannon 2004; Faith 2008, 2011c), keeping in mind the importance of the assumptions underlying regression analysis (Weaver et al. 2011). By this we have in mind the fact that one of the drawbacks to regression-based approaches is that not all assemblages contribute equally to the regression. Leverage points are those that influence the slope of the regression unduly (Sokal and Rohlf 1995). In the case of the Boomplaas Cave ungulates (Figure 8.3), the assemblage from member CL has the highest leverage (0.44) due to its large sample size. Were it to have yielded a few more or less ungulate taxa, the regression slope could change considerably, translating to a change in residuals across all assemblages and potentially altering the outcome of a comparison of two or more regressions. In this particular case, member CL also has a relatively small residual (Table 8.2), implying that it reasonably fits the relationship between logNISP and NTAXA documented across all assemblages. If it were an outlier, however, it might be reasonable to recalculate the regression without member CL (see Sokal and Rohlf 1995:531–532). While the issue of leverage points is not a major problem for the Boomplaas Cave ungulates, it may be undesirable, particularly in the case of residual analysis, to have a proxy for richness in which the observed value for any given assemblage is influenced by the other assemblages included in the analysis.

Taxonomic Heterogeneity

Measures of taxonomic heterogeneity incorporate both richness and evenness into a single diversity metric. The most commonly used heterogeneity measure in ecology and paleozoology is the Shannon–Wiener index, also called the Shannon index or sometimes incorrectly the Shannon–Weaver index (Spellerberg and Fedor 2003). It is calculated as:

$$H = -\sum p_i \ln p_i$$

where p_i is the proportional abundance of taxon i. Larger values of H indicate greater heterogeneity, with values in living communities (tallied at the species

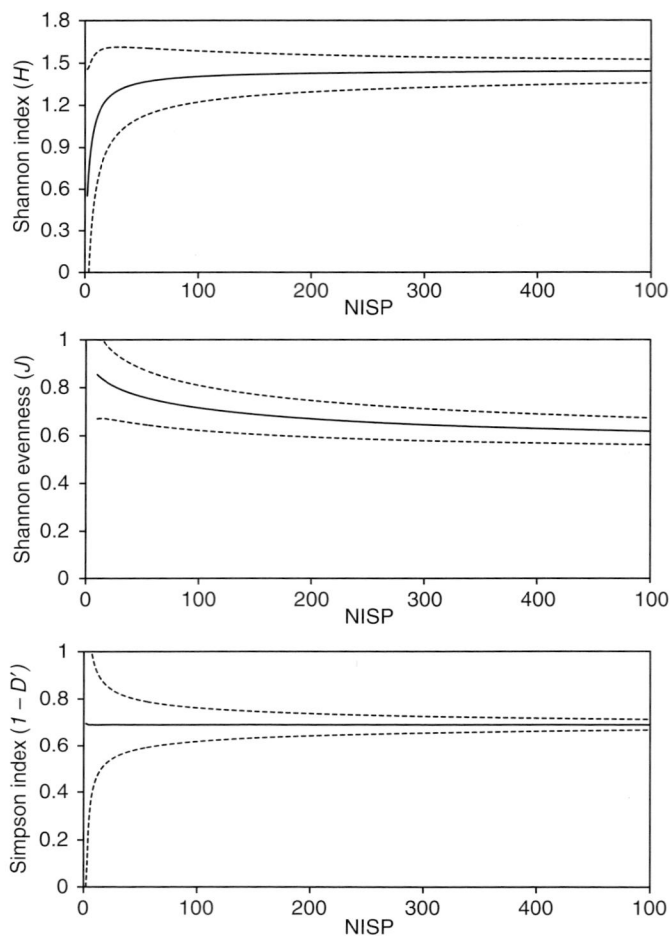

8.9. The relationship between sample size (NISP) and metrics of taxonomic heterogeneity (Shannon index) and evenness (Shannon evenness index and Simpson index) for random sub-samples of the small mammal assemblage from Homestead Cave Stratum IV (5,000 sub-samples for each NISP value from 1 to 500), with taxa aggregated at the genus level. Dashed lines indicate 95% confidence limits.

level) typically ranging from 1.5 and 3.5. For a given number of taxa, this index reaches its peak value when all taxa are equally abundant; it will also increase with greater richness.

It should come as no surprise that heterogeneity, a metric determined in part by taxonomic richness, can sometimes be influenced by sample size (Lande 1996; Lyman 2008b; Magurran 1988, 2004). If data allow, one could follow Simpson (1937) and determine whether the total collection has sampled a deposit to redundancy. Using the same set of samples from the Portland basin in northwestern Oregon as shown in Figure 8.5B, the Shannon index fluctuates considerably over the first couple of samples but then levels off after the addition of the third sample. Such an outcome would suggest the total collection provides a measure of heterogeneity not greatly influenced by sample size. Using the random sub-samples of the small mammal assemblage from Homestead Cave Stratum IV, Figure 8.9 illustrates the influence of sample size on the Shannon index. When sample size is low, the Shannon

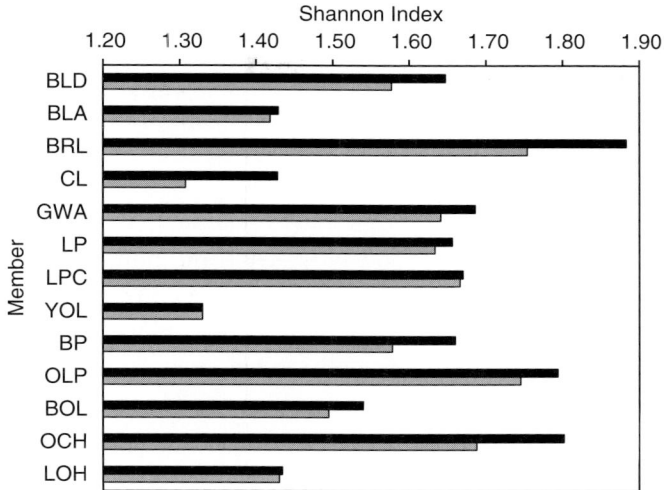

8.10. Comparison of the Shannon index for Boomplaas Cave ungulates (black bars) with values observed when assemblages are standardized to an equivalent sample size (grey bars). Taxa aggregated at the genus level.

index is considerably off the mark (i.e., it is biased), but it becomes more accurate (at a decreasing rate) as sample size increases.

Returning to the Boomplaas Cave ungulates, we find that the Shannon index is not correlated with assemblage sample size (Shannon index: $r = 0.295$, $p = 0.328$). While this means that we can reasonably compare index values across assemblages without worrying too much about the effects of sampling, it is informative to examine how values change when we standardize assemblages to equivalent sample sizes. Figure 8.10 compares the Shannon index calculated for each stratum (unstandardized) to the mean values obtained for the 5,000 randomly generated sub-samples of forty-three specimens each (standardized), equivalent to the size of the smallest assemblages (YOL and LOH). While the general patterns are quite similar, the standardized values for the larger assemblages (e.g., BRL, CL, and OCH) are considerably lower than the unstandardized values, with the ranking of some assemblages changing. The change in ranks implies that the unstandardized data are still influenced to some extent by sample size, which is certainly undesirable if the goal is to compare values across assemblages. It follows that it may be sensible to standardize assemblages to an equivalent sample size (as in Blois et al. 2010), even if there is no statistically significant correlation with sample size. In those cases where a significant correlation with sample size is observed, another alternative is to implement regression or residual analysis as described above for NTAXA (e.g., Cruz-Uribe 1988; Marean et al. 1994).

Taxonomic Evenness

While heterogeneity indices are designed to incorporate aspects of both richness and evenness into a single measure, this can make interpretation difficult. Is an increase in the Shannon index indicating a change in richness, a change in evenness, or both? A metric that captures evenness alone is therefore a useful index for describing the diversity of a faunal assemblage. One of the more widely used evenness metrics is the Shannon evenness index, often referred to by ecologists as Pielou's *J*, calculated as:

$$J = H / \ln NTAXA$$

where *H* is the Shannon index and NTAXA is the number of taxa (Pielou 1966). This is equivalent to the Shannon index divided by the maximum possible value of *H* for any value of NTAXA, which is observed when all taxa are equally abundant. The Shannon evenness index ranges from 0, in the case of an infinitely uneven assemblage dominated by a single taxon, to 1 in the case of an assemblage where all taxonomic abundances are equal.

A second evenness index commonly used by zooarchaeologists is the Simpson index (*D*), calculated as:

$$D = \sum p_i^2$$

where p_i is the proportional abundance of taxon *i*. The Simpson index represents the probability that two randomly sampled individuals (specimens) will belong to the same taxon. For finite samples, which applies to *all* zooarchaeological assemblages, an unbiased estimate of the Simpson index is provided by:

$$D' = \sum \left(\frac{n_i (n_i - 1)}{N (N - 1)} \right)$$

where n_i is the abundance of taxon *i* and *N* is the total number of individuals (specimens) in the sample. The Simpson index ranges from a minimum of $1/$ NTAXA, which is observed when all taxa are equally abundant, to ≈ 1 in the case of an assemblage dominated by a single taxon. Because *D* decreases as evenness increases (the probability that two individuals are of the same taxon declines), it is usually expressed as $1 - D$ or as $1/D$ so that larger values are associated with greater evenness (e.g., Berto et al. 2017). Note that $1 - D'$ (bias-adjusted) is equivalent to Hurlbert's (1971) probability of interspecific encounter (PIE), the calculation of which is usually given by:

$$PIE = \frac{N}{N-1} \left(1 - \sum p_i^2 \right)$$

The term PIE is used in ecological (Gotelli and Graves 1996) and paleo-ecological studies (Alroy et al. 2008), and although the formula is different, it is identical to $1 - D'$.

Most evenness indices, except for the bias-adjusted Simpson index, are highly sensitive to sample size. This is because larger samples lead to enhanced recovery of rare taxa (Grayson 1984b; Lyman 2008b), which in turn drives evenness down (Faith and Du 2018; Smith and Wilson 1996). This is apparent in Figure 8.9, which illustrates the relationship between NISP and both the Shannon evenness index (J) and Simpson index ($1 - D'$) or PIE for the random sub-samples of Homestead Cave Stratum IV. Unlike the Shannon evenness index, which declines as sample size increases, the bias-adjusted Simpson index is invariant to changing sample size. This means that it provides an unbiased estimate of the evenness of the population from which the sub-samples are drawn, even when few taxa are recovered. Its invariance to sampling artifacts, coupled with the fact that the index values represent probabilities (i.e., the values have an intuitive meaning), makes it a preferred metric for the measurement of evenness.

Which Metric Should I Choose?

From our discussion above, it should be abundantly clear there are various aspects of diversity that can be measured (richness, evenness, both), and there are different indices that can be used to measure them. Which one should be used? Of the indices discussed here, we recommend using NTAXA for richness and the bias-adjusted Simpson index for evenness. The extent to which heterogeneity indices provide useful information has been questioned (Magurran 2004), but if one is to be used, the Shannon index is the standard in paleozoological analysis. While selecting a robust metric is critical, we reiterate Magurran's (2004:101) advice to resist the temptation to calculate a suite of diversity indices, especially because many software packages will do this automatically (e.g., Hammer et al. 2001), and select the one that provides the most appealing outcome. "It is important to know in advance which aspect of biodiversity is being investigated – and why" (Magurran 2004:101). With respect to our objectives in this chapter, the answer to the "why" question is a simple one. We want a measure that is telling us something about the paleoenvironment, and how it changes through time or space. Before we can get to this issue, however, there are several additional matters that warrant attention when measuring diversity in the fossil record.

PRACTICAL CONSIDERATIONS WITH FOSSIL DATA

Faunal analysts from both archaeological and paleontological backgrounds tend to be acutely aware of the effects of sample size on diversity metrics and regularly take steps to identify and, if necessary, circumvent them. However, there are additional practical issues related to the measurement of diversity in the fossil record that, while certainly recognized in the paleozoological literature (e.g., Behrensmeyer et al. 2000; Blois et al. 2010; Faith et al. 2017; Grayson and Delpech 1998; Grayson et al. 2001; Tipper 1979), deserve greater consideration when comparing diversity across fossil assemblages. We highlight these issues here.

Taxonomic Scope

Both ecologists (Hurlbert 1971) and paleontologists (Sanders 1968; Tipper 1979) have long emphasized the importance of comparing diversity across taxonomically similar organisms. There are ecological and taphonomic reasons for doing so. The prime ecological reason is that "real and interesting differences between different trophic or taxonomic groups could cancel each other out and thus yield parameter values of little interest" (Hurlbert 1971:584). The fossil record substantiates Hurlbert's (1971) concern. Cruz-Uribe (1988), for example, notes that late Quaternary micromammal and macromammals assemblages in southern Africa show opposing trends in taxonomic diversity, a pattern that would be masked by analysis of both groups combined. Taken to the extreme, one cannot reasonably expect the diversity of distantly related taxonomic groups, such as fish, birds, mammals, and reptiles, to respond similarly to an environmental gradient. Building on Hurlbert's (1971) recommendations for ecologists, paleoecologically meaningful measures of diversity are best provided for groups of species that are similar in size, with comparable life histories, and that compete with each other for comparable resources over ecological and evolutionary timescales. From a practical perspective, we note that many paleozoological studies tend to be organized around such groupings from the outset, which is why a single site might generate separate faunal studies for the micromammals, macromammals, fish, birds, and so on.

The taphonomic reason for focusing on similar organisms is that different taxa can provide different relationships between sample size and diversity (e.g., Lyman 2015b). This arises in part from differential identifiability. Some taxa provide more identifiable skeletal parts than others, either because they have a greater number of bones or because more of the bones are diagnostic. It may also be the case that some taxa include more elements that are likely to be preserved in an identifiable state (Lyman 2015b) or that are more likely to be recovered (Gordon 1993; Nagaoka 2005). At Boomplaas Cave, for example, most sufficiently well-preserved skeletal parts of the smaller mammals, including riverine rabbit (*Bunolagus monticularis*), Cape hare (*Lepus capensis*), and rock hyrax (*Procavia capensis*), can be identified to species. In contrast, with respect to the Bovidae, for which there are numerous species present, species-level identifications are obtained primarily from teeth and (rare) horn cores because the postcrania (with few exceptions) are morphologically very similar between species of similar body size and therefore often cannot be distinguished taxonomically. When specimen counts (NISP) are used to quantify taxonomic abundances, the analysis of diversity across assemblages of species with very different numbers of identifiable elements can influence the outcome. This can be overcome by restricting analysis to those taxa that are similarly identifiable, using MNI counts, or by considering only those skeletal elements that can be identified to the same taxonomic level across all taxa (e.g., teeth only).

Recovery and Analytical Methods

Diversity can only be meaningfully compared across assemblages that have been collected and analyzed using similar methodologies. It is well documented that the taxa recovered and their relative abundances, both of which directly influence measures of diversity, vary depending on whether the bones are surface-collected or excavated, and on the types of screens used during excavation (Cannon 1999; Gordon 1993; Kowalewski and Hoffmeister 2003; Lyman 2012c; Nagaoka 2005; Shaffer and Sanchez 1994; Wolff 1975; Zohar and Belmaker 2005). This means that comparisons of diversity between assemblages collected using varied methodologies can reveal apparent differences in diversity that have little to do with paleoenvironmental differences.

By the same token, it is equally important to ensure that analytical procedures used to identify taxa are comparable between assemblages, as different analysts may use different protocols for taxonomic identification (e.g., Driver 1992, 2011; Gobalet 2001; Lyman 2002). An example is provided by the history of zooarchaeological analysis of fish remains from Pacific island archaeological sites. Several decades ago, faunal analysts used only a small subset of elements for taxonomic identification (Leach 1986), including five paired cranial elements (dentary, premaxilla, maxilla, articular, and quadrate) and the "specials" (e.g., scutes, pharyngeal grinding plates) that are not universally present across all taxa but can be diagnostic for some taxa. More recent research, fueled by development of larger comparative collections and concern for taphonomic issues, has expanded the types of bones that are typically identified to include additional cranial elements and vertebrae (Butler 1988; Lambrides and Weisler 2015; Weisler 1993). Not surprisingly, the number of kinds of elements that are identified has a considerable influence on diversity. Table 8.3 reports taxonomic abundances and diversity indices from Lambrides and Weisler's (2015) analysis of late Holocene fish remains from Henderson Island in southeast Polynesia (Test Pit 12, Layer 1B) for the five cranial elements and specials relative to an expanded sample of cranial elements and vertebrae. Not only are the taxonomic abundances significantly different ($\chi^2 = 242.551$, p < 0.001), but the diversity indices differ substantially, with the expanded sample associated with greater richness but lower evenness and heterogeneity. These differences have nothing to do with changing diversity of the sampled communities; they are solely the result of different analytical procedures.

TABLE 8.3 *Taxonomic list and diversity indices for Henderson Island fish remains according to five paired cranial elements (+ specials) and an expanded number of cranial and vertebral elements (from Lambrides and Weisler 2015).*

Taxon	Five paired cranial bones	Expanded
Acanthuridae	145	855
Carangidae	89	175
Holocentridae	1	32
Labridae	8	14
Mullidae	11	41
Scaridae	0	19
Scrombidae	33	11
Serranidae	59	739
NISP	346	1,886
NTAXA (NISP = 346)	7	7.76 (6.85 − 8.67)
H (NISP = 346)	1.45	1.20 (1.10 − 1.30)
$1 - D'$	0.72	0.63 (0.60 to 0.66)

Notes: Diversity indices for the larger sample are standardized to a sample size of 346 specimens (mean of 5,000 random sub-samples). Values in parentheses indicate 95% confidence interval of sub-sampled assemblages.

Agent of Accumulation

When comparing diversity across assemblages through time or space, it is prudent to worry about the extent to which the agent of bone accumulation is responsible for changing taxonomic diversity (e.g., Andrews 1996). For instance, one cannot reasonably expect the taxonomic diversity of mammal remains accumulated in a cave by leopards (*Panthera pardus*), which preferentially prey upon species in the 10–40 kg range (Hayward et al. 2006), to be comparable to an attritional death assemblage deposited in a fluvial system, even if both assemblages sampled species from the same habitat. A clear example of the problem is provided by the modern small mammal assemblages accumulated by three species of owl in central Oregon (Maser et al. 1970), which we introduced in Chapter 6. Despite sampling similar habitats, measures of taxonomic diversity are different between owl species (Table 8.4), with the long-eared owl (*Asio otis*) assemblage characterized by lower richness than that produced by the great horned owl (*Bubo virginianus*), and lower heterogeneity and evenness than both the great horned owl and short-eared owl (*Asio flammeus*). An uncritical comparison of diversity between these assemblages might give the impression of environmentally important changes in diversity, when all that has happened is a change in the accumulator of the faunal remains. As discussed in Chapter 3, it is sensible to only compare assemblages that are isotaphonomic, or at least to ensure that changes in diversity are not related to changing taphonomic circumstances.

TABLE 8.4 *Diversity indices for small mammal assemblages accumulated by three owl species in Oregon.*

Index	Asio otus	Asio flammeus	Bubo virginianus
NTAXA	7	6.98 (6.68 – 7.27)	8.80 (7.30 – 10.46)
H	1.18	1.68 (1.55 – 1.81)	1.80 (1.64 – 1.95)
$1 - D'$	0.57	0.79 (0.75 – 0.83)	0.81 (0.77 – 0.85)

Notes: Diversity indices for the larger samples are standardized to a sample size of 102 specimens (mean of 5,000 random sub-samples). Values in parentheses indicate 95% confidence interval of sub-sampled assemblages.

Differential Fragmentation

Mechanical (taphonomic) processes, especially differential fragmentation, have the potential to alter taxonomic diversity (Grayson and Delpech 1998; Grayson et al. 2001). If two assemblages sample the same species pool, but are subjected to differential fragmentation, then the more fragmented assemblage will provide more specimens but the same number of species (or even fewer species if fragments are so small as to not retain sufficient anatomical detail to allow identification). When abundances are quantified by NISP, this translates to lower apparent richness when sample size is taken into account. Likewise, intensified fragmentation of a single taxon can drive evenness down (if it is an abundant taxon) or up (if it is a rare taxon) by increasing the number of specimens attributed to that taxon. These hypotheticals assume that specimens remain identifiable as fragmentation increases. Reality is probably more complex, however, with increased fragmentation initially leading to an increase in the number of identified specimens, followed by an eventual decline as specimens become fragmented to the point where they are no longer identifiable (Cannon 2013; Marshall and Pilgram 1993). The use of MNI counts to quantify taxonomic abundances might appear to circumvent the fragmentation problem, but these too can become depressed at high levels of fragmentation as fewer specimens are identified (Marshall and Pilgram 1993). We are aware of few studies that attempt to control for fragmentation, but the sensible protocol is to assess the correlation between diversity and bone fragmentation (Grayson and Delpech 1998; Grayson et al. 2001), with potential indicators of fragmentation including epiphysis to shaft ratios, average fragment size, or NISP to MNE (minimum number of elements) ratios. If no significant relationship is found, the conclusion is that fragmentation is not responsible for changing diversity between assemblages. But if such a relationship exists, a reasonable workaround is residual analysis, whereby the variation in diversity that cannot be explained by fragmentation is calculated and analyzed in place of the raw diversity values.

Differential Skeletal Element Representation

Any taphonomic process that causes the (taxonomically identifiable) skeletal elements of a taxon to be differentially represented can influence taxonomic diversity. This is because differential skeletal element representation alters the distribution of identified specimens across taxa, which in turn alters diversity. Although few have attempted to control for this issue when exploring diversity metrics (Grayson and Delpech 1998; Grayson et al. 2001), there are numerous taphonomic processes well known to influence skeletal element abundances, such as differential bone transport, density-mediated attrition (e.g., carnivore ravaging), and fluvial winnowing (Lyman 1994). Fortunately, this issue is relatively easy to address. Skeletal part data are analyzed to determine whether any particular elements differ significantly across assemblages, and those that do are removed from the taxonomic counts used to calculate diversity (Grayson and Delpech 1998; Grayson et al. 2001).

Time Averaging

In Chapter 3, we noted that time averaging has the potential to obscure environmental gradients or to mix faunas from different environment and species pools. If a fossil assemblage is time averaged across a temporal environmental gradient, especially one that contributes to turnover in the faunal communities, taxonomic diversity will be affected. Richness is likely to increase because multiple species pools are sampled, although the effects on evenness or heterogeneity are less predictable and will likely vary on a case by case basis. It is therefore sensible to ensure that changing diversity, especially richness, between fossil assemblages is not the result of differential time averaging.

PALEOENVIRONMENTAL IMPLICATIONS OF TAXONOMIC DIVERSITY

Just as there are a diversity of diversity indices, there are also numerous ways in which taxonomic diversity can be (and has been) related to the paleoenvironment. Paleozoologists have interpreted taxonomic diversity in terms of precipitation (Faith 2013b; Faith et al. 2017; Grayson 1998; Schmitt and Lupo 2012), temperature (Barnosky 2001; Grayson 2000b; Grayson et al. 2001), primary productivity (Faith 2011c; Janis et al. 2000, 2004), habitat type (Cruz-Uribe 1988), habitat heterogeneity (Faith 2013b; Marean et al. 1994), tectonics (Barnosky and Carrasco 2002), elevation (Grayson 1991a), and environmental stress in relation to extremes in temperature, seasonality, or precipitation (Andrews 1996; Avery 1982, 1992). This array of proposed explanatory variables, many of which are related to each other (e.g., temperature, precipitation, and productivity), arises from the fact that taxonomic diversity is not driven by any single variable. Instead, it is influenced by a complex combination of biotic factors, including species interactions (e.g., Shurin and Allen 2001), and abiotic factors, including those mentioned above, as well as geographic area and isolation, topography, and soil fertility, to name a few others (e.g., Lomolino et al. 2010; Olff et al. 2002; Rosenzweig 1995). Added to this is the fact that diversity is also sensitive to factors operating over geological timescales (i.e., historical factors), including past climate change and long-term patterns of extinction, speciation, and dispersal (Araújo et al. 2007; Dodd and Stanton 1990; Hawkins and Porter 2003; Ricklefs and Schluter 1993; Wiens and Donoghue 2004).

Despite this complexity, paleozoologists have long focused on broader relationships between taxonomic diversity and the environment. Avery (1982:307) notes that "animal life in tropical climates tends to be more abundant and varied than that in temperate climates" and "that mammalian communities in harsh climates contain fewer species than those in milder climates." Andrews (1996:271) suggests that "[i]n harsh conditions, such as extreme aridity, extreme cold or extreme seasonality, mammalian faunas are generally dominated by a small number of species and the overall species diversity [heterogeneity] is also low." Similarly, Grayson et al. (2001:121) note "harsh abiotic conditions often create landscapes that are dominated by a small number of species, presumably because relatively few species can thrive in such contexts." In their recent review, Reed et al. (2013:23) provide an explicit summary of such generalizations: "It has been argued that a high species-diversity index is equated with an environment that is mesic, warm, and structurally varied (i.e., having animals that utilize various substrates). A low faunal-diversity index would indicate cooler or more arid, open environments with limited structure." These statements are all supported by diversity gradients observed in contemporary ecosystems (e.g., Lomolino et al. 2010; Menge and Sutherland 1987; Odum and Barrett 2005; Rosenzweig 1995; Willig et al. 2003), but, as we highlight below, reality can be far more complex.

The Varied Response of Diversity to Environmental Gradients

Most ecologists would agree that primary productivity, the rate at which energy is converted to biomass for a given area and amount of time, has a strong influence on diversity (e.g., Chase and Leibold 2002; Ricklefs and Schluter 1993; Rosenzweig 1995; Waide et al. 1999). This is why so many variables that track primary productivity, including temperature, precipitation, and evapotranspiration, are known to correlate with diversity, especially richness. However, the causal mechanism underlying these relationships, and especially the form of these relationships, is not widely agreed upon. As pointed out in meta-analyses of the literature (Mittelbach et al. 2001; Waide et al. 1999), for both plants and animals the relationship between richness and productivity is varied. In

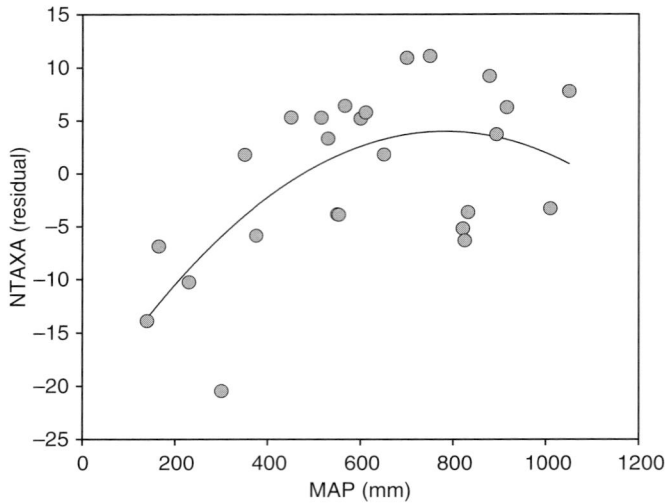

8.11. The relationship between mean annual precipitation (MAP) and species richness (NTAXA) for sub-Saharan African ungulates. Because NTAXA is also related to the size of the wildlife reserves examined here, it is presented here as the residual derived from the relationship between geographic area and richness. Data from Faith (2013b).

many cases, it is unimodal, with richness increasing from low to intermediate productivity and declining at higher productivity. The pattern is observed so frequently that some consider it to be the "true" relationship in nature (e.g., Rosenzweig 1995; Tilman and Pacala 1993). An example of the unimodal response is illustrated in Figure 8.11 for African ungulates. The increase phase of this relationship intuitively makes sense. Increased precipitation from xeric to mesic environments enhances primary productivity, which in turn translates to increased forage availability, thereby supporting a greater number of ungulate species on the landscape. There are various hypotheses to explain the decrease phase (see review in Rosenzweig 1995), which in this case may include a decline in plant nutrient content at higher precipitation, such that smaller-bodied species requiring high-quality forage cannot persist (Olff et al. 2002). Whatever the ecological explanation(s) might be, such a relationship poses interpretive challenges for paleoenvironmental reconstruction. To the extent that this par-ticular relationship between precipitation and richness can be applied to the fossil record, it implies that an increase in richness could arise from either an increase (from low to intermediate) or a decrease (from high to intermediate) in precipitation, and vice versa. This is not an insurmountable problem because we might expect other aspects of faunal composition, including the species that are present, to be very different in an arid versus a well-watered environment, in which case it should be possible to infer the direction of precipitation change. In other cases, for example, if a site is situated in an arid environment today and richness increases from recent to older strata, one could reasonably infer that this is tracking an increase in precipitation, as a shift from arid to even more arid conditions is not expected to drive an increase in richness.

8.12. The relationship between mean annual precipitation (MAP) and species richness (NTAXA) for Australian mammals (>500 g). Both axes are log-transformed because of increased heteroscedasticity at higher MAP. Data from Faith et al. (2017).

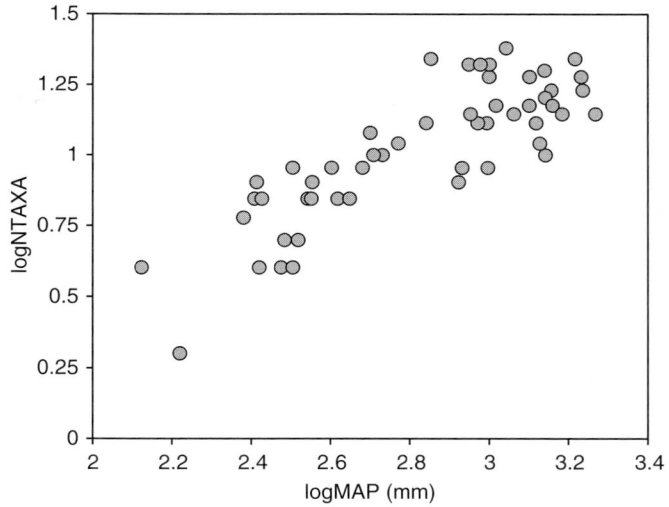

Although unimodal responses are commonly observed in nature, in other cases the relationships between richness and productivity are linear (positive or negative), u-shaped, or non-existent (Mittelbach et al. 2001; Waide et al. 1999). Differences of spatial scale might play a role in explaining some of this variation (Chase and Leibold 2002), but it cannot account for all of it (Mittelbach et al. 2001). Because there are so many factors that modulate the richness–productivity relationship, including interspecific competition, habitat structure and heterogeneity, predator–prey ratios, forage quality, and spatial scale, "it may be the cumulative or interactive effects of all such factors that determines the empirical pattern" (Waide et al. 1999:259). The empirical relationship between mammalian richness and precipitation across Australia (Figure 8.12), for example, is very different than the one documented for African ungulates (Figure 8.11). In contrast to the unimodal pattern in Africa, the relationship in Australia is positive and generally linear, although richness is more varied at higher precipitation (Faith et al. 2017). This discrepancy between the two continents may be related to the unique prevalence of extremely low-nutrient soils in Australia (Orians and Milewski 2007), such that the ecological consequences of high precipitation on forage quality are less important than for the high-nutrient soils found in parts of Africa (Olff et al. 2002). Regardless of the explanation, the varied responses of diversity across environmental gradients pose obvious challenges for paleoenvironmental reconstruction.

The paleozoological record provides further evidence for the varied ways in which environmental changes translate to changes in diversity. Our first example is provided by two faunal sequences from very different regions: the Homestead Cave small mammals (Grayson 2000b) and the late Pleistocene (~65 to 14 ka) ungulates from Grotte XVI in southwest France (Grayson et al.

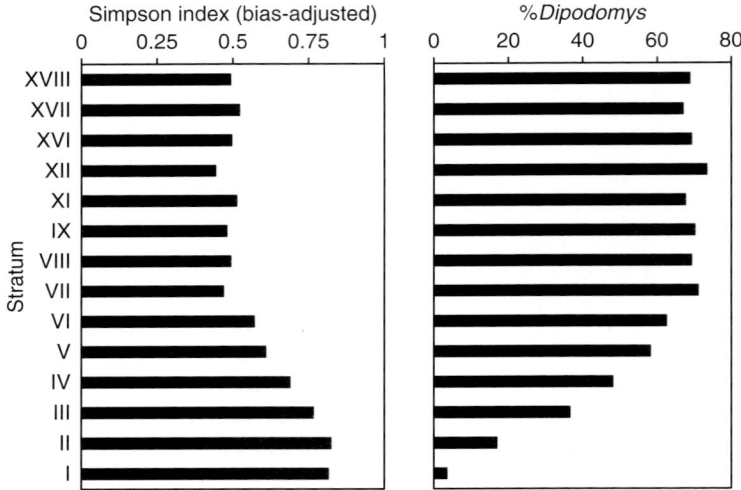

8.13. Assemblage evenness and the abundance of kangaroo rats (% of NISP attributed to *Dipodomys* spp.) through time at Homestead Cave. Assemblages arranged in stratigraphic order (oldest at the base). Taxa aggregated at the genus level for calculation of evenness. Data from Table 4.5.

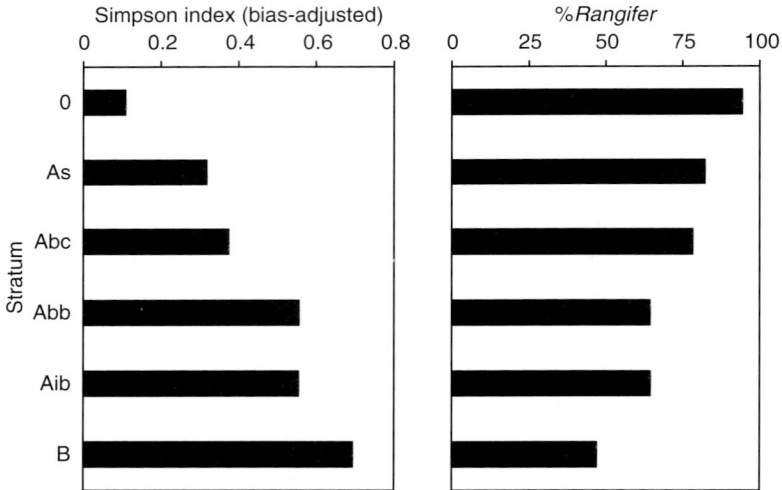

8.14. Assemblage evenness and the abundance of reindeer through time at Grotte XVI. Assemblages arranged in stratigraphic order (oldest at the base). Data from Grayson et al. (2001).

2001). Both of these assemblages demonstrate marked shifts in the evenness through time (Figures 8.13 and 8.14). At Homestead Cave, there is a decline in small mammal evenness from the base of the sequence to the middle Holocene. This decline is driven to a large extent by the increased dominance of arid-adapted kangaroo rats (*Dipodomys* spp.), and corresponds with other lines of paleoenvironmental evidence indicating warmer temperatures in the Great Basin. In contrast, at Grotte XVI, there is a steady decline in ungulate evenness from the base of the sequence (~65 ka) to the end of the Pleistocene (~14 ka), resulting from dramatic increases in the abundance of cold-adapted reindeer (*Rangifer tarandus*). Climate reconstructions from pollen cores elsewhere in France suggest that this trend is related to a decline in summer temperatures. In

8.15. Changes in evenness and richness (rarefied) in the micromammal assemblage from Boomplaas Cave. Assemblages arranged in stratigraphic order (oldest at the base). Members corresponding to the Last Glacial Maximum are indicated by grey shading. Taxa are aggregated at the species level. Data from Table 4.2.

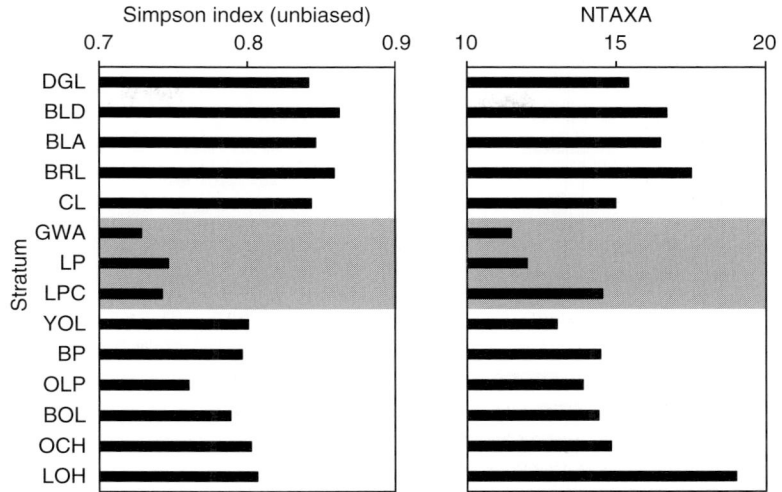

both cases, the interpretation of changing evenness is supported by other lines of paleoclimate evidence together with a paleoenvironmental assessment of those taxa whose dominance contributed to the decline in evenness. Thus, the interpretations are sensible. The important issue here, however, is that declines in evenness correspond to *both* increased and decreased temperatures. This is not at all inconsistent with the generalization that harsh abiotic landscapes tend to be dominated by a few species – very cold or very warm environments are indeed harsh – but it does illustrate the fact that there is no simple rule that can be used to infer paleoenvironment from changes in diversity.

Boomplaas Cave provides a second example of the complex relationships between diversity and paleoenvironmental change. In her examination of the Boomplaas Cave micromammals, Avery (1982) documented marked changes in taxonomic diversity through time. This is summarized here in Figure 8.15, which shows changes in NTAXA and evenness across the sequence. Both indices show a general decline from the base of the sequence to the Last Glacial Maximum (GWA to LPC), followed by high diversity during the Holocene. Based on the generalizations noted above, Avery interpreted the diversity indices as indicating a harsh Last Glacial Maximum and a milder Holocene (1982:315). She later found there was no relationship between climate and taxonomic diversity of modern micromammal assemblages derived from barn owl (*Tyto alba*) pellets in southern Africa (Avery 1999), rendering the climatic implications of diversity change in micromammals at Boomplaas Cave uncertain. What is crucial to our point here, however, is that a very different pattern is observed for the Boomplaas Cave ungulates (Figure 8.16). Evenness shows no apparent temporal trends, while richness shows a pattern that is the inverse of the one documented for micromammals, with the highest values observed during the Last Glacial Maximum and lowest during the

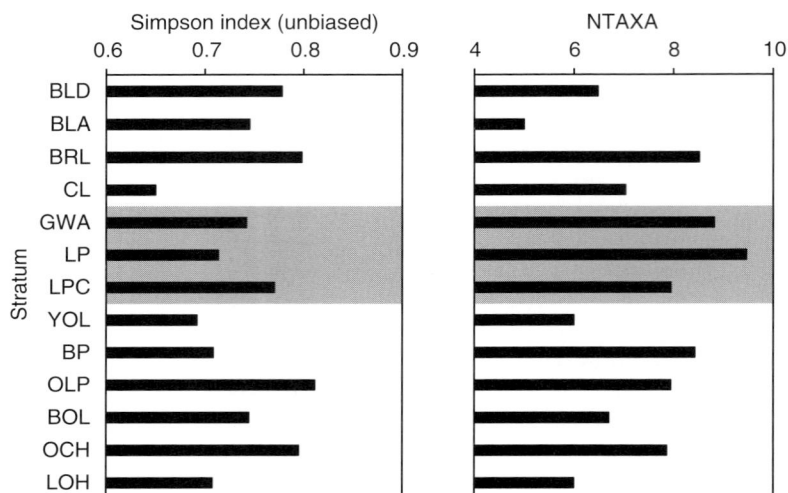

8.16. Changes in evenness and richness (rarefied) in the ungulate assemblage from Boomplaas Cave. Assemblages arranged in stratigraphic order (oldest at the base). Members corresponding to the Last Glacial Maximum are indicated by grey shading. Taxa are aggregated at the genus level. Data from Table 4.3.

Holocene. There is a weakly significant *inverse* correlation between rarefied NTAXA of micromammals and ungulates ($r = -0.581$, $p = 0.070$), implying that as richness in one taxonomic group increases, it declines in the other. This pattern cannot be brushed aside as a peculiarity specific to Boomplaas Cave, as it is also documented at other sites in the region (Cruz-Uribe 1988). Indeed, the differential response of larger versus smaller mammals to environmental gradients is also precisely why – as we discussed in Chapter 7 – cenogram analyses of modern communities indicate that different size classes track different environmental variables. The implication of this, together with our previous discussion, is that a change in diversity *on its own* tells us little about how the paleoenvironment changed in the absence of other evidence.

Paleoenvironmental Inferences

Given the many factors that influence taxonomic diversity, how can one possibly begin to infer the nature of paleoenvironmental change from changes in taxonomic diversity? For once we can provide a simple answer: with great caution. One might start with attempting to identify the relationship between environmental variables and the taxonomic diversity (richness, evenness, and heterogeneity) of faunas from contemporary ecosystems (e.g., Figures 8.11 and 8.12). This provides the essential uniformitarian link, allowing for patterns observed in the past to be more confidently related to those observed in the present. The link is obviously stronger when it is demonstrated in the same geographic region as the fossil sample (e.g., because Australia and Africa provide different patterns) and for the same taxonomic group that is the focus of analysis (e.g., because small and large mammals may respond differently). We recognize that this can pose a major setback, as such data from modern contexts may not be readily available in the zoological literature, especially with respect to abundance data necessary for determining evenness or heterogeneity (lists of taxonomic presences being more often available).

In the absence of robust links from contemporary ecosystems, one might instead attempt to identify those taxa responsible for changes in diversity, as in the case of Homestead Cave (Grayson 2000b) and Grotte XVI (Grayson et al. 2001), where changes in evenness are largely driven by changing abundances of kangaroo rats and reindeer, respectively. Knowledge of the ecology of these taxa can provide some indication as to why diversity is changing. But when this is done, from a logical perspective we lose the ability to treat diversity as an independent proxy for the environment. The environmental interpretation is instead derived from an assessment of taxonomic abundances, not from the change in diversity itself. This is perhaps why so many paleozoological studies examine taxonomic diversity to understand the *response* of faunal communities to, in some cases previously documented, paleoenvironmental change (e.g., Blois et al. 2010; Grayson 1998, 2000b; Grayson et al. 2001; Lyman 2014a; Schmitt and Lupo 2012), rather than to provide new information concerning the *nature* of paleoenvironmental change.

CASE STUDY

A case study in the analysis of diversity using the Boomplaas Cave ungulates will highlight many of the points we have made in this chapter. We focus on those changes documented since the Last Glacial Maximum, as the nature of precipitation change over this time in this part of southern Africa has been the subject of substantial discussion, with the Last Glacial Maximum variably interpreted as wetter or drier than the present (see Chapter 4). Given the observed relationship between ungulate richness and precipitation in sub-Saharan Africa today (Figure 8.11), an analysis of richness through time has potential to inform on how annual precipitation has changed (Faith 2013a, 2013b).

We begin with the observation that when sample size is taken into account by way of rarefaction (Figure 8.16) and residual analysis (Table 8.2), richness is highest in member LP (Last Glacial Maximum) and broadly declines thereafter (rarefaction: r_s = -0.733, p = 0.039; residuals: r_s = 0.867, p = 0.015). While this could be a reflection of precipitation change, there are numerous issues we must first take into account. From the outset, we can rule out variation in recovery and analytical methods, as those are equivalent for all levels in the site. However, there is reason to be concerned with the potential confounding effects of differential identifiability, fragmentation, skeletal element representation, time averaging, and the agent of bone accumulation.

Differential Identifiability

Differential identifiability is not likely to be an issue for the majority of ungulate taxa (Bovidae), because identifications are based primarily on teeth across all
taxa. One possible exception includes zebras (*Equus* spp.), as highly fragmented
zebra teeth can usually be identified as belonging to *Equus*, there being no
other known genus to which they could reasonably belong; however, similarly
fragmented bovid teeth are likely to be identified to family only (Bovidae).
Inclusion of fragmented *Equus* teeth in our counts could potentially drive
richness down in those assemblages where zebras are common by inflating the
number of specimens relative to the number of species. This possibility can be
ruled out in this case, however, as zebras tend to be more abundant in the taxonomically richer Last Glacial Maximum and Late-glacial levels compared with
the Holocene (Chapter 6). In addition, if we exclude *Equus* from the rarefaction analysis, the temporal trends in richness remain tightly correlated with the
original values calculated for all taxa (r_s = 0.857, p = 0.012). Thus, differential
identifiability is not responsible for the changes in richness observed here.

Differential Fragmentation

The majority of ungulate specimens identified to genus or species from
these levels of Boomplaas Cave are skull elements (89%), particularly those
that include teeth (mandibles, maxillae, or isolated teeth). While most highly
fragmented teeth are unlikely to be identified to genus or species (except
for *Equus* – dealt with above), it is possible that differential fragmentation of
mandibles and maxillae, which can result in the presence of multiple identifiable but isolated teeth, could depress apparent richness by increasing the
NISP while leaving NTAXA unchanged. At Boomplaas Cave, the number
of isolated teeth relative to mandibular or maxillary specimens with multiple
teeth differs significantly across stratigraphic levels (χ^2 = 86.617, p < 0.001).
While this is precisely the sort of taphonomic bias that *could* alter the relationship between sample size and richness, the relative abundance of isolated
teeth is not correlated with rarefied richness (r_s = 0.464, p = 0.302) or with
the residuals (r_s = 0.500, p = 0.267) (Figure 8.17). It follows that changes in
richness cannot be explained by changing abundances of isolated teeth.

Skeletal Element Representation

Ethnographic observations of hunter-gatherer carcass transport behavior demonstrate that for larger-bodied ungulates – size class 3 (84 to 296 kg body
mass) and larger using Brain's (1981) size divisions – skulls are unlikely to be
transported long distances (Bunn et al. 1988; O'Connell et al. 1988; Schoville

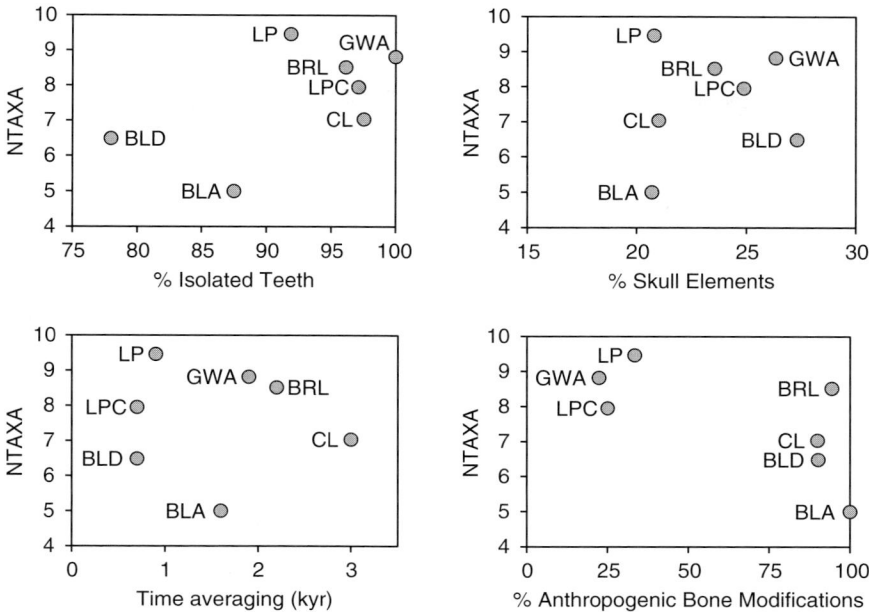

8.17. The relationship between richness (rarefied) and the abundance of isolated teeth, abundance of skull elements, time averaging, and anthropogenic bone modifications in the ungulate assemblage from Boomplaas Cave. Taxa aggregated at the genus level.

and Otárola-Castillo 2014). Because skull elements represent the majority of taxonomically identified specimens, differential transport of the skulls of larger taxa will alter the abundance of identified large-bodied ungulates (size class 3 and larger) relative to smaller-bodied ungulates (size classes 1 [0 to 23 kg] and 2 [23 to 84 kg]), which in turn can change the relationship between sampling effort and richness (Grayson and Delpech 1998). We provide an assessment of whether this could account for the changes in richness by examining the abundance of skull elements (crania, mandibles, and teeth) relative to high-survival postcranial elements, which include the long bones (femur, tibia, humerus, radius, metapodials) when mid-shafts are included in the analysis (Cleghorn and Marean 2004, 2007; Marean and Cleghorn 2003). Although there is evidence for differential representation of skull elements across assemblages (χ^2 = 12.708, p = 0.048), there is no relationship between their abundance and richness (rarefaction: r_s = 0.133, p = 0.776; residuals: r_s = -0.171, p = 0.713; Figure 8.17). We can infer the changes in richness are not the result of differential transport of skulls through time.

Time Averaging

If the decline in richness since the Last Glacial Maximum was the result of differential time averaging, we would expect the richer assemblages to be averaged over longer periods of time, as these are most likely to sample multiple species pools from different environments. The amount of time sampled by each of the stratigraphic units considered here is variable (from 0.7 to 3.0 kyr; see Table 4.1), but uncorrelated with richness (rarefaction: $r_s = 0.108$, $p = 0.822$; residuals: $r_s = 0.198$, $p = 0.667$; Figure 8.17). Differential time averaging does not account for the trends in richness.

Agent of Bone Accumulation

The taphonomic history of the Boomplaas Cave ungulates is characterized by a complex mixture of bone accumulation by people, carnivores (probably leopards), and raptors (Faith 2013a). Rather than tracking changing paleoenvironments, it is possible that changes in richness are instead related to changes in the proportion of assemblages accumulated by different taphonomic agents. To evaluate this possibility, we examine the relationship between richness and bone surface modifications related to human (cut and percussion marks) and non-human bone accumulation (carnivore tooth marks and gastric etching). Richness is not significantly correlated with the proportion of anthropogenic bone modifications (rarefaction: $r_s = -0.643$, $p = 0.110$; residuals: $r_s = -0.571$, $p = 0.167$; Figure 8.17), carnivore tooth marks (rarefaction: $r_s = 0.595$, $p = 0.170$; residuals: $r_s = 0.523$, $p = 0.235$), or gastric etching (rarefaction: $r_s = 0.649$, $p = 0.126$; residuals: $r_s = 0.559$, $p = 0.206$). This implies that changes in richness are not the result of changes in the bone accumulator.

Interpretation

Changes in richness since the Last Glacial Maximum at Boomplaas Cave cannot be explained by sampling artifacts or by any of several taphonomic variables. It follows that paleoenvironmental change remains a plausible mechanism, especially considering that other sites (presumably taphonomically distinct from Boomplaas Cave) in the region demonstrate parallel trends (Faith 2013b). To the extent that the contemporary relationship between ungulate richness and precipitation (Figure 8.11) can be extended into the past, a likely driver of the change in richness is precipitation change. Given contemporary annual rainfall of ~400 mm, below the peak (~750 mm) beyond which richness declines as precipitation increases (Figure 8.11), the increase in richness from the upper levels (member BLD) to the Last Glacial Maximum (LP) is consistent with enhanced precipitation through time, an interpretation substantiated by evidence from nearby paleoenvironmental archives (Chase et al. 2018).

SUMMARY

The measurement of diversity has strong appeal because the concept comprises a set of indices (richness, evenness, heterogeneity) seemingly straightforward to calculate and that allow the taxonomic composition of a faunal assemblage to be summarized in ways that are potentially informative of paleoenvironmental conditions. Because the diversity indices condense complex data into a single or several metrics, it comes as little surprise that the analysis of diversity is one of the more frequently used tools in our array of analytical approaches. However, its frequent use belies the complexity involved in translating a faunal assemblage to a paleoenvironmental inference.

The first set of challenges we face is related to sampling issues and taphonomic problems. These issues are hardly disastrous – they simply require that we be aware of them and take appropriate steps to ensure they have not obscured the patterns we hope to identify and interpret. One example of how this might be done is provided by our analysis of Boomplaas Cave (see Grayson and Delpech [1998] and Grayson et al. [2001] for additional examples). The second and more difficult set of challenges concerns the interpretation of diversity. While there are broad-scale relationships between diversity and environmental gradients today, these should not be uncritically applied to the fossil record. There are numerous factors that influence diversity, leading to different diversity–environment relationships for different taxa in different regions. Paleoenvironmental inferences are therefore more robust when contemporary relationships between diversity and an environmental gradient have been established for the set of taxa under consideration and in the same geographic area. When such evidence is unavailable, as may often be the case, diversity on its own may tell us about the response of faunal communities to environmental change, but it may not unambiguously tell us about the types of environmental changes that occurred.

NINE

TRANSFER FUNCTIONS AND QUANTITATIVE PALEOENVIRONMENTAL RECONSTRUCTION

Most analytical techniques we have discussed in previous chapters allow paleozoologists to relate faunal assemblages to a particular habitat type or to make relative (ordinal-scale) statements about an environmental variable of interest, be it temperature, rainfall, vegetation structure, or something else. But many of us desire a high-resolution paleoenvironmental reconstruction that reveals, for example, a 3.5°C temperature decline coupled with a reduction in annual rainfall of 100 mm. The appeal of such reconstructions is obvious – they provide much greater resolution than "cooler and drier" – and it should come as little surprise that efforts to provide them have been around for some time (e.g., Hokr 1951).

Besides promising more detailed insight to what the environment was like and how it changed in the past, quantitative paleoenvironmental reconstructions potentially allow us to overcome some of the subjectivity inherent to other analytical approaches. The types of analyses discussed in Chapters 5 and 6, for example, require the faunal analyst to assess the paleo-environmental implications of taxa in a faunal assemblage. As we saw in those earlier chapters, this assessment is informed by knowledge of the ecology of the taxa in question. It is therefore open to problems arising from variation in ecological knowledge or varied and subconscious weighting of that know-ledge. This results in some degree of interpretive subjectivity, which in turn creates the potential for lack of reproducibility and disagreement between two faunal analysts examining the same data. For example, on one hand Avery

(1982) interprets high frequencies of certain rodent taxa (*Otomys saundersiae* and *O. unisulcatus*) during the Last Glacial Maximum at Boomplaas Cave as evidence for arid conditions. Faith (2013a), on the other hand, interprets the same pattern as indicative of cool but not arid conditions. An agreed-upon technique that could accurately translate the Boomplaas Cave rodent assemblage into a numerical estimate of rainfall and temperature would resolve this issue. As we shall see in this chapter, there are a handful of methods advertised as able to do just that. Whether they accurately do so is another matter altogether, and one that we discuss in this chapter.

The major revolution in quantitative paleoecology occurred in the early 1970s when databases had become sufficiently large and computing technology had become sufficiently powerful (Webb and Clark 1977). Several papers describing quantitative techniques for converting ancient biological data into, in particular, climatic data appeared within the span of a few months (e.g., Fritts et al. 1971; Imbrie and Kipp 1971; Webb and Bryson 1972). In a now classic paper, Imbrie and Kipp (1971) developed a procedure for transforming the taxonomic abundances of planktonic foraminifera from late Quaternary ocean cores in the Caribbean into quantitative estimates of past sea surface temperatures (winter and summer) and salinity. Rather than interpreting, for example, a given assemblage as indicating warmer or cooler winter sea surface temperatures, the transformation process allowed a numerical or quantitative reconstruction: 100,000 years ago, winter sea surface temperature was 25.1°C. Since then, similar approaches have been implemented by Quaternary scientists to provide quantitative paleoenvironmental interpretations for a variety of fossil data, including pollen (e.g., Guiot et al. 1989; Peyron et al. 1998), chironomids (e.g., Chang et al. 2015; Larocque et al. 2001; Olander et al. 1997), diatoms (e.g., Gasse and Tekaia 1983; Gasse et al. 1995), and mammals (e.g., Hernández-Fernández and Peláez-Campomanes 2005; Liu et al. 2012; Thackeray 1987). In his early effort, Hokr (1951) used taxonomic presences in assemblages to build what would today be thought of as mutual climatic range or coexistence-based climatic envelopes (see Chapter 5) to derive quantitative estimates of ancient seasonal temperatures, precipitation, and so on. His work had remarkably little influence on subsequent analysts, likely because the reasoning was in some ways subjective and not mathematically rigorous. The latter is, of course, what many researchers want and can (seemingly) attain with modern computing power.

HOW IT WORKS

There are a bewildering array of mathematical procedures that have been implemented to generate quantitative paleoenvironmental inferences (see Birks [1995] for an overview), but the general premise is as follows. Following Birks (1995:161), "the primary aim throughout is to express the value of an environmental variable (e.g., lake-water pH) as a function of biological data (e.g., diatom assemblages)." The first step usually (Birks 1995; Webb and Clark 1977), but not always (e.g., Mix et al. 1999; Thackeray 1987), involves developing a *training set* of modern biotic communities from various sites with associated environmental variables. The relationships between the taxonomic composition of those communities and the environmental variables are modeled quantitatively and the resulting function – referred to as a *transfer function* (Sachs et al. 1977) – is used to translate fossil data into a quantitative estimate of the environmental variable. In short, once the modern relationship is established, the taxonomic data from fossil assemblages are simply "plugged in" to the transfer function to generate a quantitative paleoenvironmental reconstruction. The primary assumptions underlying quantitative paleoenvironmental reconstruction are highlighted below (after Birks 1995:168), many of which correspond to those outlined in Chapter 3.

(1)   The taxa in the modern training set are systematically related to the environment in which they are found. In other words, it is assumed that the taxonomic composition of contemporary biotic communities tracks the environment.

(2)   The target environmental variable(s) is an important determinant of taxonomic composition. For example, if taxonomic composition is sensitive to precipitation but invariant relative to temperature, then quantitative reconstructions can be developed for precipitation but not temperature.

(3)   The taxa included in the modern training set are the same as those found in the fossil assemblages (unless taxon-free approaches are used [see Chapter 7]), and their responses to changing environments through time are the same as those observed across contemporary spatial environmental gradients (i.e., taxonomic uniformitarianism). In other words, modern spatial variation is (hopefully) validly transmogrified into (ancient) temporal variation.

(4)   The statistical methods used to model the relationship between contemporary biotic communities and the environment generate transfer functions that provide accurate and unbiased reconstructions of the target environmental variable. A model that poorly predicts the response of contemporary biotic communities to environmental gradients will provide a poor estimate of the paleoenvironment.

(5)   Environmental variables other than the target variable have negligible influence on taxonomic composition. Depending on the statistical methods used, for example, if the target variable is temperature, but some species respond to precipitation, this can incorporate analytical noise or systematic bias that may influence the accuracy of the transfer function.

Development of a robust training set is an important requirement of quantitative paleoenvironmental reconstruction, and one that poses several challenges. First, the spatial extent of the training set should be sufficiently large to encompass the likely range of taxonomic and environmental variation documented in the analyzed fossil assemblages. If the fossil data fall beyond this range, either because the taxonomic composition or associated environments differ from those in the training set, the pitfalls are equivalent to extrapolating a regression beyond the limits of the given data; it is unknown whether the transfer function will provide valid estimates beyond the range of the training set because that relationship has not been determined in the present, potentially leading to biased or skewed estimates of the paleoenvironment. This does not imply, however, that global or continental training sets are the gold standard. If the geographic extent of the training set is too large, then large-scale environment–fauna relationships (e.g., latitudinal gradients) may swamp and mute small-scale relationships relevant to the fossil sample(s) of interest. A second requirement in developing a robust training set is that the geographic coverage of modern communities in the training set should overlap with the fossil assemblages. The relationship between modern biotic communities and the environment may be variable in different geographic contexts; recall we mention the interaction of variables in the section on Ecological Tolerances in Chapter 2. The significance of the preceding is that, for example, a transfer function developed from a low-latitude training set may have poor predictive power at higher latitudes, even if the taxa involved are the same.

Lastly, the training set should be collected in the same manner as the fossil data (e.g., similar spatial scale, sampling methods, and counting procedures). This is not a serious problem for quantitative paleoenvironmental reconstructions based on fossil assemblages from ocean or lake cores (e.g., pollen, diatoms, foraminifera), because the modern data are collected from core tops using similar procedures as the fossil data. Core-top assemblages represent the most recently (~modern) accumulated and deposited biotic remains. In the case of fossil animal remains from archaeological or paleontological sites, however, this is a major limiting factor, given that taxonomic data derived from modern skeletal remains equivalent to those found in the fossil record are rarely available. Wildlife census data could potentially be used to develop a training set, but the assumption is that counts of species derived from censuses of living animals, which can make use of any number of techniques (e.g., trapping, aerial census), are equivalent to those observed in the fossil sample, which often reflect a smaller spatial scale and are based on measures of abundance (e.g., NISP, MNI, or biomass) not equivalent to counts of living organisms. A more viable alternative may be to use recent (e.g., within the past 1,000 years) assemblages from fossil sites to develop training sets, which would be comparable to the use of time-averaged core-top assemblages of biotic communities sampled from ocean or lake cores. Increasing numbers of fidelity studies (Miller 2011; Miller et al. 2014; Terry 2010a, 2010b; Western and Behrensmeyer 2009) suggest training sets of modern data may not be too difficult to assemble. The challenge of building a robust training set has likely contributed to the restricted use of quantitative paleoenvironmental reconstructions based on mammals compared with the proliferation of literature concerned with other organisms. However, paleozoologists have come up with unique ways of dealing with this problem, and we highlight these ways below.

THE ANALYTICAL TOOLKIT

Here we outline a handful of techniques that have been developed to provide quantitative paleoenvironmental reconstructions based on mammalian fossil data. We apply each of the techniques to the Boomplaas Cave faunal assemblage so that analytical results can be compared and contrasted. For reasons discussed below, in not all cases are the Boomplaas Cave data ideally suited to the method at hand. But rather than seeking alternative faunal assemblages, we see this as a useful opportunity to simultaneously demonstrate how the methods work and highlight potential pitfalls. We believe an emphasis on the latter is important, not for the sake of being critical, but rather because these methods – more so than any others described in this book – are ripe for misuse if they are not judiciously applied. This is because they allow paleozoologists to generate seemingly precise (e.g., ratio-scale) but potentially inaccurate reconstructions without requiring any detailed knowledge of the taxa involved or consideration of taphonomic or sampling issues. To be clear, these issues are just as important with these methods as with any other (see Chapter 8 for discussion of techniques to analytically contend with taphonomic issues).

Thackeray's Method

Francis Thackeray published a series of papers that represent the first efforts to apply some of the quantitative techniques developed in the 1970s to terrestrial paleozoological contexts (Sénégas and Thackeray 2008; Thackeray 1987, 1990, 1992, 2002; Thackeray and Avery 1990). In the original paper that provided the methodological framework for the subsequent studies, Thackeray (1987) developed a modified version of Imbrie and Kipp's (1971) approach to reconstruct late Quaternary temperature changes from the abundances (NISP) of sixteen micromammal (rodent and insectivore) taxa found in 124 assemblages spanning nine South African sites, including Boomplaas Cave. Following Imbrie and Kipp (1971), his (1987) approach is based on factor analysis, an ordination method related to principal components analysis, which aims to resolve an assemblage of numerous taxa (variables) into a small number of components (factors). Unlike Imbrie and Kipp (1971), Thackeray lacked a set of modern training data required to generate a transfer function, but he did know the mean annual temperature (MAT) across the range of each taxon. For

TABLE 9.1 *Factor 1 (F1) loadings and mean annual temperature (MAT) of South African micromammal taxa from Thackeray (1987).*

Taxon	F1	MAT	p_i	$p_i \times$ F1
Aethomys namaquensis	0.248	16.89	0.08	0.0203
Crocidura flavescens	−0.159	15.02	0.14	−0.0216
Crocidura hirta	0.413	17.72	0.00	0.0000
Cryptomys hottentotus	0.022	16.85	0.19	0.0041
Dendromys melanotis	0.078	16.62	0.00	0.0002
Mus minutoides	0.100	16.98	0.01	0.0007
Myosorex varius	−0.374	13.64	0.06	−0.0214
Mystromys albicaudatus	0.265	14.53	0.07	0.0195
Otomys irroratus	−0.069	14.79	0.30	−0.0207
Otomys laminatus	−0.013	15.07	0.00	−0.0001
Otomys saundersiae	−0.415	13.16	0.09	−0.0354
Otomys unisulcatus	−0.262	13.19	0.02	−0.0056
Praomys natalensis	0.257	17.25	0.00	0.0000
Saccostomus campestris	0.418	17.37	0.02	0.0069
Steatomys krebsi	0.143	15.56	0.02	0.0032
Suncus varilla	0.104	14.56	0.01	0.0006
Σ			1.00	−0.0493

Note: For stratum BLD at Boomplaas Cave, p_i indicates the proportion of specimens assigned to taxon i.

the first three factors, Thackeray examined the factor loadings of each taxon in relation to these MAT data. He noted that those taxa with extreme loadings on the first factor (F1) tend to favor warmer (positive values) versus cooler (negative values) climates, and that factor loadings and MAT across a taxon's range are strongly correlated ($r = 0.811$, $p < 0.001$), leading to the inference that F1 is primarily tracking MAT (Table 9.1). To generate a factor score (FS) that could be used as an index of MAT for a given assemblage, Thackeray multiplied the proportional abundance of a taxon by its factor loading, and summed the values for all taxa in an assemblage (see Table 9.1 for an example from Boomplaas Cave assemblage BLD). Thackeray rescaled all FS values from 0 to 100, an arbitrary procedure, but one that facilitates interpretation by increasing the quantitative separation between values. Table 9.2 provides raw and rescaled values for the Boomplaas Cave sequence, based on the factor loadings from Thackeray's (1987) analysis (Table 9.1). These values differ from those reported by Thackeray (1987) because he examined finer-scale stratigraphic subdivisions than we do here.

The final step in Thackeray's (1987) analytical protocol was to convert the FS values to temperature. To do so, he inferred that the FS value from the top of the Boomplaas Cave sequence (member DGL) corresponds to modern MAT (16.5°C), and that the lowest FS value (member GWA), which dates to the Last Glacial Maximum, signals a 5°C drop relative to the present (11.5°C), based on

TABLE 9.2 *Reconstructions of past climate (MAT = mean annual temperature; MAP = mean annual precipitation) derived from the Boomplaas Cave micromammals using Thackeray's (1987) method and Hernández-Fernández and Peláez-Campomanes' (2005) quantitative bioclimatic model.*

Stratum	Thackeray's method			Quantitative bioclimatic model	
	Factor score (FS)	Rescaled (FS)	MAT (°C)	MAT (°C)	MAP (mm)
DGL	−0.053	98.6	16.5	21.46	565.11
BLD	−0.049	100	16.57	21.53	478.03
BLA	−0.058	96.8	16.41	21.66	472.87
BRL	−0.05	99.8	16.56	21.51	503.36
CL	−0.097	81.3	15.62	21.26	500.88
GWA	−0.307	0	11.5	21.16	570.91
LP	−0.294	4.8	11.74	21.15	596.81
LPC	−0.303	1.2	11.56	21.58	450.1
YOL	−0.269	14.7	12.24	21.29	571.32
BP	−0.25	21.9	12.61	21.19	540.34
OLP	−0.252	21.2	12.57	21.26	500.88
BOL	−0.253	20.8	12.55	21.23	551.36
OCH	−0.246	23.6	12.7	21.15	596.81
LOH	−0.248	22.6	12.65	21.71	483.44

(at that time) the best estimate of Last Glacial Maximum temperatures in the region. These two observations, along with another recent fossil sample from a separate site, were used to develop a calibration curve relating FS values to temperature. Table 9.2 provides MAT values from Boomplaas Cave, following Thackeray's assumption that the FS value from the upper level corresponds to a MAT of 16.5°C and that the lowest value signals a 5°C drop. The assumption of a 5°C temperature difference between the present and the Last Glacial Maximum is problematic because it assumes that which we wish to know in the first place – specifically, the past temperature at Boomplaas Cave.

Even if the 5°C temperature difference is accurate, the temperature values must be interpreted with extreme caution. The primary reason, one that has important implications for all quantitative paleoenvironmental interpretations, is that these indices are derived from NISP counts, which, as discussed in Chapter 6, provide only an ordinal-scale measure of taxonomic abundances (temperature in °C represents an interval-scale measurement). It follows that paleotemperature (Table 9.2), *or any other environmental variable derived from an ordinal-scale measure*, is best interpreted in ordinal-scale terms. In other words, the ranking of temperatures across the Boomplaas Cave sequence is important, but not the magnitude of difference between observations. For example, we can be reasonably confident that the assemblage from member BRL indicates warmer conditions compared with underlying member CL, but not that it indicates temperatures that are 1°C warmer.

9.1. The ranking of temperature reconstructions from coolest to warmest for Boomplaas Cave micromammals following Thackeray's (1987) method. Members corresponding to the Last Glacial Maximum are indicated by grey shading.

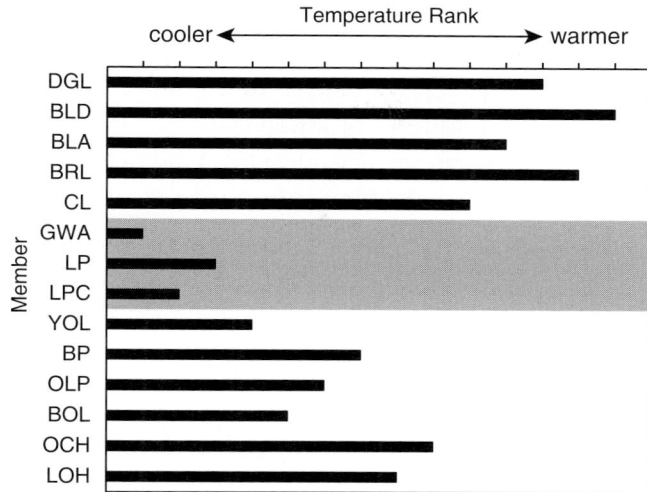

Awareness of the interpretive limitations of quantitative paleoenvironmental reconstructions derived from ordinal-scale data does not imply that quantitative reconstructions of the sort generated by Thackeray (1987) are a hopeless pursuit. Instead, it requires only that we pay careful attention to the questions we ask of our data, questions concerning (for instance) the scale of the target variable and the scale of the measured variable. While we may not be able to confidently identify temperatures at any given moment in time, we are able to address the following question: how has temperature changed through time? Ordinal-scale data are suitable for such a question because they allow interpretations in relative terms (e.g., warmer versus cooler). This is admittedly less appealing than the quantitative interval-scale statements (e.g., temperature at time X is 10°C) that we are looking for in this chapter, but it nevertheless provides useful paleoenvironmental information. At Boomplaas Cave, for example, if our temperature proxy is conceptualized in rank-order terms (Figure 9.1), there is a clear trend of declining temperature from the base of the sequence to the Last Glacial Maximum, with Holocene temperatures greater than at any other time in the sequence. These trends are consistent with local (Talma and Vogel 1992) and global climate proxies (Jouzel et al. 2007; Petit et al. 1999; see also Thackeray 1990) and imply that Thackeray's (1987) method – at least in this case – is providing reasonable ordinal-scale results.

Thackeray's (1987) method differs from the sort of quantitative paleoenvironmental reconstructions more commonly used by Quaternary scientists because he lacked a modern training set, meaning that he could not simply feed the fossil data into a transfer function to generate his temperature reconstructions. Rather, his approach can be distilled into three steps: (1) ordinate the fossil data, (2) identify environmental correlates of the ordination, and (3) interpret the ordination in terms of the environment. So despite the math involved, when viewed in this way Thackeray's method can be considered a quantitatively rigorous analysis of taxonomic abundances (of the sort discussed in Chapter 6) conducted in a manner that eliminates some of the subjectivity inherent to most interpretations. To our knowledge, Thackeray's method has not been applied outside southern Africa, but the increased availability of databases on the ecology – including climate preferences – of mammalian species across the globe (e.g., the PanTHERIA Database; Jones et al. 2009) means that applications in other contexts should not be too hard to provide.

Quantitative Bioclimatic Models

A very different sort of technique for quantitative paleoenvironmental reconstruction is provided by Hernández-Fernández and Peláez-Campomanes (2005). Their approach – which they call *quantitative bioclimatic models* – expands on the qualitative bioclimatic models discussed in Chapter 7 (Hernández-Fernández 2001; Hernández-Fernández and Peláez-Campomanes 2003), providing transfer functions that generate numerical reconstructions for a broad range of climatic variables. Recall from Chapter 7 that within the bioclimatic framework developed by Hernández-Fernández (2001), a species is assigned a climate restriction index (CRI) based on its occurrence across ten different bioclimatic zones (see Table 7.3), and the average CRI for all species in a community is calculated to provide a bioclimatic spectrum made up of ten values (the scores for each of the ten zones). In the quantitative bioclimatic model, the bioclimatic spectra derived from taxonomic presences for fifty modern faunal communities, including five for each of the ten bioclimatic zones (Table 7.3), were examined relative to more than a dozen climatic variables associated with each modern community, such as mean annual temperature (MAT), temperature annual range, and mean annual precipitation (MAP). Multiple linear regression was used to establish the relationship between each climatic variable (dependent variable) and the bioclimatic components (independent variables) comprising the bioclimatic spectra of each modern community. The resulting regression equations can be used as transfer functions to estimate past climate from fossil data.

The approach of Hernández-Fernández and Peláez-Campomanes (2005) differs from traditional quantitative applications, because rather than relying on the taxonomic identity of a set of organisms to infer the environment, individual species are converted to taxon-free CRIs (i.e., niche models). This means their transfer functions can be applied to faunas from anywhere. There is need for caution, however, given their modern training set comprises only fifty animal communities, which is woefully small in relation to the entire globe. By comparison, Kipp (1976) provides transfer functions for North Atlantic planktonic foraminifera based on a training set of 191 core tops, while Woodward and Shulmeister (2006) generate transfer functions for New Zealand chironomids

TABLE 9.3 *A comparison of observed mean annual temperatures (MAT) with values reconstructed using Hernández-Fernández and Peláez-Campomanes' (2005) transfer functions for rodents.*

Locality	Country	MAT (°C)		
		Observed	Reconstructed	Difference
Ust Kamchatsk	Russia	−0.9	−9.5	8.6
Saskatoon	Canada	2.0	1.9	0.1
Riga	Latvia	5.6	5.5	0.1
Budapest	Hungary	11.2	10.0	1.2
Dalian (Dairen)	China	10.3	6.8	3.5
Portland	USA	12.6	10.9	1.7
Charleston	USA	18.3	13.6	4.7
Nice	France	14.8	11.6	3.2
Guilin (Kweilin)	China	19.4	19.9	−0.5
Smara	Western Sahara	21.7	21.9	−0.2
Puerto Ayacucho	Venezuela	27.2	26.3	0.9
Tabou	Ivory Coast	26.0	26.8	−0.8
Kuala Lumpur	Malaysia	26.2	26.4	−0.2

Note: Locality data from Hernández-Fernández and Peláez-Campomanes (2003).

based on forty-six modern lake assemblages. Importantly, they note the transfer functions they derive are only applicable to New Zealand!

The extent to which a training set of fifty faunas captures the relationship between taxa and the environment on a near-global scale is uncertain. To partially address this issue, Hernández-Fernández and Peláez-Campomanes (2005) show that for thirteen modern localities not included in the training set, climate values predicted from faunal composition are strongly correlated with observed values, providing some confidence that the transfer functions work the way they are supposed to. However, this correlation does not necessarily mean that the predicted values are accurate, only that the general trends are maintained (e.g., warmer localities are associated with warmer reconstructions, and cooler localities with cooler reconstructions). For example, a direct comparison of observed temperatures relative to reconstructed values based on the rodent faunas from the thirteen modern localities (Table 9.3) shows in some cases the transfer functions provide highly accurate reconstructions (within 0.1°C of the observed value), but in four cases the reconstructions differ by more than 3°C from the observed value. It follows that, as is also the case with Thackeray's (1987) method, we can probably use Hernández-Fernández and Peláez-Campomanes' (2005) transfer functions to estimate ordinal-scale differences in climate parameters between fossil assemblages, but not always to provide accurate interval- or ratio-scale estimates of climate parameters at a given moment in time.

How well do the Hernández-Fernández and Peláez-Campomanes (2005) transfer functions perform with fossil data? Once again we turn to the Boomplaas Cave micromammals to highlight the promise and pitfalls of the technique. This time we consider rodents only (i.e., no insectivores) to correspond with the set of taxa included in their rodent transfer function. In Chapter 7, we previously generated CRIs for the Boomplaas Cave rodents and bioclimatic spectra for each assemblage (see Table 7.4). Hernández-Fernández and Peláez-Campomanes (2005) found the bioclimatic spectra of rodents are strongly correlated with MAT ($r^2 = 0.930$) and MAP ($r^2 = 0.746$) in their training set, so we use their rodent transfer function for both variables to translate bioclimatic spectra (see Table 7.4) into quantitative paleoenvironmental reconstructions (Table 9.2). The rodent transfer functions are as follows:

$$MAT = 26.686 + 0.024(II) - 0.029(II/III) - 0.024(III) - 0.074(IV)$$
$$- 0.120(V) - 0.135(VI) - 0.217(VII) - 0.404(VIII) - 0.386(IX)$$

$$MAP = 2978.195 - 21.237(II) - 27.563(II/III) - 33.050(III) - 32.648(IV)$$
$$- 6.678(V) - 5.076(VI) - 28.400(VII) - 33.109(VIII) - 25.980(IX)$$

where the values in parentheses refer to the bioclimatic spectra for a given assemblage (see Table 7.4 for the Boomplaas Cave bioclimatic spectra).

We turn first to temperature. Across the Boomplaas sequence, MAT values average 21.4°C, with a limited range of between 21.1°C and 21.7°C (Table 9.2). These values are well above those that characterize Boomplaas today (16.5°C according to Thackeray 1987), being more similar to temperatures from the humid subtropical climate on the northeast coast of South Africa (e.g., Durban) (Hijmans et al. 2005). There are many likely factors underlying this discrepancy, one of which is that the nearest modern locality in the training set is from Lüderitz (Namibia), 1,000 km from Boomplaas Cave. One cannot reasonably expect to reconstruct accurate temperatures from a South African fossil fauna using transfer functions developed from a training set lacking modern South African localities. Perhaps more importantly, many of the modern localities in the training set with temperatures comparable to Boomplaas Cave today include at least some taxa whose range extends to the cooler bioclimatic zones (e.g., steppe, boreal forest). The absence of such taxa from Boomplaas Cave – or indeed anywhere in sub-Saharan Africa – implies the relationship between faunal communities and climate in this part of the world is going to differ from the global relationship. This is why the reconstructed temperatures based on the Boomplaas rodents are too high. To obtain more accurate reconstructions, we would need to develop transfer functions specific to southern Africa. The extent to which this geographic problem applies to faunal communities from elsewhere in the world is unknown, but it is certainly something to consider when applying the technique.

Equally troublesome is that temperatures are reconstructed as very stable throughout the Boomplaas rodent sequence, with portions of the Last Glacial

Maximum (members GWA and LP) indicating temperatures only minimally cooler than the Holocene. This contrasts with results obtained from a previous study of Plio-Pleistocene rodent faunas in Iberia (Hernández-Fernández et al. 2007), which shows dramatic temperature variations that broadly track global trends. The reason the transfer function fails to indicate substantial temperature declines during the Last Glacial Maximum at Boomplaas Cave is that for the functions to generate cooler temperatures, taxa inhabiting the cooler climatic zones need to be documented in the assemblage. Whereas cold-adapted species from high latitudes may disperse southward into the Iberian Peninsula during cool phases of the Pleistocene, the same cannot happen at Boomplaas Cave; the site is positioned near the southern tip of the continent (there are no higher-latitude faunas to disperse equatorward during cold phases) and there simply are no cold-adapted species typical of tundra or boreal forest (i.e., the cooler zones) on the continent. This highlights an important limitation of the method. Because the climatic zones underpinning the analysis broadly track latitude, the method is likely to perform better in geographic contexts where warmer or cooler (or wetter and drier) faunas can disperse north and south in response to Quaternary climate changes.

Finally, we also note the presence of two prominent outliers in our reconstruction of temperatures at Boomplaas Cave (LPC and LOH: Figure 9.2). For example, Last Glacial Maximum assemblage LPC provides one of the highest temperature estimates for the sequence, a very unrealistic outcome. However, these anomalies are easy to explain. Because the bioclimatic spectra are determined from taxonomic presences, which are in turn influenced by assemblage sample size (recall discussion in Chapter 6), it is not surprising reconstructed MAT values demonstrate a weakly significant correlation with sample size (r_s = -0.482, p = 0.081), such that smaller assemblages provide higher MAT values. Thus, LPC appears warmer than the overlying LP assemblage because its smaller sample size (rodent MNI: LPC = 69, LP = 1,242) translates to an absence of taxa that would otherwise drive down reconstructed MAT (e.g., *Otomys laminatus*). When those assemblages with the smallest sample sizes (LPC and LOH) are excluded, the correlation between sample size and MAT weakens substantially (r_s = -0.210, p = 0.553). More optimistically, a significant positive ordinal-scale correlation between MAT values reconstructed using Hernández-Fernández and Peláez-Campomanes' (2005) transfer functions and Thackeray's (1987) factor analysis (Table 9.2) emerges (r_s = 0.716, p = 0.009). While the absolute temperature values reconstructed from the transfer functions are clearly erroneous, if the temperature reconstructions are conceptualized in ordinal-scale (rank-order) terms, then a more realistic outcome (based on independent data) is obtained (Figure 9.2).

Turning now to precipitation (Table 9.2), we first note there is no correlation between assemblage sample size and MAP (r_s = 0.196, p = 0.502). But the two smallest Pleistocene assemblages (those excluded above: LPC and LOH) also provide two of the lowest MAP estimates so we exclude these from further

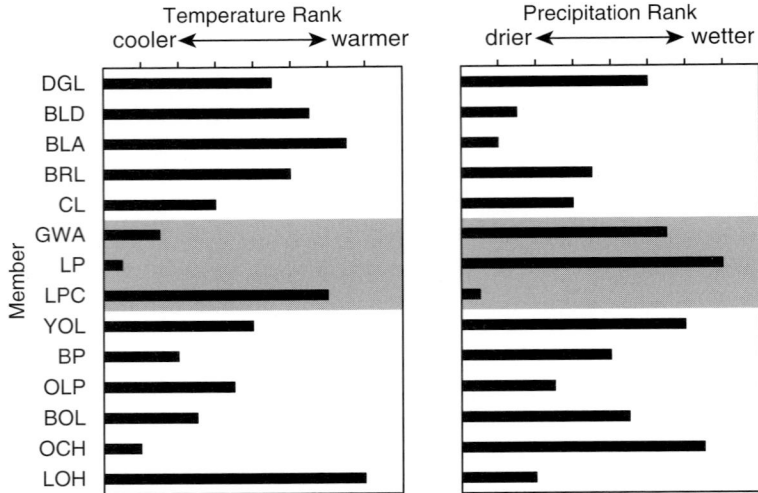

9.2. The ranking of temperature (left) and precipitation (right) reconstructions from coolest/driest to warmest/wettest for Boomplaas Cave rodents based on Hernández-Fernández and Peláez-Campomanes' (2005) transfer functions. Members corresponding to the Last Glacial Maximum are indicated by grey shading. Note that reconstructions for LPC and LOH are biased because of their small sample sizes (see text).

consideration. All reconstructed MAP values are above the ~400 mm annual precipitation observed today. For example, reconstructed MAP for the late Holocene assemblage from member DGL is ~40% higher than present, even though we might expect a climate quite similar to the present. Once again, this can be attributed to a training set that is not specific to our region of interest. In particular, the lack of taxa from bioclimatic zones VII and above – zones with substantial influence on the MAP reconstructions (see regression coefficients above) – at Boomplaas Cave leads to anomalously high estimates of precipitation. Given this and the problems with the above temperature reconstructions, it is sensible to also consider the MAP reconstructions in ordinal-scale terms. When this is done (Figure 9.2), we see a general increase in MAP from the base of the sequence to the Last Glacial Maximum – barring a distinct peak in member OCH – followed by a decline into the Holocene, a pattern broadly consistent with some interpretations from Boomplaas Cave (Chase et al. 2018; Faith 2013a) (but recall from Chapter 4 the precipitation history is debated). Once again, the ratio-scale paleoenvironmental reconstructions we are hoping for are seemingly beyond our grasp, but we do have what appear to be reasonable ordinal-scale reconstructions.

Dental Ecometrics

In Chapter 7 we discussed a handful of related studies that document relationships between climate and the distribution of large herbivore dental traits across modern ecosystems (e.g., Eronen et al. 2010b; Liu et al. 2012). These studies provide a framework for quantitative paleoenvironmental reconstruction based on taxon-free dental ecometrics. We focus here on the method developed by Liu et al. (2012), which presents an updated version of the technique used by Eronen et al. (2010b). Recall from Chapter 7 that Liu et al. (2012) showed that across modern large herbivore communities (orders Artiodactyla, Perissodactyla, Primates, Proboscidea), the mean hypsodonty score (brachydont = 1, mesodont = 2, hypsodont = 3) and lophedness (number of cutting edges) of the taxa present are correlated with mean annual temperature (r^2 = 0.67), mean annual precipitation (r^2 = 0.63), and net primary productivity (r^2 = 0.73) (see Chapter 7 for a more detailed discussion of these dental traits). Given the strength of the relationships, Liu et al. (2012) reasoned that it should be possible to predict these environmental parameters from fossil data, and they provided regression equations for doing just that:

$$MAT = 24.7 + 13.8(HYP) - 25.1(LOP)$$
$$MAP = 2727.7 - 411.9(HYP) - 859.7(LOP)$$
$$NPP = 2957.8 - 304.3(HYP) - 1043.7(LOP)$$

where HYP refers to the mean hypsodonty score and LOP refers to the mean lophedness score in a given assemblage. In theory, these regression equations should allow us to translate the large mammal species lists from virtually any reasonably well-sampled fossil assemblage into a quantitative estimate of the paleoenvironment.

We turn once again to Boomplaas Cave – this time focusing on the large herbivores (ungulates and primates) – to reconstruct MAT and MAP through the sequence. We do not explore productivity, but the analytical steps are no different than for the other parameters. Table 9.4 reports the hypsodonty score and lophedness of the ungulate taxa, and Table 9.5 reports the mean values across the sequence tallied across all non-overlapping taxa (see Chapter 8 for a discussion of non-overlapping taxa). Because the means are derived from taxonomic presences, we do have to worry about sampling effects similar to those discussed in Chapter 6. In this case, mean hypsodonty and mean lophedness are uncorrelated with sample size (hypsodonty: r_s = -0.138, p = 0.652; lophedness: r_s = -0.108, p = 0.725). It is important to note the majority of taxa at Boomplaas Cave are hypsodont and include two longitudinal cusps (Table 9.4), meaning that changing taxonomic presences are unlikely to translate to much variation in the ecometric variables. Indeed, Kruskal–Wallis tests show there are no significant differences in the medians of either variable across the sequence (hypsodonty score: p = 0.903; lophedness: p = 0.989). In terms of dental ecometrics, this means community structure is effectively stable through the sequence and temporal shifts in reconstructed climate parameters are not necessarily meaningful. But for sake of illustration, we proceed as if they are.

Turning first to MAT (Table 9.5), we observe a few good things. Several of the Holocene reconstructions are reasonably close (within ~1°C) to the present

TABLE 9.4 *Hypsodonty score (HYP) and lophedness (LOP) for Boomplaas Cave primates and ungulates.*

Taxon	HYP	LOP
Papio ursinus	1	0
Equus capensis	3	2
Equus zebra/quagga	3	2
Potamochoerus larvatus	1	0
Taurotragus oryx	2	2
Tragelaphus strepsiceros	2	2
Tragelaphini indet.	2	2
Hippotragus leucophaeus	3	2
Hippotragus equinus	3	2
Hippotragus sp.	3	2
Redunca fulvorufula	3	2
Redunca arundinum	3	2
Redunca sp.	3	2
Alcelaphus buselaphus	3	2
Connochaetes cf. *taurinus*	3	2
Connochaetes cf. *gnou*	3	2
Connochaetes/Alcelaphus	3	2
Damaliscus cf. *dorcas*	3	2
Alcelaphini indet.	3	2
Extinct caprin	3	2
Pelea capreolus	3	2
Antidorcas cf. *marsupialis*	3	2
Oreotragus oreotragus	3	2
Raphicerus melanotis	3	2
Raphicerus campestris	3	2
Raphicerus sp.	3	2
Syncerus antiquus	3	2
Syncerus caffer	3	2
Syncerus sp.	3	2

(16.5°C), and the 4–5.5°C reduction in temperatures during the Last Glacial Maximum roughly matches the 6°C decline documented in other nearby paleoclimate records (Talma and Vogel 1992). However, there are also some problems. Most of the pre-Last Glacial Maximum temperature reconstructions are within or above the range of temperatures documented in the Holocene – these values are too high based on global temperature reconstructions – and the basal assemblage (member LOH) provides the highest temperature for the sequence (18.7°C). This assemblage also provides the lowest average lophedness score, largely because it is the only one lacking *Equus* (Table 4.3). Given the small sample size of the LOH assemblage, the absence of *Equus* is very likely a reflection of sampling error; had it been recovered, reconstructed MAT would decline to 15.6°C. While it is reassuring to know this would provide a temperature reconstruction more closely aligned with other pre-Last Glacial Maximum assemblages, it is worrisome that the addition of a

TABLE 9.5 *Mean hypsodonty score (HYP) and lophedness (LOP) of large mammalian herbivores across the Boomplaas Cave sequence and reconstructions of past climate derived from them.*

Assemblage	Dental ecometrics		Climate reconstruction	
	HYP	LOP	MAT	MAP
BLD	2.67	1.89	14.1	3.1
BLA	2.71	1.86	15.4	12.4
BRL	2.65	1.82	15.6	71.5
CL	2.80	2.00	13.1	−145.0
GWA	2.91	2.18	10.1	−345.1
LP	2.85	2.00	13.8	−165.6
LPC	2.70	2.00	11.8	−103.8
YOL	2.71	1.86	15.4	12.4
BP	2.77	1.92	14.7	−63.9
OLP	2.80	1.90	15.7	−59.1
BOL	2.80	1.90	15.7	−59.1
OCH	2.79	1.93	14.8	−80.7
LOH	2.75	1.75	18.7	90.5

Note: Mean annual temperature (MAT) and mean annual precipitation (MAP) reconstructed from equations in Liu et al. (2012).

single taxon translates to a greater than 3.0°C temperature difference. This highlights a critical issue with the technique. Like other methods based on taxonomic presences, sample size issues of the sort discussed in Chapter 6 are going to be pervasive; faunal assemblages will need to be well sampled to avoid the risk of generating problematic results. Also worrisome is that the ranking of reconstructed MAT values is not correlated with that provided by Thackeray's (1987) method (r_s = 0.320, p = 0.287) or the quantitative bio-climatic models (r_s = 0.327, p = 0.275). This is because the pre-Last Glacial Maximum temperatures are unreasonably high.

Things are even worse if we consider MAP, for which the reconstructed values are nonsensical. Negative values are observed in many assemblages, and in no case do they come anywhere reasonably close to the ~400 mm/yr observed in the present. If we consider these reconstructions in ordinal-scale terms, we see a pattern that might be sensible (Figure 9.3). MAP values decline into the Last Glacial Maximum, and reach peak levels in the Holocene. But then again, this is *opposite* the MAP trends documented using the quantitative bioclimatic models (Figure 9.2). We do not attempt to sort out which reconstruction is correct here, but the discrepancy between various methods is important because it implies the ordinal-scale trends documented by at least one (if not both) of the methods has to be wrong. The implication is that the seemingly precise outputs provided by quantitative paleoenvironmental reconstructions should not be taken to imply the results are necessarily correct or even somewhat accurate!

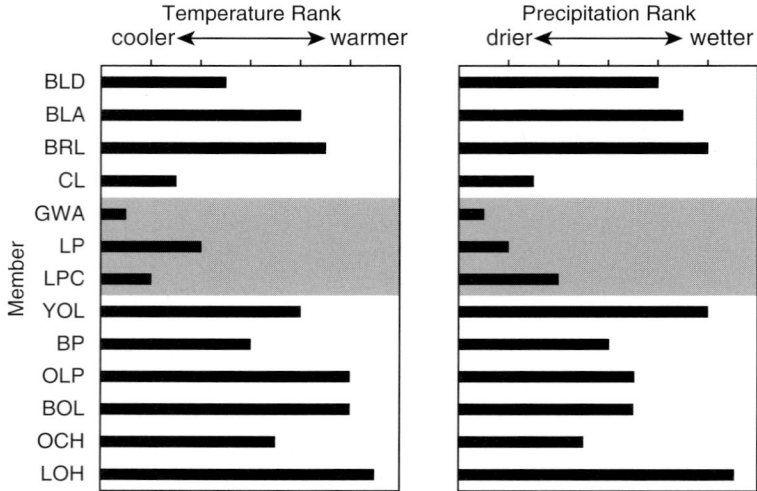

9.3. The ranking of temperature (left) and precipitation (right) reconstructions from coolest/driest to warmest/wettest for Boomplaas Cave large herbivores based on the transfer functions for dental ecometrics provided by Liu et al. (2012). Members corresponding to the Last Glacial Maximum are indicated by grey shading.

If hypsodonty and lophedness track MAT and MAP so closely in modern ecosystems across the globe (Liu et al. 2012), then why do these proxies perform poorly at Boomplaas Cave? The first problem is that the mean hypsodonty values for many of the Boomplaas Cave assemblages fall above the range of modern ecosystems examined by Liu et al. (2012). Thus, we are extrapolating beyond the limits of their regression (beyond the known measurements), which is why, for instance, negative MAP values are observed in some assemblages. More importantly, this means we are failing to meet a critical requirement of quantitative paleoenvironmental reconstruction, namely the range of taxonomic (and environmental) variation in the training set must (but in this case does not) encompass that observed in the fossil record. This failure could be due to shortcomings in the training set or taphonomic biases that have caused the mean ecometric values at Boomplaas Cave to fall outside the range of modern communities.

A second problem is highlighted by Fortelius et al. (2016:3), who note "[w]e have observed that modern-day distributions of [these] traits substantially differ across continents owing to paleobiogeographic effects." The relationship between dental ecometrics and climate varies in different parts of the world, which means the global patterns cannot be uncritically applied to the fossil record. Awareness of this problem led Fortelius et al. (2016) to develop new regression equations specific to their region of focus (equatorial Africa). What is needed in our case are regression equations relevant to southern Africa. This might also remedy the first problem (extrapolating beyond the transfer function regression) if it were to include modern communities simi-larly dominated by hypsodont ungulates. But even if we were to develop such equations, we must remain cognizant of the fact that – as demonstrated by our previous examples – interpretations of reconstructed climate parameters are best restricted to ordinal-scale statements (e.g., warmer vs. cooler; wetter vs. drier) at the risk of generating interpretations that seem precise but are (poten-tially) wildly inaccurate.

Tree Cover from Taxon-Free Characterizations

Paleoecologists often think of vegetation structure in terms of discrete habitat types, such as grassland, woodland, or forest. These categories represent a useful but arbitrary partitioning of what is really a complex multidimensional spec-trum of vegetation structure. Louys et al. (2015) present a technique designed to capture some of this complexity by providing a quantitative estimate of tree canopy cover based on faunal composition (see also Louys et al. 2011a). To do so, they recorded taxonomic presences of mammals for sixty-three wild-life areas across Central and South America (n = 8), Africa (n = 23), and Asia (n = 32). Species were characterized using a taxon-free framework according to body size, trophic level, and locomotion. Size categories include small (B: 1–10 kg), medium (C: 10–45 kg), large (D: 45–180 kg), and very large (E: >180 kg); smaller taxa (A: <1 kg) were not considered in their analysis because of concerns over potential taphonomic biases in the fossil record. Trophic levels include primary consumers (P: taxa that are strictly herbiv-orous) and secondary consumers (S: taxa that include any amount of animal tissue in the diet). Locomotor categories include terrestrial (T) or arboreal (A), with the latter broadly defined to include taxa that spend any amount of time in the trees, even if only rarely. These three ecological variables were used to define fifteen groups (Table 9.6) – note that not all possible ecological categories are observed in modern communities (e.g., very large arboreal pri-mary consumers) – and for each modern community the percentage of species falling within a given group was determined. In order to characterize tree cover, Louys et al. (2011a) used satellite imagery to capture a 25 km by 25 km square centered on the geographic coordinates of each wildlife area. GIS software was used to quantify tree cover according to the proportion of the landscape with no tree cover (%absent), light tree cover (%light), moderate tree cover (%moderate), and heavy tree cover (%heavy) (Figure 9.4). These data provide a framework for establishing relationships between faunal community composition and tree cover, which can then be used to predict tree cover in the past.

   Louys et al. (2015) developed two quantitative models for predicting tree cover from the ecological structure of the modern faunal communities. The

TABLE 9.6 *Ecological groups for mammals defined by Louys et al. (2015).*

Group	Abbreviation
Small, arboreal primary consumer	BAP
Small, arboreal secondary consumer	BAS
Small, terrestrial primary consumer	BTP
Small, terrestrial secondary consumer	BTS
Medium, arboreal primary consumer	CAP
Medium, arboreal secondary consumer	CAS
Medium, terrestrial primary consumer	CTP
Medium, terrestrial secondary consumer	CTS
Large, arboreal primary consumer	DAP
Large, arboreal secondary consumer	DAS
Large, terrestrial primary consumer	DTP
Large, terrestrial secondary consumer	DTS
Very large, arboreal secondary consumer	EAS
Very large, terrestrial primary consumer	ETP
Very large, terrestrial secondary consumer	ETS

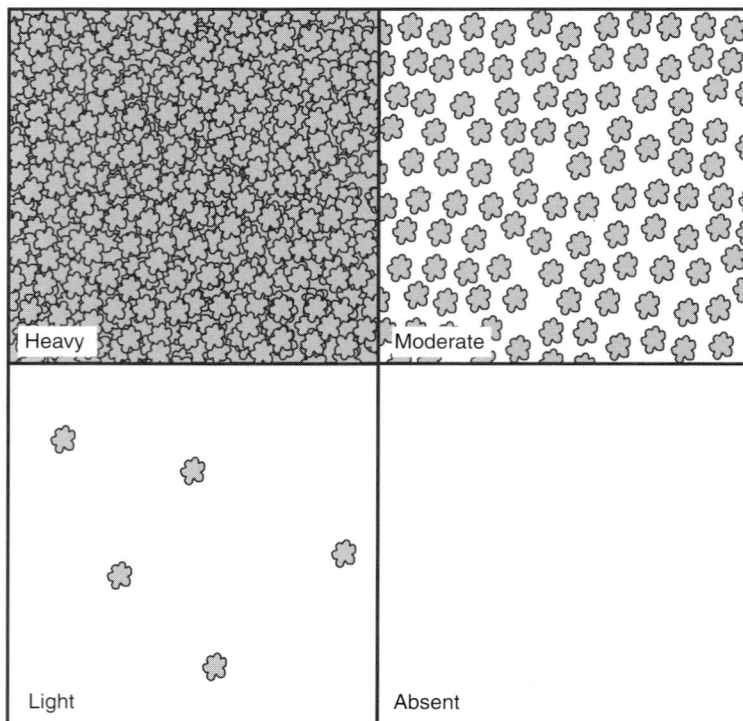

9.4. Schematic representation of tree cover categories examined by Louys et al. (2015). Redrawn from Louys et al. (2015).

first is based on a linear regression (LR) of the primary principal components derived from two principal components analyses (PCAs) of the sixty-three wildlife areas, one for the four tree cover categories and another for the fifteen taxonomic categories. The second model is based on multivariate multiple regression (MMR) with the vegetation types as dependent variables and the taxonomic categories as independent variables. Across their training set, these models do a good job predicting (from faunal composition) the observed amount of heavy tree cover (LR: $r^2 = 0.743$, $p < 0.001$; MMR: $r^2 = 0.786$, $p < 0.001$), but predictive power is weaker for moderate tree cover (LR: $r^2 = 0.050$, $p = 0.078$; MMR: $r^2 = 0.303$, $p < 0.001$), light tree cover (LR: $r^2 = 0.316$, $p < 0.001$; MMR: $r^2 = 0.412$, $p < 0.001$), and absent tree cover (LR: $r^2 = 0.265$, $p < 0.001$; MMR: $r^2 = 0.401$, $p < 0.001$). Importantly, they showed that the two models provide slightly different results, with the MMR, for example, tending to overestimate the amount of heavy tree cover in the modern communities. Perhaps not surprisingly, it follows that reconstructions generated using their technique are best interpreted in ordinal-scale terms. This is made apparent when we examine their reconstructions of tree cover for the Pliocene faunas from the Upper Laetolil Beds of Laetoli in Tanzania. Louys et al. (2015) generated complementary but still varied reconstructions using the LR (%heavy = 0.52%, %moderate = 7.79%, %light = 31.7%, %absent = 59.98%) and the MMR model (%heavy = 0.78%, %moderate = 4.21%, %light = 9.75%, %absent = 96.51%). The differences imply we cannot accept these numbers at face value. We can, however, be reasonably confident the Laetoli paleoenvironment was dominated by habitats with little or no heavy tree cover.

The technique developed by Louys et al. (2015) requires fossil samples that provide reasonable representations of the entire large mammal community (>1 kg). We suspect it is going to be difficult in many cases to obtain adequate samples. Turning once again to Boomplaas Cave, Table 9.7 shows clear discrepancies between the distributions of taxa in the fossil sample compared with the training set. There are too many large-bodied (size B and up) terrestrial primary consumers at Boomplaas, and this is driven by the dearth of secondary consumers (range: 1 to 9) compared with the modern communities examined by Louys et al. (2015) (range: 17 to 45). These biases render the Boomplaas Cave data incomparable with the modern training set, violating the same requirement of quantitative paleoenvironmental reconstruction we violated in the above discussion of dental ecometrics. It is informative, nevertheless, to proceed with an application of the method.

We first note that Louys et al.'s (2015) technique is best suited for tracking the amount of heavy cover, as this is the only category for which the modern faunal data predict more than 50% of the variance in habitat structure. Louys et al. (2015) show the LR model is more accurate and less biased than the MMR model, so we focus on a modified version of the former.

TABLE 9.7 *The percentage of large mammal species (>1 kg) across ecological categories in the modern communities examined by Louys et al. (2015) (minimum and maximum value) compared with the fossil assemblage from Boomplaas Cave.*

Category	Modern		Boomplaas Cave												
	min	max	BLD	BLA	BRL	CL	GWA	LP	LPC	YOL	BP	OLP	BOL	OCH	LOH
BAP	0.0	14.6	4.5	7.1	4.5	5.0	7.7	6.7	7.7	10.0	5.6	5.9	6.3	5.0	8.3
BAS	6.3	42.9	13.6	7.1	0.0	0.0	0.0	0.0	0.0	0.0	0.0	0.0	0.0	5.0	0.0
BTP	0.0	11.4	9.1	**14.3**	9.1	10.0	7.7	6.7	**15.4**	**20.0**	11.1	**11.8**	**12.5**	10.0	8.3
BTS	2.3	24.4	9.1	7.1	0.0	5.0	0.0	0.0	0.0	10.0	0.0	5.9	0.0	0.0	0.0
CAP	0.0	3.1	0.0	0.0	0.0	0.0	0.0	0.0	0.0	0.0	0.0	0.0	0.0	0.0	0.0
CAS	2.0	25.5	13.6	14.3	13.6	15.0	7.7	6.7	7.7	10.0	11.1	11.8	12.5	10.0	16.7
CTP	0.0	13.9	**22.7**	**21.4**	**18.2**	**20.0**	**15.4**	**26.7**	**30.8**	**30.0**	**16.7**	**23.5**	**25.0**	**20.0**	**33.3**
CTS	0.0	25.0	0.0	0.0	0.0	0.0	0.0	0.0	0.0	0.0	0.0	0.0	6.3	5.0	0.0
DAP	0.0	2.4	0.0	0.0	0.0	0.0	0.0	0.0	0.0	0.0	0.0	0.0	0.0	0.0	0.0
DAS	0.0	9.4	4.5	7.1	0.0	0.0	0.0	0.0	0.0	10.0	5.6	5.9	6.3	5.0	8.3
DTP	0.0	21.1	13.6	14.3	**27.3**	15.0	**30.8**	**26.7**	7.7	0.0	**22.2**	17.6	18.8	**25.0**	**25.0**
DTS	0.0	8.7	0.0	0.0	4.5	0.0	0.0	0.0	0.0	0.0	0.0	0.0	0.0	0.0	0.0
EAS	0.0	3.1	0.0	0.0	0.0	0.0	0.0	0.0	0.0	0.0	0.0	0.0	0.0	0.0	0.0
ETP	0.0	28.9	9.1	7.1	22.7	30.0	30.8	26.7	30.8	10.0	27.8	17.6	12.5	15.0	0.0
ETS	0.0	6.3	0.0	0.0	0.0	0.0	0.0	0.0	0.0	0.0	0.0	0.0	0.0	0.0	0.0

Note: Bold values fall outside the range of the modern communities. See Table 9.6 for abbreviations of ecological categories.

9.5. The relationship between faunal community composition (faunal PC1) and %heavy tree cover for the sixty-three modern communities examined by Louys et al. (2015). The solid line represents the least-squares regression [%heavy = 2.408 PC1 + 58.307; r^2 = 0.743, p < 0.001].

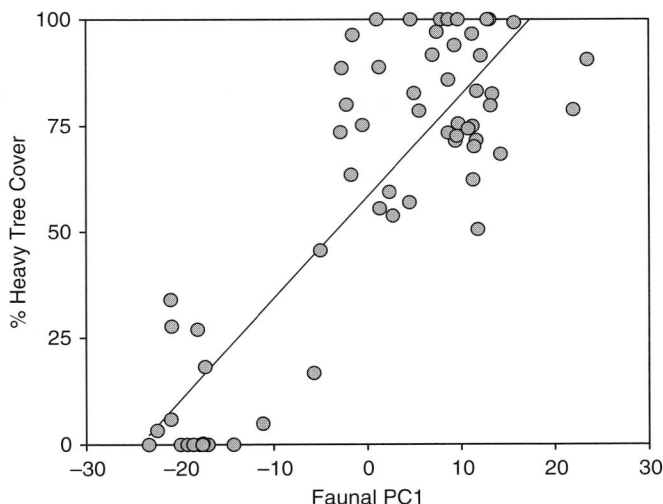

We streamlined their analytical protocol to eliminate an unnecessary analytical step (conducting a PCA on the vegetation data) for the reconstruction of %heavy from fossil data. Our modified protocol requires fewer analytical steps, but provides estimates that are perfectly correlated (r^2 = 1) with those of Louys et al. (2015), though our %heavy estimates are slightly higher. We first conducted a PCA of their modern faunal data (fifteen ecological categories across sixty-three modern communities). As shown in Figure 9.5, the first principal component (PC1) score of each community is strongly correlated with the percentage of heavy tree cover (r^2 = 0.743, p < 0.001). We then ordinated the Boomplaas Cave assemblages (data from Table 9.7) in the same multivariate space as the modern data to derive PC1 scores for the fossil data (as in Louys et al. 2015). And lastly, those Boomplaas PC1 scores were plugged in to the least-squares regression describing the relationship between PC1 and %heavy tree cover in modern communities (%heavy = 2.408 PC1 + 58.307; see Figure 9.5) to generate a quantitative environmental reconstruction.

The PC1 scores for the Boomplaas Cave fauna are illustrated in Figure 9.6. As this method is based on taxonomic presences, it is important to first check whether the PC1 scores are correlated with sample size. They are not (r_s = 0.297, p = 0.324), which means we can proceed with interpretation without worrying too much about sampling artifacts. The PC1 scores for some assemblages are more negative than any of the modern communities in Figure 9.5, an outcome related to the dominance of large-bodied terrestrial primary consumers at Boomplaas Cave. This skews the Boomplaas Cave data in the direction of (or even beyond) the more open modern communities, which is why, when translated to an estimate of %heavy using the regression equation provided above, we obtain negative values for some assemblages (Figure 9.6). However,

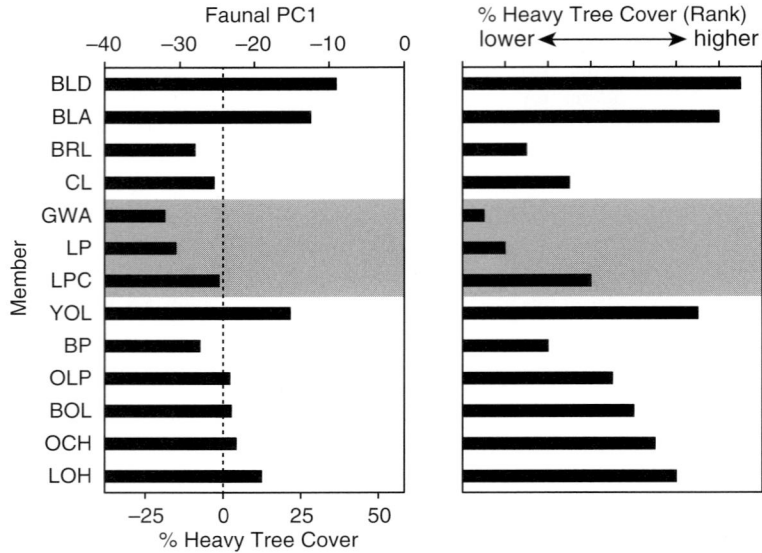

9.6. Faunal PC1 and %heavy tree cover (left) and the ordinal-scale ranking of %heavy tree cover (right) for Boomplaas Cave large mammals based on the method of Louys et al. (2015). Member corresponding to the Last Glacial Maximum are indicated by grey shading.

if we consider these reconstructions in ordinal-scale terms (Figure 9.6), we see what appears to be a reasonable pattern. Heavy tree cover decreases from the base of the sequence to the Last Glacial Maximum and increases into the Holocene. This trend is broadly consistent with the rise and decline of grassland habitats inferred from the more conventional analytical techniques outlined in previous chapters. It also complements the Boomplaas Cave charcoal record, which documents various shrubs during the Last Glacial Maximum but no trees (Deacon et al. 1983; Scholtz 1986).

Our analysis suggests it might be possible to generate reasonable ordinal-scale paleoenvironmental reconstructions using the Louys et al. (2015) technique even with biased fossil samples. It all comes down to whether or not bias has overprinted paleoenvironmental differences between assemblages. The problem of course is we have no way of knowing whether or not this has happened, which means the best we can hope for (when dealing with such biased samples) is correspondence between results generated by other types of analyses or (preferably) independent paleoenvironmental proxies.

Future Prospects

The four techniques outlined above provide a glimpse into the potential and pitfalls of quantitative paleoenvironmental reconstruction in paleozoology. Though the literature that makes use of these or related techniques to generate paleoenvironmental inferences based on vertebrate faunas pales in comparison to that for other taxonomic groups (e.g., foraminifera, diatoms, chironomids), we expect this to change considerably in the near future. Online maps of geographic ranges for most vertebrate taxa (keep in mind that such maps often vary from source to source), such as those available from the IUCN (www. iucnredlist.org) or the World Wildlife Fund WildFinder Database (www. worldwildlife.org/phages/wildfinder-database), can now be used to obtain taxonomic presence data at varying resolution (e.g., 0.5° latitude and longitude grid cells) for virtually any region of interest. Associated climatic or environmental data can be extracted from any number of online sources, including the global WorldClim database (Fick and Hijmans 2017; Hijmans et al. 2005; see www.worldclim.org) or the North American PRISM database (PRISM Climate Group 2004; see http://prism.oregonstate.edu/). This means it is now possible to assemble modern training sets without ever leaving the office! Eronen et al. (2010b) and Liu et al. (2012) did just this in their studies of the relationships between dental ecometrics and climate (see also Barr 2017), and there is abundant potential to use this approach to explore how species (or their taxon-free traits) vary as a function of any number of environmental parameters, with the aim of generating transfer functions for paleoenvironmental reconstruction. This could lead to exciting innovations in the way paleozoologists approach paleoenvironmental reconstruction, but if we are correct that such research is likely to proliferate in the future, then what we have to say next will be all the more relevant.

THE NO-ANALOG PROBLEM

Methods of quantitative paleoenvironmental reconstruction assume relationships between biotic communities and environmental parameters observed in the present can be directly applied to the past, the general uniformitarian assumption we discuss in Chapter 3. But what if the ancient ecosystems are unlike those observed anywhere on the planet today? What if relationships between fauna and certain environmental parameters differed in the past? These are not hypothetical "what if?" sorts of questions. There is ample evidence from climate and vegetation models (e.g., Harrison and Prentice 2003; Kutzbach et al. 1998), paleobotanical archives (e.g., Overpeck et al. 1992; Williams et al. 2001), and mammalian fossil records (e.g., Graham 2005; Janis et al. 2004; Stafford et al. 1999) that imply ancient ecosystems without modern analogs (see reviews in Jackson and Williams 2004; Williams and Jackson 2007). As we outline here, there is good reason to believe no-analog ecosystems can cause problems for quantitative reconstructions.

No-analog climates and faunal communities have long been recognized as problematic for quantitative paleoenvironmental reconstruction in the marine micropaleontology literature (e.g., Hutson 1977; Mix et al. 1999; Ter Braak et al. 1993). Hutson (1977) used the term "no-analog conditions" to describe those cases where the fossil record samples a set of environmental or biological conditions outside of the modern training set. No-analog conditions could include climates that are more or less seasonal than today or faunal communities characterized by atypical taxonomic abundances or associations of presently allopatric taxa. (It is important, we think, to note that we follow Graham [2005] and use the term "*non-analog*" for faunas that include species in contexts indicating sympatry when today the species are allopatric to distinguish them from other sets of species such as indicated by Hutson's [1977] use of the term no-analog. We use the latter term in a generic sense to indicate either kind of species aggregate or kind of climate.) Hutson also considered insufficiently sampled training sets or taphonomic biases that skew taxonomic representation as drivers of no-analog conditions because both lead to fossil communities unlike those in the training set.

To explore how no-analog conditions impacted the accuracy of quantitative paleoenvironmental reconstruction (sea-surface temperatures) derived from planktonic foraminifera, Hutson (1977) developed transfer functions based on modern ocean core-top assemblages, and applied those functions to the same modern assemblages after subjecting them to high levels of dissolution. Because some species are more resistant to dissolution than others, the process alters taxonomic abundances in a way that creates no-analog taxonomic compositions. Not surprisingly, he showed most techniques produced systematically biased temperature reconstructions when applied to the no-analog samples, leading him to conclude "the accuracy of a transfer function within a [modern training] set does not necessarily ensure accurate temperature reconstructions when the transfer function is applied to a no-analog data set" (Hutson 1977:364). Later studies relying on simulated no-analog conditions reached the same conclusion (Ter Braak et al. 1993). Because of the no-analog problem, Hutson (1977:364) emphasized "[e]very effort should be made to ensure that the data set on which a transfer function is calibrated is representative of the range of biological and environmental conditions which may occur [in the fossil sample]." This of course is a fundamental requirement of quantitative paleoenvironmental reconstruction (Birks 1995), one we are forced to ignore whenever the methods are applied to no-analog ecosystems.

The problem for quantitative paleoenvironmental reconstructions in paleozoology is that no-analog climates, floras, and faunas are common in the not-so-distant past (e.g., Jackson and Williams 2004; Williams and Jackson 2007). To provide one example, we consider the late Quaternary of eastern North America. Toward the end of the Pleistocene and the earliest Holocene (roughly 16,000 to 10,000 years ago), no-analog plant communities were especially widespread in the eastern portion of the continent (Williams et al. 2001). These communities were characterized by, among other things, co-occurrence

9.7. The range of summer and winter temperatures in which trees of the genus *Fraxinus* (ash) are found in North America today. The position of the realized niche of *Fraxinus* at the edge of the modern climate envelope implies that its fundamental niche extends into more seasonal climates (warmer summer temperatures and cooler winter temperatures) than exist today, but which did exist at times during the late Quaternary. Redrawn from Williams and Jackson (2007).

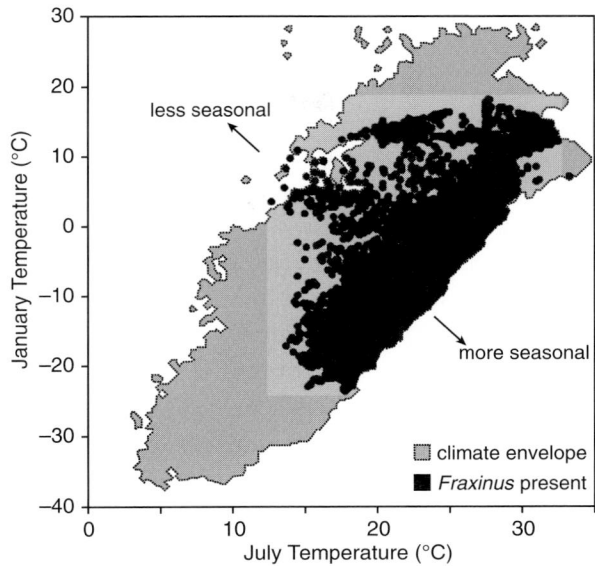

at high frequencies of taxa that are nearly allopatric today. As illustrated in Figure 9.7, the realized niches for some of the tree taxa (e.g., *Fraxinus*), whose high abundances render these no-analog communities so unusual (relative to the present), are found along the more seasonal edge of the modern North American climate envelope (i.e., the current configuration of summer and winter temperatures). This implies their fundamental niches extend into more seasonal climates characterized by warmer summers and cooler winters (Jackson and Williams 2004; Williams and Jackson 2007). Some climate models predict higher-than-present annual ranges of temperature seasonality (Kutzbach et al. 1998), broadly tracking orbitally driven increases in summer *insolation* (the amount of solar radiation reaching the earth's surface) coupled with decreases in winter insolation (Berger and Loutre 1991). For these reasons, the thought among paleobotanists is that no-analog plant communities probably have something to do with climate regimes that are more seasonal than the present (Jackson and Williams 2004; Williams and Jackson 2007).

Quaternary mammalian fossil archives from eastern North America and elsewhere document non-analog communities (e.g., Graham 2005; Graham and Mead 1987; Semken et al. 2010; Stafford et al. 1999). These communities are characterized by associations of species that are today allopatric, oftentimes involving combinations of species that are currently northern and southern in distribution. In contrast to the hypothesis outlined above, the prevailing explanation among paleozoologists is that this reflects more equable (i.e., less seasonal) climates, with cooler summers allowing northern species to disperse south and warmer winters allowing southern species to disperse north. Whatever the correct answer may be – our goal is not to resolve this discrepancy here – the point is that both no-analog faunas and floras existed in the past and were likely an outcome of no-analog climates.

What we do not know is precisely how these no-analog conditions influence the quantitative analytical techniques outlined in this chapter. This places paleozoologists at a disadvantage compared with our colleagues working in marine micropaleontological contexts, for which there is a vast literature that critically evaluates the quantitative paleoenvironmental techniques commonly applied in their discipline (e.g., Hutson 1977; Le 1992; Le and Shackleton 1994; Ortiz and Mix 1997; Telford et al. 2004; Telford and Birks 2009; Webb and Clark 1977). Not so for the techniques outlined here. However, the reconstructions provided above for the Boomplaas Cave mammals give us an indication of what we can expect. On numerous occasions we noted instances where limitations with the modern training sets or taphonomic biases created mismatches between the modern and fossil data. In all of these cases we generated quantitative paleoenvironmental reconstructions that were clearly inaccurate, even if sensible ordinal-scale trends were retained. Thus, it is reasonable to expect no-analog climates or fossil assemblages, both of which lead to incompatibility between the training sets and the fossil data, can also lead to problematic results because they require us to extrapolate beyond the limits of modern ecosystems.

Further Cause for Concern

As we discussed in Chapter 2, it is generally thought plants are more sensitive to climate than animals, because the distribution and abundance of the latter are often more directly linked to vegetation structure than to climate parameters such as temperature or precipitation. We also pointed out in Chapter 2 this is not a new idea.

To illustrate the point, let us consider the blue wildebeest (*Connochaetes taurinus*), an antelope found in grasslands and open woodlands in sub-Saharan Africa. Based on zoological observations (e.g., Kingdon and Hoffman 2013; Skinner and Chimimba 2005), it is clear an essential ecological requirement for this species is access to open, grassy habitats, grass being its preferred food. Blue wildebeest tend to not be found in regions that receive much more than ~1,000 mm annual rainfall – the average annual precipitation across its range is ~600 mm (Jones et al. 2009) – but few would argue this is due to some sort of physiological intolerance to too much rain (though having both lived in Seattle we can sympathize!). Rather, when rainfall increases, grasslands and woodlands often give way to forests (e.g., Good and Caylor 2011; Sankaran et al. 2005), which means the habitat and forage blue wildebeest depend on are not usually found in areas with high precipitation. On the other hand, we would never expect to find blue wildebeest in a dense shrubland simply because annual rainfall was similar to that observed across most of its geographic range. So although there may be relationships between the distribution of blue wildebeest and various climate parameters, these are mediated to a large extent by vegetation structure.

The point we wish to make is (to at least some extent) correlations between mammal communities and climate used for quantitative paleoenvironmental reconstruction exist insofar as there are correlations between vegetation structure and climate. Recall that earlier in this chapter we noted one of the handful of assumptions underpinning quantitative paleoenvironmental reconstructions is environmental variables other than the target variable have negligible influence on taxonomic composition (Chapter 3, assumption 5). If the target variable of interest concerns the climate (e.g., MAP or MAT), but the main determinant of taxonomic composition is vegetation structure, then we have violated this assumption and there is ample scope for problematic reconstructions. This is especially true if relationships between certain climate parameters and vegetation structure – and faunal communities by extension – differed in the past. As we discuss next, there is good reason to believe those relationships did occasionally differ.

Perhaps one of the more important variables that can alter the relationship between climate parameters such as MAP or MAT and vegetation structure is the amount of CO_2 in the atmosphere (Figure 9.8). Ice cores and other geological evidence indicate atmospheric CO_2 concentrations fluctuated widely throughout the Quaternary (e.g., Lüthi et al. 2008; Petit et al. 1999) and deeper in time (e.g., Berner and Kothavala 2001; Zhang et al. 2013), with important implications for vegetation communities (e.g., Ehleringer et al. 1991, 1997; Prentice et al. 2011). When atmospheric CO_2 concentrations are low, C_4 plants, primarily warm-season grass species (monocots), have a competitive advantage over C_3 plants, which include cool season-grasses (monocots) and most trees, shrubs, and herbaceous plants (dicots) (Ehleringer et al. 1997). The opposite is true during phases of elevated CO_2 concentrations. The effects of CO_2 on plant communities are not likely trivial. It is believed by some, for example, that changing CO_2 concentrations played a pivotal role in driving the global expansion of C_4 grasses at the end of the Miocene (e.g., Cerling et al. 1997; Ehleringer et al. 1997), vegetation change across Quaternary glacial–interglacial cycles (e.g., Ehleringer et al. 1997; Harrison and Prentice 2003), and the expansion of woody vegetation in parts of the world in recent decades (e.g., Polley 1997; Polley et al. 1997). The implication is that relationships observed between climate and vegetation (and animals) today would have differed in the past when CO_2 concentrations were either lower (e.g., glacial Pleistocene) or higher (e.g., the Miocene).

An example of this is provided in Figure 9.9, which shows the predicted dominance of C_4 and C_3 plants as a function of temperature and atmospheric

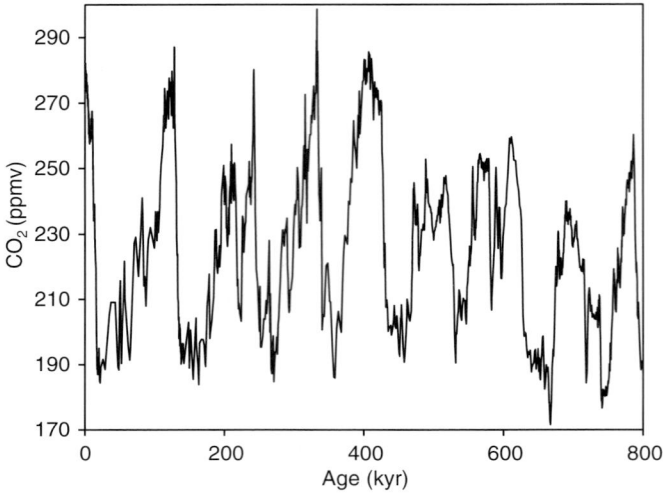

9.8. Antarctic ice core records of changes in atmospheric CO_2 over the past 800,000 years. Data from Lüthi et al. (2008).

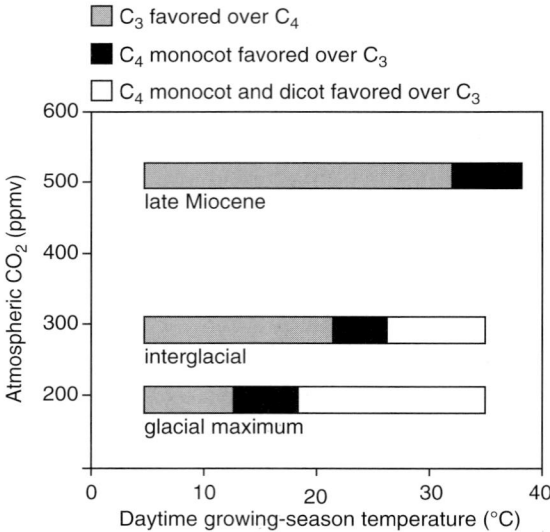

9.9. The predicted superiority of C_3 and C_4 photosynthetic pathways as a function of day-time growing-season temperature (°C) and atmospheric CO_2 concentrations (ppmv). Redrawn from Ehleringer et al. (1997).

CO_2 (after Ehleringer et al. 1997). The important point is that for any given temperature, the favored vegetation types change as a function of CO_2. To the extent vegetation determines faunal composition, this means quantitative paleoclimate reconstructions are potentially off the mark whenever CO_2 concentrations are substantially higher or lower than at present (the source of transfer functions). For example, while C_4 grasslands and their attendant faunas are restricted today (an interglacial) to relatively warm temperatures, they can occur at much cooler temperatures when CO_2 concentrations are lower. Indeed, there is abundant evidence that C_4 grasslands were more widespread during glacial phases of the Pleistocene when CO_2 concentrations were lower (Ehleringer et al. 1997). But because our transfer functions are necessarily restricted to present-day observations, it would be difficult (if not impossible) to reconstruct a cool-temperature C_4 grassland fauna during a Pleistocene glacial as indicating cool temperatures because there are no cool C_4 grasslands to be found today.

Changing CO_2 concentrations also have potential to influence precipitation estimates. On a global scale, precipitation is an important determinant of tree cover (Staver et al. 2011). Especially in the tropics, however, vegetation models show that lower CO_2 concentrations during glacial phases of the Pleistocene limit tree cover to a greater extent than predicted by precipitation change alone because of the physiological advantage of C_4 grasses at reduced CO_2 (Harrison and Prentice 2003; Prentice et al. 2011). Such models are supported by empirical paleoenvironmental archives (e.g., Sinninghe Damsté et al. 2011). This can be compounded by the effects of fire because the growth rate of trees is reduced under conditions of low CO_2, which prevents them from reaching a fire-proof size between burn intervals and allows faster-growing grasses to expand (Bond and Midgley 2000; Bond et al. 2003). It follows that when CO_2 concentrations are reduced (e.g., glacial Pleistocene) we may need *more* rainfall than at present to increase tree cover. Turning to methods of quantitative paleoclimate reconstruction, modern training sets for which relationships between tree cover and precipitation translate to relationships between fauna and precipitation can systematically underestimate rainfall whenever CO_2 concentrations are lower than present. We show next the opposite is also true whenever CO_2 concentrations are higher than at present.

Christine Janis et al. (2004) provide an important paleozoological case study illustrating how changing CO_2 concentrations can lead to problematic quantitative paleoclimate reconstructions. They were interested in explaining the tremendous species richness (relative to the present) of mid-Miocene (~18 to 12 Ma) browsing ungulates in North America. Especially relevant to our point here is they took advantage of clearly erroneous quantitative paleoenvironmental reconstructions to demonstrate the unusual composition of the Miocene faunas. They used a metric called *per-species mean hypsodonty* (PMH) – the average hypsodonty score (see Chapter 7 for a discussion of the terminology) of the ungulate fauna divided by the number of mammalian species >1 kg present – to reconstruct past precipitation (for other applications see Damuth and Fortelius 2001; Damuth et al. 2002; Eronen 2006; Eronen and Rook 2004; Fortelius 2003). PMH is reported to explain more than 60% of the variance in (log-transformed) MAP in modern communities, although full details concerning this relationship remain to be published. The transfer function relating PMH to MAP is provided by the following:

$$logMAP = -1.27(logPMH) + 1.36$$

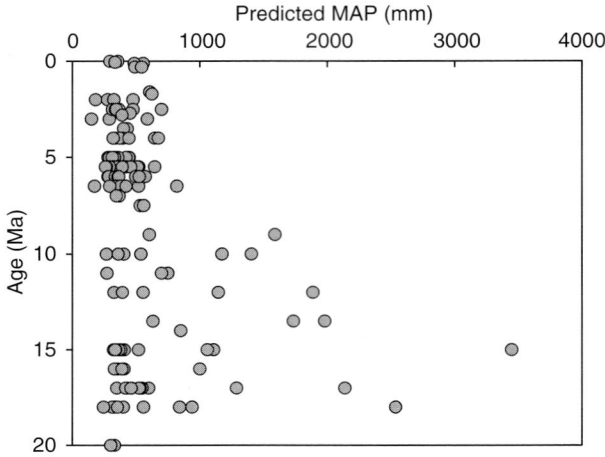

9.10. Reconstructed mean annual precipitation (mm) for the Great Plains of North America over the past ~25 Ma based on PMH values for ungulates. Raw data courtesy of Christine Janis and John Damuth.

PMH declines as rainfall increases, partly because higher rainfall is associated with closed habitats (e.g., forests) dominated by brachydont (low-crowned) browsing ungulates (Damuth and Janis 2011) and partly because higher rainfall often (but not always; see Chapter 8) drives an increase in richness.

The MAP estimates generated by Janis et al. (2004) for the Great Plains of North America are shown in Figure 9.10. Many of the values are absurdly high for localities older than 9 Ma. Not only do they substantially exceed reconstructions based on paleosols and ancient floras, but they approach levels more typical of tropical forests, which is inconsistent with any previous paleo-environmental reconstruction for the region. Janis and colleagues recognized this as anomalous and noted "[t]he exceptionally high number of brachydont species … is off-scale and causes the PMH regression – which is based solely on patterns among extant species – to predict unrealistically high values for the mid Miocene" (Janis et al. 2004:381). Among other possibilities, Janis et al. (2004) reasoned that elevated CO_2 concentrations were at least partially responsible for the high number of brachydont species (see also Janis et al. 2000). This is because when CO_2 concentrations are high, C_3 plants show increased rates of photosynthesis and growth, translating to an increase in primary productivity in C_3-dominated ecosystems (Campbell and Smith 2000). Recall from Chapter 8 that primary productivity is thought to be an important determinant of taxonomic richness in mammal communities. A CO_2-driven increase in primary productivity, Janis et al. (2004) argued, is likely responsible for the high number of (C_3) browsers in North America during the mid-Miocene, and by extension the anomalous precipitation estimates. An acknowledged weakness of their argument is that CO_2 concentrations this far back in time are notoriously contentious, but there are recent reconstructions consistent with their hypothesis (Zhang et al. 2013).

In short, changing CO_2 concentrations alter the relationship between climate, vegetation, and fauna relative to the present. For those cases where modern relationships between zoological communities and climate are mediated by vegetation structure – something likely to be the case for many mammal communities – quantitative paleoclimate reconstructions are likely to be biased whenever CO_2 concentrations are lower or higher than the present. The work by Janis et al. (2004) is the only study we are aware of that explores the magnitude of this bias, and it clearly demonstrates the bias can be severe. Until we have a better handle on these issues, we suspect the best potential for quantitative paleoenvironmental reconstruction in paleozoology (particularly those based on terrestrial mammals) is in applications that aim to quantify vegetation structure as opposed to climate (e.g., Louys et al. 2015). The good news is that with the proliferation of remote sensing data that allow us to quantify various aspects of vegetation structure, there is ample scope for the development of such techniques in the future.

SUMMARY

Quantitative paleoenvironmental reconstructions based on paleozoological data are alluring because they promise to generate definitive quantitative (interval- or ratio-scale) statements about past climate and environment, rather than limiting interpretations to qualitative (ordinal-scale) statements. However, at least with respect to examples examined here, limitations posed by the fossil data, the analytical techniques, and no-analog ecosystems mean that at best we are limited to low-resolution, ordinal-scale statements such as warmer or cooler. At worst we are left with reconstructions that are incorrect, as implied by the conflicting MAP trends documented at Boomplaas Cave (Figures 9.2 and 9.3).

Although we have been critical in this chapter, we do not wish to imply that quantitative approaches are a waste of time. Provided that sufficient caution is exercised, these approaches can enhance our confidence that observed patterns of variation in faunal composition are related to the environmental variable of interest. Thus, we can be reasonably confident, for example, that changes in micromammal composition at Boomplaas Cave are tracking changes in temperature. Avery's (1982) study of the Boomplaas Cave micromammals indicates that it is certainly possible to reach the same conclusion through conventional analyses, but quantitative paleoenvironmental approaches can help strengthen the link between (fossil) pattern and (environmental) process. Ultimately, we must remain alert to the numerous potential pitfalls that attend quantitative paleoenvironmental reconstruction; and as is the case in paleoenvironmental reconstruction in general, confidence in the outcome requires an assessment of multiple lines of evidence.

TEN

SIZE CLINES AS PALEOENVIRONMENTAL INDICATORS

In Chapter 2 we discussed the three ways in which a species might respond to environmental change. Provided that the magnitude of change is sufficient to warrant a response, populations of a species can shift their distributions to track preferred environments, they can adapt to the novel conditions, or they can die out (extirpation and extinction). These options need not be mutually exclusive (e.g., Davis and Shaw 2001). The paleoenvironmental reconstruction techniques outlined in previous chapters are sensitive to the first and third responses because they are ultimately based on the presence or abundance of taxa in paleozoological assemblages. Our focus in this chapter is on the second response, the ways in which species might adapt to changing environments. In particular, we are interested in phenotypic changes that are detectable in the paleozoological record and what they tell us about paleoenvironments.

The so-called ecogeographic rules are empirical generalizations that describe variation in a species' phenotype across environmental gradients (Mayr 1956). These include Bergmann's rule – "races from cooler climates in species of warm-blooded vertebrates tend to be larger than races of the same species living in warmer climates" (Mayr 1970:197) – and Allen's rule – "in warm-blooded animals protruding body parts like tail and ears are shorter in cooler than in warmer climates" (Mayr 1970:199). The traditional causal explanation for both, Bergmann's rule being the better known, is related to thermoregulation. The volume of a body increases as the cube whereas surface area increases as the square of a linear dimension (Figure 10.1). The larger the body, the

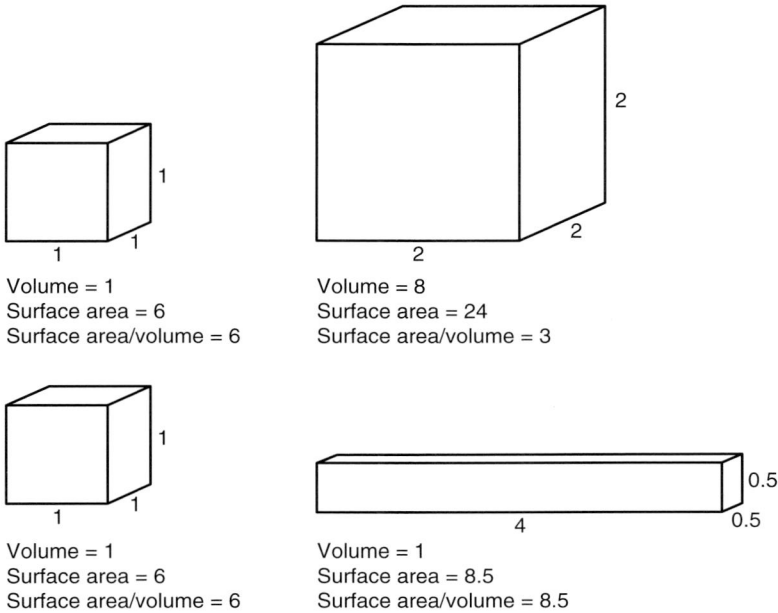

10.1. (Top) The geometric basis of Bergmann's rule: increasing volume decreases the surface area to volume ratio. (Bottom) The geometric basis of Allen's rule: increasing surface area increases the surface area to volume ratio even if volume is held constant. Adapted from Ruff (1994).

smaller the ratio of surface area to volume (Bergmann's rule). The shorter the appendage (forelimb, hindlimb, ear), the smaller the surface area (Allen's rule). In cool climates, larger bodies and shorter appendages translate to smaller surface area (from which heat is radiated) relative to volume (from which heat is generated), enhancing heat retention. And in warm climates, smaller bodies with longer appendages will have relatively larger surface areas and greater loss of (unnecessary) metabolic body heat.

Though there are other ecogeographic rules besides these (Lomolino et al. 2010; Rensch 1938), we focus our discussion in this chapter on ecogeographic variation in body size. Not only has this topic produced a massive literature, but it is also readily studied in the fossil record. All that is needed is a trusty set of calipers (more on this below). Thus, the study of *chronoclines* − a character gradient (e.g., size) expressed within a taxon over time (Simpson 1943a:174) − has provided insight to paleoenvironments by assuming documented size clines reflect the operation of an ecological principle such as Bergmann's rule. Allen's rule also has received attention in the paleozoological literature, where it is typically invoked to explain the adaptive significance of differences in body proportions between closely related species, such as those between Neanderthals and modern humans (Ruff 1991, 1994; Trinkaus 1981). However, we are unaware of examples where Allen's rule has provided the basis for paleoenvironmental reconstruction, probably because the necessary metrics (e.g., limb proportions relative to an independent measure of body size) are not readily available from most (typically disarticulated, fragmented, and scattered) fossil samples. This does not mean there isn't potential to use Allen's rule as a framework to generate paleoenvironmental reconstruction. But because the analytical and inferential steps as well as potential pitfalls would be much the same as those involved in the analysis of size clines, we focus on the latter in this chapter.

SIZE CLINES IN MODERN ORGANISMS

We first turn our attention to size clines in modern organisms, as this provides the basis for reading the paleoenvironmental significance of size clines (chronological or spatial) in the paleozoological record. Here we provide a brief overview of Bergmann's rule and the many other factors known to mediate body size.

Bergmann's Rule: More than Just Temperature

In the middle of the nineteenth century, German biologist Carl Bergmann (1847:648) observed that within genera of mammals and birds, "the larger species live farther north [i.e., high latitudes] and the smaller ones farther south [i.e., low latitudes]" (translated in James 1970). He reasoned this was because of the reduction in surface area relative to volume as size increases (Figure 10.1), allowing larger-bodied species to more effectively conserve body heat in cooler high-latitude climates. Specifically, Bergmann (1847:600–601) observed "for the amount of warmth, of which an animal can raise itself above its surrounding, the relation of its volume to its surface is of course of great importance," and "it is thus clear, that animals have to create less warmth in relation to their size the larger they are, to gain a certain increase in temperature above the one of their surrounding. This law must be of great importance on the mode of life of warm-blooded animals." And a few pages later he noted, "if there would be genera, which species are distinguished as much as possible only by size, the smaller species would all need a warmer climate" (Bergmann 1847:638; translation from Watt et al. 2010).

In the decades after Bergmann published, sufficient empirical support for his observations (both among and within species) had been amassed that it came to be known as Bergmann's rule (e.g., Rensch 1938). Although debate still exists as to whether Bergmann intended his "law" to pertain to size variation between species of the same genus or between populations of the same species (Meiri 2011; Watt et al. 2010), Rensch (1938) and Mayr (1954, 1956) formulated the rule to refer exclusively to intraspecific variation. Their conceptualization of Bergmann's rule is adopted here and in most modern paleozoological work on the matter.

10.2. Changes in adult body mass of moose (*Alces alces*) individuals across a latitudinal gradient in Sweden. Data from Sand et al. (1995).

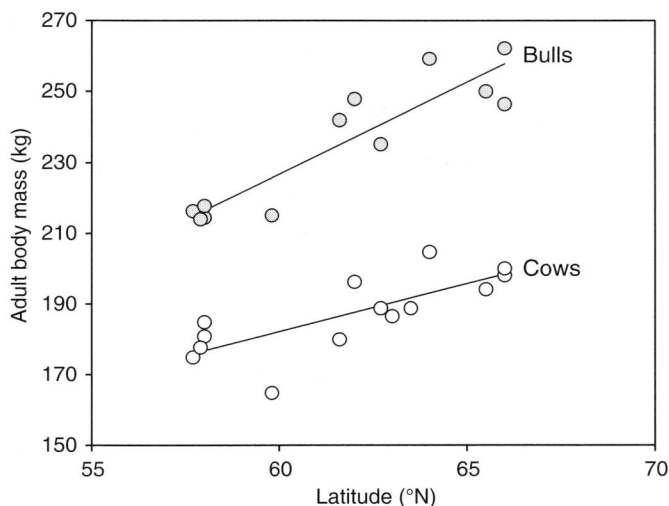

A massive body of literature has explored the extent to which Bergmann's rule holds for vertebrates (see reviews in Ashton 2002; Ashton et al. 2000; Meiri and Dayan 2003; Millien et al. 2006). Correlations between body size and latitude (or temperature) are commonly observed for most vertebrate groups, except for fishes, snakes, and lizards (Millien et al. 2006). Traditionally these relationships are assessed by examining a measure of body size (e.g., mass, length, dental metrics) across populations of a species spanning a latitudinal gradient, latitude taken as a proxy for temperature. An example is illustrated in Figure 10.2, which shows the correlation between body mass and latitude for moose (*Alces alces*) from Sweden (Sand et al. 1995). Others have examined size clines in relation to observations from local weather stations or interpolated climate data (e.g., Avery 2004; Blois et al. 2008; James 1970; Klein and Cruz-Uribe 1996; Yom-Tov and Nix 1986). Global climate data are now increasingly available (e.g., the WorldClim global climate data; Fick and Hijmans 2017; Hijmans et al. 2005) for analysis using GIS (geographic information system) software, allowing for evaluation of size clines relative to temperature and numerous other climatic parameters (e.g., Faith et al. 2016b).

There has been much discussion and debate concerning the empirical generality and the presumed adaptive basis of Bergmann's rule. To some, exceptions to the rule or the existence of other potentially more effective thermoregulatory mechanisms (e.g., fur and plumage, vascularization, fat deposits, behavior) provide sufficient reason to doubt its validity (e.g., Geist 1987; McNab 1971, 2010; Sholander 1955). One of many examples where Bergmann's rule does not apply is provided by Wasserman and Nash (1979), who showed that the body mass of deer mice (*Peromyscus maniculatus*) in central Colorado did not vary as a function of altitude (and therefore temperature; ambient air temperature tends to decrease 2°C [3.6°F] for every 305 m [1,000 ft] increase in elevation), but that the length and density of their hair did. Likewise, some carnivores are known to display size clines in complete opposition to the predictions of Bergmann's rule (Dayan et al. 1991). But as Mayr (1956:106) prudently noted more than half a century ago, exceptions are to be expected because "[t]he need for heat conservation is only one of many possible selection pressures affecting absolute or relative body size." He also argued that the existence of alternative means of temperature regulation does not completely eliminate any selective advantage of changing body mass: "Multiple solutions for biological needs are the general rule in evolution" (Mayr 1956:107). Today, biogeographers are comfortable viewing Bergmann's rule as "a general tendency exhibited by most of the species or races" (Lomolino et al. 2010:587).

Even if Bergmann's rule seems to hold for many species, we do need to be concerned that mechanisms other than (or in addition to) temperature can mediate size clines. Indeed, there are plenty of examples that illustrate this point. In her analysis of North American birds, James (1970) found that body size was better explained by a combination of temperature and humidity than by temperature alone (see also Wiggington and Dobson 1999). Rosenzweig (1968) argued that primary productivity and forage availability – variables that are often correlated with latitude – are more important than temperature in explaining the body size of North American carnivores (see also Geist 1987, 1998; Kolb 1978). Consistent with this argument, Wolverton et al. (2009) showed that size clines in white-tailed deer (*Odocoileus virginianus*) can be explained by food availability; larger deer are found in areas with greater productivity or lower deer population densities, the latter translating to reduced competition and more forage available for any given individual (more on this below). The *productivity hypothesis* is also supported by observations indicating that precipitation, a key determinant of primary productivity (see Chapter 8), is an important correlate of body size in some species (e.g., Avery 2004; Blois et al. 2008; Klein 1991; Klein and Cruz-Uribe 1996; Yom-Tov and Geffen 2006; Yom-Tov and Nix 1986). For example, Klein and Cruz-Uribe (1996) showed that the skull length of rock hyraxes (*Procavia capensis*) in South Africa varies as a function of annual precipitation but not temperature, presumably because rainfall determines availability of the foods preferred by hyraxes. Others have proposed that rather than reflecting a need for heat conservation, larger body size at high latitudes may instead be driven by greater seasonality and its effects on resource availability (Boyce 1978, 1979; Lindstedt and Boyce 1985). This is because larger individuals are better equipped to endure periods of forage scarcity, a phenomenon known as *fasting endurance* (Millar and Hickling 1990).

Yet another complicating factor that emerges in discussions of Bergmann's rule and size clines is *character displacement* – the situation where differences between two species (especially potential competitors or closely related taxa) with partially overlapping ranges are accentuated in the zone of sympatry, presumably to enhance the ecological distance between them and reduce competition (Brown and Wilson 1956). Thus, the extent to which size clines are observed within one species may have something to do with the presence or absence of another. For instance, McNab (1971) noted that martens (*Martes americana*) in western Canada increase in size (length) only in latitudes above 62° N, which corresponds to the northern limit of the larger-bodied fisher (*Martes pennanti*). But below that latitude, marten body size is stable. Dayan et al. (1989) showed that the red fox (*Vulpes vulpes*) displays a typical Bergmannian response except in the warmer parts of its range where it overlaps with the smaller Ruppell's sand fox (*Vulpes ruppelli*). In both of these cases, character displacement resulting from the presence of a closely related competitor seems to prevent populations from responding to environmental gradients via changes in body size (see also Chetboun and Tchernov 1983; Tchernov 1979). In a later study, Dayan et al. (1991) argued that character displacement might explain many cases where carnivore populations exhibit correlations between temperature and size in some regions but not in others. On the other hand, there are cases where character displacement has been explicitly examined and deemed to not be very important (e.g., Avery 2004; Blois et al. 2008; Marean et al. 1994; Purdue 1980), but it remains an important ecological mechanism to keep in mind.

Although Bergmann's rule is traditionally studied by exploring size clines across space, important insights have been gleaned from longitudinal studies spanning years to decades. For example, Smith et al. (1998) documented a steady ~30 g decline in the average body mass of white-throated woodrats (*Neotoma albigula*) in New Mexico from 1989 to 1996. They showed that the decline corresponded to increasingly warmer summer and winter temperatures, consistent with Bergmann's rule. Meanwhile butterflies in Greenland have become smaller over the past two decades as summer temperatures have increased (Bowden et al. 2015). There are numerous other examples besides these (e.g., Caruso et al. 2015; Teplitsky et al. 2008; Yom-Tov et al. 2006; Yom-Tov et al. 2010), and as global climates continue to change we can expect to see more in the future (Sheridan and Bickford 2011).

The fact that Bergmannian size clines can be established rapidly – in a geological instant from the paleozoological perspective – provides reason to be optimistic about the use of body size as a paleoenvironmental proxy. We do not need long temporal sequences to detect changes in size. But then again, observations of size clines through recent time also reinforce the complications that may attend interpreting the fossil record. Gardner et al. (2011) provide an important critical review of recent literature on temporal clines in historic times. They found that not only are the clines variable – some species are getting bigger, some smaller, and some don't seem to be changing at all – but forage quality and availability are frequently implicated as a driver of change (we turn to this topic later in this chapter). At high latitudes, for example, many species are becoming larger as temperatures increase, a pattern attributed to longer growing seasons and greater food availability during the winter months (e.g., Yom-Tov and Yom-Tov 2005; Yom-Tov et al. 2008). Such results all too clearly support Mayr's (1956) observation that temperature is but one of many variables that can influence body size.

The Island Rule

The *island rule* deserves some attention here, and not just for our readers working on island faunas. Among insular vertebrates, the island rule is the tendency for smaller species to become larger and for larger species to become smaller (Foster 1964; Lomolino 1985, 2005; Van Valen 1973). There are spectacular examples of the latter from the fossil record, with well-known examples including the pygmy mammoths (*Mammuthus exilis*) from California's Channel Islands (Agenbroad 2009, 2012), the dwarf mammoths from Wrangel Island in the Chukchi Sea (Vartanyan et al. 1993), and the so-called hobbits (*Homo floresiensis*) from Flores in Indonesia (P. Brown et al. 2004). The island rule stems from J. Bristol Foster's (1964) observations that insular rodents from western North America and Europe tend to be larger than their mainland conspecifics (island gigantism), whereas lagomorphs, carnivores, and artiodactyls tend to be smaller (island dwarfing). Van Valen (1973) later argued that the pattern was sufficiently generalizable that it deserved to be considered a rule. Although there are probably several mechanisms at work, the typical explanation includes a combination of *ecological release* stemming from a lack of predators and intraspecific competitors (species richness being lower on islands) coupled with resource limitations (e.g., Lomolino 2005; Lomolino et al. 2010). On one hand, smaller species (e.g., rodents) on islands need not rely much on their diminutive size for predator avoidance, but they do need to compete with their conspecifics and larger size may facilitate exploitation of a broader range of resources. And on the other hand, island dwarfing in otherwise larger species (e.g., artiodactyls) may be a response to limited resource availability, together with the fact that large size is no longer a necessary predator deterrence strategy.

Like Bergmann's rule, the island rule has produced a vast literature (see reviews in Lomolino 2005; Millien et al. 2006), some of which documents important exceptions (e.g., Meiri et al. 2005) or questions its generality (e.g., Meiri et al. 2008). It suffices to note here that the island rule need not apply to all species or to all populations of a species, as is also the case with Bergmann's rule. But this does not detract much from its potential as a paleoenvironmental indicator. For example, Lister (1989) showed that red deer (*Cervus elephas*) on what is today the island of Jersey near France's Normandy coast rapidly dwarfed (from ~200 to 36 kg) during the Last Interglacial, when sea levels rose and the island became isolated from the mainland. So at least for insular faunas, size clines may have potential to inform on changes in sea level, especially because the strength of the island effect seems to be related to island size (Heaney 1978).

The island rule need not apply only to islands. For example, Schmidt and Jensen (2003, 2005) documented island rule-like size shifts in Danish mammals and birds over the past two centuries, a pattern they attributed to anthropogenic habitat fragmentation – the fragmented habitat patches being comparable (in a biogeographic sense) to oceanic islands. People are but one of several factors (e.g., climate, fire, herbivory) that might drive habitat fragmentation over geological timescales, and the observations on Danish vertebrates pose interesting possibilities for future paleozoological studies, even if they make things a little more complicated for those interested in Bergmannian size clines.

Summary

If there is one thing to take from our brief overview, it is that size clines cannot be uncritically interpreted as the outcome of a direct influence of temperature or any other environmental variable. It is true Bergmann's rule is frequently exhibited by many species, but it would be dangerous to assume (in the absence of other evidence) that those size clines – or clines observed in the fossil record – necessarily reflect an adaptation to temperature. We agree with Hill et al. (2008:1755), who noted "Bergmann's Rule should be treated as a generalized empirical finding or pattern (with many exceptions) but not as an explanation." There are multiple variables in play – some of which are likely to be correlated with each other (e.g., temperature, seasonality, productivity, forage availability) – and these may be more or less important between different species and even between different populations within a species (today and in the past). So even when size clines across environmental gradients are clearly established for a species, such correlations may be spurious and need not reflect the operative mechanism responsible for (body) size. Of course, this is not the first time in this volume we have encountered a case where general patterns in nature become more complex when examined in detail. This does not mean size clines are a problematic paleoenvironmental indicator, only that we need to keep such complexity in mind when we begin to interpret patterns found in the fossil record. But before we turn to that topic, we need to briefly review some of the basics of gathering and analyzing metric data on fossils.

ASSUMPTIONS, WHY MEASURE, AND METHOD BASICS

Much like the analytical assumptions requisite to paleoenvironmental reconstruction we summarize in Chapter 3 of this volume, the requisite assumptions and basic procedures of paleozoological metric analyses are not always mentioned in the literature we are familiar with, which is not to imply the procedures are unimportant. In fact, they are extremely important if valid analytical results and well-supported interpretations are to be produced. We therefore begin this section by stating the assumptions required of analyses seeking to reconstruct paleoenvironments on the basis of metric data. We then turn to a brief consideration of the several reasons, other than paleoenvironmental reconstruction, why paleozoologists have collected metric data from animal fossils. That discussion identifies factors that render interpretation of metric data in terms of paleoenvironments less than straightforward. Finally, we describe the basics of collecting osteometric data from both modern and fossil bones and teeth.

Assumptions

If a chronocline documented by ostemetric data recorded among fossils is to be interpreted in terms of changing paleoenvironments, then several assumptions are necessary. One of these (after Martin 1990) is that the size change documented in the fossil record is a function of change in one or more environmental variables. This assumption can be analytically evaluated, and in later pages of this chapter we present examples of how to do so. Another assumption (after Martin 1990) is that the relationship between the bone or tooth dimension that has been measured and one or more environmental variables documented among modern individuals was the same in the past as it is today. This second assumption is simply the fundamental assumption of uniformitarianism described in Chapter 3 but here phrased in terms of size.

Another assumption, or something of a rephrasing of the second assumption immediately above, is that the dimension(s) chosen for measurement reflects phenotypic change rather than genetic change (e.g., Hadly 1997; Jass et al. 2015). That is, the cause of a size change in the fossil record could be evolution (defined as genetic change), which might or might not be driven by environmental change. For instance, one more or less genetically distinct population (a subspecies, say) might be displaced by an immigrating population of conspecific subspecies. An intriguing example concerns North American gophers. On the basis of a study of modern individuals, Hadly (1997) argued that the mandibular alveolar length of northern pocket gophers (*Thomomys talpoides*) was genetically controlled and thus of taxonomic significance at the subspecies level, whereas the mandibular diastema length correlated with body size and was an ecophenotypic trait (influenced by environment but not genetics). She was concerned the replacement of one subspecies of pocket gopher by another could result in what might otherwise be (mis)interpreted as a chronocline that reflected environmental change. In an analysis of fossils deposited over the past

3,200 years in a cave in northwestern Wyoming, she was able to show that diastemal length fluctuated basically in harmony with other environmental records whereas alveolar length remained stable. Hadly concluded that the chronocline was not the result of genetic replacement of the resident pocket gopher population, but rather a phenotypic response to changing environments. She subsequently analyzed ancient DNA from some of the prehistoric gopher mandibles and showed there had been genetic stasis in the measured mandibles (Hadly et al. 1998; see also Jass et al. 2015).

A related issue worthy of consideration is genetic drift – the sampling error that sometimes attends biological reproduction and associated phenotypic change. Particularly in small populations, genetic drift can lead to phenotypic changes (e.g., in an organism's size) through time that are random and unrelated to paleoenvironmental changes. With this in mind, a variant of the first assumption noted above is when interpreting size clines in terms of the paleoenvironment, the paleozoologist assumes genetic drift is unimportant. We are not aware of any paleozoologist who has considered drift in this context, but this need not imply such an effort could (or should) not be made. For those readers interested in this issue, we note paleontologists have long considered genetic drift in attempts to determine whether evolutionary changes in an organism's phenotype are the result of (random) drift or (non-random) selection (e.g., Clyde and Gingerich 1994; Gingerich 1993; Raup 1977). The traditional procedure for doing so is to model phenotypic change resulting from drift as a random walk (i.e., a stochastic process), which serves as a null model that can be compared with the fossil data. If the fossil data are inconsistent with the null model, the inference is that some sort of non-random process (e.g., selection) is responsible for the observed phenotypic changes. The limitation to this approach is that it is often difficult to reject the null model (i.e., it lacks statistical power) even when there is good evidence that non-random evolutionary processes are in play (Bell et al. 2006; Sheets and Mitchell 2001). A sophisticated solution to this problem involves fitting the fossil data to various evolutionary models (e.g., drift, stasis, selection) and using information statistics such as the Akaike Information Criterion (AIC) to determine which model is the better fit (Hunt 2006, 2007; Hunt et al. 2008). We recommend this approach for determining whether drift or paleoenvironmental change is responsible for size clines observed in the fossil record.

A final assumption that should be mentioned is that the analyst must assume the dimensions measured have been measured with sufficient *accuracy* and resolution to capture size change across an environmental gradient. This is where the search for a correlation between the measurements of modern specimens and an environmental variable comes into play (e.g., Olcott and Barry 2000). If no correlation is found, then perhaps a dimension was not measured with sufficient resolution, or a dimension that does not vary across environmental gradients was measured.

Why Measure?

There are several reasons in modern paleozoology to take measurements of dimensions of fossils, where a *dimension* is the theoretical distance between two anatomical landmarks, where a *landmark* is an easily defined and located fixed point on a bone or tooth. A *measurement* (in this case) is the measured distance between two landmarks on an individual bone or tooth. Metric data have been collected from fossils to reconstruct past environments and for numerous other reasons as well (Boessneck and Driesch 1978; Orton 2014).

Osteometric data are sometimes used to identify the taxon of animals represented by a bone or tooth, whether domestic species or "breeds" (e.g., Eisenmann 1986; Grigson 1969; Higham 1969; Payne 1969; Reitz and Ruff 1994), or wild forms (e.g., Harris 1984; Livingston 1987; Plug 2005; Stahl 2005). Some analysts think such use of metric data is ill-advised when seeking to establish definitive criteria that distinguish a new species from a closely related well-known species (e.g., Geist 1992; Ryder 1992; for an excellent paleozoological example, see Mead et al. 2000; Graham 2001; Mead and Spiess 2001). One of the hallmark (though not universal) features of the process of animal domestication is diminution of body size, so detection of such requires bone measurements (e.g., Grigson 1989; Tchernov and Horwitz 1991; Zeder 2001; Zeder and Hesse 2000). Given the two preceding uses, it is not at all surprising metric data are often used to distinguish wild from domestic forms of congeners in a collection of fossils (e.g., Evin et al. 2014; Walker and Frison 1982). Because bones increase in size as an animal grows, osteometric data have been used to assess the ontogenetic age of individuals represented by fossils (e.g., Etnier 2002, 2004). Finally, many species display sexually dimorphic body sizes that might be recognized in the fossil record. Among other things, this feature has prompted a tremendous amount of research on how to determine the sex of the individual represented by bones of the iconic mammal of North America, the bison (*Bison* spp.) (e.g., Bedford 1978; Duffield 1973; Kooyman and Sandgathe 2001; Lewis et al. 2005; Morlan 1991; Roberts 1982; Todd 1986; Walde 2004). Knowing the sex makeup of a fossil collection representing multiple individuals may reveal something about the behavior of the animal group, or of the predator that accumulated the bones of that animal. Finally, metric data have been used to assist with anatomical refitting (determining, say, which left tibia goes with which right tibia such that both can be argued to be from the same animal) (e.g., Lyman 2006a; Todd 1987). Distances between anatomically refitted specimens could reveal something about the disarticulation and dispersal history of the represented carcasses.

Method Basics

To record osteometric data, one must take measurements. Well and good, but things are seldom so simple. Replicability of measurements is preferable, and in two senses (Breslawski and Byers 2015). The analyst should be able to measure precisely the same dimension on specimen after specimen, and also different analysts should be able to measure exactly the same dimension. The former requirement is necessary if a series of measurements are to show what the analyst hopes they show. The latter requirement allows one researcher to compare her data with another's. One thing that has facilitated meeting both requirements is the use of standard, well-defined dimensions, that is, ones that are specified by easily located anatomical landmarks. Toward that end, many zooarchaeologists and some paleontologists measure the skeletal dimensions defined by Driesch (1976) for mammals and birds; Morales and Rosenlund (1979) have defined a number of dimensions of fish bones. Dimensions other than those specified by these researchers can be defined and used, but in doing so the analyst should make sure that they are well defined and replicable – the dimensions should be between two easily located (that is, replicable) landmarks.

Skeletal dimensions to be measured should be well preserved and, hopefully, abundant in that portion of the fossil record under study. Given preceding chapters, it should be clear that larger samples will provide more statistically robust results and lessen the chance of, say, the average of a dimension among fossils differing from the average of that dimension among the comparative sample because of a non-representative sample of one or the other population (i.e., random noise). If one or both landmarks defining a dimension tends to be poorly preserved (e.g., subject to abrasion, fracture, or exfoliation; Breslawski and Byers 2015), measurements likely will not reflect the actual size of the dimension. This aspect of measuring bones and teeth concerns *accuracy*, or how close a measurement is to the true value (size) of a dimension of a particular specimen (Lyman and VanPool 2009).

Related to accuracy, as *resolution* increases, so too does accuracy (usually), up to the limits of the measuring tool. A measurement taken to the nearest 0.01 mm provides more resolution than one to the nearest 0.1 mm, though this does not mean faunal analysts can reliably replicate measurements to the nearest 0.01 mm. How much resolution, then, should the analyst seek? As paleontologist George Gaylord Simpson indicated some years ago, dimensions should be logically related to a research problem, adequately reported, well defined, and standardized (Simpson et al. 1960). Because they are analytical tools, and are taken by humans using tools, resolution depends on both the skill of the human and the resolution built into the tool. So an analytically useful resolution level will depend on the measurer's skill, the tool's inherent resolution limits, and the research question. A good rule of thumb is to measure dimensions in which the number of unit steps from the smallest to the largest measurement is between 30 and 300 units (e.g., centimeters, millimeters, tenths of millimeters) (VanPool and Leonard 2009). For instance, if one is measuring the length of specimens (say, femora) that range from 11.2 cm to 15.7 cm, measuring specimens to the nearest cm produces six unit steps (11, 12, 13, 14, 15, 16 cm). Such will obscure variation that likely would be analytically meaningful. Measuring these specimens to the nearest 0.01 cm (which produces 450 unit steps) will produce data at a higher resolution than likely is necessary and probably also to some degree inaccurate (some degree of measurement error is incorporated). Measuring the femora specimens to the nearest 0.1 cm will result in 45 unit steps and likely be of sufficient resolution and accuracy to serve one's analytical purposes. But please recognize there are no hard-fast-universal rules here. Common sense is good to exercise in this regard.

When it comes to using common sense, if one wishes to compare measurements taken from fossil specimens with those taken from modern specimens, one needs to choose modern specimens wisely. First, do not measure the skeletal remains of what were zoo or captive animals as their behavior patterns and diets were likely different than those of the wild animals represented by the fossils, and this could influence the size of their bones and teeth (O'Regan and Turner 2004). Second, measure only one side of each modern comparative skeleton to avoid interdependence of measurements. This part of the protocol follows the assumption that animals are bilaterally symmetrical, which, while we know it to be false to some degree (e.g., Lyman 2006a), serves to underscore that metric data taken from both sides of an animal result in non-independent observations and could thus skew results. Typically, both sides of fossil specimens are measured because it is usually impossible to determine whether, say, a particular left and particular right tibia represent the same individual animal (Lyman 2006a; but see Todd 1987). Third, measure only bones of skeletally mature adult animals among both modern skeletons and fossil remains. This will avoid the potentiality that one or the other sample provides smaller measurements because the represented individuals were not fully grown. Fourth, measure bones of both sexes in the comparative collection to avoid skewing results for sexually dimorphic taxa. And finally, insofar as possible, measure modern specimens from the same general geographic area as the fossil collection in order to avoid geographic influences on size.

SIZE CLINES AS A PALEOENVIRONMENTAL PROXY

Because ecogeographic variation in body size has long been a well-known (albeit complex) phenomenon, it should come as little surprise that paleozoologists have long explored size clines in the fossil record for purposes of deciphering their paleoenvironmental implications (e.g., Guilday et al. 1964; Hibbard 1963;

Hooijer 1947; Kurtén 1957, 1960, 1965). Some of that literature is summarized in Table 10.1, which provides a selection of paleozoological studies that have explicitly examined size clines across the modern range of a species to facilitate interpretation of the environmental significance of size clines in the fossil record. While Table 10.1 is by no means exhaustive, it shows that the best-studied taxa are carnivorans – in large part due to Klein's (1986) efforts – and rodents. The rarity of large herbivores on our list should not be taken to imply that they do not display size clines across environmental gradients through space or time. As we discuss below, they most certainly do (e.g., Ashton et al. 2000; S. Davis 1981; Guthrie 1984a, 2003; Hill et al. 2008; Klein 1976a; Purdue 1989; Reynolds 2007).

In addition to the studies in Table 10.1, many paleozoologists have explored size clines through time in the absence of detailed knowledge about modern spatial variation (e.g., Avery 1982; Guthrie 2003; Hill et al. 2008; Kurtén 1957, 1965; Weinstock 1997; Wilson 1978). Such studies are no less interesting. They provide insight into how species respond to (usually) independently documented paleoenvironmental changes. But this is not the same as using size clines to infer environmental change. As Avery (1982:277) noted, in order to do that "basic data must be available for modern representatives of the species concerned." In light of our previous discussion (see also Table 10.1), the reason should now be clear – there are numerous factors that mediate body size. One might assume that observed size clines are a reflection of an ecological principle such as Bergmann's rule, but that assumption is tenuous at best. And even if observed clines appear to be consistent with Bergmann's rule (or any other environmental correlate), one risks a breakdown in logic if that pattern is used as the basis for paleoenvironmental interpretation. Consider a species that decreases in size from the Last Glacial Maximum through the Holocene, a timespan characterized globally by an increase in temperatures (among other things). We can reasonably suggest that this species is responding in a manner consistent with Bergmann's rule (e.g., Guilday et al. 1964), even if we cannot be certain that factors other than temperature are responsible. But we cannot then infer from the size clines that the Last Glacial Maximum was cooler. That would be circular reasoning.

Before delving into a few paleozoological case studies, we first outline what we take to be a typical strategy for using size clines in paleoenvironmental reconstruction. The process may seem fairly straightforward at first glance, but there are enough considerations and potential complications to warrant some discussion.

One Way to Go about It

The first step begins with selection of a target species. Which species should be studied? This of course depends on our research questions, but we are

TABLE 10.1 *A selection of mammalian taxa for which size clines across spatial environmental gradients have been used to infer paleoenvironmental change.*

Order	Species	Size proxy	Environmental correlate	Source
Artiodactyla	Odocoileus virginianus	astragalus	forage availability	Wolverton et al. 2007, 2009
	Sus scrofa	m3	summer temperature	S. Davis 1981
Carnivora	Acinonyx jubatus	m1	latitude	Klein 1986
	Aonyx capensis	P4	latitude	Klein 1986
	Atilax paludinosus	m1	latitude	Klein 1986
	Canis lupus	m1	summer temperature; winter temperature	S. Davis 1981
	Canis mesomelas	m1	latitude	Klein 1986; Klein et al. 1999, 2007
	Caracal caracal	m1	latitude	Klein 1986
	Crocuta crocuta	m1	latitude	Klein 1986; Klein and Scott 1989
	Felis silvestris	P4	latitude	Klein 1986
	Galerella pulverulenta	m1	latitude	Klein 1986; Klein and Cruz-Uribe 2000
	Herpestes ichneumon	m1	latitude	Klein 1986
	Ictonyx striatus	m1	latitude	Klein 1986
	Leptailurus serval	m1	latitude	Klein 1986
	Lycaon pictus	P4	latitude	Klein 1986
	Mellivora capensis	P4	latitude	Klein 1986
	Panthera leo	m1	latitude	Klein 1986
	Parahyaena brunnea	m1	latitude	Klein 1986
	Vulpes chama	m1	latitude	Klein 1986
	Vulpes vulpes	m1; mandibular molar row	summer temperature; winter temperature	Davis 1977

Order	Species	Measurement	Environmental correlate	Reference
Hyracoidea	*Procavia capensis*	cranial length; distal humerus	annual precipitation	Klein and Cruz-Uribe 1996, 2000
Lagomorpha	*Sylvilagus floridanus*	p4; distal humerus; astragalus	longitude	Purdue 1980
Rodentia	*Bathyergus suillus*	distal humerus	annual precipitation	Klein 1991; Klein and Cruz-Uribe 2000
	Cryptomys hottentotus	mandibular alveolar row	%winter rainfall; summer aridity	Avery 2004
	Neotoma cinerea	fecal pellets	summer temperature; winter temperature	Smith et al. 1995, 2009; Smith and Betancourt 1998, 2003
	Neotoma lepida	fecal pellets	summer temperature; winter temperature	Smith and Betancourt 2003
	Ondatra spp.	m1 (length/width)[a]	latitude	Nelson and Semken 1970
	Tachyoryctes splendens	mandibular alveolar row	annual temperature	Faith et al. 2016b
	Thomomys talpoides	mandibular diastema	elevation	Hadly 1997
	Sciurus carolinensis	distal humerus; proximal radius; astragalus	latitude; longitude	Purdue 1980
	Otospermophilus beecheyi	mandibular diastema	annual precipitation; winter precipitation	Blois et al. 2008

Notes: For species with multiple significant environmental correlates, only the strongest are indicated here.

[a] By dividing m1 length by width, this becomes a measure of shape rather than size.

also constrained by those species that are reasonably abundant in the fossil assemblage(s) of interest. And in the case of species for which the environmental correlates of body size (e.g., temperature, precipitation, productivity) are poorly understood in modern populations, it also pays to choose one that is reasonably well represented in modern zoological collections, as these will provide the foundation for identifying how size varies across modern environmental gradients (more on this below). Keeping in mind these constraints, the decision often can be informed by previous work hinting at the possibility of environmentally mediated size clines. For example, Klein's (1986) influential study of size clines in modern and Quaternary carnivorans from southern Africa was motivated by Hendey's (1974) observation that some carnivorans in the region varied markedly in size through time. And Avery's (2004) work on size variation in the common mole-rat (*Cryptomys hottentotus*) from southern Africa stems from her earlier observations that late Quaternary specimens fluctuated in size in a manner suggesting an environmental driver (Avery 1982).

What is the appropriate proxy for size? When zoologists think of an animal's size, it is usually in terms of body mass, length, or height — mass being the variable of greatest physiological and ecological importance (J. Brown et al. 2004). Paleozoologists are faced with typically fragmentary skeletal remains from which they hope to provide a reasonable proxy of an individual's size. Fortunately, the relationships between various cranial, dental, and postcranial measurements and size (mass) are well documented for some taxonomic groups (e.g., Damuth and MacFadden 1990; Hopkins 2008; Legendre and Roth 1988; Purdue 1987). Table 10.1 illustrates the diversity of proxies used to measure size for paleoenvironmental reconstruction. The decision of which one to choose is often influenced by what is preserved and identifiable in the fossil record. Klein (1991:246), for example, measured the distal humerus breadth of Cape dune mole-rats (*Bathyergus suillus*) for the very sensible reason that it "is very durable and thus the most abundant skeletal element in the fossil samples." Ideally we also want a metric that is known to track body size at the intraspecific level (e.g., Blois et al. 2008; Hill et al. 2008; Purdue 1989; Smith et al. 1995; Wolverton et al. 2007). In practice, however, this is not always possible (e.g., Avery 1982; S. Davis 1981; Faith et al. 2016b; Klein 1986) because modern zoological specimens on which skeletal metrics are taken are not always associated with independent body size data (e.g., body mass). Complete tooth rows or diastema lengths are typically measured on rodents (e.g., Hadly 1997; Jass et al. 2015), for which mandibles tend to be well preserved, though for larger species isolated teeth may be all that is available (Table 10.1). The reliance on teeth or tooth rows is not unproblematic, in part because they may be subject to selective pressures (e.g., those related to diet) independent of an animal's body size (Dayan et al. 1991). And although tooth size tends to be tightly correlated with body size when a broad range of small and large species are considered together (e.g., Creighton 1980; Damuth and MacFadden 1990) – Gould's (1975:351) so-called "mouse-to-elephant" plots – the same may not be true at the intraspecific level (e.g., Wood 1979). This means that dental metrics may not track changes in body size across environmental gradients, even if other skeletal elements do. But we can at least be confident that evidence of size clines in dental metrics in modern populations of a species means that we should be able to detect meaningful patterns (for the same species) in the past.

While many paleozoological studies explore size clines in a single skeletal element (e.g., Blois et al. 2008; Klein 1986; Klein and Cruz-Uribe 1996) or in a combination of elements examined separately (e.g., Hill et al. 2008; Purdue 1980), it is also possible to consider multiple elements simultaneously (see Albarella [2002] and Meadow [1999] for critical reviews of several techniques for doing this). In a study of size clines in fossil reindeer (*Rangifer tarandus*) from Europe, Weinstock (1997) used an index called the variability size index (VSI) to pool multiple elements from an assemblage into a single measure of size (see also Boudadi-Maligne and Escarguel 2014; Weinstock 2002). The VSI is based on comparison of measurements of paleozoological specimens with those of a standard population (e.g., a population of modern individuals) and is calculated as

$$VSI = \frac{x_i - \mu_i}{2\sigma_i} \times 50$$

where x_i is the value of measurement i on a fossil specimen, μ is the mean of that measurement in the standard population, and σ is the standard deviation of that measurement in the standard population. A fossil specimen that is larger than the standard population mean will generate a positive VSI, whereas a specimen that is smaller than the population mean will generate a negative VSI. Those readers who took (and can still remember!) introductory statistics may note that the VSI is similar to the more familiar z-score transformation:

$$z = \frac{x_i - \mu_i}{\sigma_i}$$

This means that either the VSI or the z-score transformation can be used to generate comparable results (VSI = 25 × z). If multiple measurements are taken on the same specimen, then the VSI (or z-score) values can be averaged to generate a mean VSI for that specimen. And in turn, the mean VSI of multiple specimens from a paleozoological assemblage can be used to generate an index of their overall size relative to the standard population. The benefit to this approach is that it can greatly enhance sample size (and statistical power) because multiple elements can be used. One drawback to pooling the VSI from multiple elements, however, is that this assumes all skeletal elements respond similarly to changes in an individual's body size (Albarella 2002; Meadow

1999). In other words, if the length of the astragalus in an individual is one standard deviation above the mean of the standard population, then the same should be true for the breadth of the distal humerus or any other dimension. This may not be the case. Different elements or portions thereof change in size at different rates (e.g., Geist 1989; Jungers and German 1981), a phenomenon known as *allometry* (Gould 1966). This means that differential skeletal representation between assemblages could introduce noise, obscure, or even drive apparent size clines between assemblages. Fortunately it is possible to evaluate this potential complication by examining whether mean VSI values across assemblages are related to differential element representation. This can be accomplished in much the same way as we outlined in Chapter 8.

The next step is to sort out how the proxy for size varies across spatial environmental gradients in modern populations. For some species those relationships are already well documented (e.g., Olcott and Barry 2000), but in many cases it is up to the paleozoologist to do the necessary groundwork – this is why paleozoologists have contributed substantially to our understanding of size clines in modern organisms (Millien et al. 2006). Typically this step involves measuring large samples of zoological specimens in museum collections, and recording when and where measured specimens were collected. As we noted earlier, when measuring those specimens (and the fossil specimens), it is standard to focus on adults – usually determined based on epiphyseal fusion for postcrania – to minimize ontogenetic (age-related) size differences. Preferably the sample of modern organisms should be from the same part of the world as the fossil sample because size within a widely distributed species will likely differ over its range. Also important is to obtain a sample that spans a large environmental gradient; this is where knowing the collection locations of measured specimens is important. Such information is necessary to detect geographic size clines and increases the likelihood that environmental conditions sampled by the modern dataset will correspond to those that existed in the past, reducing the risk of extrapolating into the unknown (see discussions in Chapter 9).

Examining size clines across a narrow environmental gradient can be problematic because in some species the size cline is not linear. For instance, Klein and Cruz-Uribe (1996) showed that the relationship between modern rock hyrax skull length and precipitation in southern Africa is quadratic. Skull length increases as rainfall increases from 0 to ~800 mm/year, but it declines as rainfall increases beyond that. This poses an obvious challenge for interpreting size clines in this species – small size could indicate low or high rainfall – an issue that would not have been apparent had their comparative sample included only hyraxes from the drier portions of their range.

Armed with a (hopefully) large database of measurements, the sizes of modern specimens are then examined across gradients in environmental variables such

as temperature, rainfall, productivity, seasonality, or any number of alternative parameters suspected or known to influence body size. Environmental
data are usually derived from local weather stations, GIS models that inter-
polate measurements from those stations, and other datasets obtained from
direct observations or remote sensing. Because different environmental
parameters often co-vary across space, it can be difficult to determine which
is more important in determining body size. A detailed discussion of statistical
approaches is beyond our scope here, but Yom-Tov and Geffen (2006) and
Blois et al. (2008) provide good examples of methods that can be used to tease
things apart.

Once the relationships between size and the environment are established,
attention can be directed to the paleozoological record. When interpreting
paleozoological size clines, the assumption is that the environmental variables
underpinning size clines in the present are the same ones responsible for
size clines in the past (i.e., the uniformitarian assumption 1 of Chapter 3).
Thus, an increase through time in the size of a species known to conform to
Bergmann's rule, for example, might be interpreted as evidence for a decline
in temperatures (but see Purdue 1989; Wolverton et al. 2009). But before the
size of fossil specimens can be confidently interpreted in environmental terms,
there are a few taphonomic issues worth considering. For species that are sexually dimorphic, differential representation of males and females can alter size
distributions in a manner that has nothing to do with the environment. This
could arise, for example, as a result of behavioral differences that might cause
one of the sexes to be more frequently captured by the predator responsible
for accumulating a bone assemblage (e.g., Marean et al. 1994). For example,
Grayson (1983b) compared a sample of 125 mandibular alveolar lengths of
modern bushy-tailed woodrats (*Neotoma cinerea*) that included 58 females and
65 males (2 were unsexed) with the distribution of a sample of 131 mandibular
alveolar lengths of bushy-tailed woodrats from sediments dating to various
time periods of the Holocene from Gatecliff Shelter in Nevada (Figure 10.3).
Because the modern and the prehistoric distributions differed for this sexually
dimorphic species, Grayson (1983b) worried that perhaps the Gatecliff sample
contained more females than males, though he doubted this was the case in
part because of the diminution in alveolar lengths that occurred over time
in the prehistoric sample (Figure 10.4). The latter might be a Bergmannian
response to locally documented climatic warming, though Grayson (1983b)
found that explanation unlikely because the chronocline in the number of
measured mandibles per time period and the mean size per time period
are correlated (Spearman's *rho* = −0.86, *p* < .10). Later work by Lyman and
O'Brien (2005) suggested that the chronocline in woodrat alveolar lengths in
the area was in fact a function of climatic warming. Not only did a second
sample of mandibles from a different site (in western Utah) display exactly the

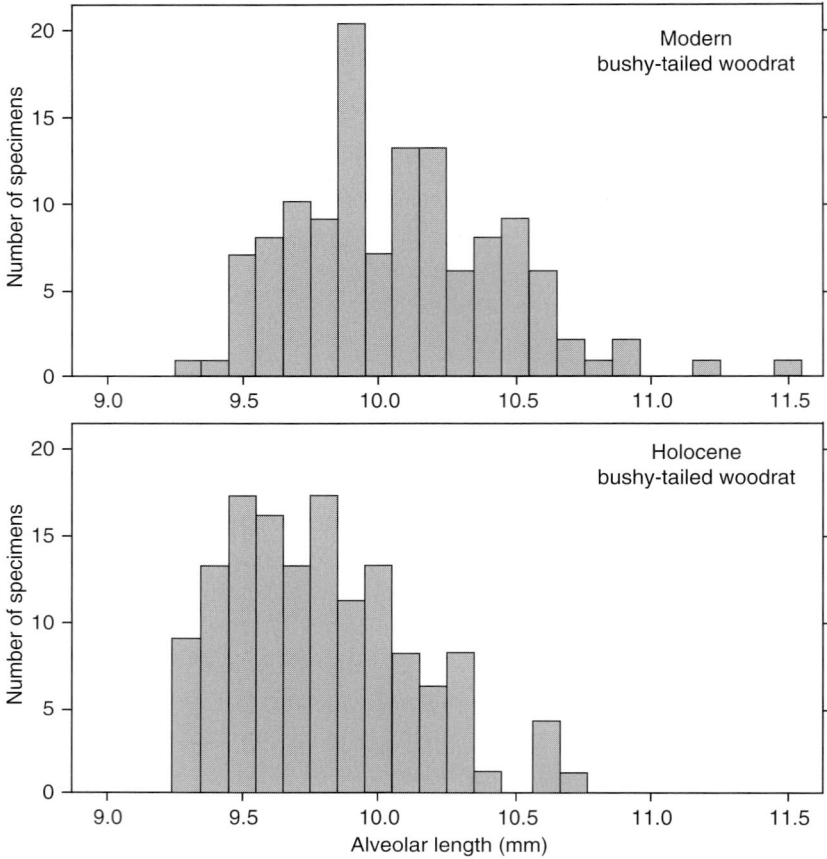

10.3. Frequency distribution of mandibular alveolar lengths in a sample of modern bushy-tailed woodrats (*Neotoma cinerea*) comprising 58 females, 65 males, and 2 of unknown sex (upper) compared with the frequency distribution of alveolar lengths of 131 bushy-tailed woodrat mandibles from Holocene sediments in Gatecliff Shelter, Nevada. Redrawn from Grayson (1983b).

same decrease in mean size, but the diversity of size classes over eleven strata spanning the terminal Pleistocene and entire Holocene fluctuated just as was predicted based on ecological principles.

A bias similar to one created by unequal samples of males and females of a sexually dimorphic species could be introduced by any taphonomic process biased against larger or smaller individuals. The good news is that it is often possible to get around these problems. For example, Klein (1991) recognized that the large size of Cape dune mole-rats (*Bathyergus suillus*) – a sexually dimorphic species whose body size varies as a function of rainfall – in the late Pleistocene at Die Kelders Cave in South Africa could arise from a fossil sample skewed toward larger males. However, he was able to show that the degree of variability in the measured fossil specimens was not unusually

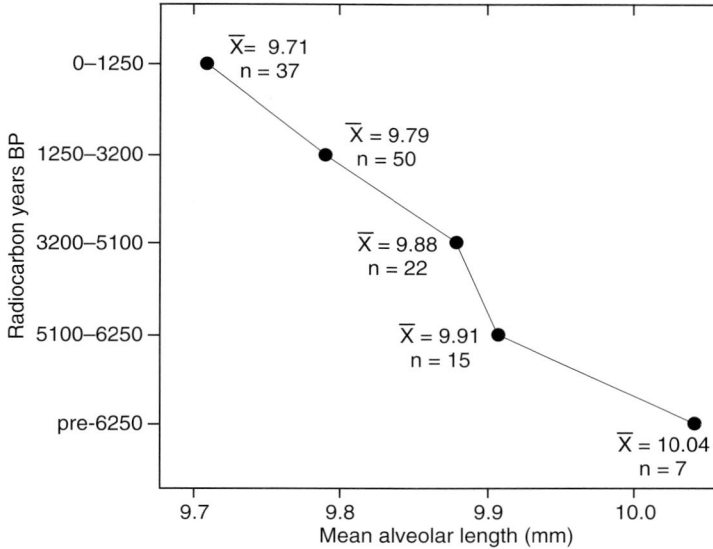

10.4. Mean mandibular alveolar length of bushy-tailed woodrats (*Neotoma cinerea*) per Holocene stratum at Gatecliff Shelter, Nevada. Note that mean length decreases over time, but so does sample size (Spearman's *rho* = −0.86, *p* < .10). Redrawn from Grayson (1983b).

low — as expected if females or smaller individuals were poorly represented — and that there was a bimodal distribution to the measurements (distal humerus breadth) that implied roughly equal numbers of males and females. Klein (1991) reasoned that taphonomic bias cannot account for the large size of the late Pleistocene Cape dune mole-rats, leaving paleoenvironmental change (altered precipitation in this case) as the more likely causal mechanism.

Provided that taphonomic or ecological processes such as character displacement can be ruled out, it is possible to develop paleoenvironmental inferences from fossil size clines. Of course, as we noted in our discussion of size clines in modern organisms, this may not be as straightforward as it sounds. Whenever sufficient zoological observations are available, it often becomes apparent that there are many variables that influence the body size of a species. We provide some examples of this later in this section. It may be necessary to explore a long list of mechanisms — sometimes drawing from alternative lines of paleoenvironmental evidence — to get at the paleoenvironmental variable of significance (e.g., Avery 2004; Faith et al. 2016b; Hill et al. 2008; Marean et al. 1994). Often the paleoenvironmental interpretation is provided in ordinal-scale terms (e.g., warmer/cooler or wetter/drier), though some have used regression equations describing the relationship between size and an environmental variable to generate numerical reconstructions. Efforts to do the latter have produced mixed results. Smith et al. (1995) provided very sensible temperature reconstructions (in light of independent proxies) based on size clines in the fecal pellets of late Quaternary woodrats (*Neotoma*) from the southwestern United States (see also Smith and Betancourt 1998, 2003; Smith et al. 1998). Then again, S. Davis (1981) provided temperature reconstructions based on the large tooth size of red foxes (*Vulpes vulpes*), wolves (*Canis lupus*), and boars (*Sus scrofa*) from the terminal Pleistocene of Israel that were ~10–11°C cooler than accepted estimates. The likely reason is that some variable in addition to temperature – S. Davis (1981) suggested forage availability – contributed to the large size of these species. Because there is no *a priori* way of knowing whether numerical reconstructions are likely to be accurate, the most conservative protocol is to interpret size clines in an ordinal-scale way.

In the following pages we outline a handful of case studies that illustrate the application of these methods in paleozoological contexts. In deciding which case studies to include, our main priority was to illustrate some of the analytical and interpretive challenges that can arise, and the ways in which they might be dealt with.

Bergmann's Rule and Past Temperature Change

We begin with a case study of a small mammal that conforms to Bergmann's rule: the East African mole-rat (*Tachyoryctes splendens*). This is a large-bodied (for a rodent) subterranean species that inhabits well-drained soils across a range of habitats spanning grasslands to forests in East Africa. East African mole-rats spend much of their time foraging for plant foods, especially roots and tubers, in subterranean burrows, but they occasionally forage on the ground surface, where they are susceptible to predation by raptors and other carnivores. Predation by raptors such as barn owls (*Tyto alba*) means that their skeletal remains are often incorporated into the deposits of caves and rockshelters in which owls roost. One such rockshelter is Enkapune ya Muto (EYM), found along the western escarpment (the Mau Escarpment) of the East African Rift Valley in Kenya. Archaeological excavations at EYM in the 1980s (Ambrose 1998) uncovered a moderate sample of *T. splendens* spanning the middle Holocene from ~7,200 to 3,000 cal yrs BP (Faith et al. 2016b; Marean et al. 1994), a time that in this part of the world is characterized by dramatic environmental changes as the African Humid Period (~14.5 to 5.5 ka) – a climatic phase that saw mesic savanna habitats expand across the Sahara desert – came to a close (e.g., DeMenocal et al. 2000; Street-Perrott and Perrott 1993).

In their study of the small mammals from EYM, Marean et al. (1994) documented a decline in the mandibular alveolar lengths – a measurement that tracks body mass in other rodent species (Martin 1984) – moving up the stratigraphic sequence (Figure 10.5A). Based on previous paleozoological studies (Avery 1982; S. Davis 1977, 1981; Klein 1986, 1991), Marean et al. (1994) suggested that changes in temperature or precipitation were possible

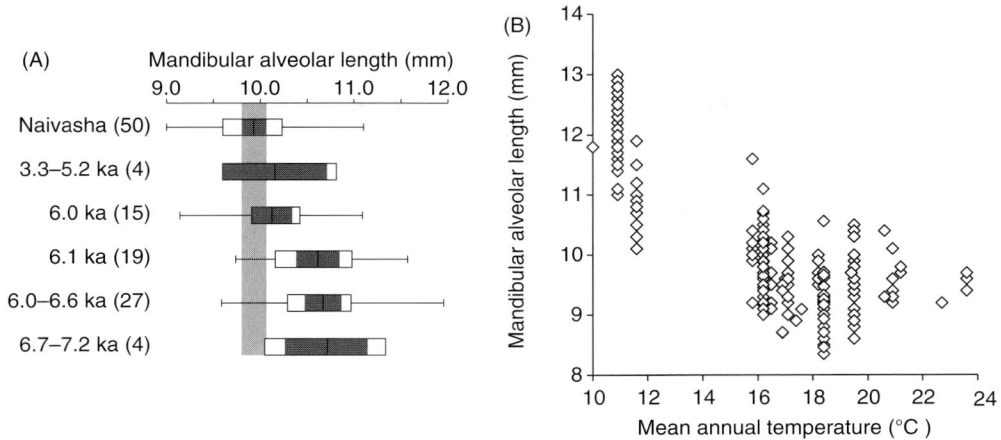

10.5. (A) Box plots illustrating the mandibular alveolar lengths of modern East African mole-rats from Lake Naivasha and the Holocene assemblages from Enkapune ya Muto. The dark grey horizontal bar indicates 95% confidence limits for the mean, and the light grey vertical bar indicates 95% confidence limits for the mean of the modern Lake Naivasha sample. Sample size in parentheses. (B) The relationship between mandibular alveolar length and mean annual temperature. Redrawn from Faith et al. (2016b).

explanations, but they were quick to note that they lacked the necessary data concerning size clines in contemporary *T. splendens* to pin down the environmental implications of the size cline.

Such data were provided more recently by Faith et al. (2016b), who demonstrated that the mandibular alveolar length of modern mole-rats in museum collections – most of their sample is from Kenya within ~300 km of EYM – varies across temperature gradients. They showed that alveolar length declined as mean annual temperature increased from 10 to 17.3°C, but remained stable thereafter (Figure 10.5B). This pattern is consistent with Bergmann's rule even if the size cline is not observed at higher temperatures, perhaps because – assuming the size cline is related to thermoregulation – heat conservation is less important at higher temperatures. Faith et al. (2016b) also showed that alveolar length did not vary as a function of mean annual precipitation, seasonality in precipitation or temperature, or net primary productivity, ruling out several alternative mechanisms known to underpin size clines in other mammalian species (Table 10.1). Although we previously noted that forage availability and quality is often an important correlate of body size (see also below), the absence of a correlation between mole-rat size and primary productivity (a proxy for forage availability) or annual precipitation – a variable known to influence plant nutrient content in East Africa (Olff et al. 2002) – leaves temperature (and thermoregulation) as a probable explanation for the size cline. The largest mole-rats, for example, belong to a subspecies known as *Tachyoryctes splendens rex*, which is found at high altitudes (sometimes in excess of 4,000 m) on the slopes of Mt. Kenya. The task of ruling out alternatives was made easier in this case by the fact that temperature gradients across the modern sample are related to elevation (correlation between mean annual temperature and elevation: r = -0.982, p < 0.001) rather than latitude, limiting co-variation in environmental parameters (e.g., temperature, seasonality, productivity) often seen when populations of a species are examined across broad latitudinal gradients.

EYM is today located in an area characterized by a mean annual temperature of ~16.3°C. Because the alveolar length of *T. splendens* is sensitive to temperature changes from 10 to 17.3°C, we should be able to track temperatures that are cooler than present, though we may not be able to detect temperatures much warmer than the present (Figure 10.5B). With this in mind, the large size of mole-rats from the base of the sequence is consistent with temperatures that are cooler than today. But as we noted above, it is worth considering taphonomic problems that could generate such a pattern. Marean et al. (1994) called attention to one such possibility. East African mole-rats are susceptible to predation by raptors when they leave their burrows. This usually occurs while foraging or when adult males are in search of mates. However, juveniles of both sexes will also leave their mothers' burrows and travel overland to establish their own burrows whenever mole-rat densities are particularly high, rendering them susceptible to predation as well. Marean et al. (1994) showed that the size decline tracks an increase in mole-rat relative abundance in the EYM small mammal assemblage, which they assumed to reflect an increase in mole-rat densities on the landscape. Because mole-rat alveolar length is sexually dimorphic (Faith et al. 2016b) and the extent to which it varies as a function of age and tooth wear is uncertain, the size cline may reflect greater predation of juveniles of both sexes in the upper layers at EYM compared with a focus on larger males in the lower layers.

Faith et al. (2016b) were able to rule out this taphonomic scenario by examining the size distribution of modern mole-rats collected from nearby Lake Naivasha (10 km east of EYM). They noted that many of the mole-rats from the older deposits are substantially larger than any of the modern specimens from Lake Naivasha (Figure 10.5A). While differential access to juveniles and adults could alter the mean size of fossil mole-rats at EYM, it cannot account for the presence of oversized individuals in the older deposits. Larger sample size could translate to the recovery of larger individuals, but the modern sample from Lake Naivasha is greater than any of the fossil samples. So it seems that in this case, taphonomic processes do not readily account for the size clines at EYM.

As we reviewed above, yet another confounding variable that could drive the size cline at EYM is character displacement. It is possible that the arrival or disappearance of a competitor species could contribute to size variation in the

EYM mole-rats (Dayan et al. 1991). However, as Marean et al. (1994) observed, character displacement can be ruled out here because the East African mole-rat is the only small mammal species found at EYM that specializes in roots and tubers. There are no obvious competitor species that would drive character displacement.

Taking all the evidence into account, rising temperatures are a sensible interpretation for the reduction in mole-rat size at EYM. We can be reasonably confident in this interpretation because (1) size clines in modern mole-rats track temperature to the exclusion of other variables that likely influence body size, (2) a likely taphonomic bias can be ruled out, and (3) character displacement is unlikely. And in addition – something that enhances confidence in all paleoenvironmental reconstructions derived from faunal remains – the temperature change suggested by the EYM mole-rats is consistent with inferences derived from other regional paleoenvironmental proxies (e.g., Berke et al. 2012; Street-Perrott and Perrott 1993).

The example provided by the EYM mole-rats is but one of many where size clines have been used to infer past changes in an environmental parameter (Table 10.1). We need not dwell on other studies available to us, however, because the analytical procedures and interpretive steps are similar to those we have just described. Instead we turn to a series of examples where conformation of a species to Bergmann's rule belies the many factors underpinning body size variation in modern and fossil populations.

Forage Availability and Predation in Large Herbivores

If the East African mole-rat represents a straightforward example of how size clines can be used to infer paleoenvironmental change, then the white-tailed deer (*Odocoileus virginianus*) illustrates the complexities of such analyses. Populations of white-tailed deer, which span a massive latitudinal range from central Canada to southern Peru, conform to Bergmann's rule; the largest individuals are found at high latitudes where temperatures are low (e.g., Koch 1986; Paterson 1990). Just as we demonstrated above for the East African mole-rat, it seems this relationship should provide the basis for inferring temperature change from size clines in the fossil record. So it may come as a surprise to learn that several investigations of modern and fossil white-tailed deer discount temperature change as a determinant of body size (Huston and Wolverton 2011; Purdue 1989; Wolverton 2008; Wolverton et al. 2007, 2009). Instead, these studies conclude that forage availability is the primary determinant of size clines in this species.

To understand why this is so we need to delve into the white-tailed deer's ecology. Because their digestive tract cannot efficiently handle high-fiber plants such as grasses or mature dicots, white-tailed deer are concentrate feeders

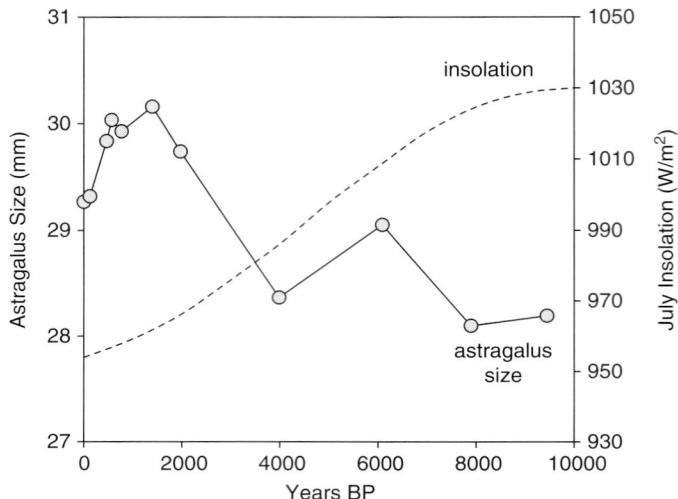

10.6. Changes in astragalus size (geometric mean of six measurements from Purdue 1989) in inferred females of white-tailed deer from Illinois (jagged line connecting dots) relative to changes in summer (July) insolation (dashed line) over the past 10,000 years. Ages of paleozoological samples differ slightly from those in Purdue (1989) because they have been calibrated to facilitate comparison with the insolation curve. Astragalus data from Purdue (1989) and insolation data from Berger and Loutre (1991).

that preferentially browse on the highest-quality parts of a diverse array of plant species, favoring new growth that is richer in protein and lower in fiber compared with old growth (Demarais et al. 2000; Hofmann 1989). Seasonal variation in availability of such foods translates to seasonal variability in growth of white-tailed deer (Lesage et al. 2001) and other cervids (e.g., Bergerud 2000; Ferguson 2002). During the winter, most plant foods have high fiber content and the calories consumed are dedicated to energetic maintenance; reliance on fat stores is essential to supplement this low-quality diet. During the plant growing season (summer), when high-quality foods are abundant, white-tailed deer replace body mass lost during the winter, store fat, and – in individuals that have not reached maturity – resume ontogenetic growth. The implication is that access to high-quality foods during the growing season determines how much energy can be diverted to ontogenetic growth and how big individuals will get (Huston and Wolverton 2011; McNab 2010). In short, greater access to high-quality foods during the growing season means bigger individuals. Though our primary concern here is with white-tailed deer, access to high-quality forage – especially during the plant growing season – is thought to be important in determining whether many ungulate species reach their potential maximum size (e.g., Geist 1989; Guthrie 1984a, 1984b; see also McNab 2010).

Purdue (1989) turned to the Holocene paleozoological record from central Illinois to explore these relationships in white-tailed deer (see also Purdue 1986). Across a series of samples spanning the past ~8,500 radiocarbon years BP, he found that the size of the astragalus – a good proxy for body size in white-tailed deer (Purdue 1987; see also Emerson 1978) – varies substantially through time (Figure 10.6). Barring a decline in his modern sample and historic samples – an issue that we turn to later – the white-tailed deer from central

Illinois have been getting bigger through the Holocene. Purdue (1989) noted that the size of white-tailed deer is inversely correlated with summer insolation, the amount of solar radiation reaching the earth's surface. He reasoned that this correlation may have something to do with the effects of insolation on the plant growing season. Climate models and paleoclimate archives available to Purdue at the time suggested that high summer insolation during the middle Holocene translated to hotter and drier summers (Bartlein et al. 1984; Kutzbach and Guetter 1986), a phenomenon well documented in more recent literature (e.g., Diffenbaugh et al. 2006; Williams et al. 2010). Purdue (1989) also noted white-tailed deer size was positively correlated with winter insolation, but there was no evidence suggesting this translated to cooler winter temperatures as might be expected if size were related to thermoregulation (i.e., Bergmann's rule). So to explain the size cline, Purdue (1989) argued the combination of hot and dry summers would have caused plants to mature more rapidly, shortening the period during which high-quality fresh growth was available to white-tailed deer (see also Guthrie 1984a). The outcome was relatively small deer during the middle Holocene (Figure 10.6). Purdue (1989) expanded on these observations to suggest that conformation of white-tailed deer to Bergmann's rule across latitudinal gradients was not due to temperature, but rather to greater access to high-quality forage in the northern parts of its range (see also Guthrie 1984b).

Much the same argument was advanced more recently by Wolverton et al. (2009). Building on work by Purdue (1986, 1989) and others (Geist 1987, 1998), they aimed to further test the hypothesis that forage availability underpins body size variation in white-tailed deer. They assumed that "body size ... in animals with determinate growth is a function of the ontogenetic growth rate during the time period prior to reaching maturity" and that "ontogenetic growth rate is a function of food availability per individual animal" (Wolverton et al. 2009:404). If this is so, then body size should vary as a function of *ecologically relevant net primary productivity* (eNPP), a measure of the amount of plant biomass produced during the growing season (Huston and Wolverton 2009). An important departure from Purdue's (1986, 1989) work, however, is that they also reasoned that body size should be influenced by white-tailed deer population densities, which influences forage availability for individual animals; greater densities means more competition and less available forage (see also Wolverton 2008). Their model outlining the influence of eNPP and population density on body size is illustrated in Figure 10.7. Growth rate and body mass should be highest when eNPP is high and population density low, and lowest when eNPP is low and population density high.

To test the eNPP component of their model (Figure 10.7), Wolverton et al. (2009) turned to astragalus measurements of modern and prehistoric white-tailed deer from central Missouri, central Texas, and southeast Texas. Although

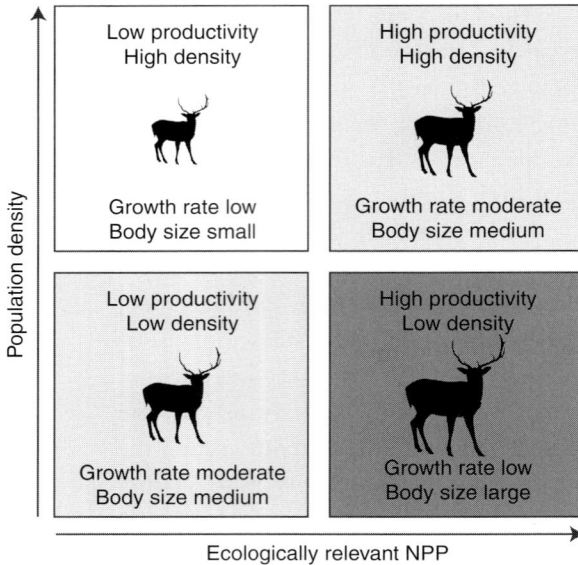

10.7. A heuristic model of how ecologically relevant net primary productivity (eNPP) and population density influence white-tailed deer growth rates and body size. Adapted from Wolverton et al. (2009).

they were unable to address temporal trends in their time-averaged fossil dataset (primarily over the past 5,000 years), they were able to demonstrate that spatial increases in astragalus size track increases in agricultural crop productivity. Wolverton et al. (2009) recognized that crop productivity is not the same as the productivity of the natural environments in which deer populations forage, but they reasoned that the combination of climatic and soil properties that influence crop productivity should influence eNPP in a similar manner. Their observations derived from paleozoological specimens are weakened by the fact that forage quality may have varied in the past due to changing climate (Purdue 1986, 1989), but it remains the case that the biggest deer are found in areas with greatest productivity, consistent with the arguments forwarded by Purdue (1986, 1989). And as Purdue also did, Wolverton et al. (2009:414) extrapolated these relationships to a broader taxonomic and geographic scale, suggesting that "the latitudinal distribution of body size in species that conform to Bergmann's rule [may] actually relate to the macrogeographic distribution of food availability per animal through the response of ontogenetic growth rates to ecologically relevant NPP." In other words, white-tailed deer and perhaps other species are not necessarily bigger at higher latitudes because they have adapted to cooler temperatures, but rather because enhanced eNPP at higher latitudes allows them to achieve greater size (e.g., Guthrie 1984b; Huston and Wolverton 2009).

As the model presented in Figure 10.7 suggests, environmental productivity is only part of the story. To explore how population density mediates white-tailed deer body size, Wolverton et al. (2009) then turned to white-tailed deer

10.8. Changes in average dressed weight (lbs) of 1.5 year old bucks and does harvested from 1971 to 2005 at Fort Hood, Texas (after Wolverton et al. 2009). Solid (bucks) and dashed (does) lines are the least-squares regression. Raw data courtesy of Steve Wolverton.

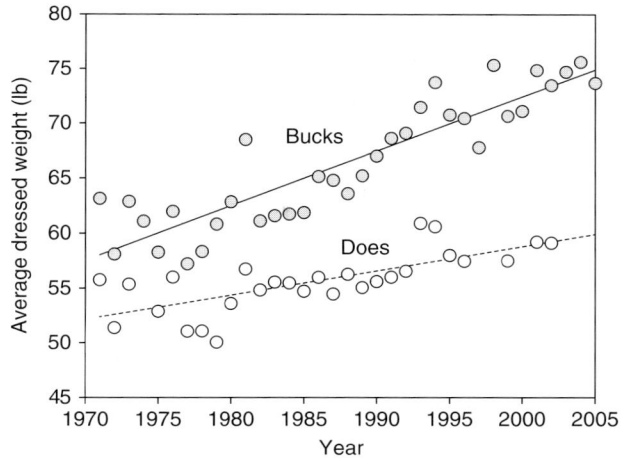

collected during controlled harvests from Fort Hood (Texas) from 1971 to 2005 (see also Wolverton 2008). While census data for the Fort Hood population are not complete for this time period, surveys from 1981 to 1991 and from 1997 to 2005 indicate a roughly 30% decline in deer population density, from 36.03 individuals/1,000 acres to 25.36 deer/1,000 acres. And as shown in Figure 10.8, as deer populations declined their body mass increased substantially. Wolverton et al. (2009) note that the large size of the more recent Fort Hood white-tailed deer contrasts starkly with that of unmanaged deer populations elsewhere in Texas where populations are high in density and the deer are relatively small (Wolverton et al. 2007). These observations provide good reason to believe that white-tailed deer population density and its effects on forage availability is an important determinant of body size.

The broader implication of these observations is that ecological factors external to the sort of environmental variables we are primarily interested in in this volume (e.g., climate and habitat), including predation pressure by humans or other carnivores, can play an important role in driving size clines (Abrams and Rowe 1996). Support for this idea was provided in a study by Wolverton et al. (2007), who noted that white-tailed deer – an important prey item in the diets of Native Americans – from Holocene zooarchaeological assemblages in central Texas are similar in size to those from the managed population at Fort Hood in 2005, but are significantly larger than unmanaged, high-density populations from elsewhere in Texas. The extent to which the prehistoric sample is also influenced by changing climate is uncertain, but they reasoned that prehistoric harvesting pressure translated to lower white-tailed deer population densities and larger individuals (as a result of reduced intra-specific competition).

Reduced predation pressure may also explain the size decline observed in the modern and historic (mid-1800s) sample of white-tailed deer examined by Purdue (1986, 1989; Figure 10.6). While the prehistoric populations Purdue examined were potentially subject to regular predation from Native American foragers, this would not have been the case after the decimation of indigenous populations following European colonization of North America. If declining predation pressure translated to elevated deer densities, then the deer may be smaller because of increased competition and reduced forage availability (Abrams and Rowe 1996). But regardless of whether this is the correct explanation for recent size decline in white-tailed deer from Illinois (Figure 10.6), the mere possibility that forage availability and quality – mediated by (bottom-up) climate or top-down ecological processes such as predation – may have such an influence poses an obvious challenge for reading the environmental significance of size clines in the fossil record. As we noted before, we may not be able to safely assume that a relationship between size and a given environmental parameter (e.g., temperature) in modern organisms implies that this environmental parameter is responsible for size change in the present or past – the correlations may be spurious.

To further complicate the picture, we cannot uncritically assume increases in predation pressure will translate to an increase in body size for all species. In some cases, intensified predation pressure can translate to a decrease in size (e.g., Klein and Cruz-Uribe 1983b; Klein et al. 2004; Stiner et al. 2000). This is because increased predation on a population influences its demographics such that there are fewer older individuals and a greater proportion of juveniles (e.g., Lyman 1987; Munro 2004; Taber et al. 1982; Wolverton 2008). So in vertebrate (e.g., tortoises) and invertebrate (e.g., shellfish) species that grow more or less continuously through life, intensified predation translates to a younger population made up of smaller individuals. Although he had few control data from modern contexts, Broughton (1997) presented a compelling case that human predation rather than climate drove a reduction in the average size of sturgeon (*Acipenser* sp.) caught by humans occupying the shore of San Francisco Bay during the late Holocene. The same effect might also be observed in larger herbivores (e.g., ungulates) if size were reconstructed from a skeletal element that takes a long time to mature, in which case a younger population (under intense predation) may appear in the paleozoological record as a smaller-bodied population, even if reduced competition for forage means a greater potential adult body size. This possibility highlights the importance of controlling for ontogeny when using size clines as a paleoenvironmental indicator.

If you are feeling that size clines are becoming complicated at this point, then do not despair! With a bit of analytical work it is possible to tease things apart (e.g., Hill et al. 2008; Wolverton 2008). To illustrate how this can be done, we turn to an excellent paleozoological study provided by Hill et al. (2008). Recognizing that body size may be tracking predation pressure or environmental change, they set out to explain size clines in late Quaternary

10.9. Changes in Great Plains bison calcaneal length over the past 37,000 years. Data from Hill et al. (2008).

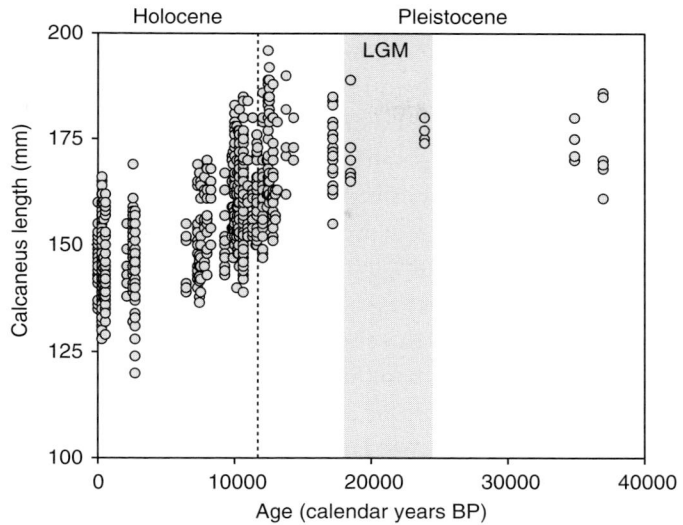

bison (*Bison* spp.) – a taxon well known to have been much larger during the late Pleistocene (e.g., Guthrie 1970; Lyman 2004; McDonald 1981; Wilson 1978) – from the Great Plains of North America. Drawing on paleozoological assemblages from fifty-seven sites spanning the past 37,000 years, they documented substantial declines in the size of the calcaneal tuber and distal humerus breadth through time. As shown in Figure 10.9, bison size is stable for most of the late Pleistocene but diminution occurs toward the very end of the Pleistocene (~13,000 BP) and well into the Holocene.

From twenty-four archaeological kill/butchery sites providing remains of multiple animals (from 20 to 211 individuals based on dental counts), Hill et al. (2008) also examined the frequencies of juveniles, prime adults, and old adults (after Stiner 1990) based on patterns of dental eruption and wear. They showed the age structures of these samples are broadly consistent with those characterizing living populations, suggesting human hunters were likely killing multiple animals at the same time rather than selectively targeting the oldest and youngest individuals (e.g., Steele 2003). And as we noted above, Hill et al. (2008) reasoned increases in human predation pressure should translate to bison populations dominated by a greater proportion of juveniles (e.g., Lyman 1987; Munro 2004; Taber et al. 1982; Wolverton 2008), but they were unable to detect any diachronic changes in the representation of juvenile bison in these kill sites. In particular, they showed the proportion of juvenile bison in Great Plains kill sites remained relatively stable across Early Paleoindian (>12,000 cal yrs BP), Late Paleoindian (12,000 to 9,000 cal yrs BP), Archaic (9,000 to 2,000 cal yrs BP), and Late Prehistoric/Protohistoric (<2,000 cal yrs BP) samples. It follows that the decline in bison size is not associated with predation-driven changes in population structure.

So what might these results be telling us about the mechanisms driving the size decline in Great Plains bison? The mortality data indicate altered human predation pressure is not responsible for the size cline, leaving one or several environmental or ecological factors as plausible alternatives. Turning first to Bergmann's rule and thermoregulation, Hill et al. (2008) note the decline in bison size broadly tracks the transition from cool Pleistocene climates to warmer Holocene climates. But they also argued temperature alone is insufficient to explain the size cline, as we would expect to see the largest bison during the Last Glacial Maximum and the smallest bison during the early Holocene. We would also expect to see a short-lived size increase during the cool Younger Dryas (12,800 to 11,500 BP). However, none of these predictions are supported by the bison data (Figure 10.9). Noting the late Pleistocene faunal community – one that includes horses (*Equus* spp.), camels (Camelidae), mammoths (*Mammuthus* spp.), and saber-tooth cats (*Smilodon* sp., *Homotherium* sp.), among others – was very different to that of the Holocene. Hill et al. (2008) also entertained the possibility that the large size of late Pleistocene bison was a predator avoidance strategy or a response to increased interspecific competition with other large herbivores. While the massive extinctions of carnivores and herbivores that occurred in North America toward the end of the Pleistocene (e.g., Grayson 2016) may have translated to ecological release that favored smaller body size, this scenario does not clearly explain the changes that occurred through the Holocene (after 11,700 BP).

After ruling out the possibilities noted above, Hill et al. (2008) proposed a reorganization of grassland composition and its effects on forage quality was the primary driver for the bison size decline. They observed the two prominent phases of bison diminution, from ~14,000 to 11,000 BP and again after ~8,000 BP (Figure 10.9), correspond to previously documented evidence for climate-driven increase in C_4 grasses (warm-season grasses) and decrease in C_3 grasses (cool-season grasses) on the landscape. Hill et al. (2008) argued these changes in grassland composition would have dramatically altered forage quality and availability. First, whereas fresh growth in C_4 grasses is limited to the summer, C_3 grasses sprout during the spring and fall rains (under ideal precipitation), translating to a longer period of the year during which protein-rich fresh growth is available. And second, C_3 grasses tend to have higher protein content and are more digestible than their C_4 counterparts. So Hill et al. (2008) argued the expansion of C_4 grasses toward the end of the Pleistocene and during the middle Holocene translated to a reduction in bison size because of climate-driven decrease in forage quality. Their inferences concerning the relationship between body size and forage quality in late Quaternary bison complement those provided for white-tailed deer (Purdue 1989; Wolverton et al. 2009) and other ungulates (e.g., Guthrie 1984a, 1984b, 2003; Weinstock 1997). For instance,

Dale Guthrie (1984a, 1984b) argued the large size of many North American ungulates during the late Pleistocene, including various cervids (*Rangifer, Cervus,* and *Alces*) and bovids (*Ovis, Ovibos,* and *Bison*), resulted from a longer growing season relative to the Holocene. His arguments were supported in part by controlled feeding experiments in Dall sheep (*Ovis dalli*), which grew to impressive sizes when their diets were supplemented with high-quality foods immediately before and after the plant growing season (effectively lengthening the growing season) (see also accounts in Geist [1989]).

It is important to note in the studies reviewed in this subsection, the primary objective of their authors was to account for the mechanisms driving size clines in modern and fossil individuals. This is very different from using those size clines to generate a paleoenvironmental reconstruction. So to what extent might we be able to use size clines in large ungulates as paleoenvironmental indicators? For various reasons outlined throughout this chapter, it would be unwise to assume size clines in paleozoological samples are the outcome of a single mechanism – we may have to explore several competing mechanisms to get to the ultimate environmental driver (e.g., Hill et al. 2008). But to the extent the inferences summarized here are correct and more broadly applicable, then size clines in ungulate species whose growth potential is determined by forage availability during the growing season may be best suited for tracking forage availability in the past, an environmental parameter mediated by a combination of climatic factors that influence plant community composition or the length of the growing season in addition to ecological factors (predation and competition) that determine the amount of food available for individual animals.

SUMMARY

Intraspecific variability in body size across environmental gradients is well documented in contemporary populations of many species, and these relationships have long been examined by paleozoologists to provide paleoenvironmental reconstructions. But just as we noted in our previous discussions concerning the relationships between taxonomic diversity and the environment (Chapter 8), we must be cautious in applying broad-scale species–environmental relationships – even for well-documented phenomena such as Bergmann's rule – to the paleozoological record. As we highlighted in this chapter, numerous ecological and environmental variables can influence size. This means correlations between size and any single environmental variable in contemporary populations of a species need not imply the latter is necessarily the operative one responsible for size clines today or in the past. The challenge is that to provide environmental inferences from ancient size clines we have to assume just this.

Fortunately, it is possible to strengthen our confidence in this critical assumption. Doing so requires that we explore multiple variables to identify the one(s) most likely responsible for driving body size variation in contemporary populations, keeping in mind the balance between competing mechanisms that influence size (e.g., temperature, rainfall, productivity, or forage quality) may differ over portions of a species' range and could also vary through time. As is clear from our discussion of white-tailed deer, knowing a bit about the ecology of the species we are dealing with can facilitate the task. In addition, we can enhance confidence in our paleoenvironmental interpretations by taking the time to address taphonomic processes, ecological factors (e.g., character displacement, competition, and predation), and alternative environmental variables that could influence size. We recognize this idealized strategy sets the bar quite high! Limitations posed by insufficient ecological knowledge for some species or by the nature of our fossil samples mean it may not always be possible to do all of this. But the more we can do, the more confident we can be in paleoenvironmental inferences derived from size clines in the paleozoological record.

ELEVEN

SOME FINAL THOUGHTS

Long recognizing that each animal species has more or less unique ecological tolerances (Chapter 2), paleozoologists have for many years analyzed ancient faunal remains to determine the nature of past environments. Knowledge of this potential had been around for nearly two centuries. Charles Lyell (1832:141), for instance, called upon environmental change – broadly construed as "not merely temperature, humidity, soil, elevation, and other circumstances of like kind, but also the existence or non-existence, the abundance or scarcity, of a particular assemblage of other plants and animals in the same region" – as a major cause of what were then becoming understood as terminal Pleistocene extinctions (see Grayson [1980, 1984a] for histories of debates over causes of these extinctions). The implication of such an attribution was that a driver or cause of faunal change over time was environmental change.

In summarizing what was then known, in the mid-nineteenth century European archaeologist Adolphe von Morlot (1861) spelled out clearly the fundamental analytical assumption of environmental reconstruction, and also described an example. Attributing the fundamental assumption of uniformitarianism to Lyell (see Chapter 3), Morlot (1861:285) wrote "the laws which govern organic creation and the inorganic world are as invariable as the results of their combinations and permutations are infinitely varied, science revealing to us everywhere the perfect stability of causes with the diversity of forms." Thus, Morlot (1861:296) observed that changes in taxonomic abundances and composition of shellfish faunas (as well as the size of individual shells) found in shell middens near Copenhagen were the result not just of human exploitation, but of changes in salinity of the marine water in which the species lived. His archaeological interests overshadowed his discussion of bird and mammal remains recovered from shell midden deposits; changes in those taxa were, Morlot thought, the result of human exploitation, and he did not consider their possible paleoenvironmental implications.

In earlier chapters of this volume we have presented other glimpses of the history of analyses of faunal remains for purposes of environmental reconstruction, and we have described in some detail many of the techniques that have been developed to undertake such analyses. In Chapter 1 we indicated knowing the environmental history of an area was important to understanding the evolutionary history of organisms, including humans and their cultures. The "environment" is, after all, the context in which evolution takes place and thus is a source of various kinds of selective processes (in the sense of Charles Darwin's and Alfred R. Wallace's theory of natural selection). We also mentioned in Chapter 1 that knowing the environmental history of an area may be important to modern conservation biology and wildlife management. We promised to return to these two topics in the final chapter of the book. This is the final chapter, so you know what is next. But before fulfilling your expectations, we think it important to first provide a context and a warrant for the paragraphs that follow.

Lyman has long had an interest in using paleozoological data to inform conservation biology and wildlife management (e.g., Lyman 1996, 2006b, 2012e; Lyman and Cannon 2004; Wolverton and Lyman 2012). An increasing number of paleontologists display similar interests (e.g., Barnosky et al. 2017; Dietl and Flessa 2011; Dietl et al. 2015; Louys 2012). Conservation and concerns about the state of the earth's ecology is a topic of increasing importance to us given that we both have children and Lyman has grandchildren. If we can slow the rate at which we are fouling our own nest, or perhaps even reverse the trend because of what we have learned by using the analytical techniques described in earlier pages of this volume, then we both will be pleased.

Faith has long had deep interests in using paleozoological data to help understand human biological and cultural evolution (e.g., Faith 2008, 2013a; Faith et al. 2009, 2016a). This is an important topic among not only biological anthropologists, but archaeologists and many of those working in other human sciences as well. For instance, when and why did early hominins first appear? What drove the development of human social and political organization, and why did some complex societies collapse? What selective environments pushed these and other early faltering steps along? And can our insights to those events and processes help us understand the modern human condition, both its current state and its potential future(s)? If this book helps researchers get closer to definitive answers to these sorts of questions, again we will both be pleased.

We do not mean to imply by the preceding that Faith has had no interest in conservation biology or that Lyman has little interest in paleoenvironmental reconstruction. Both of us have long had interests in these topics as well as in zoology and the biology and ecology of animals; those are some of the reasons we became paleozoologists in the first place. And the references we cite in this book should make it clear we have both used paleofaunal data for both purposes. We do intend the preceding couple of paragraphs to provide a bit of explanation as to why we have written this chapter the way we have, particularly why it covers the topics it does. Hopefully as well it will help the reader grasp why we discuss what we do in this chapter rather than continue in more direct terms the central theme of this volume – how to reconstruct paleoenvironments on the basis of ancient faunal remains.

CONSERVATION PALEOZOOLOGY

Seventy years ago, North American zoologist Raymond Gilmore (1949:167) argued that paleozoological research could provide a robust "picture of former ecology and subsequent ecologic successions for certain regions." Importantly in the context of this volume, he went on to note "this [picture] may be of importance in future judgments on land management involving problems of agriculture, fish, game, and general plant cover" (Gilmore 1949:167). A growing number of paleozoologists have, during the past ~30 years, acted on Gilmore's vision. Paleontologists have, for example, initiated efforts to develop what they refer to as "conservation paleobiology" (Dietl and Flessa 2009, 2011; Dietl et al. 2012, 2015; Louys 2012). A few zooarchaeologists have similarly argued for what some of them refer to as "applied zooarchaeology" (Lyman 1996, 2006b, 2012e; Lyman and Cannon 2004; Wolverton and Lyman 2012; Wolverton et al. 2016). In both cases, paleozoologists have argued the ancient archives of animal remains they study represent temporal spans unprecedented in duration relative to what biologists study. The long duration of the fossil record reveals unique evolutionary and ecological processes and biotic responses to a variety of shifting environments, some of which are unheard of today, such as the non-analog biotas mentioned in Chapter 2. A similar situation is found in modern ecology concerning so-called "novel ecosystems" (e.g., Young 2014; see also Hobbs et al. 2009, 2013; Murcia et al. 2014; Williams et al. 2007). The latter are communities or biotas previously undocumented and seemingly at least partially the result of human actions, such as translocating (intentionally or unintentionally) exotic species into areas where they would not occur without human intervention. Further, the fossil record provides insight to pre-industrial biotas and ecological baselines that still guide some biological conservation decisions, though it is at the same time becoming clear that it is impossible to recreate those baselines, hence the growing interest in so-called novel ecosystems. Will those ecosystems provide the ecological resources and services humanity requires? Finally, the paleozoological record displays how faunas (and floras) respond to various modes and tempos of environmental change that can be used to inform conservation decisions or to anticipate and predict outcomes of conservation applications.

As we approach the third decade of the third millennium, the human population of earth is racing from its current 7.5 billion individuals to about 10 billion by ~2050 (Barnosky and Hadly 2016). Environmental change is a hot (no pun intended, unless you wish it so) topic of discussion among ecologists, conservationists, politicians, and others. "Global warming" (see Barnosky [2009] for a very readable, informative, and entertaining account) is a major concern because at least some of its causes are anthropogenic. And the latter has resulted in much discussion of a possible new geological era referred to as the Anthropocene, the age of humans (e.g., Davies 2016). Although not yet officially recognized by the International Commission on Stratigraphy or ratified by the International Union of Geological Sciences, the Anthropocene (ratified or not) as a concept can be conceived as having two purposes. First, it underscores the magnitude of human influences on earth's ecological processes and products (Ruddiman 2013), and second, it serves as a wake-up call that humans must become more ecologically conscientious else we may find ourselves in the midst of an unfavorable (to us and other life forms on which we depend) ecological maelstrom of our own making.

Whatever one thinks of the general notion of an "Anthropocene," and there is much to think about with respect to the concept above and beyond what we have mentioned (e.g., Lorimer 2015), knowing something about past ecologies, ecosystems, and environments can be valuable to all sorts of modern conservation issues. Other paleoecologists (Seddon et al. 2014) have compiled a series of fifty questions they and their colleagues believe their field should pursue in the future. Of direct pertinence to our discussion here are the following questions:

- Why are some ecosystems and species more sensitive to environmental change than others and therefore respond first or to a greater degree than others?
- Why do some species and ecosystems experience varying degrees of time-lag in responding to environmental change?
- What influence has Holocene landscape modification (especially fragmentation) had on the ability of biotas to respond to environmental change?
- How can paleofaunal data be used to inform ecological restoration, recovery, and reintroduction efforts?
- Which variables make some ecosystems or biotas more resilient to environmental change than others?

It should be clear that answers to these questions could be critically important to the human species as we move into the future. We do not mean to imply paleozoological data underpinning paleoenvironmental reconstruction will provide definitive answers to these questions. We do believe, however, thoughtful application of the analytical techniques we have described in earlier chapters can produce at least some insights to what answers to these questions will look like. It is clear, for example, some of the more specific questions asked by modern conservation biologists can be answered with paleozoological data (Faith 2012b; Lyman 2012e). Why not these more general questions posed by paleoecologists?

ENVIRONMENTS OF HUMAN BIOLOGICAL AND CULTURAL EVOLUTION

Given how the fossil record samples spans of time (time averaging of greater or lesser magnitude) and chunks of geographic space (spatial or habitat averaging of greater or lesser magnitude), scales vary tremendously from sample to sample (Chapter 3). That is, both the amount of space and of time included (sometimes referred to as extent) and the resolution or grain of variables (DPI, if you will) vary given a diversity of taphonomic and analytical histories. Although such can result in misleading implications of faunas, recognizing variability in extent and grain across multiple samples can provide varied scales of ecological contexts in which humans evolved biologically and culturally. These can range from, say, a site-specific location involving a morphometric shift (e.g., Bergmannian response) of a taxon to an environmental change, to a shift in abundances of taxa or in the taxonomic composition of a local fauna, to regional weather patterns implicated by shifts in the taxonomic composition of multiple local faunas.

Knowledge of site-specific environmental contexts and ecological conditions might reveal potential catalysts for shifts in local hominin populations, their tool assemblages, or both. A well-known example of this concerns a long-term discussion over the replacement in Europe of Neanderthals by modern humans and the more or less concomitant change from Middle Paleolithic stone tool industries to Upper Paleolithic industries (e.g., Grayson and Delpech 2008; Hodgkins et al. 2016; Morin 2012). What exactly happened roughly forty thousand years ago when these shifts took place has baffled archaeologists for decades; well, maybe not "baffled," as some researchers think they know precisely what happened, but not everyone agrees with any particular scenario. Importantly in the context of this volume, analysts of faunal remains from this time period have examined those remains not only for evidence of variability in who ate what, but also for evidence of whether faunal change was driven by hominin predation or climate change or both (e.g., Grayson and Delpech 2003,

2006). Grappling with such issues is enhanced by close study of faunal remains because those remains provide critical data on the environmental context of hominin biological and cultural evolution (as well as holding implications for modern conservation biology [Grayson and Delpech 2005]).

Another example, similar to the one regarding Neanderthals just described, concerns the transition from the Middle Stone Age to the Late Stone Age in southern Arica. There, a somewhat smaller (fewer people are involved, and less has been published) but nevertheless important debate concerns the hunting proficiency of Middle Stone Age people relative to Late Stone Age peoples. In a long series of papers Richard Klein (e.g., 1994, 1995, 1998, 2000) suggests that the former were less proficient hunters than the latter; he believes the transition from one cultural industry to the other and a concomitant shift to greater hunting proficiency was the result of a major neurological reorganization. Faith (2008, 2011c) suggested on the basis of large samples of faunal assemblages from both eras that a more parsimonious explanation for observed changes in the faunal assemblages was a combination of (climate-driven) declining environmental productivity together with an increase in human population density, both resulting in reduced availability of large game. The debate that followed has so far been brief (Weaver et al. 2011; Faith 2011b), with both sides agreeing that environmental change more or less coincident with the perceived changes in both stone tool industries and mammalian faunas is an important consideration. As with the European discussion of the shift from Neanderthals to modern humans, the faunal remains have played a role in detecting what could be significant environmental shifts.

Concerning the initial appearance of a human ancestor – a hominin – and subsequent diversification of what would be the human lineage, Behrensmeyer et al. (2007b:335) indicate correlating the appearance and diversification events (as revealed by the fossil record) with "well-documented paleoenvironmental events could shed light on our understanding of the role of ecological factors shaping hominin evolution." Faunal remains associated with either ancient artifacts or remains of the hominins themselves can provide evidence of the kinds of habitats and climates the hominins lived in and if and how the hominins interacted with fauna. Further, faunal remains associated with early stone tools, hominin fossils, or both may reveal whether evolutionary changes in the hominin lineage tracked those of other animal taxa (e.g., Vrba 1995). These questions can be addressed to any specific period (not just the era of the earliest hominins) of human biological or cultural evolution. And the faunal remains provide insight to the local environmental contexts of the evolutionary processes and environmental events involved.

Building regional scenarios requires correlation of faunal assemblages of different resolution and grain (how much time and space, how many different communities are represented), variable taphonomic skewing, and sometimes imprecise chronological placement (exactly how contemporary are two assemblages assigned to the same temporal period?). These are problems beyond our scope in this volume, but here they emphasize how the paleozoologist must often have collaborators from rather disparate fields (e.g., archaeologists, geomorphologists, geochronologists). And we believe that is often a very good thing for reconstructing ancient environments, regardless of the purpose for doing so.

WHAT NEXT?

There are other reasons besides the ones discussed above why one might want to decipher the paleoenvironmental meaning of a faunal assemblage. And though there is no limit to the interesting and important questions that could be asked, our ability to answer them is only as good as the methods at hand. Together with the temporal and spatial resolution of our fossil samples, the methods we use determine the resolution of our inferences and of the questions we ask of the data (more on this below). We dedicated the majority of space in this volume to what we view as the main types of techniques in the methodological toolkit. We also noted some instances where there is room for improvement, and it is likely that some readers will perceive other gaps that we did not. So we would be pleased if this volume prompted the development of new methods, providing techniques that might, for example, offer enhanced resolution or reveal environmental parameters otherwise difficult to reconstruct.

From our previous discussions of the history and development of various analytical techniques, it is clear that all are unified by the same fundamental approach, one that is grounded in uniformitarianism and that would also guide the formulation of new techniques. The fundamental procedure is as follows. We first perceive some variable or pattern among living faunas that relates to an environmental gradient. We then interpret that pattern in the paleozoological record (assuming that it is visible, or that we have a measurable proxy for it) as reflecting that environmental gradient. Or conversely, we might first perceive a pattern in the fossil record and then look for that pattern among contemporary animals, hoping that we can identify an environmental correlate. Either approach will do the trick, but here it is important to re-emphasize a point we first made in Chapter 3. Besides simply recognizing correlations between faunas and the environment, robust inferences require that we identify the causal (ecological) mechanism responsible for those correlations. This is because, as we encountered several times in this volume (see especially Chapters 8 and 10), a correlation between faunal variable X and environmental variable Y does not mean that Y is the driver of X – such correlations are not infrequently spurious. But if we have an ecological understanding of *why* such a correlation exists in the first place, then we can be more confident that changes in fossil faunas reflect changes in the target environmental variable.

With this fundamental approach in mind, there is good reason to be optimistic about the potential for novel additions to the paleoenvironmental toolkit. Electronic databases on the ecology and environmental preferences of some animal groups are now widely available (e.g., the PanTHERIA database: Jones et al. 2009), as are databases detailing species ranges in the present and recent past (e.g., Galster et al. 2007; Sauer et al. 2017). The same is also true of a diverse array of climatic and environmental parameters (e.g., Giglio et al. 2013; Hengl et al. 2014; Running et al. 2004), including especially the WorldClim global climate data (Fick and Hijmans 2017; Hijmans et al. 2005). So for those readers faced with paleoenvironmental problems that cannot be addressed using the methods described in previous chapters, such databases provide plenty of grist for the development of new techniques. Should there be a need for a second edition of this volume – and provided we have the fortitude to tackle it – we suspect there may be many new methods to discuss.

FINAL THOUGHTS

Some of the analytical techniques we have described in earlier chapters might seem quantitatively rigorous; some of them indeed are, so we have tried to describe at least the basics of the particular techniques. Given the statistical rigor involved in some of those techniques, a caveat that requires emphasis here is that "Even the most sophisticated and rigorous methods of quantitative analysis cannot overcome deficiencies in the data or inherent limitations of the fossil [samples available]" (Behrensmeyer et al. 2007a:4). We have touched on how to evaluate sample adequacy (e.g., sample to redundancy) and sample representativeness in terms of taphonomic skewing (e.g., does variability in a taphonomic variable such as percent of carnivore-gnawed bones correlate with shifts in taxonomic abundances?). It is beyond our scope to do more here along these lines as the volume of literature on such topics is large. It suffices to say that some of that literature should be consulted as analysis proceeds along the lines described in earlier chapters.

In various places in this book we use the term "time averaged," a useful concept that warrants two final comments. First, it should be clear that the sample(s) of faunal remains under study may not be a true "average" (in a statistical sense) of the temporal period represented given how remains were accumulated, preserved, recovered, and identified. All of the latter processes tend to be non-random, hence the analyzed data may not represent an average of the biotic community(ies) represented. Once again issues of grain or resolution come to the forefront. The second comment is that although time averaging may be difficult to contend with analytically, we suggest that thoughtfulness in phrasing of research questions in terms of their resolution and how that resolution compares with the resolution provided by the faunal sample(s) under study should lessen the treachery of time averaging. In our view, both of these comments should always be on your mind as you apply the analytical techniques we have described.

When we first thought about writing this book, one of the things that occurred to us was that no such book existed. Books on paleoenvironmental reconstruction that did exist concerned plant remains (e.g., pollen, seeds), stable isotopes, geomorphological data, insect remains, and other, largely non-vertebrate faunal materials. Analytical techniques designed to be applied to these other data sources are seldom applicable to vertebrate remains, so we perceived a tremendous gap in the literature. Further, the value of having a book ready to hand that focused on vertebrate remains seemed to be increasing as research interest in issues such as those discussed in this chapter – conservation paleozoology, and contextualizing human biological and cultural evolution – has increased. In both cases the emphasis has been on vertebrate remains, though not to the exclusion of invertebrate or plant remains. After writing the book, we both feel like we have learned a lot in the process. We hope readers will feel the same after reading it, and we hope readers will apply some of what they have learned to samples of paleozoological remains for the reasons we have discussed in this chapter.

GLOSSARY

accuracy: the closeness of a measurement value to the true value

actualism: the research method of inferring the nature of past events and identity of past processes by analogy with events and processes observed in action in the present

adaptation: a trait or feature of an organism shaped by natural selection and that promotes survival and reproduction (*see also*: ecophenotypic plasticity)

Allen's rule: in warm-blooded animals, protruding body parts such as the tail and ears are shorter in cooler than in warmer climates

allometry: the growth of different body parts at different rates such that their proportions vary as a function of size

allopatric species: two or more species that do not co-occur in an area or habitat; their geographic ranges do not overlap (*see also*: sympatric species)

area of sympatry: the geographic area within which all (or most) species represented in a fossil collection are today found, the interpretive assumption being that this area represents the climate and/or habitat that existed in the location where the fossils were found at the time they were accumulated and deposited (*see also*: coexistence approach, mutual climatic range technique)

assemblage: a collection of faunal remains whose aggregateness is a result of an analytical decision; the decision might be to aggregate (analytically) all fossils recovered from a depositional unit such as a geological stratum, or to aggregate all the remains of a particular taxon recovered from a particular multi-stratum site or locality; the aggregateness, or spatio-temporal–taxonomic boundaries of an assemblage are an analytical decision, sometimes facilitated by stratigraphic boundaries, though the latter are not necessary (*syn*: collection)

autecology: the study of an individual organism or species and its relationship to its environment (*see also*: synecology)

Bergmann's rule: races from cooler climates in species of warm-blooded vertebrates tend to be larger than races of the same species living in warmer climates

bioclimatic model: models that use associations between climatic variables and species occurrences to estimate climates suitable for the presence of species

biodiversity: the diversity (usually number of kinds) of life forms in a specific spatio-temporal context, typically but not always measured as species (*see also*: diversity)

biogeography: study of the geographical distribution of organisms, both patterns and processes

biome: a particular combination of kinds of plants and animals found over large and multiple areas as a result of similar climatic regimes

bottom-up ecology: the structure and composition of a community is dictated by the bottom level of the trophic pyramid such that changes in that level influence higher levels (*see also*: top-down ecology)

brachydont: a tooth form with a low crown; enamel does not extend on to the root area

browser: a type of herbivore that feeds on dicots, including leaves, shoots, or fruits of generally woody plants

cenogram: a graph of the log (usually $\log_{10}$) of each taxon's mean adult body mass for taxa present in a community or fossil fauna, rank ordered from greatest to least (left to right)

central–marginal hypothesis: a species will display an abundance or frequency distribution with few individuals near range boundaries and numerous individuals near the center of the species' range; this presumed geographic frequency distribution is idealized and reality tends to be more complex with more or less isolated populations varying in size, density, and isolation such that the total population may or may not increase as one moves from the center of the species' range to the range margin

character displacement: a divergence in one or more characters or attributes of two taxa such that they are not in competition and can coexist in the same habitat (*see also*: ecological release)

chronocline: a character gradient (e.g., increasing size) expressed within a taxon over time (*see also*: cline)

climate restriction index: in bioclimatic models, a measure of the climatic zones in which a species occurs

climatograph: a graph of two climatic variables (typically temperature and precipitation) defining the climatic envelope within which the species is found, based on the margins of the species' geographic distribution

climax community: the theoretical final or mature stage of a biological community, supposedly developmentally static (*see also*: succession, ecological)

cline: a character gradient (e.g., increasing size) expressed within a taxon, usually over geographic space (*see also*: chronocline)

cliseral shift hypothesis: proposal that biotic communities responded uniformly to past environmental changes through synchronized shifts in their distribution; influenced by plant ecologist Frederic E. Clements, who proposed communities are immutable and organism-like because constituent species are functionally interlinked (*see also*: individualistic hypothesis)

coexistence approach: determination of the climatic parameters under which all (or as many as possible) of the species in a fossil assemblage coexist today (*see also*: area of sympatry, mutual climatic range technique)

collection: *see* assemblage

community: a set of organisms that live together and together form a more or less discrete entity; organisms comprising a community may, or may not, be functionally interrelated and linked through competition or some other process (*see also*: distal community, proximal community)

cultural filter: the human behaviors responsible for the accumulation, modification, and deposition of animal remains in archaeological deposits (Reed 1963; Reed and Braidwood 1960); a concept labeling the process(es) thought to jeopardize attempts to reconstruct paleoenvironments from zooarchaeological remains

dental mesowear: a method of paleodietary reconstruction based on development of wear facets on the teeth

dental microwear: a method of paleodietary reconstruction based on microscopic wear patterns on a tooth's surface

dimension: the theoretical distance between two anatomical landmarks

disharmonious fauna: *see* non-analog fauna

disjunct distribution: a discontinuous distribution of a population or species

distal community: one or more biological communities that provided remains from a distant location(s) to a fossil collection locality (Shotwell 1955, 1958) (*see also*: proximal community)

diversity: a complex term with a plethora of definitions; often it is used as a synonym for richness (*see also*: biodiversity, evenness, heterogeneity)

  alpha diversity: diversity at the local scale, usually within a single habitat

  beta diversity: change in diversity between communities, also referred to as turnover, usually across adjacent habitats on a landscape scale

  gamma diversity: variably defined as the species diversity of large geographic areas (e.g., biomes, continents), equivalent to alpha diversity writ large, or as the change in diversity across such areas, equivalent to beta diversity at a larger spatial scale

ecological diversity analysis: the comparison of modern and fossil communities based on the distribution of taxon-free traits such as body mass, diet, and locomotion

ecological diversity spectra: in ecological diversity analysis, the distribution of taxa in a community across taxon-free ecological categories

ecological release: freeing a species or population from an environmental limiting factor such as removal of a competitor or increase in exploitable resources (*see also*: character displacement)

ecological threshold: an abrupt change in ecosystem quality, property or phenomenon, or a case in which small changes in an environmental driver produce large responses in ecosystems (*see also*: tipping point)

ecological tolerances: the biotic and abiotic factors that control the presence/absence and abundance of a species

ecologically relevant net primary productivity: the amount of solar energy converted to plant biomass during the plant growing season

ecology: study of the relationships of organisms to one another and to their abiotic environment

ecometrics: the analysis of measurable, ecologically relevant morphological (anatomical) traits across the taxa in a community

ecomorphology: study of the relationship between an animal's morphology and the ecological context of its function

ecophenotypic plasticity: the ability of a species or population to alter its phenotype in response to environmental change without requiring genetic change (*see also*: adaptation)

ecotone: a transition between two or more diverse communities or habitats that contains plants and animals representative of the adjacent communities

edge effect: the tendency that the variety and density of organisms are greater at community junctions (ecotones) than in the communities flanking the junctions

environment: the total of all physical (abiotic) and biological factors impinging on a particular organismic unit

eurybiomic: species that occur in a wide variety of biomes (*see also*: stenobiomic)

eurytopic: species with wide tolerance limits (*see also*: stenotopic)

evenness: how individual organisms are distributed across species; equal numbers of individuals per species is even, unequal numbers of individuals per species is uneven

extralimital: a biogeographic record of a species beyond or outside of its range

fasting endurance: the ability to withstand periods of forage scarcity, usually found among larger-bodied individuals

faunule: an assemblage of associated animal remains recovered from one or several continuous strata and dominated by members of one biological community (Tedford 1970:677) (*see also*: local fauna)

fidelity study: actualistic assessment of the quantitative faithfulness of a fossil record of phenotypes, age classes, species richness, species abundance, trophic structure, etc., to the original biota

fossil (as used in this volume): any ancient remain of an animal, regardless of the specimen's age or fossilization (mineralization) condition

grazer: an herbivore that feeds primarily on monocots (grasses)

habitat: description of a kind of a spatio-temporally bounded place where an organism lived, usually focusing on vegetation

habitat metric: techniques that relate the habitat associations of the taxa in a faunal assemblage to a particular habitat type; e.g., ecological diversity analysis, niche models, taxonomic habitat index (THI)

habitat spectrum (habitat spectrum index): the distribution of taxa in a community according to their modern habitat associations

habitat tracking: dispersal of a species in tandem with shifts in the distribution of the habitat to which it is adapted

heterogeneity: a combination of both richness (number of kinds) and evenness (how individuals are distributed across kinds); the more difficult it is to predict the identity (kind or species) of an individual randomly chosen from a fauna, the more heterogeneous the fauna

hypsodont: a tooth form that has a high crown; enamel extends toward the root

hypsodonty index: a quantitative measurement of dental hypsodonty, usually taken as the unworn crown height of a tooth divided by its length or width

hypsodonty score: an ordinal scale ranking of dental hypsodonty: brachydont = 1, mesodont = 2, hypsodont = 3

indicator species: species that has well-known, relatively narrow ecological tolerances, whose presence in a fossil assemblage (relatively unambiguously) indicates the presence of a particular habitat or environment

individualistic hypothesis: ecologist H. A. Gleason's model of communities and biotic provinces indicating each species of organism responds individually to environmental variables given that each species has its own unique ecological tolerance limits and is not functionally interlinked with other species (*see also*: cliseral shift hypothesis)

insolation: the amount of solar radiation reaching the earth's surface

island rule: the tendency for insular (island-occupying) populations of smaller-bodied species to become larger and larger-bodied species to become smaller

isotaphonomy: two or more assemblages of biotic remains are said to be "isotaphonomic" if analysis indicates they had similar taphonomic histories; the implication is that assemblages that are isotaphonomic can be compared without fear that differences are an artifact of different taphonomic histories

keystone species: a species whose presence and role in an ecosystem has a disproportionate effect on other organisms in the system; often but not always a dominant predator occupying a high level of the trophic pyramid and whose removal allows prey species populations to irrupt thereby influencing nutrient cycling and ecosystem structure and function

landmark: an easily defined and located fixed point on a bone or tooth

law of tolerance: proposed by Victor Shelford, each species is adapted to both minimum and maximum values of environmental parameters such that too little or too great a value of that parameter precludes the occurrence of that organism (also known as Shelford's law of tolerance) (*see also*: limiting factors)

Liebig's law of the minimum: the ecological variable necessary to an organism's occurrence that is least abundant in an area will limit the organism's abundance in that area; other necessary resources may be more abundant than necessary, but the least available resource will be the limiting one

life zone: a geographic region (or altitudinal belt) characterized by a distinct set of plants and animals more or less limited to that region, proposed by biologist C. Hart Merriam

limiting factors: values of an environmental variable at the extremes of a species' tolerance curve (*see also*: law of tolerance)

local fauna: an assemblage of faunal remains from one locality or site, or several closely grouped localities or sites that are stratigraphically equivalent or nearly so, thus the represented taxa are close in space and time (Tedford 1970:678) (*see also*: faunule)

measured variable: the variable (e.g., length, frequency of occurrence) that is measured (*see also*: target variable)

measurement: the measured distance between two landmarks on an individual bone or tooth

   accuracy: how close a measurement is to the true value (size) of a dimension of a particular specimen

   resolution: the ability of a measurement system to detect and indicate small changes in a measurement result

mesodont: a tooth form with an intermediate crown height

mixed feeder: an herbivore that feeds on both grasses and dicots

mutual climatic range technique: determination of the climatic parameters under which all (or as many as possible) of the species in a fossil assemblage coexist today (*see also*: area of sympatry, coexistence approach)

natural (historical) range of variation: habitats, communities, and ecosystems are not literally static but rather dynamic, undergoing all sorts of changes of multiple kinds and scales; those that exceed this natural range are those that are of interest in paleoenvironmental reconstruction

nearest living relative (NLR): the assumption that an extinct species had the same ecological tolerances as its phylogenetically nearest living relative

niche: a multidimensional hypervolume of environmental conditions and resources defining the requirements for a species to exist (after G. Evelyn Hutchinson)

   fundamental niche: the particular combination of environmental variables in which a species can occur

   realized niche: incorporates the effects of factors that preclude a species occurrence such that it only occurs in a portion of its fundamental niche

niche conservatism: a species represented by fossils had the same ecological tolerances in the past that it has today; the species' niche has not evolved or changed over time but instead remained static

no-analog: climates or environments without modern analogs, or biotic communities characterized by association of presently allopatric taxa or taxonomic abundances that are atypical relative to the present (*see also*: non-analog fauna)

non-analog fauna: faunas made up of species that are stratigraphically and temporally associated and inferred to represent biological communities of species that were in the past biogeographically sympatric but which today are allopatric; also referred to (incorrectly) as a disharmonious fauna (*see also*: no-analog)

ordination: putting phenomena in order; usually refers to a family of statistical techniques (e.g., principal components analysis, correspondence analysis) that arrange phenomena such as faunal assemblages or communities along one or more axes according to their taxonomic composition

paleoautecology: the study of a fossil organism or species and its relationship to its environment

paleosynecology: the paleoecological study of a fossil group of organisms, species, or communities and their relationships to their environment

paleozoology: the study of ancient animal remains collected from geological (non-artifact bearing) deposits or from archaeological (artifact-bearing) deposits

palimpsest: in paleozoology (particularly zooarchaeology), a fossil collection representing multiple communities or biomes

primary productivity: the rate at which solar energy is converted to plant biomass for a given area and amount of time

productivity hypothesis: larger-bodied individuals of an animal species will be found in areas of greater productivity or lower population density resulting in greater availability of forage

proximal community: the biological community from which remains of animals originated and which is essentially geographically coincident with the location from which the remains were collected (Shotwell 1955, 1958) (*see also*: distal community)

rarefaction: probabilistic reduction of all large samples to the smallest of the samples under study, such that variability in sample size across assemblages is eliminated; it assumes statistical independence of the quantitative units being rarified, specifically that each tally of "1" represents a different organism

resolution: a property of the fineness of scale at which we perceive something, similar to grain or DPI (dots per inch) in photographs; the temporal or spatial resolution of a paleozoological assemblage is related to both taphonomic factors (e.g., depositional environment) and analytical decisions (e.g., how assemblages are aggregated)

richness: number of taxa (usually species) in a community, biome, or other unit

sampling radius: the area surrounding a fossil deposit from which fossils were derived (*syn*: accumulation radius)

sampling to redundancy: increasing sample size while simultaneously monitoring the value of the variable of interest (e.g., taxonomic richness) until it stabilizes over cumulatively larger sample sizes; stability indicates one has sampled to redundancy

scale: in ecology, *extent* concerns the total size of the area or total duration of time under study, and *grain* concerns the sampling strategy, including size and frequency of samples; in paleoecology, scale concerns both extent and grain and also resolution

stenobiomic: species that occur in only one or a few biomes (*see also*: eurybiomic)

stenotopic: species with narrow tolerance limits (*see also*: eurytopic)

succession, ecological: changes in or development of a community over time as plants and animals influence biomass production and nutrient cycling theoretically toward a stable state known as a *climax*

sweepstakes dispersal: a random long-distance colonization often by a single species

sympatric species: species that co-occur in a region or habitat; their geographic ranges overlap (*see also*: allopatric species)

synecology: the ecological study of a group of organisms, species, or communities and their relationships to their environment (*see also*: autecology)

taphonomy: the study of the transition of biological materials from the biosphere to the lithosphere (Efremov 1940) (*see also*: isotaphonomy)

target variable: the variable (e.g., length, frequency of occurrence) of analytical and interpretive interest that the analyst seeks to measure or estimate (*see also*: measured variable)

taxon free: in paleozoology, kinds of variables (e.g., richness) and analyses (e.g., comparison of distributions of body sizes) of biota that focus on non-taxonomic properties; species identities are less important than the variable measured (and usually not considered)

time averaging: the process of accumulating and mixing remains of organisms from different habitats, usually over extended time periods but potentially also distributed over large geographic areas (Behrensmeyer 1982); a time-averaged assemblage of animal (or plant) remains is one that represents a span of time longer than the ecological or climatic event of interest (*see also*: time perspectivism)

time perspectivism: being cognizant of the fact that the analyst must align the temporal scale (duration) of the event or process of interest – the hypothesis being tested or research question answered, and the analytical techniques used – with the scale and temporal resolution of the assemblage(s) under investigation (*see also*: time averaging)

tipping point: a particular moment in time when a small change can have long-term consequences for an ecosystem, or a time when an ecosystem shifts radically and potentially irreversibly into a different state (*see also*: ecological threshold)

tolerance curve: response of a species' total fitness measured as population size across an environmental gradient

tooth wear, abrasion: caused by food on tooth contact

tooth wear, attrition: caused by tooth on tooth contact that occurs when tooth surfaces slide past one another

top-down ecology: the structure and composition of a community is dictated by a high level of the trophic pyramid such that changes in that level influence lower levels (*see also*: bottom-up ecology)

training set: a set of modern biotic communities from various location that are systematically and mathematically related to one or more environmental or climatic variables to produce a transfer function

transfer function: an established mathematical/statistical relationship between biological data and (usually) one or more climate variables (e.g., seasonal temperature, annual precipitation) that is used to hindcast paleoclimatic conditions from paleobiological data

trophic cascade: occurs when predators at a high level in an ecosystem's trophic pyramid influence the abundance or behavior of their prey such that reciprocal changes occur in lower trophic levels, thereby causing changes in ecosystem structure and nutrient cycling

trophic pyramid: a model in the shape of a triangle or pyramid showing biomass or bioproductivity decreasing from bottom (producers) to top (carnivores) and energy flow through trophic or feeding levels

uniformitarianism: the assumption that the laws of nature (how earth and natural processes work) are invariant in time and space

vicariance event: geological or climate-driven event that causes populations to become geographically isolated from one another

zooarchaeology: the study of animal remains recovered from archaeological deposits (deposits containing artifacts)

REFERENCES

Abrams, Peter A., and Locke Rowe. 1996. The Effects of Predation on the Age and Size of Maturity of Prey. *Evolution* 50:1052–1061.

Agenbroad, Larry D. 2009. *Mammuthus exilis* from the California Channel Islands: Height, Mass, and Geologic Age. In *Proceedings of the 7th California Islands Symposium*, edited by C. C. Damiani and D. K. Garcelon, pp. 15–19. Institute for Wildlife Studies, Arcata, CA.

—— 2012. Giants and Pygmies: Mammoths of Santa Rosa Island, California (USA). *Quaternary International* 255:2–8.

Ager, Derek V. 1979. Paleoecology. In *Encyclopedia of Paleontology*, edited by Rhodes W. Fairbridge and David Jablonski, pp. 530–541. Dowden, Hutchinson and Ross, Stroudsburg, PA.

Albarella, Umberto. 2002. "Size Matters": How and Why Biometry is Still Important in Zooarchaeology. In *Bones and the Man: Studies in Honour of Don Brothwell*, edited by Keith Dobney and Terry O'Connor, pp. 51–62. Oxbow Books, Oxford.

Alemseged, Zeresenay. 2003. An Integrated Approach to Taphonomy and Faunal Change in the Shungura Formation (Ethiopia) and its Implications for Hominid Evolution. *Journal of Human Evolution* 44:451–478.

Alemseged, Z[eresenay], R. Bobe, and D. Geraads. 2007. Comparability of Fossil Data and its Significance for the Interpretation of Hominin Environments: A Case Study in the Lower Omo Valley, Ethiopia. In *Hominin Environments in the East African Pliocene: An Assessment of the Faunal Evidence*, edited by R[ené] Bobe, Z[eresenay] Alemseged, and A[nna] K. Behrensmeyer, pp. 159–181. Springer, Dordrecht.

Alfimov, A. V., and D. I. Berman. 2009. Possible Errors of the Mutual Climatic Range (MCR) Method in Reconstructing the Pleistocene Climate of Beringia. *Entomological Review* 89:487–499.

Allan, Robert S. 1948. Geological Correlation and Paleoecology. *Geological Society of America Bulletin* 59:1–10.

Allègre, Claude J. 2008. *Isotope Geochemistry*. Cambridge University Press.

Allison, Peter A., and David J. Bottjer (editors). 2011. *Taphonomy: Process and Bias through Time*. Topics in Geobiology 32. Springer, Dordrecht.

Alroy, John. 2000. New Methods for Quantifying Macroevolutionary Patterns and Processes. *Paleobiology* 26:707–733.

Alroy, John, Martin Aberhan, David J. Bottjer et al. 2008. Phanerozoic Trends in the Global Diversity of Marine Invertebrates. *Science* 321:97–100.

Ambrose, Stanley H. 1998. Chronology of the Later Stone Age and Food Production in East Africa. *Journal of Archaeological Science* 25:377–392.

Ambrose, Stanley H., and Michael J. DeNiro. 1986. The Isotopic Ecology of East African Mammals. *Oecologia* 69:395–406.

Ambrose, Stanley H., and M. Anne Katzenberg (editors). 2000. *Biogeochemical Approaches to Paleodietary Analysis*. Kluwer Academic, New York.

Ambrose, Stanley H., and John Krigbaum. 2003. Bone Chemistry and Bioarchaeology. *Journal of Anthropological Archaeology* 22:193–199.

Ambrose, Stanley H., and Lynette Norr. 1993. Experimental Evidence for the Relationship of the Carbon Isotope Ratios of Whole Diet and Dietary Protein to Those of Bone Collagen and Carbonate. In *Prehistoric Human Bone: Archaeology at the Molecular Level*, edited by Joseph B. Lambert and Gisela Grupe, pp. 1–37. Springer, Berlin.

Anderson, Alexander C. 1875. Further Remarks on the Reindeer; and on its Assumed Coexistence with the Hippopotamus. In *Reliquiae Aquitanicae; Being Contributions to the Archaeology and Paleontology of Périgord and the Adjoining Provinces of Southern France, by Edourd Lartet and Henry Christy*, edited by Thomas R. Jones, pp. 153–160. Williams and Norgate, London.

Anderson, Elaine. 1968. Fauna of the Little Box Elder Cave, Converse County, Wyoming: The Carnivora. *University of Colorado Studies, Series in Earth Sciences* 6:1–59.

Andrews, Peter. 1990. *Owls, Caves and Fossils: Predation, Preservation, and Accumulation of Small Mammal Bones in Caves, with an Analysis of the Pleistocene Cave Faunas from Westbury-sub-Mendip, Somerset, UK*. University of Chicago Press.

1992. Community Evolution in Forest Habitats. *Journal of Human Evolution* 22:423–438.

1995. Mammals as Palaeoecological Indicators. *Acta Zoologica Cracoviensia* 38:59–72.

1996. Palaeoecology and Hominoid Palaeoenvironments. *Biological Reviews* 71:257–300.

2006. Taphonomic Effects of Faunal Impoverishment and Faunal Mixing. *Palaeogeography, Palaeoclimatology, Palaeoecology* 241:572–589.

Andrews, Peter, and Sylvia Hixson. 2014. Taxon-Free Methods of Palaeoecology. *Annales Zoologici Fennici* 51:269–284.

Andrews, Peter, and Louise Humphrey. 1999. African Miocene Environments and the Transition to Early Hominins. In *African Biogeography, Climate Change, and Human Evolution*, edited by Timothy G. Bromage and Friedemann Schrenk, pp. 282–300. Oxford University Press.

Andrews, Peter, J. M. Lord, and Elisabeth M. Nesbit Evans. 1979. Patterns of Ecological Diversity in Fossil and Modern Mammalian Faunas. *Biological Journal of the Linnean Society* 11:177–205.

Anyonge, William. 1996. Microwear on Canines and Killing Behavior in Large Carnivores: Saber Function in *Smilodon fatalis*. *Journal of Mammalogy* 77:1059–1067.

Araújo, Miguel B., and A. Townsend Peterson. 2012. Uses and Misuses of Bioclimatic Envelope Modeling. *Ecology* 93:1527–1539.

Araújo, Miguel B., David Nogués-Bravo, José Alexandre F. Diniz-Filho et al. 2007. Quaternary Climate Changes Explain Diversity among Reptiles and Amphibians. *Ecography* 31:8–15.

Ashton, Kyle G. 2002. Patterns of Within-Species Body Size Variation of Birds: Strong Evidence for Bergmann's Rule. *Global Ecology and Biogeography* 11:505–523.

Ashton, Kyle G., Mark C. Tracy, and Alan de Queiroz. 2000. Is Bergmann's Rule Valid for Mammals? *American Naturalist* 156:390–415.

Assefa, Zelalem, Solomon Yirga, and Kaye E. Reed. 2008. The Large-Mammal Fauna from the Kibish Formation. *Journal of Human Evolution* 55:501–512.

Atkinson, T. C., K. R. Briffa, G. R. Coope, M. J. Joachim, and D. W. Perry. 1986. Climatic Calibration of Coleopteran Data. In *Handbook of Holocene Palaeoecology and Palaeohydrology*, edited by B. E. Bergland, pp. 851–858. John Wiley and Sons, Chichester.

Atkinson, T. C., K. R. Briffa, and G. R. Coope. 1987. Seasonal Temperatures in Britain during the Past 22,000 Years, Reconstructed Using Beetle Remains. *Nature* 325:587–592.

Austin, Mike. 2007. Species Distribution Models and Ecological Theory: A Critical Assessment and Some Possible New Approaches. *Ecological Modeling* 200:1–19.

Avery, D. M. 1982. Micromammals as Palaeoenvironmental Indicators and an Interpretation of the Late Quaternary in the Southern Cape Province, South Africa. *Annals of the South African Museum* 85:183–374.

———. 1988. Micromammals and Paleoenvironmental Interpretation in Southern Africa. *Geoarchaeology* 3:41–52.

———. 1990. Holocene Climatic Change in Southern Africa: The Contribution of Micromammals to its Study. *South African Journal of Science* 86:407–412.

———. 1992. The Environment of Early Modern Humans at Border Cave, South Africa: Micromammalian Evidence. *Palaeogeography, Palaeoclimatology, Palaeoecology* 91:71–87.

———. 1999. A Preliminary Assessment of the Relationship between Trophic Variability in Southern African Barn Owls *Tyto alba* and Climate. *Ostrich* 70:179–186.

———. 2004. Size Variation in the Common Molerat *Cryptomys hottentotus* from Southern Africa and its Potential for Palaeoenvironmental Reconstruction. *Journal of Archaeological Science* 31:273–282.

———. 2007. Micromammals as Palaeoenvironmental Indicators of the Southern African Quaternary. *Transactions of the Royal Society of South Africa* 62:17–23.

Ayliffe, Linda, and A. R. Chivas. 1990. Oxygen Isotope Composition of the Bone Phosphate of Australian Kangaroos: Potential as a Palaeoenvironmental Recorder. *Geochimica et Cosmochimica Acta* 54:2603–2609.

Badgley, Catherine. 1986. Counting Individuals in Mammalian Fossil Assemblages from Fluvial Environments. *Palaios* 1:328–338.

Bailey, I. W., and E. W. Sinnott. 1915. A Botanical Index of Cretaceous and Tertiary Climates. *Science* 41:831–834.

———. 1916. The Climatic Distribution of Certain Types of Angiosperm Leaves. *American Journal of Botany* 3:24–39.

Baker, G., L. H. P. Jones, and I. D. Wardrop. 1959. Cause of Wear in Sheeps' Teeth. *Nature* 184:1583–1584.

Balkwill, Darlene McCuaig, and Stephen L. Cumbaa. 1992. *A Guide to the Identification of Post-Cranial Bones of Bos taurus and Bison bison.* Syllogeus 71. Canadian Museum of Nature, Ottawa.

Bañuls-Cardona, Sandra, J. M. López-García, Hugues-Alexandre Blain, and A. C. Salomó. 2012. Climate and Landscape during the Last Glacial Maximum in Southwestern Iberia: The Small-Vertebrate Association from the Sala De Las Chimeneas, Maltravieso, Extremadura. *Comptes Rendus Palevol* 11:31–40.

Bañuls-Cardona, Sandra, J. M. López-García, Hugues-Alexandre Blain, I. Lozano-Fernández, and Gloria Cuenca-Bescós. 2014. The End of the Last Glacial Maximum in the Iberian Peninsula Characterized by the Small-Mammal Assemblages. *Journal of Iberian Geology* 40:19–27.

Barnosky, Anthony D. 1998. What Causes "Disharmonious" Mammal Assemblages? In *Quaternary Paleozoology in the Northern Hemisphere*, edited by Jeffrey J. Saunders, Bonnie W. Styles, and Gennady F. Baryshnikov, pp. 173–186. Illinois State Museum Scientific Papers 27. Illinois State University, Springfield.

———. 2001. Distinguishing the Effects of the Red Queen and Court Jester on Miocene Mammal Evolution in the Northern Rocky Mountains. *Journal of Vertebrate Paleontology* 21:172–185.

(editor) 2004. *Biodiversity Response to Climate Change in the Middle Pleistocene: The Porcupine Cave Fauna from Colorado*. University of California Press, Berkeley.

2009. *Heatstroke: Nature in an Age of Global Warming*. Island Press, Washington, DC.

Barnosky, Anthony D., and Marc A. Carrasco. 2002. Effects of Oligo-Miocene Global Climate Changes on Mammalian Species Richness in the Northwestern Quarter of the USA. *Evolutionary Ecology Research* 4:811–841.

Barnosky, Anthony D., and Elizabeth A. Hadly. 2016. *Tipping Points for Planet Earth*. Thomas Dunne Books, St. Martin's Press, New York.

Barnosky, Anthony D., Elizabeth A. Hadly, Patrick Gonzalez et al. 2017. Merging Paleobiology with Conservation Biology to Guide the Future of Terrestrial Ecosystems. *Science* 355:eaah4787.

Barr, W. Andrew. 2014. Functional Morphology of the Bovid Astragalus in Relation to Habitat: Controlling Phylogenetic Signal in Ecomorphology. *Journal of Morphology* 275:1201–1216.

2015. Paleoenvironments of the Shungura Formation (Plio-Pleistocene: Ethiopia) Based on Ecomorphology of the Bovid Astragalus. *Journal of Human Evolution* 88:97–107.

2017. Bovid Locomotor Functional Trait Distributions Reflect Land Cover and Annual Precipitation in Sub-Saharan Africa. *Evolutionary Ecology Research* 18:253–269.

Barr, W. Andrew, and Robert S. Scott. 2014. Phylogenetic Comparative Methods Complement Discriminant Function Analysis in Ecomorphology. *American Journal of Physical Anthropology* 153:663–674.

Barry, John C., Michèle Morgan, Lawrence J. Flynn et al. 2002. Faunal and Environmental Change in the Late Miocene Siwaliks of Northern Pakistan. *Paleobiology* 28 (special issue 3):1–71.

Bartlein, P. J., T. Webb III, and E. Fleri. 1984. Holocene Climatic Change in the Northern Midwest: Pollen-Derived Estimates. *Quaternary Research* 22:361–374.

Bate, D. M. A. 1937. Palaeontology: The Fossil Fauna of the Wady El-Mughara Caves. In *The Stone Age of Mt. Carmel*, vol. 1: *Excavations at the Wady El-Mughara*, by D. A. E. Garrod and D. M. A. Bate, pp. 135–240. Oxford University Press.

Baxter, M. J. 2001. Methodological Issues in the Study of Assemblage Diversity. *American Antiquity* 66:715–725.

Bedford, Jean N. 1978. A Technique for Sex Determination of Mature Bison Metapodials. *Plains Anthropologist* 14:40–43.

Begon, Michael, Colin R. Townsend, and John L. Harper. 2006. *Ecology: From Individuals to Ecosystems*, fourth edition. Blackwell, Malden, MA.

Behrensmeyer, Anna K. 1982. Time Resolution in Fluvial Vertebrate Assemblages. *Paleobiology* 8:211–227.

2006. Climate Change and Human Evolution. *Science* 311:476–478.

Behrensmeyer, Anna K., and Ralph E. Chapman. 1993. Models and Simulations of Time-Averaging in Terrestrial Veretebrate Accumulations. In *Taphonomic Approaches to Time Resolution in Fossil Assemblages*, edited by Susan M. Kidwell and Anna K. Behrensmeyer, pp. 125–149. Short Courses in Paleontology 6, The Paleontological Society, Knoxville, TN.

Behrensmeyer, Anna K., and Robert W. Hook. 1992. Paleoenvironmental Contexts and Taphonomic Modes. In *Terrestrial Ecosystems through Time: Evolutionary Paleoecology of Terrestrial Plants and Animals*, edited by Anna K. Behrensmeyer, John D. Damuth, William A. DiMichele, Richard Potts, Hans-Dieter Sues, and Scott L. Wing, pp. 15–36. University of Chicago Press.

Behrensmeyer, Anna K., Susan M. Kidwell, and Robert A. Gastaldo. 2000. Taphonomy and Paleobiology. In *Deep Time: Paleobiology's Perspective*, edited by Douglas H. Erwin and Scott W. Wing, pp. 103–147. *Paleobiology* (Supplement) 26.

Behrensmeyer, Anna K., C. Tristan Stayton, and Ralph E. Chapman. 2003. Taphonomy and Ecology of Modern Avifaunal Remains from Amboseli Park, Kenya. *Paleobiology* 29:52–70.

Behrensmeyer, A[nna] K., R[ené] Bobe, and Z[eresenay] Alemseged. 2007a. Approaches to the Analysis of Faunal Change during the East African Pliocene. In *Hominin Environments in the East African Pliocene: An Assessment of the Faunal Evidence*, edited by R[ené] Bobe, Z[eresenay] Alemseged, and A[nna] K. Behrensmeyer, pp. 1–24. Springer, Dordrecht.

Behrensmeyer, A[nna] K., Z[eresenay] Alemseged, and R[ené] Bobe. 2007b. Finale and Future: Investigating Faunal Evidence for Hominin Paleoecology in East Africa. In *Hominin Environments in the East African Pliocene: An Assessment of the Faunal Evidence*, edited by R[ené] Bobe, Z[eresenay] Alemseged, and A[nna] K. Behrensmeyer, pp. 333–345. Springer, Dordrecht.

Bell, Christopher J., Jacques A. Gauthier, and Gabe S. Bever. 2010. Covert Biases, Circularity, and Apomorphies: A Critical Look at the North American Quaternary Herpetofaunal Stability Hypothesis. *Quaternary International* 217:30–36.

Bell, Michael A., Matthew P. Travis, and D. Max Blouw. 2005. Inferring Natural Selection in a Fossil Threespine Stickleback. *Paleobiology* 32:562–577.

Belyea, Lisa R. 2007. Revealing the Emperor's New Clothes: Niche-Based Palaeoenvironmental Reconstruction in the Light of Recent Ecological Theory. *The Holocene* 17:683–688.

Berger, A[ndre], and M. F. Loutre. 1991. Insolation Values for the Climate of the Last 10 Million Years. *Quaternary Science Reviews* 10:297–317.

Bergerud, Arthur T. 2000. Caribou. In *Ecology and Management of Large Mammals in North America*, edited by Stephen Demarais and Paul R. Krausman, pp. 658–693. Prentice Hall, Upper Saddle River, NJ.

Bergmann, Carl. 1847. Ueber die Verhältnisse der Wärmeökonomie der Thiere zu Ihrer Grösse. *Gottinger studien* 3:595–708.

Berke, Melissa A., Thomas C. Johnson, Josef P. Werne, Stefan Schouten, and Jaap S. Sinninghe Damsté. 2012. A Mid-Holocene Thermal Maximum at the End of the African Humid Period. *Earth and Planetary Science Letters* 351–352:95–104.

Berner, Robert A., and Zavareth Kothavala. 2001. Geocarb III: A Revised Model of Atmospheric CO_2 over Phanerozoic Time. *American Journal of Science* 301:182–204.

Berto, Claudio, Paolo Boscato, Francesco Boschin, Elisa Luzi, and Annamaria Ronchitelli. 2017. Paleoenvironmental and Paleoclimatic Context during the Upper Palaeolithic (Late Upper Pleistocene) in the Italian Peninsula: The Small Mammal Record from Grotta Paglicci (Rignano Garganico, Foggia, Southern Italy). *Quaternary Science Reviews* 168:30–41.

Bininda-Emonds, Olaf R. P., Marcel Cardillo, Kate E. Jones et al. 2007. The Delayed Rise of Present-Day Mammals. *Nature* 446:507–512.

Birch, L. C. 1957. The Role of Weather in Determining the Distribution and Abundance of Animals. *Cold Spring Harbor Symposium on Quantitative Biology* 22:203–218.

Birks, H. J[ohn] B. 1995. Quantitative Palaeoenvironmental Reconstructions. In *Statistical Modelling of Quaternary Science Data*, edited by Darrel Maddy and John S. Brew, pp. 161–254. Technical Guide 5. Quaternary Research Association, Cambridge, UK.

Birks, H. John B., Oliver Hein, Heikki Seppa, and Anne E. Bjune. 2010. Strengths and Weaknesses of Quantitative Climate Reconstructions Based on Late-Quaternary Biological Proxies. *Open Ecology Journal* 3:68–110.

Blaauw, Maarten. 2012. Out of Tune: The Dangers of Aligning Proxy Archives. *Quaternary Science Reviews* 36:38–49.

Blain, Hugues-Alexandre, Salvador Bailon, Gloria Cuenca-Bescós et al. 2009. Long-Term Climate Record Inferred from Early-Middle Pleistocene Amphibian and Squamate Reptile Assemblages at the Gran Dolina Cave, Atapuerca, Spain. *Journal of Human Evolution* 56:55–65.

Blain, Hugues-Alexandre, Salvador Bailon, Gloria Cuenca-Bescós et al. 2010. Climate and Environment of the Earliest West European Hominins Inferred from the Amphibian and Squamate Reptile Assemblages: Sima del Elefante Lower Red Unit, Atapuerca, Spain. *Quaternary Science Reviews* 29:3034–3044.

Blain, Hugues-Alexandre, Iván Lozano-Fernández, Andreu Ollé, Jesus Rodríguez, Manuel Santonja, and Alfredo Pérez-González. 2015. The Continental Record of Marine Isotope Stage 11 (Middle Pleistocene) on the Iberian Peninsula Characterized by the Herpetofaunal Assemblages. *Journal of Quaternary Science* 30:667–678.

Blair, W. Frank. 1958. Distributional Patterns of Vertebrates in the Southern United States in Relation to Past and Present Environments. In *Zoogeography*, edited by Carl L. Hubbs, pp. 433–468. Publication 51. American Association for the Advancement of Science, Washington, DC.

Blois, Jessica L., Robert S. Feranec, and Elizabeth A. Hadly. 2008. Environmental Influences on Spatial and Temporal Patterns of Body-Size Variation in California Ground Squirrels (*Spermophilus beecheyi*). *Journal of Biogeography* 35:602–613.

Blois, Jessica L., Jenny L. McGuire, and Elizabeth A. Hadly. 2010. Small Mammal Diversity Loss in Response to Late-Pleistocene Climatic Change. *Nature* 465:771–774.

Blois, Jessica L., Phoebe L. Zametske, Matthew C. Fitzpatrick, and Seth Finnegan. 2013. Climate Change and the Past, Present, and Future of Biotic Interactions. *Science* 341:499–504.

Blumenthal, Scott A., Naomi E. Levin, Francis H. Brown et al. 2017. Aridity and Hominin Environments. *Proceedings of the National Academy of Sciences USA* 114:7331–7336.

Bobe, René. 2006. The Evolution of Arid Ecosystems in Eastern Africa. *Journal of Arid Environments* 66:564–584.

Bobe, René, and Anna K. Behrensmeyer. 2004. The Expansion of Grassland Ecosystems in Africa in Relation to Mammalian Evolution and the Origin of the Genus *Homo*. *Palaeogeography, Palaeoclimatology, Palaeoecology* 207:399–420.

Bobe, René, and Gerald G. Eck. 2001. Responses of African Bovids to Pliocene Climatic Change. *Paleobiology* 27 (special issue 2):1–48.

Bobe, René, and Meave G. Leakey. 2009. Ecology of Plio-Pleistocene Mammals in the Omo-Turkana Basin and the Emergence of *Homo*. In *The First Humans: Origin and Early Evolution of the Genus Homo*, edited by Frederick E. Grine, John G. Fleagle, and Richard E. Leakey, pp. 173–184. Springer, Dordrecht.

Bobe, René, Anna K. Behrensmeyer, and Ralph E. Chapman. 2002. Faunal Change, Environmental Variability and Late Pliocene Hominin Evolution. *Journal of Human Evolution* 42:475–497.

Bobe, R[ené], Z[eresenay] Alemseged, and A[nna] K. Behrensmeyer (editors). 2007. *Hominin Environments in the East African Pliocene: An Assessment of the Faunal Evidence*. Springer, Dordrecht.

Boessneck, Joachim, and Angela von den Driesch. 1978. The Significance of Measuring Animal Bones from Archaeological Sites. In *Approaches to Faunal Analysis in the Middle*

East, edited by Richard H. Meadow and Melinda A. Zeder, pp. 25–39. Peabody Museum of Archaeology and Ethnology Bulletin 2. Harvard University, Cambridge, MA.

Boivin, Nicole L., Melinda A. Zeder, Dorian Q. Fuller et al. 2016. Ecological Consequences of Human Niche Construction: Examining Long-Term Anthropogenic Shaping of Global Species Distributions. *Proceedings of the National Academy of Sciences USA* 113:6388–6396.

Bökönyi, Sándor. 1982. The Climatic Interpretation of Macrofaunal Assemblages in the Near East. In *Palaeoclimates, Palaeoenvironments and Human Communities in the Eastern Mediterranean Region in Later Prehistory*, edited by John L. Bintliff and Willem Van Zeist, pp. 149–163. British Archaeological Reports, International Series 133. BAR, Oxford.

Bond, William J., and Guy F. Midgley. 2000. A Proposed CO_2–Controlled Mechanism of Woody Plant Invasion in Grasslands and Savannas. *Global Change Biology* 6:865–869.

Bond, W[illiam] J., G. F. Midgley, and F. I. Woodward. 2003. The Importance of Low Atmospheric CO_2 and Fires in Promoting the Spread of Grasslands and Savannas. *Global Change Biology* 9:973–982.

Boshoff, Andre F., and Graham I. H. Kerley. 2001. Potential Distributions of the Medium- to Large-Sized Mammals in the Cape Floristic Region, Based on Historical Accounts and Habitat Requirements. *African Zoology* 36:245–273.

Boshoff, Andre F., Graham I. H. Kerley, and Richard M. Cowling. 2001. A Pragmatic Approach to Estimating the Distributions and Spatial Requirements of the Medium- to Large-Sized Mammals in the Cape Floristic Region, South Africa. *Diversity and Distributions* 7:29–43.

Bottjer, David J., Kathleen A. Cambell, Jennifer K. Schubert, and Mary I. Droser. 1995. Palaeoecological Models, Non-Uniformitarianism, and Tracking the Changing Ecology of the Past. In *Marine Palaeoenvironmental Analysis from Fossils*, edited by Dan W. J. Bosence and Peter A. Allison, pp. 7–26. Geological Society of America Special Paper 83. Geological Society of America, Boulder, CO.

Boudadi-Maligne, Myriam, and Gilles Escarguel. 2014. A Biometric Re-Evaluation of Recent Claims for Early Upper Palaeolithic Wolf Domestication in Eurasia. *Journal of Archaeological Science* 45:80–89.

Bowden, Joseph J., Anne Eskildsen, Rikke R. Hansen et al. 2015. High-Arctic Butterflies Become Smaller with Rising Temperature. *Biology Letters* 11:20150574.

Boyce, Mark S. 1978. Climatic Variability and Body Size Variation in the Muskrats (*Ondatra zibethicus*) of North America. *Oecologia* 36:1–19.

1979. Seasonality and Patterns of Natural Selection for Life Histories. *American Naturalist* 114:569–583.

Bradley, Raymond S. 1985. *Quaternary Paleoclimatology: Methods of Paleoclimatic Reconstruction*. Allen and Unwin, Boston.

2015. *Paleoclimatology: Reconstructing Climates of the Quaternary*, third edition. Elsevier, Amsterdam.

Brain, C. K. 1981. *The Hunters or the Hunted? An Introduction to African Cave Taphonomy*. University of Chicago Press.

Bray, P. J., S. P. E. Blockley, G. R. Coope et al. 2006. Refining Mutual Climatic Range (MCR) Quantitative Estimates of Palaeotemperature Using Ubiquity Analysis. *Quaternary Science Reviews* 25:1865–1876.

Breslawski, Ryan P., and David A. Byers. 2015. Assessing Measurement Error in Paleozoological Osteometrics with Bison Remains. *Journal of Archaeological Science* 53:235–242.

Brink, James S. 1999. Preliminary Report on a Caprine from the Cape Mountains, South Africa. *Archaeozoologia* 10:11–26.

Bronk Ramsey, Christopher. 2009. Bayesian Analysis of Radiocarbon Dates. *Radiocarbon* 51:337–360.

Brook, Barry W., Erle C. Ellis, Michael P. Perring, Anson W. Mackay, and Linus Blomqvist. 2013. Does the Terrestrial Biosphere Have Planetary Tipping Points? *Trends in Ecology and Evolution* 28:396–401.

Brothwell, D., and R. Jones. 1978. The Relevance of Small Mammal Studies to Archaeology. In *Research Problems in Zooarchaeology*, edited by D. R. Brothwell, K. D. Thomas, and Juliet Clutton-Brock, pp. 47–57. Occasional Publication 3. Institute of Archaeology, University of London.

Broughton, Jack M. 1997. Widening Diet Breadth, Declining Foraging Efficiency, and Prehistoric Harvest Pressure: Ichthyofaunal Evidence from the Emeryville Shellmound, California. *Antiquity* 71:845–862.

———. 2000. Terminal Pleistocene Fish Remains from Homestead Cave, Utah, and Implications for Fish Biogeography in the Bonneville Basin. *Copeia* 2000:645–656.

Broughton, Jack M., and Michael D. Cannon (editors). 2010. *Evolutionary Ecology and Archaeology: Applications to Problems in Human Evolution and Prehistory*. University of Utah Press, Salt Lake City.

Broughton, Jack M., David B. Madsen, and Jay Quade. 2000. Fish Remains from Homestead Cave and Lake Levels of the Past 13,000 Years in the Bonneville Basin. *Quaternary Research* 53:392–401.

Broughton, Jack M., Virginia I. Cannon, Shannon Arnold, Raymond J. Bogiatto, and Kenya Dalton. 2006. The Taphonomy of Owl-Deposited Fish Remains and the Origin of the Homestead Cave Ichthyofauna. *Journal of Taphonomy* 4:69–95.

Broughton, Jack M., David A. Byers, Reid A. Bryson, William Eckerle, and David B. Madsen. 2008. Did Climatic Seasonality Control Late Quaternary Artiodactyl Densities in Western North America? *Quaternary Science Reviews* 27:1916–1937.

Brown, Barnum. 1908. The Conrad Fissure, a Pleistocene Bone Deposit in Northern Arkansas: With Descriptions of Two New Genera and Twenty New Species of Mammals. *American Museum of Natural History Memoir* 9:155–208.

Brown, James H. 1984. On the Relationship between Abundance and Distribution of Species. *American Naturalist* 124:255–279.

Brown, James H., James F. Gillooly, Andrew P. Allen, Van M. Savage, and Geoffrey B. West. 2004. Toward a Metabolic Theory of Ecology. *Ecology* 85:1771–1789.

Brown, P., T. Sutikna, M. J. Morwood et al. 2004. A New Small-Bodied Hominin from the Late Pleistocene of Flores, Indonesia. *Nature* 431:1055–1061.

Brown, W. L., Jr., and E. O. Wilson. 1956. Character Displacement. *Systematic Zoology* 5:49–64.

Buckley, Michael. 2018. Zooarchaeology by Mass Spectrometry (ZooMS) Collagen Fingerprinting for the Species Identification of Archaeological Bone Fragments. In *Zooarchaeology in Practice*, edited by Christina M. Giovas and Michelle J. LeFebvre, pp. 227–247. Springer, Cham.

Buckley, Michael, Matthew Collins, Jane Thomas-Oates, and Julie C. Wilson. 2009. Species Identification by Analysis of Bone Collagen Using Matrix-Assisted Laser Desorption/Ionisation Time-of-Flight Mass Spectrometry. *Rapid Communications in Mass Spectrometry* 23:3843–3854.

Bulinski, Katherine V. 2007. Analysis of Sample-Level Properties along a Paleoenvironmental Gradient: The Behavior of Evenness as a Function of Sample Size. *Palaeogeography, Palaeoclimatology, Palaeoecology* 253:490–508.

Bunn, Henry T., Laurence E. Bartram, and Ellen M. Kroll. 1988. Variability in Bone Assemblage Formation from Hadza Hunting, Scavenging, and Carcass Processing. *Journal of Anthropological Archaeology* 7:412–457.

Burgman, Jenny H. E., Jennifer Leichliter, Nico L. Avenant, and Peter S. Ungar. 2016. Dental Microwear of Sympatric Rodent Species Sampled across Habitats in Southern Africa: Implications for Environmental Influence. *Integrative Zoology* 11:111–127.

Burnham, Robyn J. 2008. Hide and Go Seek: What Does Presence Mean in the Fossil Record? *Annals of the Missouri Botanical Garden* 95:51–71.

Burt, William H. 1958. The History and Affinities of the Recent Land Mammals of Western North America. In *Zoogeography*, edited by Carl L. Hubbs, pp. 131–154. Publication 51. American Association for the Advancement of Science, Washington, DC.

Burt, William H., and Richard P. Grossenheider. 1964. *A Field Guide to the Mammals.* Houghton Mifflin, Boston.

Burton, James H., and T. Douglas Price. 1990. The Ratio of Barium to Strontium as a Paleodietary Indicator of Consumption of Marine Resources. *Journal of Archaeological Science* 17:547–557.

Burton, James H., T. Douglas Price, and W. D. Middleton. 1999. Correlation of Bone Ba/Ca and Sr/Ca due to Biological Purification of Calcium. *Journal of Archaeological Science* 26:609–616.

Butler, B. Robert. 1969. More Information on the Frozen Ground Features and Further Interpretation of the Small Mammal Sequence at the Wasden Site (Owl Cave), Bonneville County, Idaho. *Tebiwa: Journal of the Idaho State University Museum* 12(1):58–63.

——— 1972a. The Holocene in the Desert West and its Cultural Significance. In *Great Basin Cultural Ecology: A Symposium*, edited by Don D. Fowler, pp. 5–12. Publications in the Social Sciences. 8. Desert Research Institute, Reno, NV.

——— 1972b. The Holocene or Postglacial Ecological Crisis on the Eastern Snake River Plain. *Tebiwa: Journal of the Idaho State University Museum* 15(1):49–63.

——— 1976. The Evolution of the Modern Sagebrush–Grass Steppe Biome on the Eastern Snake River Plain. In *Holocene Environmental Change in the Great Basin*, edited by Robert Elston, pp. 4–39. Research Paper 6. Nevada Archeological Survey, Reno.

——— 1978. *A Guide to Understanding Idaho Archaeology: The Upper Snake and Salmon River Country*. Idaho Museum of Natural History, Pocatello.

Butler, Kaylene, Julien Louys, and Kenny Travouillon. 2014. Extending Dental Mesowear Analyses to Australian Marsupials, with Applications to Six Plio-Pleistocene Kangaroos from Southeast Queensland. *Palaeogeography, Palaeoclimatology, Palaeoecology* 408:11–25.

Butler, P. M. 1952. The Milk Molars of Perissodactyla, with Remarks on Molar Occlusion. *Proceedings of the Zoological Society of London* 121:777–817.

Butler, Virginia L. 1988. Fish Feeding Behaviour and Fish Capture: The Case for Variation in Lapita Fishing Strategies. *Archaeology in Oceania* 29:81–90.

Byers, John A. 1997. *American Pronghorn: Social Adaptations and the Ghosts of Predators Past.* University of Chicago Press.

Calaby, J. H. 1971. Man, Fauna, and Climate in Aboriginal Australia. In *Aboriginal Man and Environment in Australia*, edited by D. J. Mulvaney and J. Golson, pp. 80–93. Australian National University Press, Canberra.

Calandra, Ivan, and Gildas Merceron. 2016. Dental Microwear Texture Analysis in Mammalian Ecology. *Mammal Review* 46:215–228.

Campbell, B. D., and D. M. Smith. 2000. A Synthesis of Recent Global Change Research on Pasture and Rangeland Production: Reduced Uncertainties and their Management Implications. *Agriculture, Ecosystems and Environments* 82:39–55.

Campbell, Timothy L., Patrick J. Lewis, and Justin K. Williams. 2011. Analysis of the Modern Distribution of South African *Gerbilliscus* (Rodentia: Gerbillinae) with Implications for Plio-Pleistocene Palaeoenvironmental Reconstruction. *South African Journal of Science* 107:Art. #497.

Campbell, Timothy L., Patrick J. Lewis, Monte L. Thies, and Justin K. Williams. 2012. A Geographic Information Systems (GIS)-Based Analysis of Modern South African Rodent Distributions, Habitat Use, and Environmental Tolerances. *Ecology and Evolution* 2:2881–2894.

Cannon, Michael D. 1999. A Mathematical Model of the Effects of Screen Size on Zooarchaeological Relative Abundance Measures. *Journal of Archaeological Science* 26:205–214.

——— 2003. A Model of Central Place Forager Prey Choice and an Application to Faunal Remains from Mimbres Valley, New Mexico. *Journal of Anthropological Archaeology* 22:1–25.

——— 2004. Geographic Variability in North American Mammal Community Richness during the Terminal Pleistocene. *Quaternary Science Reviews* 23:1099–1123.

——— 2013. NISP, Bone Fragmentation, and the Measurement of Taxonomic Abundance. *Journal of Archaeological Method and Theory* 20:397–419.

Caporale, Salvatore S., and Peter S. Ungar. 2016. Rodent Incisor Microwear as a Proxy for Ecological Reconstruction. *Palaeogeography, Palaeoclimatology, Palaeoecology* 446:225–233.

Cartmill, Matt. 1967. The Early Pleistocene Mammalian Microfaunas of Sub-Saharan Africa and their Ecological Significance. *Quaternaria* 9:169–198.

Caruso, Nicholas M., Michael D. Sears, Dean C. Adams, and Karen R. Lips. 2015. Widespread Rapid Reductions in Body Size of Adult Salamanders in Response to Climate Change. *Global Change Biology* 20:1751–1759.

Casanovas-Vilar, Isaac, and Jordi Agusti. 2007. Ecogeographical Stability and Climate Forcing in the Late Miocene (Vallesian) Rodent Record of Spain. *Palaeogeography, Palaeoclimatology, Palaeoecology* 248:169–189.

Case, E. C. 1921. Criteria for the Determination of the Climatic Environment of Extinct Animals. *Geological Society of America Bulletin* 32:333–338.

——— 1936. Paleoecology of the Vertebrates. In *Report of the Committee on Paleoecology, 1935–1936*, edited by W. H. Twenhofel, pp. 10–21. National Research Council, Division of Geology and Geography, Washington, DC.

Cassiliano, Michael L. 1994. Paleoecology and Taphonomy of Vertebrate Faunas from the Anza-Borrego Desert of California. Unpublished Ph.D. dissertation, Department of Geosciences, University of Arizona, Tucson.

Castaños, Jone, Pedro Castaños, Xabier Murelaga et al. 2014. Osteometric Analysis of the Scapula and Humerus of *Rangifer tarandus* and *Cervus elaphus*: A Contribution to the Discrimination of Late Pleistocene Cervids. *Acta Palaeontologica Polonica* 59:779–786.

Cerling, Thure E., and John M. Harris. 1999. Carbon Isotope Fractionation between Diet and Bioapatite in Ungulate Mammals and Implications for Ecological and Paleoecological Studies. *Oecologia* 120:347–363.

Cerling, Thure E., Yang Wang, and Jay Quade. 1993. Expansion of C_4 Ecosystems as an Indicator of Global Ecological Change in the Late Miocene. *Nature* 361:344–345.

Cerling, Thure E., John M. Harris, Bruce J. MacFadden et al. 1997. Global Vegetation Change through the Miocene–Pliocene Boundary. *Nature* 389:153–158.

Cerling, Thure E., John M. Harris, and Benjamin H. Passey. 2003. Diets of East African Bovidae Based on Stable Isotope Analysis. *Journal of Mammalogy* 84:456–470.

Cerling, Thure E., Samuel A. Andanje, Scott A. Blumenthal et al. 2015. Dietary Changes of Large Herbivores in the Turkana Basin, Kenya from 4 to 1 Ma. *Proceedings of the National Academy of Sciences USA* 112:11467–11472.

Chaline, J. 1977. Rodents, Evolution, and Prehistory. *Endeavor* n.s. 1(2):44–51.

Chang, Jie Christine, James Shulmeister, and Craig Woodward. 2015. A Chironomid Based Transfer Function for Reconstructing Summer Temperatures in Southeastern Australia. *Palaeogeography, Palaeoclimatology, Palaeoecology* 423:109–121.

Chapin, F. Stuart, III, and Anthony M. Starfield. 1997. Time Lags and Novel Ecosystems in Response to Transient Climatic Change in Alaska. *Climate Change* 35:449–461.

Chase, Brian M., and M. E. Meadows. 2007. Late Quaternary Dynamics of Southern Africa's Winter Rainfall Zone. *Earth-Science Reviews* 84:103–138.

Chase, Brian M., J. Tyler Faith, Alex Mackay et al. 2018. Climatic Controls on Later Stone Age Human Adaptation in Africa's Southern Cape. *Journal of Human Evolution* 114:35–44.

Chase, Jonathan M., and Matthew A. Leibold. 2002. Spatial Scale Dictates the Productivity–Biodiversity Relationship. *Nature* 416:427–430.

Cheatum, E. P., and Don Allen. 1964. Limitations in Paleoecological Reconstruction Utilizing Data from Fossil Non-Marine Molluscs. In *The Reconstruction of Past Environments: Proceedings*, assembled by James J. Hester and James Schoenwetter, pp. 31–33. Publication 3. Fort Burgwin Research Center, Taos, NM.

Chetboun, R., and Etian Tchernov. 1983. Temporal and Spatial Morphological Variation in *Meriones tristrami* (Rodentia: Gerbillidae) from Israel. *Israel Journal of Zoology* 32:63–90.

Churcher, C. S., and M[ichael] C. Wilson. 1990. Methods in Quaternary Ecology #12: Vertebrates. *Geoscience Canada* 17:59–78.

Clark, William C. 1985. Scales of Climate Impacts. *Climatic Change* 7:5–27.

Cleghorn, Naomi, and Curtis W. Marean. 2004. Distinguishing Selective Transport and *In Situ* Attrition: A Critical Review of Analytical Approaches. *Journal of Taphonomy* 2:43–67.

——— 2007. The Destruction of Skeletal Elements by Carnivores: The Growth of a General Model for Skeletal Element Destruction and Survival in Zooarchaeological Assemblages. In *Breathing Life into Fossils: Taphonomic Studies in Honor of C. K. (Bob) Brain*, edited by Travis R. Pickering, Kathy Schick, and Nicholas Toth, pp. 37–66. Publication Series. 2. Stone Age Institute Press, Gosport, IN.

Cleland, Carol E. 2001. Historical Science, Experimental Science, and the Scientific Method. *Geology* 29:987–990.

——— 2011. Prediction and Explanation in Historical Natural Science. *British Journal for the Philosophy of Science* 62:551–582.

Cleland, Charles E. 1966. *The Prehistoric Animal Ecology and Ethnozoology of the Upper Great Lakes Region*. Anthropological Papers 29. Museum of Anthropology, University of Michigan, Ann Arbor.

Clements, Frederic E. 1936. Nature and the Structure of Climax. *Journal of Ecology* 24:252–284.

Clements, Frederic E., and Victor Shelford. 1939. *Bio-Ecology*. Wiley, New York.

Clyde, William C., and Philip D. Gingerich. 1994. Rates of Evolution in the Dentition of Early Eocene *Cantius*: Comparison of Size and Shape. *Paleobiology* 20:506–522.

Codding, Brian F., Douglas W. Bird, and Rebecca Bliege Bird. 2010a. Interpreting Abundance Indices: Some Zooarchaeological Implications of Martu Foraging. *Journal of Archaeological Science* 37:3200–3210.

Codding, Brian F., Judith F. Porcasi, and Terry L. Jones. 2010b. Explaining Prehistoric Variation in the Abundance of Large Prey: A Zooarchaeological Analysis of Deer and Rabbit Hunting along the Pecho Coast of Central California. *Journal of Anthropological Archaeology* 29:47–61.

Cole, Kenneth L. 1995. Equable Climates, Mixed Assemblages, and the Regression Fallacy. In *Late Quaternary Environments and Deep History: A Tribute to Paul S. Martin*, edited by David W. Steadman and Jim I. Mead, pp. 131–138. Scientific Papers 3. Mammoth Site of Hot Springs, Hot Springs, SD.

Colwell, Robert K., and Thiago F. Rangel. 2009. Hutchinson's Duality: The Once and Future Niche. *Proceedings of the National Academy of Sciences USA* 106 (Supplement 2):19651–19658.

Colwell, Robert K., Anne Chao, Nicholas J. Gotelli et al. 2012. Models and Estimators Linking Individual-Based and Sample-Based Rarefaction, Extrapolation and Comparison of Assemblages. *Journal of Plant Ecology* 5:3–21.

Coope, G. R. 1959. A Late Pleistocene Insect Fauna from Chelford, Cheshire. *Proceedings of the Royal Society B* 151:70–86.

——— 1986. Coleoptera Analysis. In *Handbook of Palaeoecology and Palaeohydrology*, edited by B. E. Bergland, pp. 703–713. John Wiley and Sons, Chichester.

Coppens, Yves, F. Clark Howell, Glynn Ll. Isaac, and Richard E. F. Leakey (editors). 1976. *Earliest Man and Environments in the Lake Rudolf Basin: Stratigraphy, Paleoecology, and Evolution*. University of Chicago Press.

Cormie, A. B., B. Luz, and H. P. Schwarcz. 1994. Relationship between the Hydrogen and Oxygen Isotopes of Deer Bone and their Use in the Estimation of Relative Humidity. *Geochimica et Cosmochimica Acta* 58:3439–3449.

Costeur, Loïc. 2004. Cenogram Analysis of the Rudabánya Mammalian Community: Palaeonvironmental Interpretations. *Paleontographica Italica* 90:303–307.

Costeur, Loïc, and Serge Legendre. 2008. Mammalian Communities Document a Latitudinal Environmental Gradient during the Miocene Climatic Optimum in Western Europe. *Palaios* 23:280–288.

Craig, G. Y. 1961. Palaeozoological Evidence of Climate: (2) Invertebrates. In *Descriptive Palaeoclimatology*, edited by A. E. M. Nairen, pp. 207–226. Interscience Publishers, New York.

Creighton, G. Ken. 1980. Static Allometry of Mammalian Teeth and the Correlation of Tooth Size and Body Size in Contemporary Mammals. *Journal of Zoology* (London) 191:435–443.

Croft, Darin A. 2001. Cenozoic Environmental Change in South America as Indicated by Mammalian Body Size Distributions (Cenograms). *Diversity and Distributions* 7:271–287.

Croft, Darin A., and Deborah Weinstein. 2008. The First Application of the Mesowear Method to Endemic South American Ungulates (Notoungulata). *Palaeogeography, Palaeoclimatology, Palaeoecology* 269:103–114.

Cruz-Uribe, Kathryn. 1983. The Mammalian Fauna from Redcliff Cave, Zimbabwe. *South African Archaeological Bulletin* 38:7–16.

——— 1988. The Use and Meaning of Species Diversity and Richness in Archaeological Faunas. *Journal of Archaeological Science* 15:179–196.

Cuenca-Bescós, G[loria], J. Rofes, and J. Garcia-Pimienta. 2005. Environmental Change across the Early–Middle Pleistocene Transition: Small Mammalian Evidence from the Trinchera Dolina Cave, Atapuerca, Spain. In *Early–Middle Pleistocene Transitions: The Land–Ocean Evidence*, edited by M. J. Head and P. L. Gibbard, pp. 277–286. Special Publication 247. Geological Society of London.

Cuenca-Bescós, Gloria, Lawrence G. Straus, Manuel R. González Morales, and Juan C. García Pimienta. 2009. The Reconstruction of Past Environments through Small

Mammals: From the Mousterian to the Bronze Age in El Mirón Cave (Cantabria, Spain). *Journal of Archaeological Science* 36:947–955.

Curran, Sabrina C. 2012. Expanding Ecomorphological Methods: Geometric Morphometric Analysis of Cervidae Post-Crania. *Journal of Archaeological Science* 39:1172–1182.

Cutler, Alan H., Anna K. Behrensmeyer, and Ralph E. Chapman. 1999. Environmental Information in a Recent Bone Assemblage: Roles of Taphonomic Processes and Ecological Change. *Palaeogeography, Palaeoclimatology, Palaeoecology* 149:359–372.

Cuvier, G. 1796. Notice sur le Squelette d'une Très-Grande Espèce de Quadrupède Inconnue jusqu'à Présent, Trouvé au Paraguay, et Déposé au Cabinet d'Histoire Naturelle de Madrid. *Magasin Encyclopédique* 1:303–310.

Daams, Remmert, Albert J. van der Meulen, Pablo Peláez-Campomanes, and Maria A. Alvarez-Sierra. 1999. Trends in Rodent Assemblages from the Aragonian (Early–Middle Miocene) of the Calatayud-Daroca Basin, Aragón, Spain. In *Hominoid Evolution and Climatic Change in Europe*, vol. 1: *The Evolution of Neogene Terrestrial Ecosystems in Europe*, edited by Jordi Agustí, Lorenzo Rook, and Peter Andrews, pp. 127–139. Cambridge University Press.

Dalquest, Walter W. 1965. New Pleistocene Formation and Local Fauna from Hardeman County, Texas. *Journal of Paleontology* 39:63–79.

Daly, Patricia. 1969. Approaches to Faunal Analysis in Archaeology. *American Antiquity* 34:146–153.

Damuth, John D. 1981. Population Density and Body Size in Mammals. *Nature* 290:699–700.

—— 1982. Analysis of the Preservation of Community Structure in Assemblages of Fossil Mammals. *Paleobiology* 8:434–446.

—— 1992. Taxon-Free Characterisation of Animal Communities. In *Terrestrial Ecosystems through Time: Evolutionary Paleoecology of Terrestrial Plants and Animals*, edited by Anna K. Behrensmeyer, John D. Damuth, William A. DiMichele, Richard Potts, Hans-Dieter Sues, and Scott L. Wing, pp. 183–203. University of Chicago Press.

Damuth, John D., and Mikael Fortelius. 2001. Reconstructing Mean Annual Precipitation, Based on Mammalian Dental Morphology and Local Species Richness. In *Eeden Plenary Workshop on Late Miocene to Early Pliocene Environments and Ecosystems*, edited by J[ordi] Agustí and O. Oms, pp. 23–24. European Science Foundation, Sabadell, Spain.

Damuth, John D., and Christine M. Janis. 2011. On the Relationship between Hypsodonty and Feeding Ecology in Ungulate Mammals, and its Utility in Palaeoecology. *Biological Reviews* 86:733–758.

Damuth, John D., and Bruce J. MacFadden (editors). 1990. *Body Size in Mammalian Paleobiology: Estimation and Biological Implications*. Cambridge University Press.

Damuth, John D., Mikael Fortelius, Peter Andrews et al. 2002. Reconstructing Mean Annual Precipitation Based on Mammalian Dental Morphology and Local Species Richness. *Journal of Vertebrate Paleontology* 22 (Supplement 3):48A.

Darlington, Philip J., Jr. 1957. *Zoogeography: The Geographical Distribution of Animals*. John Wiley and Sons, New York.

Darwin, Charles. 1859. *On the Origin of Species by Means of Natural Selection, or the Preservation of Favoured Races in the Struggle for Life*. John Murray, London.

Daubenmire, Rexford F. 1938. Merriam's Life Zones of North America. *Quarterly Review of Biology* 13:317–332.

Dauphin, Y., C. Kowalski, and C. Denys. 1994. Assemblage Data and Bone and Teeth Modifications as an Aid to Paleoenvironmental Interpretations of the Open-Air Pleistocene Site of Tighenif (Algeria). *Quaternary Research* 42:340–349.

Davies, Jeremy. 2016. *The Birth of the Anthropocene*. University of California Press, Berkeley.

Davies, T. Jonathan, Andy Purvis, and John L. Gittleman. 2009. Quaternary Climate Change and the Geographic Ranges of Mammals. *American Naturalist* 174:297–307.

Davis, Edward Byrd, and Nicholas D. Pyenson. 2007. Diversity Biases in Terrestrial Mammalian Assemblages and Quantifying the Differences between Museum Collections and Published Accounts: A Case Study from the Miocene of Nevada. *Palaeogeography, Palaeoclimatology, Palaeoecology* 250:139–149.

Davis, Margaret B. 1976. Pleistocene Biogeography of Temperate Deciduous Forests. In *Ecology of the Pleistocene: A Symposium*, edited by R. C. West and W. G. Haag, pp. 13–26. Geoscience and Man 13. Louisiana State University, Baton Rouge.

——— 1981. Quaternary History and the Stability of Forest Communities. In *Forest Succession: Concepts and Application*, edited by Darrell C. West, Herman H. Shugart, and Daniel B. Botkin, pp. 132–153. Springer, New York.

Davis, Margaret B., and Ruth G. Shaw. 2001. Range Shifts and Adaptive Responses to Quaternary Climate Change. *Science* 292:637–679.

Davis, Margaret B., Ruth G. Shaw, and Julie R. Etterson. 2005. Evolutionary Responses to Changing Climate. *Ecology* 86:1704–1714.

Davis, Matt, and Silvia Pineda-Munoz. 2016. The Temporal Scale of Diet and Dietary Proxies. *Ecology and Evolution* 6:1883–1897.

Davis, Simon J. M. 1977. Size Variation of the Fox, *Vulpes vulpes* in the Palaearctic Region Today, and in Israel during the Late Quaternary. *Journal of Zoology* (London) 182:343–351.

——— 1981. The Effects of Temperature Change and Domestication on the Body Size of Late Pleistocene to Holocene Mammals of Israel. *Paleobiology* 7:101–104.

Dawkins, W. Boyd. 1869. On the Distribution of British Postglacial Mammals. *Quarterly Journal of the Geological Society of London* 25:192–217.

——— 1871. On Pleistocene Climate and the Relation of the Pleistocene Mammalia to the Glacial Period. *Popular Science Review* 10:377–397.

——— 1874. *Cave Hunting, Researches on the Evidence of Caves Respecting the Early Inhabitants of Europe*. Macmillan, London.

Dawson, Terence P., Stephen T. Jackson, Joanna I. House, Iain Colin Prentice, and Georgina M. Mace. 2011. Beyond Predictions: Biodiversity Conservation in a Changing Climate. *Science* 332:53–58.

Dayan, Tamar, Etian Tchernov, Yoram Yom-Tov and Daniel Simberloff. 1989. Ecological Character Displacement in Saharo-Arabian *Vulpes*: Outfoxing Bergmann's Rule. *Oikos* 55:263–272.

Dayan, Tamar, Daniel Simberloff, Etian Tchernov, and Yoram Yom-Tov. 1991. Calibrating the Paleothermometer: Climate, Communities, and the Evolution of Size. *Paleobiology* 17:189–199.

Deacon, H. J. 1979. Excavations at Boomplaas Cave: A Sequence through the Upper Pleistocene and Holocene in South Africa. *World Archaeology* 10:241–257.

Deacon, H. J., and Mary Brooker. 1976. The Holocene and Upper Pleistocene Sequence in the Southern Cape. *Annals of the South African Museum* 71:203–214.

Deacon, H. J., Janette Deacon, and Mary Brooker. 1976. Four Painted Stones from Boomplaas Cave, Oudtshoorn District. *South African Archaeological Bulletin* 31:141–145.

Deacon, H. J., Janette Deacon, Mary Brooker, and M. L. Wilson. 1978. The Evidence for Herding at Boomplaas Cave in the Southern Cape, South Africa. *South African Archaeological Bulletin* 33:39–65.

Deacon, H. J., A. Scholtz, and L. D. Daitz. 1983. Fossil Charcoals as a Source of Palaeoecological Information in the Fynbos Region. In *Fynbos Palaeoecology: A Preliminary Synthesis, South African National Scientific Programmes Report No. 75*, edited by H. J. Deacon, Q. B. Hendey and J. J. N. Lambrechts, pp. 174–182. Mills Litho, Cape Town.

Deacon, H. J., Janette Deacon, A. Scholtz, J. F. Thackeray, and James S. Brink. 1984. Correlation of Palaeoenvironmental Data from the Late Pleistocene and Holocene Deposits at Boomplaas Cave, Southern Cape. In *Late Cainozoic Palaeoclimates of the Southern Hemisphere*, edited by J. C. Vogel, pp. 339–351. Balkema, Rotterdam.

Deacon, Hilary J. 1995. Two Late Pleistocene–Holocene Archaeological Depositories from the Southern Cape, South Africa. *South African Archaeological Bulletin* 50:121–131.

Deacon, Janette. 1984. *The Later Stone Age of Southernmost Africa*. British Archaeological Reports, International Series 213. BAR, Oxford.

Deacon, Janette, and N. Lancaster. 1988. *Late Quaternary Palaeoenvironments of Southern Africa*. Oxford University Press.

De Graaf, G. 1961. A Preliminary Investigation of the Mammalian Microfauna in Pleistocene Deposits of Caves in the Transvaal System. *Palaeontologia Africana* 7:59–117.

DeGusta, David, and Elisabeth S. Vrba. 2003. A Method for Inferring Paleohabitats from the Functional Morphology of Astragali. *Journal of Archaeological Science* 30:1009–1022.

——— 2005a. Methods for Inferring Paleohabitats from Discrete Traits of the Bovid Postcranial Skeleton. *Journal of Archaeological Science* 32:1115–1123.

——— 2005b. Methods for Inferring Paleohabitats from the Functional Morphology of Bovid Phalanges. *Journal of Archaeological Science* 32:1099–1113.

Delcourt, Hazel R., and Paul A. Delcourt. 1988. Quaternary Landscape Ecology: Relevant Scales in Space and Time. *Landscape Ecology* 2:23–44.

Delcourt, Hazel R., Paul A. Delcourt, and Thompson Webb III. 1983. Dynamic Plant Ecology: The Spectrum of Vegetational Change in Space and Time. *Quaternary Science Reviews* 1:153–175.

Demarais, Stephen, Karl V. Miller, and Harry A. Jacobson. 2000. White-Tailed Deer. In *Ecology and Management of Large Mammals in North America*, edited by Stephen Demarais and Paul R. Krausman, pp. 601–628. Prentice Hall, Upper Saddle River, NJ.

DeMenocal, Peter, Joseph Ortiz, Tom Guilderson et al. 2000. Abrupt Onset and Termination of the African Humid Period: Rapid Climate Responses to Gradual Insolation Forcing. *Quaternary Science Reviews* 19:347–361.

Demirel, Arzu, Peter Andrews, Isin Yalcinkay, and Ayhan Ersoy. 2011. The Taphonomy and Palaeoenvironmental Implications of the Small Mammals from Karain Cave, Turkey. *Journal of Archaeological Science* 38:3048–3059.

Deng, Tao. 2009. Late Cenozoic Environmental Changes in the Linxia Basin (Gansu, China) as Indicated by Cenograms of Fossil Mammals. *Vertebrata PalAsiatica* 10:282–298.

DeNiro, Michael J., and Samuel Epstein. 1978. Influence of Diet on the Distribution of Carbon Isotopes in Animals. *Geochimica et Cosmochimica Acta* 42:495–506.

——— 1981. Influence of Diet on the Distribution of Nitrogen Isotopes in Animals. *Geochimica et Cosmochimica Acta* 45:341–351.

Denys, C., Y. Dauphin, B. Rzebik-Kowalski, and K. Kowalski. 1996. Taphonomic Study of Algerian Owl Pellet Assemblages and Differential Preservation of Some Rodents: Palaeontological Implications. *Acta Zoologica Cracoviensia* 39:103–116.

de Ruiter, Darryl J., Matt Sponheimer, and Julia A. Lee-Thorp. 2008. Indications of Habitat Association of *Australopithecus robustus* in the Bloubank Valley, South Africa. *Journal of Human Evolution* 55:1015–1030.

DeSantis, Larisa R. G. 2016. Dental Microwear Textures: Reconstructing Diets of Fossil Mammals. *Surface Topography: Metrology and Properties* 4:023002.

DeSantis, Larisa R. G., and Ryan J. Haupt. 2014. Cougars' Key to Survival through the Late Pleistocene Extinction: Insights from Dental Microwear Texture Analysis. *Biology Letters* 10:20140203.

DeSantis, Larisa R. G., Blaine W. Schubert, Jessica R. Scott, and Peter S. Ungar. 2012a. Implications of Diet for the Extinction of Saber-Toothed Cats and American Lions. *PLoS ONE* 7:e52453.

DeSantis, Larisa R. G., Rachel A. Beavins Tracy, Cassandra S. Koontz, John C. Roseberry, and Matthew C. Velasco. 2012b. Mammalian Niche Conservation through Deep Time. *PLOS One* 7:e35624.

Dietl, Gregory P., and Karl W. Flessa (editors). 2009. *Conservation Paleobiology: Using the Past to Manage for the Future.* Paleontological Society Papers 15. Paleontological Society, Boulder, CO.

——— 2011. Conservation Paleobiology: Putting the Dead to Work. *Trends in Ecology and Evolution* 26:30–37.

Dietl, Gregory P., Susan Kidwell, Mark Brenner et al. 2012. *Conservation Paleobiology: Opportunities for the Earth Sciences.* Paleontological Research Institute, Ithaca, NY.

Dietl, Gregory P., Susan M. Kidwell, Mark Brenner et al. 2015. Conservation Paleobiology: Leveraging Knowledge of the Past to Inform Conservation and Restoration. *Annual Review of Earth and Planetary Sciences* 43:79–103.

Diffenbaugh, Noah S., Moetasim Ashfaq, Bryan Shuman, John W. Williams, and Patrick J. Bartlein. 2006. Summer Aridity in the United States: Response to Mid-Holocene Changes in Insolation and Sea Surface Temperature. *Geophysical Research Letters* 33:L22712.

DiMichele, W. A., A. K. Behrensmeyer, T. D. Olszewski et al. 2004. Long-Term Stasis in Ecological Assemblages: Evidence from the Fossil Record. *Annual Review of Ecology, Evolution, and Systematics* 35:285–322.

Dodd, J. Robert, and Robert J. Stanton, Jr. 1981. *Paleoecology: Concepts and Applications.* John Wiley and Sons, New York.

——— 1990. *Paleoecology: Concepts and Applications*, second edition. John Wiley and Sons, New York.

Dodson, Peter. 1973. The Significance of Small Bones in Paleoecological Interpretation. *University of Wyoming Contributions to Geology* 12:15–19.

Domínguez-Rodrigo, Manuel. 2012. Critical Review of the MNI (Minimum Number of Individuals) as a Zooarchaeological Unit of Quantification. *Archaeological and Anthropological Sciences* 4:47–59.

Domínguez-Rodrigo, Manuel, and Charles M. Musiba. 2010. How Accurate are Paleoecological Reconstructions of Early Paleontological and Archaeological Sites? *Evolutionary Biology* 37:128–140.

Donohue, Shelly L., Larisa R. G. DeSantis, Blaine W. Schubert, and Peter S. Ungar. 2013. Was the Giant Short-Faced Bear a Hyper-Scavenger? A New Approach to the Dietary Study of Ursids Using Dental Microwear Textures. *PLoS ONE* 8:e77531.

Dorf, E. 1959. Climatic Changes of the Past and Present. *Contributions from the Museum of Paleontology* 13:181–210. University of Michigan, Ann Arbor.

Dortch, Joe, and Richard Wright. 2010. Identifying Palaeo-Environments and Changes in Aboriginal Subsistence from Dual-Patterned Faunal Assemblages, South-Western Australia. *Journal of Archaeological Science* 37:1053–1064.

Driesch, Angela von den. 1976. *A Guide to the Measurement of Animal Bones from Archaeological Sites*. Peabody Museum of Archaeology and Ethnology Bulletin 1. Harvard University, Cambridge, MA.

Driesch, Angela von den, and Hilary J. Deacon. 1985. Sheep Remains from Boomplaas Cave, South Africa. *South African Archaeological Bulletin* 40:39–44.

Driver, Jonathan C. 1992. Identification, Classification and Zooarchaeology. *Circaea* 9:35–47.

——— 2001. Paleoecological and Archaeological Implications of the Charlie Lake Cave Fauna, British Columbia. In *People and Wildlife in Northern North America: Essays in Honor of R. Dale Guthrie*, edited by S. Craig Gerlach and Maribeth S. Murray, pp. 13–22. British Archaeological Reports, International Series 944. BAR, Oxford.

——— 2011. Identification, Classification and Zooarchaeology. *Ethnobiology Letters* 2:19–39.

Driver, Jonathan C., and Joshua R. Woiderski. 2008. Interpretation of the "Lagomorph Index" in the American Southwest. *Quaternary International* 185:3–11.

Duffield, Lathel F. 1973. Aging and Sexing the Post-Cranial Skeleton of Bison. *Plains Anthropologist* 18:132–139.

Dulian, James J. 1975. Paleoecology of the Brayton Local Biota, Late Wisconsinan of Southwestern Iowa. Unpublished Master of Science thesis, University of Iowa, Iowa City.

Eastham, L. C., R. S. Feranec, and D. R. Begun. 2016. Stable Isotopes Show Resource Partitioning among the Early Late Miocene Herbivore Community at Rudabánya II: Paleoenvironmental Implications for the Hominoid, *Rudapithecus hungaricus*. *Palaeogeography, Palaeoclimatology, Palaeoecology* 454:161–174.

Eckert, C. G., K. E. Samis, and S. C. Lougheed. 2008. Genetic Variation across Species' Geographical Ranges: The Central–Marginal Hypothesis and Beyond. *Molecular Ecology* 17:1170–1188.

Efremov, J. A. 1940. Taphonomy: New Branch of Paleontology. *Pan-American Geologist* 74:81–93.

Ehleringer, James R., R. F. Sage, L. B. Flanagan, and R. W. Pearcy. 1991. Climate Change and the Evolution of C_4 Photosynthesis. *Trends in Ecology and Evolution* 6:95–99.

Ehleringer, James R., Thure E. Cerling, and Brent R. Helliker. 1997. C_4 Photosynthesis, Atmospheric CO_2, and Climate. *Oecologia* 112:285–299.

Eisenmann, Véra. 1986. Comparative Osteology of Modern and Fossil Horses, Half-Asses, and Asses. In *Equids in the Ancient World*, edited by Richard H. Meadow and Hans-Peter Uerpmann, pp. 67–116. Dr. Ludwig Reichert Verlag, Wiesbaden.

Elias, Robert W., Yoshimitsu Hirao, and Clair C. Patterson. 1982. The Circumvention of the Natural Biopurification of Calcium along Nutrient Pathways by Atmospheric Inputs of Industrial Lead. *Geochimica et Cosmochimica Acta* 46:2561–2580.

Elias, Scott A. 1997. The Mutual Climatic Range Method of Palaeoclimate Reconstruction Based on Insect Fossils: New Applications and Interhemispheric Comparisons. *Quaternary Science Reviews* 16:1217–1225.

——— 2001. Mutual Climatic Range Reconstructions of Seasonal Temperatures Based on Late Pleistocene Fossil Assemblages in Eastern Beringia. *Quaternary Science Reviews* 20:77–91.

Elias, Scott A., and Nicki J. Whitehouse. 2014. G. Russell Coope: Papers Honouring his Life and Career. *Quaternary International* 341:1–5.

Elias, Scott A., John T. Andrews, and Katherine H. Anderson. 1999. Insights on the Climatic Constraints on the Beetle Fauna of Coastal Alaska, USA, Derived from the Mutual Climatic Range Method of Palaeoclimate Reconstruction. *Arctic and Alpine Research* 31:94–98.

Elith, Jane, and John R. Leathwick. 2009. Species Distribution Models: Ecological Explanation and Prediction across Space and Time. *Annual Review of Ecology, Evolution, and Systematics* 40:677–697.

Elton, Charles S. 1927. *Animal Ecology.* Sedgwick and Jackson, London.

Emerson, Thomas E. 1978. A New Method for Calculating the Live Weight of the Northern White-Tailed Deer from Osteoarchaeological Material. *Mid-Continental Journal of Archaeology* 3:35–44.

Emery, Kitty F., and Erin Kennedy Thornton. 2008. Zooarchaeological Habitat Analysis of Ancient Maya Landscape Changes. *Journal of Ethnobiology* 28:154–178.

——— 2012. Using Animal Remains to Reconstruct Ancient Landscapes and Climate in Central and Southern Maya Lowlands. In *Proceedings of the General Session of the 11th International Council for Archaeozoology Conference,* edited by Christine Lefèvre, pp. 203–225. British Archaeological Reports, International Series 2354. BAR, Oxford.

——— 2014. Tracking Climate Change in the Ancient Maya World through Zooarchaeological Habitat Analysis. In *The Great Maya Droughts in Cultural Context: Case Studies in Resilience and Vulnerability,* edited by Gyles Iannone, pp. 301–331. University Press of Colorado, Boulder.

Emslie, Steven D. 1982. Osteological Identification of Long-Eared and Short-Eared Owls. *American Antiquity* 47:155–157.

Endler, John A. 1982. Problems in Distinguishing Historical from Ecological Factors in Biogeography. *American Zoologist* 22:441–452.

Erasmus, Barend F. N., Albert S. Van Jaarswel, Steven L. Chown, Mrigesh Kshatriya, and Konrad J. Wessels. 2002. Vulnerability of South African Animal Taxa to Climate Change. *Global Change Biology* 8:679–693.

Eronen, Jussi T. 2006. Eurasian Neogene Large Herbivorous Mammals and Climate. *Acta Zoologica Fennica* 216:1–72.

Eronen, Jussi T., and Lorenzo Rook. 2004. The Mio-Pliocene Primate Fossil Record: Dynamics and Habitat Tracking. *Journal of Human Evolution* 47:323–341.

Eronen, Jussi T., Majid M. Ataabadi, Arne Micheels et al. 2009. Distribution History and Climatic Controls of the Late Miocene Pikermian Chronofauna. *Proceedings of the National Academy of Sciences USA* 106:11867–11871.

Eronen, Jussi T., P. David Polly, Marianne Fred et al. 2010a. Ecometrics: The Traits that Bind the Past and Present Together. *Integrative Zoology* 5:88–101.

Eronen, J[ussi] T., K. Puolamäki, L. Liu et al. 2010b. Precipitation and Large Herbivorous Mammals I: Estimates from Present-Day Communities. *Evolutionary Ecology Research* 12:217–233.

Eronen, J[ussi] T., K. Puolamäki, L. Liu et al. 2010c. Precipitation and Large Herbivorous Mammals II: Application to Fossil Data. *Evolutionary Ecology Research* 12:235–248.

Eronen, J[ussi] T., C. Janis, C. P. Chamberlain, and A. Mulch. 2015. Mountain Uplift Explains Differences in Palaeogene Patterns of Mammalian Evolution and Extinction between North America and Europe. *Proceedings of the Royal Society B* 282:20150136.

Ervynck, Anton. 1999. Possibilities and Limitations of the Use of Archaeozoological Data in Biogeographical Analysis: A Review with Examples from the Benelux Region. *Belgian Journal of Zoology* 129:125–138.

Escarguel, Gilles, Emmanuel Fara, Arnaud Brayard, and Serge Legendre. 2011. Biodiversity is Not (and Never Has Been) a Bed of Roses! *Comptes Rendus Biologies* 334:351–359.

Estes, James A. 1996. Predators and Ecosystem Management. *Wildlife Society Bulletin* 24:390–396.

Estes, Richard, and Paul Berberian. 1970. Paleoecology of a Late Cretaceous Vertebrate Community from Montana. *Breviora* 343:1–35.

Etheridge, Richard. 1958. Pleistocene Lizards of the Cragin Quarry Fauna of Meade County, Kansas. *Copeia* 1958:94–101.

Etnier, Michael A. 2002. The Effects of Human Hunting on Northern Fur Seal (*Callorhinus ursinus*) Migration and Breeding Distributions in the Late Holocene. Unpublished Ph.D. dissertation, Department of Anthropology, University of Washington, Seattle.

——— 2004. Reevaluating Evidence of Density-Dependent Growth in Northern Fur Seals (*Callorhinus ursinus*) Based on Measurements of Archived Skeletal Specimens. *Canadian Journal of Fisheries and Aquatic Sciences* 61:1616–1626.

Evans, Alistair R. 2013. Shape Descriptors as Ecometrics in Dental Ecology. *Hystrix* 24:133–140.

Evin, Allowen, Thomas Cucchi, Gilles Escarguel et al. 2014. Using Traditional Biometrical Data to Distinguish West Palearctic Wild Boar and Domestic Pigs in the Archaeological Record: New Methods and Standards. *Journal of Archaeological Science* 43:1–8.

Fagerstrom, J. A. 1964. Fossil Communities in Paleoecology: Their Recognition and Significance. *Geological Society of America Bulletin* 75:1197–1216.

Faith, J. Tyler. 2008. Eland, Buffalo, and Wild Pigs: Were Middle Stone Age Humans Ineffective Hunters? *Journal of Human Evolution* 55:24–36.

——— 2011a. Late Quaternary Dietary Shifts of the Cape Grysbok (*Raphicerus melanotis*) in Southern Africa. *Quaternary Research* 75:159–165.

——— 2011b. Ungulate Biogeography, Statistical Methods, and the Proficiency of Middle Stone Age Hunters. *Journal of Human Evolution* 60:315–317.

——— 2011c. Ungulate Community Richness, Grazer Extinctions, and Human Subsistence Behavior in Southern Africa's Cape Floral Region. *Palaeogeography, Palaeoclimatology, Palaeoecology* 306:219–227.

——— 2012a. Conservation Implications of Fossil Roan Antelope (*Hippotragus equinus*) in Southern Africa's Cape Floristic Region. In *Paleontology in Ecology and Conservation*, edited by Julien Louys, pp. 239–251. Springer, Heidelberg.

——— 2012b. Palaeozoological Insights into Management Options for a Threatened Mammal: Southern Africa's Cape Mountain Zebra (*Equus zebra zebra*). *Diversity and Distributions* 18:438–447.

——— 2013a. Taphonomic and Paleoecological Change in the Large Mammal Sequence from Boomplaas Cave, Western Cape, South Africa. *Journal of Human Evolution* 65:715–730.

——— 2013b. Ungulate Diversity and Precipitation History since the Last Glacial Maximum in the Western Cape, South Africa. *Quaternary Science Reviews* 68:191–199.

——— 2014. Late Pleistocene and Holocene Mammal Extinctions on Continental Africa. *Earth-Science Reviews* 128:105–121.

——— 2018. Paleodietary Change and its Implications for Aridity Indices Derived from $\delta^{18}O$ of Herbivore Tooth Enamel. *Palaeogeography, Palaeoclimatology, Palaeoecology* 490:571–578.

Faith, J. Tyler, and Andrew Du. 2018. The Measurement of Taxonomic Evenness in Zooarchaeology. *Archaeological and Anthropological Sciences* 10:1419–1428.

Faith, J. Tyler, and Anna K. Behrensmeyer. 2013. Climate Change and Faunal Turnover: Testing the Mechanics of the Turnover-Pulse Hypothesis with South African Fossil Data. *Paleobiology* 39:609–627.

Faith, J. Tyler, and Adam D. Gordon. 2007. Skeletal Element Abundances in Archaeofaunal Assemblages: Economic Utility, Sample Size, and Assessment of Carcass Transport Strategies. *Journal of Archaeological Science* 34:872–882.

Faith, J. Tyler, and James F. O'Connell. 2011. Revisiting the Late Pleistocene Mammal Extinction Record at Tight Entrance Cave, Southwestern Australia. *Quaternary Research* 76:397–400.

Faith, J. Tyler, Manuel Domínguez-Rodrigo, and Adam D. Gordon. 2009. Long-Distance Carcass Transport at Olduvai Gorge? A Quantitative Examination of Bed I Skeletal Element Abundances. *Journal of Human Evolution* 56:247–256.

Faith, J. Tyler, Jonah. N. Choiniere, Christian A. Tryon, Daniel J. Peppe, and David L. Fox. 2011. Taxonomic Status and Paleoecology of *Rusingoryx atopocranion* (Mammalia, Artiodactyla), an Extinct Pleistocene Bovid from Rusinga Island, Kenya. *Quaternary Research* 75:697–707.

Faith, J. Tyler, Richard Potts, Thomas W. Plummer et al. 2012. New Perspectives on Middle Pleistocene Change in the Large Mammal Faunas of East Africa: *Damaliscus hypsodon* sp. nov. (Mammalia, Artiodactyla) from Lainyamok, Kenya. *Palaeogeography, Palaeoclimatology, Palaeoecology* 361–362:84–93.

Faith, J. Tyler, Christian A. Tryon, Daniel J. Peppe, and David L. Fox. 2013. The Fossil History of Grévy's Zebra (*Equus grevyi*) in Equatorial East Africa. *Journal of Biogeography* 40:359–369.

Faith, J. Tyler, Christian A. Tryon, Daniel J. Peppe et al. 2015. Paleoenvironmental Context of the Middle Stone Age Record from Karungu, Lake Victoria Basin, Kenya, and its Implications for Human and Faunal Dispersals in East Africa. *Journal of Human Evolution* 83:28–45.

Faith, J. Tyler, Christian A. Tryon, and Daniel J. Peppe. 2016a. Environmental Change, Ungulate Biogeography, and their Implications for Early Human Dispersals in Equatorial East Africa. In *Africa from MIS 6–2: Population Dynamics and Paleoenvironments*, edited by Sacha C. Jones and Brian A. Stewart, pp. 233–245. Springer, Dordrecht.

Faith, J. Tyler, David B. Patterson, Nick Blegen et al. 2016b. Size Variation in *Tachyoryctes splendens* (East African Mole-Rat) and its Implications for Late Quaternary Temperature Change in Equatorial East Africa. *Quaternary Science Reviews* 140:39–48.

Faith, J. Tyler, Joe Dortch, Chelsea Jones, James Shulmeister, and Kenny J. Travouillon. 2017. Large Mammal Species Richness and Late Quaternary Precipitation Change in Southwestern Australia. *Journal of Quaternary Science* 32:760–769.

Falk, Carl R., and Holmes A. Semken, Jr. 1998. Taphonomy of Rodent and Insectivore Remains in North American Archaeological Sites: Selected Examples and Interpretations. In *Quaternary Paleozoology in the Northern Hemisphere*, edited by Jeffrey J. Saunders, Bonnie W. Styles, and Gennady F. Baryshnikov, pp. 285–321. Illinois State Museum Scientific Papers 27. Illinois State Museum, Springfield.

Faure, Gunter, and Teresa M. Mensing. 2004. *Isotopes: Principles and Applications*, third edition. Wiley, Hoboken, NJ.

Ferguson, Steven H. 2002. The Effects of Productivity and Seasonality on Life History: Comparing Age at Maturity among Moose (*Alces alces*) Populations. *Global Ecology and Biogeography* 11:303–312.

Fernández-García, Mónica, and Juan Manuel López-García. 2013. Palaeoecology and Biochronology Based on the Rodents Analysis from the Late Pleistocene/Holocene of Toll Cave (Moià, Barcelona). *Spanish Journal of Paleontology* 28:227–238.

Fernández-García, Mónica, Juan Manuel López-García, and Carlos Lorenzo. 2016. Palaeoecological Implications of Rodents as Proxies for the Late Pleistocene–Holocene Environmental and Climatic Changes in Northeastern Iberia. *Comptes Rendus Palevol* 15:707–719.

Fernández-Jalvo, Yolanda, and Peter Andrews. 2016. *Atlas of Taphonomic Identifications: 1001+ Images of Fossil and Recent Mammal Bone Modification.* Springer, Dordrecht.

Fernández-Jalvo, Yolanda, Christiane Denys, Peter Andrews, Terry Williams, Yanicke Dauphin, and Louise Humphrey. 1998. Taphonomy and Palaeoecology of Olduvai Bed-I (Pleistocene, Tanzania). *Journal of Human Evolution* 34:137–172.

Fernández-Jalvo, Y[olanda], L. Scott, and P. Andrews. 2011. Taphonomy in Palaeoecological Interpretations. *Quaternary Science Reviews* 30:1296–1302.

Fichman, Martin. 1977. Wallace: Zoogeography and the Problem of Landbridges. *Journal of the History of Biology* 10:45–63.

Fick, Stephen E., and Robert J. Hijmans. 2017. WorldClim 2: New 1-km Spatial Resolution Climate Surfaces for Global Land Areas. *International Journal of Climatology* 37:4302–5315

Figueirido, Borja, Paul Palmqvist, and J. A. Pérez-Claros. 2009. Ecomorphological Correlates of Craniodental Variation in Bears and Paleobiological Implications for Extinct Taxa: An Approach Based on Geometric Morphometrics. *Journal of Zoology* (London) 277:70–80.

Findley, James S. 1964. Paleoecological Reconstruction: Vertebrate Limitations. In *The Reconstruction of Past Environments: Proceedings*, assembled by James J. Hester and James Schoenwetter, pp. 23–25. Publication 3. Fort Burgwin Research Center, Taos, NM.

Fisher, Jacob L. 2012. Shifting Prehistoric Abundances of Leporids at Five Finger Ridge, a Central Utah Archaeological Site. *Western North American Naturalist* 72:60–68.

Fisher, Jacob L., and Benjamin Valentine. 2013. Resource Depression, Climate Change, and Mountain Sheep in the Eastern Great Basin of Western North America. *Archaeological and Anthropological Sciences* 5:145–157.

Flannery, Kent V. 1967. Vertebrate Fauna and Hunting Patterns. In *The Prehistory of the Tehuacan Valley*, vol. 1: *Environment and Subsistence*, edited by Douglas S. Byers, pp. 132–177. University of Texas Press, Austin.

Fleagle, John G. 1978. Size Distributions of Living and Fossil Primate Faunas. *Paleobiology* 4:67–76.

Fleming, Theodore H. 1973. Number of Mammal Species in North and Central American Forest Communities. *Ecology* 54:555–563.

Flynn, Lawrence J. 2003. Small Mammal Indicators of Forest Paleo-Environment in the Siwalik Deposits of the Potwar Plateau, Pakistan. In *Distribution and Migration of Tertiary Mammals in Eurasia: A Volume in Honour of Hans de Bruun*, edited by Jelle W. F. Reumer and Wilma Wessels, pp. 183–196. Deinsea 10. Natural History Museum, Rotterdam.

Fortelius, Mikael 1985. Ungulate Cheek Teeth: Developmental, Functional, and Evolutionary Interrelations. *Acta Zoologica Fennica* 180:1–76.

——— 2003. Evolution of Dental Capability in Western Eurasian Large Mammal Plant Eaters 22–2 Million Years Ago: A Case for Environmental Forcing Mediated by Biotic Processes. In *The New Panorama of Animal Evolution*, edited by A. Legakis, S. Sfenthourakis, R. Polymeni, and M. Thessalou-Legaki, pp. 61–67. Pensoft, Sofia and Moscow.

Fortelius, Mikael, and Nikos Solounias. 2000. Functional Characterization of Ungulate Molars Using the Abrasion–Attrition Wear Gradient: A New Method for Reconstructing Paleodiets. *American Museum Novitates* 3301:1–36.

Fortelius, Mikael, Jussi T. Eronen, Jukka Jernvall et al. 2002. Fossil Mammals Resolve Regional Patterns of Eurasian Climate Change over 20 Million Years. *Evolutionary Ecology Research* 4:1005–1016.

Fortelius, Mikael, Jussi T. Eronen, Liping Liu et al. 2003. Continental-Scale Hypsodonty Patterns, Climatic Palaeobiogeography, and Dispersal of Eurasian Neogene Large

Mammal Herbivores. In *Distribution and Migration of Tertiary Mammals in Eurasia*, edited by Jelle W. F. Reumer and Wilma Wessels, pp. 1–11. Deinsea 10. Natural History Museum, Rotterdam.

Fortelius, Mikael, Jussi T. Eronen, Liping Liu et al. 2006. Late Miocene and Pliocene Large Land Mammals and Climatic Changes in Eurasia. *Palaeogeography, Palaeoclimatology, Palaeoecology* 238:219–227.

Fortelius, Mikael, Indrė Žliobaitė, Ferhat Kaya et al. 2016. An Ecometric Analysis of the Fossil Mammal Record of the Turkana Basin. *Philosophical Transactions of the Royal Society B* 371:20150232.

Foster, J. Bristol. 1964. Evolution of Mammals on Islands. *Nature* 202:234–235.

Fowler, Melvin L., and Paul W. Parmalee. 1959. Ecological Interpretation of Data on Archaeological Sites: The Modoc Rock Shelter. *Transactions of the Illinois State Academy of Science* 52(3–4):109–119.

Francis, C. M. 2008. *A Guide to the Mammals of Southeast Asia*. Princeton University Press.

Franz-Odendaal, Tamara A., and Thomas M. Kaiser. 2003. Differential Mesowear in the Maxillary and Mandibular Cheek Dentition of Some Ruminants (Artiodactyla). *Annales Zoologici Fennici* 40:395–410.

Fraser, Danielle, and Jessica M. Theodor. 2010. The Use of Gross Dental Wear in Dietary Studies of Extinct Lagomorphs. *Journal of Paleontology* 84:720–729.

Frazier, Michael K. 1977. New Records of *Neofiber leonardi* (Rodentia: Cricetidae) and the Paleoecology of the Genus. *Journal of Mammalogy* 58:368–373.

Fricke, Henry C., William C. Clyde, and James R. O'Neill. 1998. Intra-Tooth Variations in $\delta^{18}O$ (PO_4) of Mammalian Tooth Enamel as a Record of Seasonal Variations in Continental Climate Variables. *Geochimica et Cosmochimica Acta* 62:1839–1850.

Fritts, Harold C., Terence J. Blasing, Bruce P. Hayden, and John E. Kutzbach. 1971. Multivariate Techniques for Specifying Tree-Growth and Climate Relationships and for Reconstructing Anomalies in Paleoclimate. *Journal of Applied Meteorology* 10:845–864.

Frost, S. R. 2007. African Pliocene and Pleistocene Cercopithecid Evolution and Global Climate Change. In *Hominin Environments in the East African Pliocene: An Assessment of the Faunal Evidence*, edited by R[ené] Bobe, Z[eresenay] Alemseged and A[nna] K. Behrensmeyer, pp. 51–76. Springer, Dordrecht.

Fukami, Tadashi. 2015. Historical Contingency in Community Assembly: Integrating Niches, Species Pools, and Priority Effects. *Annual Review of Ecology, Evolution, and Systematics* 46:1–23.

Gailer, Juan Pablo, Ivan Calandra, Ellen Schulz-Kornas, and Thomas M. Kaiser. 2016. Morphology is Not Destiny: Discrepancy between Form, Function and Dietary Adaptation in Bovid Cheek Teeth. *Journal of Mammalian Evolution* 23:369–383.

Galster, S., N. D. Burgess, J. Fjeldså, L. A. Hansen, and C. Rahbek. 2007. *One Degree Resolution Databases of the Distribution of 1085 Mammals in Sub-Saharan Africa*. Zoological Museum, University of Copenhagen.

Gamble, Clive. 1978. Optimising Information from Studies of Faunal Remains. In *Sampling in Contemporary British Archaeology*, edited by John F. Cherry, Clive Gamble, and Stephen Shennan, pp. 321–353. British Archaeological Reports, British Series 50. BAR, Oxford.

Gandiwa, Edson. 2013. Top-Down and Bottom-Up Control of Large Herbivore Populations: A Review of Natural and Human-Induced Influences. *Tropical Conservation Science* 6:493–505.

García-Alix, Antonio, Raef Minwer-Barakat, Elvira Martín Suárez, Matthijs Freudenthal, and José M. Martín. 2008. Late Miocene–Early Pliocene Climatic Evolution of the

Granada Basin (Southern Spain) Deduced from the Paleoecology of the Micromammal Associations. *Palaeogeography, Palaeoclimatology, Palaeoecology* 265:214–225.

García-Alix, Antonio, Raef Minwer-Barakat, Elvira Martín Suárez, and Matthijs Freudenthal. 2009. Small Mammals from the Early Pleistocene of the Granada Basin, Southern Spain. *Quaternary Research* 72:265–274.

García Yelo, B. A., A. R. Gómez Cano, J. L. Cantalapiedra et al. 2014. Palaeoenvironmental Analysis of the Aragonian (Middle Miocene) Mammalian Faunas from the Madrid Basin Based on Body-Size Structure. *Journal of Iberian Geology* 40:129–140.

Gardner, A. L. (editor). 2007. *Mammals of South America*, vol. 1: *Marsupials, Xenarthans, Shrews, and Bats.* University of Chicago Press.

Gardner, Janet L., Anne Peters, Michael R. Kearney, Leo Joseph, and Robert Heinsohn. 2011. Declining Body Size: A Third Universal Response to Warming? *Trends in Ecology and Evolution* 26:285–291.

Garrard, Andrew N. 1982. The Environmental Implications of a Re-analysis of the Large Mammal Fauna from the Wadi El-Mughara Caves, Palestine. In *Palaeoclimates, Palaeoenvironments and Human Communities in the Eastern Mediterranean Region in Later Prehistory*, edited by John L. Bintliff and Willem Van Zeist, pp. 165–187. British Archaeological Reports, International Series 133(i). BAR, Oxford.

Garrett, Nicole D., David L. Fox, Kieran P. McNulty et al. 2015. Stable Isotope Paleoecology of Late Pleistocene Middle Stone Age Humans from Equatorial East Africa, Lake Victoria Basin, Kenya. *Journal of Human Evolution* 82:1–14.

Gasse, F., and F. Tekaia. 1983. Transfer Functions for Estimating Paleoecological Conditions (Ph) from East African Diatoms. *Hydrobiologia* 103:85–90.

Gasse, F., S. Juggins, and L. Ben Khelifa. 1995. Diatom-Based Transfer Functions for Inferring Past Hydrochemical Characteristics of African Lakes. *Palaeogeography, Palaeoclimatology, Palaeoecology* 117:31–54.

Gaston, Kevin J. 1996. Species Richness: Measure and Measurement. In *Biodiversity: A Biology of Numbers and Difference*, edited by Kevin J. Gaston, pp. 77–113. Blackwell, Oxford.

———. 2007. Latitudinal Gradient in Species Richness. *Current Biology* 17(15):R574.

Gauthreaux, Sidney A., Jr. 1980. The Influences of Long-Term and Short-Term Climatic Changes on the Dispersal and Migration of Organisms. In *Animal Migration, Orientation and Navigation*, edited by Sidney A. Gauthreaux, Jr., pp. 103–174. Academic Press, New York.

Gavin, Daniel G., W. Wyatt Oswald, Eugene R. Wahlf, and John W. Williams. 2003. A Statistical Approach to Evaluating Distance Metrics and Analog Assignments for Pollen Records. *Quaternary Research* 60:356–367.

Geist, Valerius. 1987. Bergmann's Rule Is Invalid. *Canadian Journal of Zoology* 65:1035–1038.

———. 1989. Environmentally Guided Phenotype Plasticity in Mammals and Some of its Consequences to Theoretical and Applied Biology. In *Alternative Life-History Styles of Animals*, edited by Michael N. Bruton, pp. 153–176. Kluwer Academic, Dordrecht.

———. 1992. Endangered Species and the Law. *Nature* 357:274–276.

———. 1998. *Deer of the World: Their Evolution, Behavior, and Ecology.* Stackpole Books, Mechanicsburg, PA.

George, Christian O. 2012. Alternative Approaches to the Identification and Reconstruction of Paleoecology of Quaternary Mammals. Unpublished Ph.D. dissertation, University of Texas, Austin.

George, T. Neville. 1958. The Ecology of Fossil Animals: I. Organism and Environment. *Science Progress* 46:677–680.

Georgina, Dianna M. 2001. The Small Mammals of Lime Hills Cave I. In *People and Wildlife in Northern North America: Essays in Honor of R. Dale Guthrie*, edited by S. Craig Gerlach and Maribeth S. Murray, pp. 23–31. British Archaeological Reports, International Series 944. BAR, Oxford.

Gidley, James W., and C. Lewis Gazin. 1938. *The Pleistocene Vertebrate Fauna from Cumberland Cave, Maryland*, United States Museum Bulletin 171. Smithsonian Institution, Washington, DC.

Gieke, James. 1872. On Changes of Climate during the Glacial Epoch. Fifth Paper. *Geological Magazine* 9:164–170.

—— 1881. *Prehistoric Europe, a Geological Sketch*. Edward Standford, London.

Gienapp, P., C. Teplitsky, J. S. Alho, J. A. Mills, and J. Merilä. 2008. Climate Change and Evolution: Disentangling Environmental and Genetic Responses. *Molecular Ecology* 17:167–178.

Gifford[-Gonzalez], Diane P. 1981. Taphonomy and Paleoecology: A Critical Review of Archaeology's Sister Disciplines. In *Advances in Archaeological Method and Theory*, vol. IV, edited by Michael B. Schiffer, pp. 365–438. Academic Press, New York.

Gifford-Gonzalez, Diane P. 1991a. Bones Are Not Enough: Analogues, Knowledge, and Interpretive Strategies in Zooarchaeology. *Journal of Anthropological Archaeology* 10:215–254.

Gifford-Gonzalez, Diane [P.]. 1991b. Examining and Refining the Quadratic Crown Height Method of Age Estimation. In *Human Predators and Prey Mortality*, edited by Mary C. Stiner, pp. 41–78. Westview Press, Boulder, CO.

Giglio, Louis, James T. Randerson, and Guido van der Werf. 2013. Analysis of Daily, Monthly, and Annual Burned Area Using the Fourth-Generation Global Fire Emissions Database (GFED4). *Journal of Geophysical Research* 118:317–328.

Gilmore, Raymond M. 1949. The Identification and Value of Mammal Bones from Archeological Excavations. *Journal of Mammalogy* 30:163–169.

Gingerich, Philip D. 1989. New Earliest Wasatchian Mammalian Fauna from the Eocene of Northwestern Wyoming: Composition and Diversity in a Rarely Sampled High-Floodplain Assemblage. *University of Michigan Papers in Paleontology* 28:1–97.

—— 1993. Quantification and Comparison of Evolutionary Rates. *American Journal of Science* 293A:453–478.

—— 2006. Environment and Evolution through the Paleocene-Eocene Thermal Maximum. *Trends in Ecology and Evolution* 21:246–253.

Giovas, Christina M. 2009. The Shell Game: Analytic Problems in Archaeological Mollusc Quantification. *Journal of Archaeological Science* 36:1557–1564.

Gleason, Henry A. 1926. The Individualistic Concept of the Plant Association. *American Midland Naturalist* 21:92–110.

Gobalet, Kenneth W. 2001. A Critique of Faunal Analysis: Inconsistency among Experts in Blind Tests. *Journal of Archaeological Science* 28:377–386.

Goillot, Cyrielle, Cécile Blondel, and Stéphane Peigné. 2009. Relationships between Dental Microwear and Diet in Carnivora (Mammalia): Implications for the Reconstruction of the Diet of Extinct Taxa. *Palaeogeography, Palaeoclimatology, Palaeoecology* 271: 13–23.

Goldblatt, P., and John C. Manning. 2002. Plant Diversity of the Cape Region of South Africa. *Annals of the Missouri Botanical Garden* 89:281–302.

Gómez Cano, Ana Rosa, Blanca A. García Yelo, and Manuel Hernández-Fernández. 2006. Cenogramas, Análisis Bioclimático y Muestreo en Faunas de Mamíferos: Implicaciones para la Aplicación de Métodos de Análisis Paleoecológico. *Estudios Geológicos* 62:135–144.

Gómez Cano, Ana Rosa, Juan L. Cantalapiedra, M. Àngels Álvarez-Sierra, and Manuel Hernández Fernández. 2014. A Macroecological Glance at the Structure of Late Miocene Rodent Assemblages from Southwest Europe. *Scientific Reports* 4:6557.

Good, Stephen P., and Kelly K. Caylor. 2011. Climatological Determinants of Woody Cover in Africa. *Proceedings of the National Academy of Sciences USA* 108:4902–4907.

Gordon, Elizabeth A. 1993. Screen Size and Differential Faunal Recovery: A Hawaiian Example. *Journal of Field Archaeology* 20:453–460.

Gordon, K. D. 1988. A Review of Methodology and Quantification in Dental Microwear Analysis. *Scanning Microscopy* 2:1139–1147.

Gotelli, Nicholas J., and Robert K. Colwell. 2001. Quantifying Biodiversity: Procedures and Pitfalls in the Measurements and Comparison of Species Richness. *Ecology Letters* 4:379–391.

Gotelli, Nicholas J., and Gary R. Graves. 1996. *Null Models in Ecology*. Smithsonian Institution Press, Washington, DC.

Gould, Stephen Jay. 1965. Is Uniformitarianism Necessary? *American Journal of Science* 263:223–228.

1966. Allometry and Size in Ontogeny and Phylogeny. *Biological Reviews* 41:587–640.

1970. Land Snail Communities and Pleistocene Climates in Bermuda: A Multivariate Analysis of Microgastropod Diversity. In *Proceedings of North American Paleontological Convention*, edited by Ellis L. Yochelson, pp. 486–521. Allen Press, Lawrence, KS.

1975. On the Scaling of Tooth Size in Mammals. *American Zoologist* 15:353–362.

1977. Eternal Metaphors in Palaeontology. In *Patterns of Evolution as Illustrated by the Fossil Record*, edited by Anthony Hallam, pp. 1–26. Elsevier, Amsterdam.

1987. *Time's Arrow, Time's Cycle: Myth and Metaphor in the Discovery of Geological Time*. Harvard University Press, Cambridge, MA.

1990. The Golden Rule: A Proper Scale for our Environmental Crisis. *Natural History* 90(9):24–30.

Gould, Stephen Jay, and Richard C. Lewontin. 1979. The Spandrels of San Marco and the Panglossian Paradigm: A Critique of the Adaptationist Programme. *Proceedings of the Royal Society B* 205:581–598.

Graham, Russell W. 1976. Late Wisconsin Mammalian Faunas and Environmental Gradients of the Eastern United States. *Paleobiology* 2:343–350.

1979. Paleoclimates and Late Pleistocene Faunal Provinces in North America. In *Pre-Llano Cultures of the Americas: Paradoxes and Possibilities*, edited by Robert L. Humphrey and Dennis Stanford, pp. 49–69. Anthropological Society of Washington, Washington, DC.

1981. Preliminary Report on Late Pleistocene Vertebrates from the Selby and Dutton Archeological/Paleontological Sites, Yuma County, Colorado. *University of Wyoming Contributions to Geology* 20:33–56.

1984. Paleoenvironmental Implications of the Quaternary Distribution of the Eastern Chipmunk (*Tamias striatus*) in Central Texas. *Quaternary Research* 21:111–114.

1985a. Diversity and Community Structure of the Late Pleistocene Mammal Fauna of North America. *Acta Zoologica Fennica* 170:181–192.

1985b. Response of Mammalian Communities to Environmental Changes during the Late Quaternary. In *Community Ecology*, edited by Jared Diamond and Ted J. Case, pp. 303–313. Harper and Row, New York.

1991. Interpreting Fossils (Review of Peter Andrews' *Owls, Caves and Fossils*). *Science* 253:213–214.

2001. Comment on "Skeleton of Extinct North American Sea Mink (*Mustela macrodon*)" by Mead et al. *Quaternary Research* 56:419–421.

2005. Quaternary Mammal Communities: Relevance of the Individualistic Response and Non-Analogue Faunas. In *Paleobiogeography: Generating New Insights into the Coevolution of the Earth and its Biota*, edited by Bruce S. Lieberman and Alycia L. Stigall, pp. 141–157. Paleontological Society Papers 11. Paleontological Society, Boulder, CO.

Graham, Russell W., and Ernest L. Lundelius, Jr. 1994. *FAUNMAP: A Database Documenting Late Quaternary Distributions of Mammal Species in the United States*, Illinois State Museum Scientific Papers 25. Illinois State Museum, Springfield.

Graham, Russell W., and Jim I. Mead. 1987. Environmental Fluctuations and Evolution of Mammalian Faunas during the Last Deglaciation in North America. In *North America and Adjacent Oceans during the Last Deglaciation*, edited by W. F. Ruddiman and H. E. Wright, Jr., pp. 371–402. Geology of North America K-3. Geological Society of America, Boulder, CO.

Graham, Russell W., and Holmes A. Semken, Jr. 1987. Philosophy and Procedures for Paleoenvironmental Studies of Quaternary Mammalian Faunas. In *Late Quaternary Mammalian Biogeography and Environments of the Great Plains and Prairies*, edited by Russell W. Graham, Holmes A. Semken, Jr., and Mary A. Graham, pp. 1–17. Illinois State Museum Scientific Papers 22. Illinois State Museum, Springfield.

Graham, Russell W., Holmes A. Semken, Jr., and Mary A. Graham (editors). 1987. *Late Quaternary Mammalian Biogeography and Environments of the Great Plains and Prairies*. Illinois State Museum Scientific Papers 22. Illinois State Museum, Springfield.

Grayson, Donald K. 1976. The Nightfire Island Avifauna and the Altithermal. In *Holocene Environmental Change in the Great Basin*, edited by Robert Elston, pp. 74–102. Research Paper 6. Nevada Archaeological Survey, Reno.

1977. A Review of the Evidence for Early Holocene Turkeys in the Northern Great Basin. *American Antiquity* 42:110–114.

1978. Reconstructing Mammalian Communities: A Discussion of Shotwell's Method of Paleoecological Analysis. *Paleobiology* 4:77–81.

1979a. Mount Mazama, Climatic Change, and Fort Rock Basin Archaeofaunas. In *Volcanic Activity and Human Ecology*, edited by Payson D. Sheets and Donald K. Grayson, pp. 427–457. Academic Press, New York.

1979b. On the Quantification of Vertebrate Archaeofaunas. In *Advances in Archaeological Method and Theory* vol. II, edited by Michael B. Schiffer, pp. 199–237. Academic Press, New York.

1980. Vicissitudes and Overkill: The Development of Explanations of Pleistocene Extinctions. In *Advances in Archaeological Method and Theory* vol. III, edited by Michael B. Schiffer, pp. 357–403. Academic Press, New York.

1981. A Critical View of the Use of Archaeological Vertebrates in Paleoenvironmental Reconstruction. *Journal of Ethnobiology* 1:28–38.

1983a. *The Establishment of Human Antiquity*. Academic Press, New York.

1983b. The Paleontology of Gatecliff Shelter. In *The Archaeology of Monitor Valley* ii: *Gatecliff Shelter*, by David Hurst Thomas, pp. 99–126. Anthropological Papers 59(1). American Museum of Natural History, New York.

1984a. Nineteenth-Century Explanations of Pleistocene Extinctions: A Review and Analysis. In *Quaternary Extinctions: A Prehistoric Revolution*, edited by Paul S. Martin and Richard G. Klein, pp. 5–39. University of Arizona Press, Tucson.

1984b. *Quantitative Zooarchaeology: Topics in the Analysis of Archaeological Faunas*. Academic Press, Orlando, FL.

1991a. Alpine Faunas from the White Mountains, California: Adaptive Change in the Prehistoric Great Basin? *Journal of Archaeological Science* 18:483–506.

1991b. The Small Mammals of Gatecliff Shelter: Did People Make a Difference? In *Beamers, Bobwhites, and Blue-Points: Tributes to the Career of Paul W. Parmalee*, edited by James R. Purdue, Walter E. Klippel, and Bonnie W. Styles, pp. 99–109. Illinois State Museum Scientific Papers 23. Illinois State Museum, Springfield.

1998. Moisture History and Small Mammal Community Richness during the Latest Pleistocene and Holocene, Northern Bonneville Basin, Utah. *Quaternary Research* 49:330–334.

2000a. The Homestead Cave Mammals. In *Late Quaternary Paleoecology in the Bonneville Basin*, edited by David B. Madsen, pp. 67–89. Bulletin 130. Utah Geological Survey, Salt Lake City.

2000b. Mammalian Responses to Middle Holocene Climatic Change in the Great Basin of the Western United States. *Journal of Biogeography* 27:181–192.

2006. The Late Quaternary Biogeographic Histories of Some Great Basin Mammals (Western USA). *Quaternary Science Reviews* 25:2964–2991.

2011. *The Great Basin: A Natural Prehistory*. University of California Press, Berkeley.

2016. *Giant Sloths and Sabertooth Cats: Extinct Mammals and the Archaeology of the Ice Age Great Basin*. University of Utah Press, Salt Lake City.

Grayson, Donald K, and Françoise Delpech. 1998. Changing Diet Breadth in the Early Upper Paleolithic of Southwestern France. *Journal of Archaeological Science* 25:1119–1129.

2003. Ungulates and the Middle-to-Upper Paleolithic Transition at Grotte XVI (Dordogne, France). *Journal of Archaeological Science* 30:1633–1648.

2005. Pleistocene Reindeer and Global Warming. *Conservation Biology* 19:557–562.

2006. Was There Increasing Dietary Specialization across the Middle-to-Upper Paleolithic Transition in France? In *When Neanderthals and Modern Humans Met*, edited by Nicholas J. Conard, pp. 377–417. Tübingen Publications in Prehistory. Kerns Verlag, Tübingen, Germany.

2008. The Large Mammals of Roc de Combe (Lot, France): The Châtelperronian and Aurignacian Assemblages. *Journal of Anthropological Archaeology* 27:338–362.

Grayson, Donald K, and David B. Madsen. 2000. Biogeographic Implications of Recent Low-Elevation Recolonization by *Neotoma cinerea* in the Great Basin. *Journal of Mammalogy* 81:1100–1105.

Grayson, Donald K., Stephanie D. Livingston, Eric Rickart, and Monson W. Shaver, III. 1996. Biogeographic Significance of Low-Elevation Records for *Neotoma cinerea* from the Northern Bonneville Basin, Utah. *Great Basin Naturalist* 56:191–196.

Grayson, Donald K, Françoise Delpech, Jean-Philippe Rigaud, and Jan F. Simek. 2001. Explaining the Development of Dietary Dominance by a Single Ungulate Taxon at Grotte XVI, Dordogne, France. *Journal of Archaeological Science* 28:115–125.

Green, Jeremy L. 2009. Dental Microwear in the Orthodentine of the Xenarthra (Mammalia) and its Use in Reconstructing the Palaeodiet of Extinct Taxa: The Case Study of *Nothrotheriops shastensis* (Xenarthra, Tardigrada, Nothrotheriidae). *Zoological Journal of the Linnean Society* 156:201–222.

Green, Jeremy L., and Daniela C. Kalthoff. 2015. Xenarthran Dental Microstructure and Dental Microwear Analyses, with New Data for *Megatherium americanum* (Megatheriidae). *Journal of Mammalogy* 96:645–657.

Green, Jeremy L., and Nicholas A. Resar. 2012. The Link between Dental Microwear and Feeding Ecology in Tree Sloths and Armadillos (Mammalia: Xenarthra). *Biological Journal of the Linnean Society* 107:277–294.

Greenacre, M. J., and E. S. Vrba. 1984. Graphical Display and Interpretation of Antelope Census Data in African Wildlife Areas, Using Correspondence Analysis. *Ecology* 65:984–997.

Grigson, Caroline. 1969. The Uses and Limitations of Differences in Absolute Size in the Distinction between the Bones of Aurochs (*Bos primigenius*) and Domestic Cattle (*Bos taurus*). In *The Domestication and Exploitation of Plants and Animals*, edited by Peter J. Ucko and G. W. Dimbleby, pp. 277–294. Aldine Atherton, Chicago.

————. 1989. Size and Sex: Evidence for the Domestication of Cattle in the Near East. In *The Beginnings of Agriculture*, edited by Annie Milles, Diane Williams, and Neville Gardner, pp. 77–109. British Archaeological Reports, International Series 496. BAR, Oxford.

Grimm, Guido W., and Alastair J. Potts. 2016. Fallacies and Fantasies: The Theoretical Underpinnings of the Coexistence Approach for Palaeoclimate Reconstruction. *Climate of the Past* 12:611–622.

Grine, Frederick E. 1986. Dental Evidence for Dietary Differences in *Australopithecus* and *Paranthropus*: A Quantitative Analysis of Permanent Molar Microwear. *Journal of Human Evolution* 15:783–822.

Grine, Frederick E., and Richard F. Kay. 1988. Early Hominid Diets from Quantitative Image Analysis of Dental Microwear. *Nature* 333:765–768.

Grine, Frederick E., Peter S. Ungar, and M. F. Teaford. 2002. Error Rates in Dental Microwear Quantification Using Scanning Electron Microscopy. *Scanning* 24:144–153.

Grinnell, Joseph. 1914. Barriers to Distribution as Regards Birds and Mammals. *American Naturalist* 48:248–254.

————. 1917a. Field Tests of Theories Concerning Distributional Control. *American Naturalist* 51:115–128.

————. 1917b. The Niche-Relationships of the California Thrasher. *Auk* 34:427–433.

————. 1924. Geography and Evolution. *Ecology* 5:225–229.

————. 1928. Presence and Absence of Animals. *University of California Chronicle* 30:429–450.

Groffman, Peter M., Jill S. Baron, Tamara Blett et al. 2006. Ecological Thresholds: The Key to Successful Environmental Management or an Important Concept with No Practical Application? *Ecosystems* 9:1–13.

Guilday, John E. 1962. The Pleistocene Local Fauna of the Natural Chimneys, Augusta County, Virginia. *Carnegie Museum of Natural History Annals* 36:87–122.

————. 1967. Differential Extinction during Late-Pleistocene and Recent Times. In *Pleistocene Extinctions: The Search for a Cause*, edited by P[aul] S. Martin and H. E. Wright, Jr., pp. 121–140. Yale University Press, New Haven, CT.

————. 1969. Small Mammal Remains from the Wasden Site (Owl Cave), Bonneville County, Idaho. *Tebiwa, Journal of the Idaho State University Museum* 12(1):47–57.

————. 1971. The Pleistocene History of the Appalachian Mammal Fauna. In *The Distributional History of the Biota of the Southern Appalachians*, Part III: *Vertebrates*, edited by Perry C. Holt, pp. 233–262. Research Division Monograph 4. Virginia Polytechnic Institute and State University, Blacksburg.

Guilday, John E., and Eleanor K. Adam. 1967. Small Mammal Remains from Jaguar Cave, Lemhi County, Idaho. *Tebiwa, Journal of the Idaho State University Museum* 10(1):26–37.

Guilday, John E., and Paul W. Parmalee. 1972. Quaternary Periglacial Records of Voles of the Genus *Phenacomys* Merriam (Cricetidae: Rodentia). *Quaternary Research* 2:170–175.

Guilday, John E., Paul S. Martin, and A. D. McCrady. 1964. New Paris No. 4: A Pleistocene Cave Deposit in Bedford County, Pennsylvania. *Bulletin of the National Speleological Society* 26:121–194.

Guilday, John E., Harold W. Hamilton, Elaine Anderson, and Paul W. Parmalee. 1978. *The Baker Bluff Cave Deposit, Tennessee, and the Late Pleistocene Faunal Gradient*. Bulletin of the Carnegie Museum of Natural History 11. Carnegie Museum of Natural History, Pittsburgh.

Guiot, J., Jean-Luc De Beaulieu, A. Pons, and M. Reille. 1989. A 140,000-Year Climatic Reconstruction from Two European Pollen Cores. *Nature* 338:309–313.

Gunnell, Gregg F. 1994. Paleocene Mammals and Faunal Analysis of the Chappo Type Locality (Tiffanian), Green River Basin, Wyoming. *Journal of Vertebrate Paleontology* 14:81–104.

——— 1997. Wasatchian–Bridgerian (Eocene) Paleoecology of the Western Interior of North America: Changing Paleoenvironments and Taxonomic Composition of Omomyid (Tarsiiformes) Primates. *Journal of Human Evolution* 32:105–132.

Guo, Qinfeng. 2014. Central-Marginal Population Dynamics in Species Invasions. *Frontiers in Ecology and Evolution* 2: article 23.

Guo, Qinfeng, Mark Taper, M. Schoenberger, and J. Brandle. 2005. Spatial-Temporal Population Dynamics across Species Range: From Centre to Margin. *Oikos* 108:47–57.

Gustafson, Carl E. 1972. Faunal Remains from the Marmes Rockshelter and Related Archaeological Sites in the Columbia Basin. Unpublished Ph.D. dissertation, Department of Zoology, Washington State University, Pullman.

Guthrie, R. Dale. 1968a. Paleoecology of the Large-Mammal Community in Interior Alaska during the Late Pleistocene. *American Midland Naturalist* 79:346–363.

——— 1968b. Paleoecology of a Late Pleistocene Small Mammal Community from Interior Alaska. *Arctic* 21:223–244.

——— 1970. Bison Evolution and Zoogeography in North America during the Pleistocene. *Quarterly Review of Biology* 45:1–15.

——— 1982. Mammals of the Mammoth Steppe as Paleoenvironmental Indicators. In *Paleoecology of Beringia*, edited by David M. Hopkins, John V. Matthews, Jr., Charles E. Schweger, and Steven B. Young, pp. 307–326. Academic Press, New York.

——— 1984a. Alaskan Megabucks, Megabulls, and Megarams: The Issue of Pleistocene Gigantism. In *Contributions in Quaternary Vertebrate Paleontology: A Volume in Honor of John E. Guilday*, edited by Hugh H. Genoways and Mary R. Dawson, pp. 482–510. Special Publication 8. Carnegie Museum of Natural History, Pittsburgh.

——— 1984b. Mosaics, Allochemics, and Nutrients: An Ecological Theory of Late Pleistocene Extinctions. In *Quaternary Extinctions: A Prehistoric Revolution*, edited by Paul S. Martin and Richard G. Klein, pp. 259–298. University of Arizona Press, Tucson.

——— 2003. Rapid Body Size Decline in Alaskan Pleistocene Horses before Extinction. *Nature* 426:169–171.

Haber, F. 1959. Fossils and the Idea of a Process of Time in Natural History. In *Forerunners of Darwin: 1745–1859*, edited by B. Glass, O. Temkin, and W. L. Strauss, pp. 222–261. Johns Hopkins University Press, Baltimore.

Hadly, Elisabeth A. 1997. Evolutionary and Ecological Response of Pocket Gophers (*Thomomys talpoides*) to Late-Holocene Climatic Change. *Biological Journal of the Linnean Society* 60:277–296.

——— 1999. Fidelity of Terrestrial Vertebrate Fossils to a Modern Ecosystem. *Palaeogeography, Palaeoclimatology, Palaeoecology* 149:389–409.

Hadly, Elizabeth A., Michael H. Kohn, Jennifer A. Leonard, and Robert K. Wayne. 1998. A Genetic Record of Population Isolation in Pocket Gophers during Holocene Climatic Change. *Proceedings of the National Academy of Sciences USA* 95:6893–6896.

Hadly, Elizabeth A., Uma Ramakrishnan, Yvonne L. Chan et al. 2004. Genetic Response to Climatic Change: Insights from Ancient DNA and Phylochronology. *PLoS Biology* 2(10): e290.

Hafner, David J. 1993. North American Pika (*Ochotona princeps*) as a Late Quaternary Biogeographic Indicator Species. *Quaternary Research* 39:373–380.

Hairston, Nelson G., Frederick E. Smith, and Lawrence B. Slobodkin. 1960. Community Structure, Population Control, and Competition. *American Naturalist* 44:421–425.

Hall, Brian K. 2002. Palaeontology and Evolutionary Developmental Biology: A Science of the Nineteenth and Twenty-First Centuries. *Palaeontology* 45:647–669.

Hall, E. Raymond. 1981. *The Mammals of North America*, second edition. John Wiley and Sons, New York.

Hall, E. Raymond, and Keith R. Kelson. 1959. *The Mammals of North America*. Ronald Press, New York.

Hammer, Øyvind, and David A. T. Harper. 2006. *Paleontological Data Analysis*. Blackwell, Malden, MA.

Hammer, Øyvind, David A. T. Harper, and Paul D. Ryan. 2001. Paleontological Statistics Software Package for Education and Data Analysis. *Palaeontologia Electronica* 4(1):9 pp.

Hampe, Arndt. 2004. Bioclimatic Envelope Models: What They Detect and What They Hide. *Global Ecology and Biogeography* 13:469–471.

Hanley, Torrance C., and Kimberly J. La Pierre (editors). 2015. *Trophic Ecology: Bottom-Up or Top-Down Interactions across Aquatic and Terrestrial Systems*. Cambridge University Press.

Hargrave, Lyndon L., and Steven D. Emslie. 1979. Osteological Identification of Sandhill Crane versus Turkey. *American Antiquity* 44:295–299.

Harris, Arthur H. 1963. *Vertebrate Remains and Past Environmental Reconstruction in the Navajo Reservoir District*. Papers in Anthropology 11. Museum of New Mexico, Santa Fe.

    1984. *Neotoma* in the Late Pleistocene of New Mexico and Chihuahua. In *Contributions in Quaternary Vertebrate Paleontology: A Volume in Memorial to John E. Guilday*, edited by Hugh H. Genoways and Mary R. Dawson, pp. 164–178. Special Publication 8. Carnegie Museum of Natural History, Pittsburgh.

    1985. *Late Pleistocene Vertebrate Paleoecology of the West*. University of Texas Press, Austin.

Harris, John M., and Thure E. Cerling. 2002. Dietary Adaptations of Extant and Neogene African Suids. *Journal of Zoology* (London) 256:45–54.

Harrison, J. L. 1962. The Distribution of Feeding Habits among Animals in a Tropical Rain Forest. *Journal of Animal Ecology* 31:53–63.

Harrison, Sandy P., and Colin I. Prentice. 2003. Climate and CO_2 Controls on Global Vegetation Distribution at the Last Glacial Maximum: Analysis Based on Palaeovegetation Data, Biome Modelling and Palaeoclimate Simulations. *Global Change Biology* 9:983–1004.

Haupt, Ryan J., Larisa R. G. DeSantis, Jeremy L. Green, and Peter S. Ungar. 2013. Dental Microwear Texture as a Proxy for Diet in Xenarthrans. *Journal of Mammalogy* 94:856–866.

Hawkins, Bradford A. 2001. Ecology's Oldest Pattern? *Trends in Ecology and Evolution* 16:470.

Hawkins, Bradford A., and Eric E. Porter. 2003. Relative Influence of Current and Historical Factors on Mammal and Bird Diversity Patterns in Deglaciated North America. *Global Ecology and Biogeography* 12:475–481.

Hayek, Lee-Ann C., Raymond L Bernor, Nikos Solounias, and Patricia Steigerwald. 1991. Preliminary Studies of Hipparionine Horse Diet as Measured by Tooth Microwear. *Annales Zoologici Fennici* 28:187–200.

Hayward, M. W., P. Henschel, J. O'Brien et al. 2006. Prey Preferences of the Leopard (*Panthera pardus*). *Journal of Zoology* (London) 270:298–313.

Heaney, Lawrence R. 1978. Island Area and Body Size of Insular Mammals: Evidence from Tri-Colored Squirrel (*Callosciurus prevosti*) of Southeast Asia. *Evolution* 32:29–44.

Hedges, R. E. M. 2002. Bone Diagenesis: An Overview of Processes. *Archaeometry* 44:319–328.

Helgen, Kristofer M., Rod T. Wells, Benjamin P. Kear, Wayne R. Gerdtz, and Timothy F. Flannery. 2006. Ecological and Evolutionary Significance of Sizes of Giant Extinct Kangaroos. *Australian Journal of Zoology* 54:293–303.

Hedgpeth, Joel W., and Harry S. Ladd (editors). 1957. *Treatise on Marine Ecology and Paleoecology*. Memoir 67 (vol. I: *Ecology*; vol. II: *Paleoecology*). Geological Society of America, New York.

Heisler, Leanne M., Christopher M. Somers, and Ray G. Poulin. 2014. Rodent Populations on the Northern Great Plains Respond to Weather Variation at a Landscape Level. *Journal of Mammalogy* 95:82–90.

Hendey, Q. B. 1974. The Late Cenozoic Carnivora of the South-Western Cape Province. *Annals of the South African Museum* 63:1–369.

Hengl, Tomslav, Jorge Mendes de Jesus, Robert A. MacMillan et al. 2014. Soilgrids1km – Global Soil Information Based on Automated Mapping. *PLoS ONE* 9: e105992.

Herm, D. 1972. Pitfalls in Paleoecologic Interpretation: An Integrated Approach to Avoid the Major Pits. In *International Geographic Congress, 24th Session, Section 7: Paleontology*, edited by B. E. Mamet and G. E. G. Westermann, pp. 82–88. International Geographic Union, Montreal.

Hernández-Fernández, Manuel. 2001. Bioclimatic Discriminant Capacity of Terrestrial Mammal Faunas. *Global Ecology and Biogeography* 10:189–204.

Hernández-Fernández, Manuel, and Pablo Peláez-Campomanes. 2003. The Bioclimatic Model: A Method of Palaeoclimatic Qualitative Inference Based on Mammal Associations. *Global Ecology and Biogeography* 12:507–517.

2005. Quantitative Palaeoclimatic Inference Based on Terrestrial Mammal Faunas. *Global Ecology and Biogeography* 14:39–56.

Hernández-Fernández, Manuel, M. T. Alberdi, B. Azanza et al. 2006. Identification Problems of Arid Environments in the Neogene–Quaternary Mammal Record of Spain. *Journal of Arid Environments* 66:585–608.

Hernández-Fernández, Manuel, M. A. Álvarez-Sierra, and Pablo Peláez-Campomanes. 2007. Bioclimatic Analysis of Rodent Palaeofaunas Reveals Severe Climatic Changes in Southwestern Europe during the Plio-Pleistocene. *Palaeogeography, Palaeoclimatology, Palaeoecology* 251:500–526.

Hester, James J. 1964. The Possibilities for Paleoecological Reconstruction – Archaeology. In *The Reconstruction of Past Environments: Proceedings*, assembled by James J. Hester and James Schoenwetter, pp. 19–23. Publication 3. Fort Burgwin Research Center, Taos, NM.

Hester, James J., and James Schoenwetter (assemblers). 1964. *The Reconstruction of Past Environments*. Publication 3. Fort Burgwin Research Center, Taos, NM.

Hibbard, Claude W. 1944. Stratigraphy and Vertebrate Paleontology of Pleistocene Deposits of Southwestern Kansas. *Geological Society of America Bulletin* 55:707–754.

1949. Techniques of Collecting Microvertebrate Fossils. *Contributions from the Museum of Paleontology* 8:7–19. University of Michigan, Ann Arbor.

1955. The Jinglebob Interglacial (Sangamon?) Fauna from Kansas and its Climatic Significance. *Contributions from the Museum of Paleontology* 12:179–228. University of Michigan, Ann Arbor.

1958. Summary of North American Pleistocene Mammalian Local Faunas. *Papers of the Michigan Academy of Science, Arts and Letters* 43:3–32.

1960. An Interpretation of Pliocene and Pleistocene Climates in North America. *Annual Report of the Michigan Academy of Science, Arts and Letters* 62:5–30.

1963. A Late Illinoian Fauna from Kansas and its Climatic Significance. *Papers of the Michigan Academy of Science, Arts and Letters* 48:187–221.

Hibbard, C[laude] W., C. E. Ray, D. E. Savage, D. W. Taylor, and J. E. Guilday. 1965. Quaternary Mammals of North America. In *The Quaternary of the United States*, edited by H. E. Wright, Jr., and David G. Frey, pp. 509–525. Princeton University Press.

Higgs, E. S. 1967. Faunal Fluctuations and Climate in Libya. In *Background to Evolution in Africa*, edited by Walter W. Bishop and J. Desmond Clark, pp. 149–163. University of Chicago Press.

Higham, C. F. 1969. The Metrical Attributes of Two Samples of Bovine Limb Bones. *Journal of Zoology* (London) 157:63–74.

Hijmans, Robert J., and Catherine H. Graham. 2006. The Ability of Climate Envelope Models to Predict the Effect of Climate Change on Species Distributions. *Global Change Biology* 12:2272–2281.

Hijmans, Robert J., Susan E. Cameron, Juan L. Parra, Peter G. Jones, and Andy Jarvis. 2005. Very High Resolution Interpolated Climate Surfaces for Global Land Areas. *International Journal of Climatology* 25:1965–1978.

Hill, M. O. 1973. Diversity and Evenness: A Unifying Notation and its Consequences. *Ecology* 54:427–432.

Hill, M. O., and H. G. Gauch, Jr. 1980. Detrended Correspondence Analysis: An Improved Ordination Technique. *Vegetatio* 42:47–58.

Hill, Matthew E., Jr., Matthew G. Hill, and Christopher C. Widga. 2008. Late Quaternary *Bison* Diminution on the Great Plains of North America: Evaluating the Role of Human Hunting Versus Climate Change. *Quaternary Science Reviews* 27:1752–1771.

Hobbs, Richard J., Eric Higgs, and James A. Harris. 2009. Novel Ecosystems: Implications for Conservation and Restoration. *Trends in Ecology and Evolution* 24:599–605.

Hobbs, Richard J., Eric Higgs, and Carol M. Hall (editors). 2013. *Novel Ecosystems: Intervening in the New Ecological World Order*. Wiley-Blackwell, Oxford.

Hobbs, Richard J., Eric Higgs, Carol M. Hall et al. 2014. Managing the Whole Landscape: Historical, Hybrid, and Novel Ecosystems. *Frontiers in Ecology and Environment* 12:557–564.

Hodgkins, Jamie, Curtis W. Marean, Alain Turq et al. 2016. Climate-Mediated Shifts in Neandertal Subsistence Behaviors at Pech de l'Azé and Roc de Marsal (Dordogne Valley, France). *Journal of Human Evolution* 96:1–18.

Hoefs, Jochen. 2015. *Stable Isotope Geochemistry*, seventh edition. Springer, New York.

Hoffman, Antoni. 1979. Community Paleoecology as an Epiphenomenal Science. *Paleobiology* 5:357–379.

Hoffmann, Robert S., and J. Knox Jones, Jr. 1970. Influence of Late-Glacial and Post-Glacial Events on the Distribution of Recent Mammals on the Northern Great Plains. In *Pleistocene and Recent Environments of the Central Great Plains*, edited by Wakefield Dort, Jr., and J. Knox Jones, pp. 355–394. Department of Geology, Special Publication 3. University of Kansas Press, Lawrence.

Hoffmeister, Donald F. 1964. Mammals and Life Zones: Time for Re-evaluation. *Plateau* 37:46–55.

Hofmann, R. R. 1989. Evolutionary Steps of Ecophysiological Adaptation and Diversification of Ruminants: A Comparative View of their Digestive System. *Oecologia* 78:443–457.

Hokr, Zdenek. 1951. A Method of the Quantitative Determination of the Climate in the Quaternary Period by Means of Mammal Associations. *Sborník of the Geological Survey of Czechoslovakia* 18:209–219.

Holbrook, Sally J. 1975. Prehistoric Paleoecology of Northwestern New Mexico. Unpublished Ph.D. dissertation, University of California, Berkeley.

1977. Rodent Faunal Turnover and Prehistoric Community Stability in Northwestern New Mexico. *American Naturalist* 111: 1195–1208.

1980. Species Diversity Patterns in Some Present and Prehistoric Rodent Communities. *Oecologia* 44:355–367.

1982a. Prehistoric Environmental Reconstruction by Mammalian Microfaunal Analysis, Grasshopper Pueblo. In *Multidisciplinary Research at Grasshopper Pueblo, Arizona*, edited by William A. Longacre, Sally J. Holbrook, and Michael W. Graves, pp. 73–86. Anthropological Papers of the University of Arizona 40. University of Arizona, Tucson.

1982b. The Prehistoric Local Environment of Grasshopper Pueblo, Arizona. *Journal of Field Archaeology* 9:207–215.

Holbrook Sally J., and James C. Mackey. 1976. Prehistoric Environmental Change in Northern New Mexico: Evidence from a Gallina Phase Archaeological Site. *Kiva* 41:309–317.

Holdaway, Simon, and LuAnn Wandsnider (editors). 2008. *Time in Archaeology: Time Perspectivism Revisited*. University of Utah Press, Salt Lake City.

Holdridge, L. R. 1947. Determination of World Plant Formations from Simple Climatic Data. *Science* 105:367–368.

Holling, C. S. 1992. Cross-Scale Morphology, Geometry, and Dynamics of Ecosytems. *Ecological Monographs* 62:447–502.

Holt, Ben G., Jean-Philippe Lessard, Michael K. Borregaard et al. 2013. An Update of Wallace's Zoogeographic Regions of the World. *Science* 339:74–78.

Holt, Robert D. 2003. On the Evolutionary Ecology of Species' Ranges. *Evolutionary Ecology Research* 5:159–178.

2009. Bringing the Hutchinsonian Niche into the 21st Century: Ecological and Evolutionary Perspectives. *Proceedings of the National Academy of Sciences USA* 106 (Supplement 2):19659–19665.

Holt, Robert D., and Timothy H. Keitt. 2005. Species' Borders: A Unifying Theme in Ecology. *Oikos* 108:3–6.

Holtzman, Richard C. 1979. Maximum Likelihood Estimation of Fossil Assemblage Composition. *Paleobiology* 5:77–89.

Hooijer, D. A. 1947. Pleistocene Remains of *Panthera tigris* (Linnaeus) Subspecies from Wansien, Szechwan, China, Compared with Fossil and Recent Tigers from Other Localities. *American Museum Novitates* 1346:1–17.

Hopkins, Samantha S. B. 2008. Reassessing the Mass of Exceptionally Large Rodents Using Toothrow Length and Area as Proxies for Body Mass. *Journal of Mammalogy* 89:232–243.

Hopley, Philip J., and Mark A. Maslin. 2010. Climate-Averaging of Terrestrial Faunas: An Example from the Plio-Pleistocene of South Africa. *Paleobiology* 36:32–50.

Hopley, Philip J., Alf G. Latham, and Jim D. Marshall. 2006. Palaeoenvironments and Palaeodiets of Mid-Pliocene Micromammals from Makapansgat Limeworks, South Africa: A Stable Isotope and Dental Microwear Approach. *Palaeogeography, Palaeoclimatology, Palaeoecology* 233:235–251.

Hostetler, Mark E. 1997. Avian Body-Size Clumps and the Response of Birds to Scale-Dependent Landscape Structure in Suburban Habitats. Unpublished Ph.D. dissertation, University of Florida, Gainesville.

Howard, Hildegarde. 1930. A Census of the Pleistocene Birds of Rancho La Brea from the Collections of the Los Angeles Museum. *Condor* 32:81–88.

Howell, F. Clark and François Bourlière (editors). 1963. *African Ecology and Human Evolution*. Aldine Transaction, New Brunswick, NJ.

Hubbell, Stephen P. 2001. *The Unified Neutral Theory of Biodiversity and Biogeography*. Princeton University Press.

Hughes, Susan S. 2009. Noble Marten (*Martes americana nobilis*) Revisited: Its Adaptation and Extinction. *Journal of Mammalogy* 90:74–92.

Hughes, Terry P., Stephen Carpenter, Johan Rockström, Marten Scheffer, and Brian Walker. 2013. Multiscale Regime Shifts and Planetary Boundaries. *Trends in Ecology and Evolution* 28:389–395.

Hunt, Gene. 2006. Fitting and Comparing Models of Phyletic Evolution: Random Walks and Beyond. *Paleobiology* 32:578–601.

2007. The Relative Importance of Directional Change, Random Walks, and Stasis in the Evolution of Fossil Lineages. *Proceedings of the National Academy of Sciences USA* 104:18404–18408.

Hunt, Gene, Michael A. Bell, and Matthew P. Travis. 2008. Evolution toward a New Adaptive Optimum: Phenotypic Evolution in a Fossil Stickleback Lineage. *Evolution* 62–63:700–710.

Hunter, Mark D., and Peter W. Price. 1992. Playing Chutes and Ladders: Heterogeneity and the Relative Roles of Bottom-Up and Top-Down Forces in Natural Communities. *Ecology* 73:724–732.

Huntley, Brian. 2012. Reconstructing Palaeoclimates from Biological Proxies: Some often Overlooked Sources of Uncertainty. *Quaternary Science Reviews* 31:1–16.

Huppert, Amit, and Andrew R. Solow. 2004. A Method for Reconstructing Climate from Fossil Beetle Assemblages. *Proceedings of the Royal Society B* 271:1125–1128.

Hurlbert, Stuart H. 1971. The Nonconcept of Species Diversity: A Critique and Alternative Parameters. *Ecology* 52:577–586.

Huston, Michael A., and Steve Wolverton. 2009. The Global Distribution of Net Primary Productivity: Resolving the Paradox. *Ecological Monographs* 79:343–377.

2011. Regulation of Animal Size by eNPP, Bergmann's Rule, and Related Phenomena. *Ecological Monographs* 81:349–405.

Hutchinson, G. Evelyn. 1957. Concluding Remarks. *Cold Spring Harbor Symposium in Quantitative Biology* 22:415–427.

1978. *An Introduction to Population Ecology*. Yale University Press, New Haven, CT.

Hutson, William H. 1977. Transfer Functions under No-Analog Conditions: Experiments with Indian Ocean Planktonic Foraminifera. *Quaternary Research* 8:355–367.

Imbrie, J., and Nilva G. Kipp. 1971. A New Micropaleontological Method for Quantitative Paleoclimatology: Application to a Late Pleistocene Carribean Core. In *The Late Cenozoic Glacial Ages*, edited by Karl K. Turekian, pp. 71–181. Yale University Press, New Haven, CT.

Imbrie, John, and Norman D. Newell (editors). 1964. *Approaches to Paleoecology*. John Wiley and Sons, New York.

Inkpen, Robert John. 2008. Explaining the Past in the Geosciences. *Philosophia* 36:495–507.

Inskeep, R. R. 1987. *Nelson Bay Cave, Cape Province, South Africa: The Holocene Levels*. British Archaeological Reports, International Series 357. BAR, Oxford.

Jablonski, David, and J. John Sepkoski, Jr. 1996. Paleobiology, Community Ecology, and Scales of Ecological Pattern. *Ecology* 77:1367–1378.

Jackson, Donald A. 1997. Compositional Data in Community Ecology: The Paradigm or Peril of Proportions? *Ecology* 78:929–940.

Jackson, Stephen T. 2009a. Alexander von Humboldt and the General Physics of the Earth. *Science* 324:596–597.

2009b. Introduction: Humboldt, Ecology, and the Cosmos. In *Essay on the Geography of Plants*, by Alexander von Humboldt, pp. 1–46. University of Chicago Press.

Jackson, Stephen T., and Jonathan T. Overpeck. 2000. Responses of Plant Populations and Communities to Environmental Changes of the Late Quaternary. *Paleobiology* 26 (Supplement):194–220.

Jackson, Stephen T, and John W. Williams. 2004. Modern Analogs in Quaternary Paleoecology: Here Today, Gone Yesterday, Gone Tomorrow? *Annual Review of Earth and Planetary Sciences* 32:495–537.

Jacobs, Zenobia, and Richard. G. Roberts. 2017. Single-Grain OSL Chronologies for the Still Bay and Howieson's Poort Industries and the Transition between Them. *Journal of Human Evolution* 107:1–13.

Jacobson, Jodi A. 2003. Identification of Mule Deer (*Odocoileus hemionus*) and White-Tailed Deer (*Odocoileus virginianus*) Postcranial Remains as a Means of Determining Human Subsistence Strategies. *Plains Anthropologist* 48:287–297.

2004. Determining Human Ecology on the Plains through the Identification of Mule Deer (*Odocoileus hemionus*) and White-Tailed Deer (*Odocoileus virginianus*) Postcranial Remains. Unpublished Ph.D. dissertation, Department of Anthropology, University of Tennessee, Knoxville.

Jaeger, J.-J., and H. B. Wesselman. 1976. Fossil Remains of Micromammals from the Omo Group Deposits. In *Earliest Man and Environments in the Lake Rudolf Basin: Stratigraphy, Paleoecology, and Evolution*, edited by Yves Coppen, F. Clark Howell, Glynn Ll. Isaac, and Richard E. F. Leakey, pp. 351–360. University of Chicago Press.

James, Frances C. 1970. Geographic Size Variation in Birds and its Relationship to Climate. *Ecology* 51:365–390.

Jamniczky, Heather A., Donald B. Brinkman, and Anthony P. Russell. 2003. Vertebrate Microsite Sampling: How Much is Enough? *Journal of Vertebrate Paleontology* 23:725–734.

2008. How Much is Enough? A Repeatable, Efficient, and Controlled Sampling Protocol for Assessing Taxonomic Diversity and Abundance in Vertebrate Microfossil Assemblages. In *Vertebrate Microfossil Assemblages: Their Role in Paleoecology and Paleobiogeography*, edited by Julia T. Sankey and Sven Baszio, pp. 9–16. Indiana University Press, Bloomington.

Janis, Christine M. 1988. An Estimation of Tooth Volume and Hypsodonty Indices in Ungulate Mammals, and the Correlation of These Factors with Dietary Preference. In *Teeth Revisited: Proceedings of the VIIth International Symposium on Dental Morphology*, edited by D. E. Russell, J. P. Santoro, and D. Sigogneau-Russell, pp. 367–387. Mémoires du Muséum National d'Historie Naturelle, Series C. Muséum National d'Historie Naturelle, Paris.

1995. Correlations between Craniodental Morphology and Feeding Behavior in Ungulates: Reciprocal Illuminations between Living and Fossil Taxa. In *Functional Morphology in Vertebrate Paleontology*, edited by J. J. Thomason, pp. 76–98. Cambridge University Press.

Janis, Christine M., and D. Ehrhardt. 1988. Correlation of Relative Muzzle Width and Relative Incisor Width with Dietary Preference in Ungulates. *Zoological Journal of the Linnean Society* 92:267–284.

Janis, Christine M., and Mikael Fortelius. 1988. On the Means Whereby Mammals Achieve Increased Functional Durability of their Dentitions, with Special Reference to Limiting Factors. *Biological Reviews* 63:197–230.

Janis, Christine M., John Damuth, and Jessica M. Theodor. 2000. Miocene Ungulates and Terrestrial Primary Productivity: Where Have All the Browsers Gone? *Proceedings of the National Academy of Sciences USA* 97:7899–7904.

———. 2004. The Species Richness of Miocene Browsers, and Implications for Habitat Type and Primary Productivity in the North American Grassland Biome. *Palaeogeography, Palaeoclimatology, Palaeoecology* 207:371–398.

Jardine, Phillip E., Christine M. Janis, Sarda Sahney, and Michael J. Benton. 2012. Grit Not Grass: Concordant Patterns of Early Origin of Hypsodonty in Great Plains Ungulates and Glires. *Palaeogeography, Palaeoclimatology, Palaeoecology* 365–366:1–10.

Jass, Christopher N., M. F. Poteet, and C. J. Bell. 2015. Response of Pocket Gophers (*Geomys*) to Late Quaternary Environmental Change on the Edwards Plateau of Central Texas. *Historical Biology* 27:192–213.

Jehl, Joseph R., Jr. 1966. Subspecies of Recent and Fossil Birds. *Auk* 83:306–307.

Jernvall, Jukka, John P. Hunter, and Mikael Fortelius. 1996. Molar Tooth Diversity, Disparity, and Ecology in Cenozoic Ungulate Radiations. *Science* 274:1489–1492.

Johnson, Ralph Gordon. 1960. Environmental Interpretation of Pleistocene Marine Species. *Journal of Geology* 68:575–576.

Jones, Emily Lena. 2015. Archaeofaunal Evidence of Human Adaptation to Climate Change in Upper Paleolithic Iberia. *Journal of Archaeological Science: Reports* 2:257–263.

Jones, J. Knox, Jr., David M. Armstrong, and Jerry R. Choate. 1985. *Guide to Mammals of the Plains States*. University of Nebraska Press, Lincoln.

Jones, Kate E., Jon Biebly, Marcel Cardillo et al. 2009. PanTHERIA: A Species-Level Database of Life History, Ecology, and Geography of Extant and Recently Extinct Mammals. *Ecology* 90:2648.

Jones, Terry L., Kenneth W. Gobalet, and Brian F. Codding. 2016. The Archaeology of Fish and Fishing on the Central Coast of California: The Case for an Under-Exploited Resource. *Journal of Anthropological Archaeology* 41:88–108.

Jouzel, J., V. Masson-Delmotte, O. Cattani et al. 2007. Orbital and Millennial Antarctic Climate Variability over the Past 800,000 Years. *Science* 317:793–796.

Jungers, William L., and Rebecca Z. German. 1981. Ontogenetic and Interspecific Skeletal Allometry in Nonhuman Primates: Bivariate Versus Multivariate Analysis. *American Journal of Physical Anthropology* 55:195–202.

Jungers, William L., Anthony B. Falsetti, and Christine E. Wall. 1995. Shape, Relative Size, and Size-Adjustments in Morphometrics. *American Journal of Physical Anthropology* 38:137–161.

Kaiser, Thomas M., and Mikael Fortelius. 2003. Differential Mesowear in Occluding Upper and Lower Molars: Opening Mesowear Analysis for Lower Molars and Premolars in Hyposodont Horses. *Journal of Morphology* 258:63–83.

Kaiser, Thomas M., and Nikos Solounias. 2003. Extending the Tooth Mesowear Method to Extinct and Extant Equids. *Geodiversitas* 25:321–345.

Kappelman, John. 1984. Plio-Pleistocene Environments of Bed I and Lower Bed II, Olduvai Gorge, Tanzania. *Palaeogeography, Palaeoclimatology, Palaeoecology* 48:171–196.

———. 1988. Morphology and Locomotor Adaptations of the Bovid Femur in Relation to Habitat. *Journal of Morphology* 198:119–130.

———. 1991. The Paleoenvironments of *Kenyapithecus* at Fort Ternan. *Journal of Human Evolution* 20:95–129.

Kappelman, John, Tom Plummer, Laura Bishop, Alex Duncan, and Scott Appleton. 1997. Bovids as Indicators of Plio-Pleistocene Paleoenvironments in East Africa. *Journal of Human Evolution* 32:229–256.

Karr, James R. 1971. Structure of Avian Communities in Selected Panama and Illinois Habitats. *Ecological Monographs* 41:207–233.

Kaufman, Daniel. 1998. Measuring Archaeological Diversity: An Application of the Jackknife Technique. *American Antiquity* 63:73–85.

Kay, Richard F. 1975. The Functional Adaptations of Primate Molar Teeth. *American Journal of Physical Anthropology* 43:195–215.

——— 1987. Analysis of Primate Dental Microwear Using Image Processing Techniques. *Scanning Microscopy* 1:657–662.

Kay, Richard F., and Richard F. Madden. 1997. Mammals and Rainfall: Paleoecology of the Middle Miocene at La Venta (Colombia, South America). *Journal of Human Evolution* 32:161–199.

Kay, Richard F., Sergio F. Vizcaíno, and M. Susana Bargo. 2012. A Review of the Paleoenvironment and Paleoecology of the Miocene Santa Cruz Formation. In *Early Miocene Paleobiology in Patagonia: High-Latitude Paleocommunities of the Santa Cruz Formation*, edited by Sergio F. Vizcaíno, Richard F. Kay, and M. Susana Bargo, pp. 331–365. Cambridge University Press.

Kaya, Ferhat, Nuretdin Kaymakçi, Faysal Bibi et al. 2016. Magnetostratigraphy and Paleoecology of the Hominid-Bearing Locality Çorakyerler, Tuglu Formation (Çankiri Basin, Central Anatolia). *Journal of Vertebrate Paleontology* 36:e1071710.

Kearney, Michael, and Warren Porter. 2009. Mechanistic Niche Modeling: Combining Physiological and Spatial Data to Predict Species' Ranges. *Ecology Letters* 12:334–350.

Kenagy, G. J. 1973. Adaptations for Leaf Eating in the Great Basin Kangaroo Rat, *Dipodomys microps*. *Oecologia* 12:383–412.

Kerley, Graham I. H., Robert L. Pressey, Richard M. Cowling, Andre F. Boshoff, and Rebecca Sims-Castley. 2003. Options for the Conservation of Large and Medium-Sized Mammals in the Cape Floristic Region Hotspot, South Africa. *Biological Conservation* 112:169–190.

Kidwell, Susan M. 2001. Preservation of Species Abundance in Marine Death Assemblages. *Science* 294:1091–1094.

Kidwell, Susan M., and Anna K. Behrensmeyer (editors). 1993. *Taphonomic Approaches to Time Resolution in Fossil Assemblages*. Short Courses in Paleontology 6. Paleontological Society, Knoxville, TN.

Kidwell, Susan M., and Adam Tomasovych. 2013. Implications of Time-Averaged Death Assemblages for Ecology and Conservation Biology. *Annual Review of Ecology, Evolution and Systematics* 44:539–563.

King, Frances B., and Russell W. Graham. 1981. Effects of Ecological and Paleoecological Patterns on Subsistence and Paleoenvironmental Reconstructions. *American Antiquity* 46:128–142.

Kingdon, J., and M. Hoffman (editors). 2013. *Mammals of Africa*, vol. VI: *Pigs, Hippopotamuses, Chevrotain, Giraffes, Deer and Bovids*. Bloomsbury, London.

Kingdon, J., D. Happold, T. Butynski, M. Hoffman and J. Kalina (editors). 2013. *Mammals of Africa* (vols. I–IV). Bloomsbury, London.

Kingston, John D. 2007. Shifting Adaptive Landscapes: Progress and Challenges in Reconstructing Early Hominid Environments. *Yearbook of Physical Anthropology* 50:20–58.

Kipp, Nilva G. 1976. New Transfer Function for Estimating Past Sea-Surface Conditions from Sea-Bed Distribution of Planktonic Foraminiferal Assemblages in the North Atlantic. *Geological Society of America Memoirs* 145:3–42.

Klein, Richard G. 1972. The Late Quaternary Mammalian Fauna of Nelson Bay Cave (Cape Province, South Africa): Its Implications for Megafaunal Extinctions and Environmental and Cultural Change. *Quaternary Research* 2:135–142.

1975. Paleoanthropological Implications of the Nonarcheological Bone Assemblage from Swartklip I, South-Western Cape Province, South Africa. *Quaternary Research* 5:275–288.

1976a. The Fossil History of *Raphicerus* H. Smith, 1827 (Bovidae, Mammalia) in the Cape Biotic Zone. *Annals of the South African Museum* 71:169–191.

1976b. The Mammalian Fauna of the Klasies River Mouth Sites, Southern Cape Province, South Africa. *South African Archaeological Bulletin* 31:75–98.

1978. A Preliminary Report on the Larger Mammals from the Boomplaas Stone Age Cave Site, Cango Valley, Oudtshoorn District, South Africa. *South African Archaeological Bulletin* 33:66–75.

1980. Environmental and Ecological Implications of Large Mammals from Upper Pleistocene and Holocene Sites in Southern Africa. *Annals of the South African Museum* 81:223–283.

1983. Palaeoenvironmental Implications of Quaternary Large Mammals in the Fynbos Region. In *Fynbos Palaeoecology: A Preliminary Synthesis*, edited by H. J. Deacon, Q. B. Hendey, and J. J. N. Lambrechts, pp. 116–138. South African National Scientific Programmes Report 75. Mills Litho, Cape Town.

1986. Carnivore Size and Quaternary Climatic Change in Southern Africa. *Quaternary Research* 25:153–170.

1991. Size Variation in Cape Dune Molerat (*Bathyergus suillus*) and Late Quaternary Climatic Change in the Southwestern Cape Province, South Africa. *Quaternary Research* 36:243–256.

1994. The Problem of Modern Human Origins. In *Origins of Anatomically Modern Humans*, edited by Matthew V. Nitecki and Doris V. Nitecki, pp. 3–17. Plenum Press, New York.

1995. Anatomy, Behavior, and Modern Human Origins. *Journal of World Prehistory* 9:167–198.

1998. Why Anatomically Modern People Did Not Disperse from Africa 100,000 Years Ago. In *Neandertals and Modern Humans in Western Asia*, edited by Takeru Akazawa, Kenichi Aoki, and Ofer Bar-Yosef, pp. 509–521. Plenum Press, New York.

2000. Archaeology and the Evolution of Human Behavior. *Evolutionary Anthropology* 9:17–36.

Klein, Richard G., and Kathryn Cruz-Uribe. 1983a. The Computation of Ungulate Age (Mortality) Profiles from Dental Crown Heights. *Paleobiology* 9:70–78.

1983b. Stone Age Population Numbers and Average Tortoise Size at Byneskranskop Cave 1 and Die Kelders Cave 1, Southern Cape Province, South Africa. *South African Archaeological Bulletin* 38:26–30.

1984. *The Analysis of Animal Bones from Archeological Sites*. University of Chicago Press.

1987. Large Mammal and Tortoise Bones from Eland's Bay Cave and Nearby Sites, Western Cape Province, South Africa. In *Papers in the Prehistory of the Western Cape, South Africa*, edited by J. E. Parkington and M. Hall, pp. 132–163. British Archaeological Reports, International Series 322. BAR, Oxford.

1996. Size Variation in the Rock Hyrax (*Procavia capensis*) and Late Quaternary Climatic Change in South Africa. *Quaternary Research* 46:193–207.

2000. Middle and Later Stone Age Large Mammal and Tortoise Remains from Die Kelders Cave 1, Western Cape Province, South Africa. *Journal of Human Evolution* 38:169–195.

Klein, Richard G., and Katharine Scott. 1989. Glacial/Interglacial Size Variation in Fossil Spotted Hyenas (*Crocuta crocuta*) from Britain. *Quaternary Research* 32:88–95.

Klein, Richard G., Kathryn Cruz-Uribe, and Peter B. Beaumont. 1991. Environmental, Ecological, and Paleoanthropological Implications of the Late Pleistocene Mammalian Fauna from Equus Cave, Northern Cape Province, South Africa. *Quaternary Research* 36:94–119.

Klein, Richard G., Kathryn Cruz-Uribe, David Halkett, Timothy Hart, and John E. Parkington. 1999. Paleoenvironmental and Human Behavioral Implications of the Boegoeberg 1 Late Pleistocene Hyena Den, Northern Cape Province, South Africa. *Quaternary Research* 52:393–403.

Klein, Richard G., Graham Avery, Kathryn Cruz-Uribe et al. 2004. The Ysterfontein 1 Middle Stone Age Site, South Africa, and Early Human Exploitation of Coastal Resources. *Proceedings of the National Academy of Sciences USA* 101:5708–5715.

Klein, Richard G., Graham Avery, Kathryn Cruz-Uribe, and Teresa E. Steele. 2007. The Mammalian Fauna Associated with an Archaic Hominin Skullcap and Later Acheulean Artifacts at Elandsfontein, Western Cape Province, South Africa. *Journal of Human Evolution* 52:164–186.

Koch, Paul L. 1986. Clinal Geographic Variation in Mammals: Implications for the Study of Chronoclines. *Paleobiology* 12:269–281.

Koch, Paul L., Noreen Tuross, and Marilyn L. Fogel. 1997. The Effects of Sample Treatment and Diagenesis on the Isotopic Integrity of Carbonate in Biogenic Hydroxylapatite. *Journal of Archaeological Science* 24:417–429.

Koch, Paul L., Kathryn A. Hoppe, and David S. Webb. 1998. The Isotopic Ecology of Late Pleistocene Mammals in North America. Part 1. Florida. *Chemical Geology* 152:119–138.

Kohn, Matthew J. 1996. Predicting Animal $\delta^{18}O$: Accounting for Diet and Physiological Adaptation. *Geochimica et Cosmochimica Acta* 60:4811–4829.

Kolb, H. H. 1978. Variation in the Size of Foxes in Scotland. *Biological Journal of the Linnean Society* 10:291–304.

Kooyman, Brian, and Dennis Sandgathe. 2001. Sexually Dimorphic Size Variation in Holocene Bison as Revealed by Carpals and Tarsals. In *People and Wildlife in Northern North America: Essays in Honor of R. Dale Guthrie*, edited by S. Craig Gerlach and Maribeth S. Murray, pp. 67–78. British Archaeological Reports, International Series 944. BAR, Oxford.

Kovarovic, Kris, and Peter Andrews. 2007. Bovid Postcranial Ecomorphological Survey of the Laetoli Paleoenvironment. *Journal of Human Evolution* 52:663–680.

Kovarovic, Kris, Peter Andrews, and Leslie Aiello. 2002. The Palaeoecology of the Upper Ndolanya Beds at Laetoli, Tanzania. *Journal of Human Evolution* 43:395–418.

Kovarovic, Kris, Leslie C. Aiello, Andrea Cardini, and Charles A. Lockwood. 2011. Discriminant Function Analyses in Archaeology: Are Classification Rates Too Good to Be True? *Journal of Archaeological Science* 38:3006–3018.

Kowalewski, Michal. 1996. Time-Averaging, Overcompleteness, and the Geological Record. *Journal of Geology* 104:317–326.

Kowalewski, Michal, and Alan P. Hoffmeister. 2003. Sieves and Fossils: Effects of Mesh Size on Paleontological Patterns. *Palaios* 18:460–469.

Kubo, Mugino O., and Eisuke Yamada. 2014. The Inter-Relationship between Dietary and Environmental Properties and Tooth Wear: Comparisons of Mesowear, Molar Wear Rate, and Hypsodonty Index of Extant Sika Deer Populations. *PLoS ONE* 9:e90745.

Kubo, Mugino O., Eisuke Yamada, Masaki Fujita, and Ituro Oshiro. 2015. Paleoecological Reconstruction of Late Pleistocene Deer from the Ryukyu Islands, Japan: Combined Evidence of Mesowear and Stable Isotope Analyses. *Palaeogeography, Palaeoclimatology, Palaeoecology* 435:159–166.

Kurtén, Björn. 1952. The Chinese *Hipparion* Fauna: A Quantitative Survey, with Comments on the Ecology of the Machairodonts and Hyaenids and the Taxonomy of Gazelles. *Societas Scientiarum Fennica: Commentationes Biologicae* 13(4).

    1957. The Bears and Hyenas of the Interglacials. *Quaternaria* 4:69–81.

    1960. Chronology and Faunal Evolution of the Earlier European Glaciations. *Societas Scientiarum Fennica Commentationes Biologicae* 21:1–62.

    1965. The Carnivora of the Palestine Caves. *Acta Zoologica Fennica* 107:1–74.

Kutzbach, John E., and Peter J. Guetter. 1986. The Influence of Changing Orbital Parameters and Surface Boundary Conditions on Climate Simulations for the Past 18000 Years. *Journal of Atmospheric Sciences* 43:1726–1759.

Kutzbach, J[ohn E.], R. Gallimore, S. Harrison, P. Behling, R. Selin, and F. Laarif. 1998. Climate and Biome Simulations for the Past 21,000 Years. *Quaternary Science Reviews* 17:473–506.

Ladd, Harry S. 1959. Ecology, Paleontology, and Stratigraphy. *Science* 129:69–78.

Laland, Kevin N., and Michael J. O'Brien. 2011. Cultural Niche Construction: An Introduction. *Biological Theory* 6:191–202.

Lambert, W. David. 2006. Functional Convergence of Ecosystems: Evidence from Body Mass Distributions of North American Late Miocene Mammal Faunas. *Ecosystems* 9:97–118.

Lambert, W. David, and Crawford S. Holling. 1998. Causes of Ecosystem Transformation at the End of the Pleistocene: Evidence from Mammal Body-Mass Distributions. *Ecosystems* 1:157–175.

Lambrides, A. B. J., and M. I. Weisler. 2015. Assessing Protocols for Identifying Pacific Island Archaeological Fish Remains: The Contribution of Vertebrae. *International Journal of Osteoarchaeology* 25:838–848.

Lande, Russell. 1996. Statistics and Partitioning of Species Diversity, and Similarity among Multiple Communities. *Oikos* 76:5–13.

Landres, Peter B. 1992. Temporal Scale Perspectives in Managing Biological Diversity. *Transactions of the North American Wildlife and Natural Resources Conference* 57:292–307.

Lang, Richard W., and Arthur H. Harris. 1984. *The Faunal Remains from Arroyo Hondo Pueblo, New Mexico: A Study in Short-Term Subsistence Change.* Arroyo Hondo Archaeological Series 5. School of American Research, Santa Fe, NM.

Larocque, I., R. I. Hall, and E. Grahn. 2001. Chironomids as Indicators of Climate Change: A 100-Lake Training Set from a Subarctic Region of Northern Sweden (Lapland). *Journal of Paleolimnology* 26:307–322.

Lartet, Edourd. 1867. Notes sur Deux Têtes de Carnassiers Fossiles (*Ursus* et *Felis*) et sur Quelques Débris de Rhinocéros Provenant des Découvertes Faites par M. Bourguignat dans les Cavernes du Midi de la France. *Annales des Sciences Naturelles: Zoologie et Paleontologie* 8:157–194.

    1875. Notes on the Reindeer and Hippopotamus. In *Reliquiae Aquitanicae; Being Contributions to the Archaeology and Paleontology of Périgord and the Adjoining Provinces of Southern France, by Edourd Lartet and Henry Christy*, edited by Thomas R. Jones, pp. 147–152. Williams and Norgate, London.

Lastrucci, Carlo L. 1963. *The Scientific Approach: Basic Principles of the Scientific Method.* Schenkman, Cambridge, MA.

Lavergne, Sébastien, Nicolas Mouquet, Wilfried Thuiller, and Ophélie Ronce. 2010. Biodiversity and Climate Change: Integrating Evolutionary and Ecological Responses to Species and Communities. *Annual Review of Ecology, Evolution, and Systematics* 41:321–350.

Lawing, A. Michelle, Jason J. Head, and P. David Polly. 2012. The Ecology of Morphology: The Ecometrics of Locomotion and Macroenvironment in North American Snakes. In *Paleontology in Ecology and Conservation*, edited by Julien Louys, pp. 117–146. Springer, Heidelberg.

Lawrence, Barbara. 1973. Problems in the Inter-Site Comparisons of Faunal Remains. In *Domestikationsforschung und Geschichte der Haustiere*, edited by J. Matolcsi, pp. 397–402. Akademiai Kiado, Budapest.

Lawrence, David R. 1968. Taphonomy and Information Losses in Fossil Communities. *Geological Society of America Bulletin* 79:1315–1330.

——— 1971. The Nature and Structure of Paleoecology. *Journal of Paleontology* 45:593–607.

Le, Jianning. 1992. Paleotemperature Estimation Methods: Sensitivity Test on Two Western Equatorial Pacific Cores. *Quaternary Science Reviews* 11:801–820.

Le, Jianning, and Nick J. Shackleton. 1994. Reconstructing Paleoenvironment by Transfer Function: Model Evaluation with Simulated Data. *Marine Micropaleontology* 24:187–199.

Leach, Foss. 1986. A Method for the Analysis of Pacific Island Fishbone Assemblages and an Associated Database Management System. *Journal of Archaeological Science* 13:147–159.

Lee-Thorp, Julia A. 2008. On Isotopes and Old Bones. *Archaeometry* 50:925–950.

Lee-Thorp, Julia A., and Matt Sponheimer. 2006. Contributions of Biogeochemistry to Understanding Hominin Dietary Ecology. *Yearbook of Physical Anthropology* 131:131–148.

——— 2013. Hominin Ecology from Hard-Tissue Biogeochemistry. In *Early Hominin Paleoecology*, edited by Matt Sponheimer, Julia A. Lee-Thorp, and Kaye E. Reed, pp. 281–324. University Press of Colorado, Boulder.

Lee-Thorp, Julia A., and Nikolaas J. van der Merwe. 1987. Carbon Isotope Analysis of Fossil Bone Apatite. *South African Journal of Science* 83:712–715.

Lee-Thorp, Julia A., Judith C. Sealy, and Nikolaas J. van der Merwe. 1989. Stable Carbon Isotope Ratio Differences between Bone Collagen and Bone Apatite, and their Relationship to Diet. *Journal of Archaeological Science* 16:585–599.

Le Fur, Soizic, Emmanuel Fara, and Patrick Vignaud. 2011. Effect of Simulated Faunal Impoverishment and Mixture on the Ecological Structure of Modern Mammal Faunas: Implications for the Reconstruction of Mio-Pliocene African Palaeoenvironments. *Palaeogeography, Palaeoclimatology, Palaeoecology* 305:295–309.

Legendre, L., and E. D. Gallagher. 2001. Ecologically Meaningful Transformations for Ordination of Species Data. *Oikos* 129:271–280.

Legendre, P., and L. Legendre. 2012. *Numerical Ecology*. Elsevier, New York.

Legendre, Serge. 1986. Analysis of Mammalian Communities from the Late Eocene and Oligocene of Southern France. *Paleovertebrata* 16:191–212.

——— 1987. Les Communautés de Mammifères d'Europe Occidentale de L'Eocène Supérieur et Oligocène: Structures et Milieux. *Münchner Geowissenschaftliche Abhandlungen A* 10:301–312.

——— 1989. Les Communautés de Mammifères du Paléogène (Eocène Supérieur et Oligocène) d'Europe Occidentale: Structures, Milieux et Évolution. *Münchner Geowissenschaftliche Abhandlungen A* 16:1–110.

Legendre, Serge, and Claudia Roth. 1988. Correlation of Carnassial Tooth Size and Body Weight in Recent Carnivores (Mammalia). *Historical Biology* 1:85–98.

Lenton, Timothy M., Hermann Held, Elmar Kriegler et al. 2008. Tipping Elements in the Earth's Climate System. *Proceedings of the National Academy of Sciences USA* 105:1786–1793.

Leroux, Shawn J., and Michel Loreau. 2015. Theoretical Perspectives on Bottom-Up and Top-Down Interactions across Ecosystems. In *Trophic Ecology: Bottom-Up and Top-Down Interactions across Aquatic and Terrestrial Systems*, edited by Torrence C. Hanley and Kimberly J. La Pierre, pp. 1–27. Cambridge University Press.

Lesage, Louis, Michel Crête, Jean Huot, and Jean-Pierre Ouellet. 2001. Evidence for a Trade-Off between Growth and Body Reserves in Northern White-Tailed Deer. *Oecologia* 126:30–41.

Levin, Naomi E., Thure E. Cerling, Benjamin H. Passey, John M. Harris, and James R. Ehleringer. 2006. A Stable Isotope Aridity Index for Terrestrial Environments. *Proceedings of the National Academy of Sciences USA* 103:11201–11205.

Levin, Simon A. 1992. The Problem of Pattern and Scale in Ecology. *Ecology* 73:1943–1967.

Levinson, M. 1985. Are Fossil Rodents Useful in Palaeo-Ecological Interpretations? *Annals of the Geological Survey of South Africa* 19:53–64.

Lewis, Patrick J., Briggs Buchanan, and Eileen Johnson. 2005. Sexing Bison Metapodials Using Principal Components Analysis. *Plains Anthropologist* 50:159–172.

Linder, H. P. 2003. The Radiation of the Cape Flora, Southern Africa. *Biological Reviews* 78:597–638.

Lindstedt, Stan L., and Mark S. Boyce. 1985. Seasonality, Fasting Endurance, and Body Size in Mammals. *American Naturalist* 125:873–878.

Lintulaakso, Kari, and Kris Kovarovic. 2016. Diet and Locomotion, but Not Body Size, Differentiate Mammal Communities in Worldwide Tropical Ecosystems. *Palaeogeography, Palaeoclimatology, Palaeoecology* 454:20–29.

Lister, A[drian] M. 1989. Rapid Dwarfing of Red Deer on Jersey in the Last Interglacial. *Nature* 342:539–542.

——— 1997. The Evolutionary Response of Vertebrates to Quaternary Environmental Change. In *Past and Future Rapid Environmental Changes: The Spatial and Evolutionary Responses of Terrestrial Biota*, edited by B. Huntley, W. Cramer, A. V. Prentice, and J. R. M. Allen, pp. 287–302. NATO ASI Series 147. Springer, Berlin.

Liu, Liping, Jussi T. Eronen, and Mikael Fortelius. 2009. Significant Mid-Latitude Aridity in the Middle Miocene of East Asia. *Palaeogeography, Palaeoclimatology, Palaeoecology* 279:201–206.

Liu, Liping, Kai Puolamäki, Jussi T. Eronen et al. 2012. Dental Functional Traits of Mammals Resolve Productivity in Terrestrial Ecosystems Past and Present. *Proceedings of the Royal Society B* 279:2793–2799.

Livingston, Stephanie D. 1987. Prehistoric Biogeography of White-Tailed Deer in Washington and Oregon. *Journal of Wildlife Management* 51:649–654.

Lomolino, Mark V. 1985. Body Size of Mammals on Islands: The Island Rule Reexamined. *American Naturalist* 125:310–316.

——— 2005. Body Size Evolution in Insular Vertebrates: Generality of the Island Rule. *Journal of Biogeography* 32:1638–1699.

Lomolino, Mark V., Brett R. Riddle, Robert J. Whittaker, and James H. Brown. 2010. *Biogeography*, fourth edition. Sinauer Associates, Sunderland, MA.

Longinelli, Antonio. 1984. Oxygen Isotopes in Mammal Bone Phosphate: A New Tool for Paleohydrological and Paleoclimatological Research? *Geochimica et Cosmochimica Acta* 48:385–390.

López-García, Juan Manuel, Hugues-Alexandre Blain, Gloria Cuenca-Bescós et al. 2010. Palaeoenvironmental and Palaeoclimatic Reconstruction of the Latest Pleistocene El Portalón Site, Sierra De Atapuerca, Northwestern Spain. *Palaeogeography, Palaeoclimatology, Palaeoecology* 292:453–464.

López-García, Juan Manuel, Hugues-Alexandre Blain, Maria Bennàsar, and Mónica Fernández-Garcí. 2014. Environmental and Climatic Context of Neanderthal Occupation in Southwestern Europe during MIS3 Inferred from the Small-Vertebrate Assemblages. *Quaternary International* 326–327:319–328.

López-García, Juan Manuel, Narcís Soler, Julià Maroto et al. 2015a. Palaeoenvironmental and Palaeoclimatic Reconstruction of the Latest Pleistocene of L'Arbreda Cave (Serinyà, Girona, Northeastern Iberia) Inferred from the Small-Mammal (Insectivore and Rodent) Assemblages. *Palaeogeography, Palaeoclimatology, Palaeoecology* 435:244–253.

López-García, Juan Manuel, Chiara dalla Valle, Mauro Cremaschi, and Marco Peresani. 2015b. Reconstruction of the Neanderthal and Modern Human Landscape and Climate from the Fumane Cave Sequence (Verona, Italy) Using Small-Mammal Assemblages. *Quaternary Science Reviews* 128:1–13.

Lorimer, Jamie. 2015. *Wildlife in the Anthropocene: Conservation after Nature*. University of Minnesota Press, Minneapolis.

Losos, Jonathan B., and Donald B. Miles. 1994. Adaptation, Constraint, and the Comparative Method: Phylogenetic Issues and Methods. In *Ecological Morphology: Integrative Organismal Biology*, edited by Peter C. Wainwright and Stephen M. Reilly, pp. 60–98. University of Chicago Press.

Louthan, Allison M., Daniel F. Doak, and Amy L. Angert. 2015. Where and When Do Species Interactions Set Range Limits? *Trends in Ecology and Evolution* 30:780–792.

Louys, Julien (editor). 2012. *Paleontology in Ecology and Conservation*. Springer, Heidelberg.

Louys, Julien, Carlo Meloro, Sarah Elton, Peter Ditchfield, and Laura C. Bishop. 2011a. Mammal Community Structure Correlates with Arboreal Heterogeneity in Faunally and Geographically Diverse Habitats: Implications for Community Convergence. *Global Ecology and Biogeography* 20:717–729.

2011b. Meoswear as a Means of Determining Diet in African Antelopes. *Journal of Archaeological Science* 38:1485–1495.

Louys, Julien, Shaena Montanari, Thomas Plummer, Fritz Hertel, and Laura C. Bishop. 2013. Evolutionary Divergence and Convergence in Shape and Size within African Antelope Proximal Phalanges. *Journal of Mammalian Evolution* 20:239–248.

Louys, Julien, Carlo Meloro, Sarah Elton, Peter Ditchfield, and Laura C. Bishop. 2015. Analytical Framework for Reconstructing Heterogeneous Environmental Variables from Mammal Community Structure. *Journal of Human Evolution* 78:1–11.

Lowe, J. J., and M. J. C. Walker. 1997. *Reconstructing Quaternary Environments*, second edition. Pearson, Prentice Hall, New York.

2015. *Reconstructing Quaternary Environments*, third edition. Routledge, London.

Lozek, V. 1986. Mollusca Analysis. In *Handbook of Palaeoecology and Palaeohydrology*, edited by B. E. Berglund, pp. 729–740. John Wiley and Sons, Chichester.

Lubinski, Patrick M. 2000. A Comparison of Methods for Evaluating Ungulate Mortality Distributions. *Archaeozoologia* 11:121–134.

Lucas, Peter W., Ridwaan Omar, Khaled Al-Fadhalah et al. 2013. Mechanisms and Causes of Wear in Tooth Enamel: Implications for Hominin Diets. *Journal of the Royal Society: Interface* 10:20120923.

Lundelius, Ernest L., Jr. 1960. Post Pleistocene Faunal Succession in Western Australia and its Climatic Interpretation. *Proceedings of the International Geological Congress* 21(4):142–153.

1964. The Use of Vertebrates in Paleoecological Reconstructions. In *The Reconstruction of Past Environments: Proceedings*, assembled by James J. Hester and James Schoenwetter, pp. 26–31. Publication 3. Fort Burgwin Research Center, Taos, NM.

1967. Late-Pleistocene and Holocene Faunal History of Central Texas. In *Pleistocene Extinctions: The Search for a Cause*, edited by P[aul] S. Martin and H. E. Wright, Jr., pp. 287–319. Yale University Press, New Haven, CT.

1972. Vertebrate Remains from the Gray Sand. In *Blackwater Locality No. 1: A Stratified, Early Man Site in Eastern New Mexico*, edited by James J. Hester, pp. 148–163. Publication 8. Fort Burgwin Research Center, Taos, NM.

1974. The Last Fifteen Thousand Years of Faunal Change in North America. In *History and Prehistory of the Lubbock Lake Site*, edited by Craig C. Black, pp. 141–160. The Museum Journal 15. Texas Tech University, Lubbock Lake.

1976. Vertebrate Paleontology of the Pleistocene: An Overview. In *Ecology of the Pleistocene: A Symposium*, edited by R. C. West and W. G. Haag, pp. 45–59. Geoscience and Man 13. Louisiana State University, Baton Rouge.

1979. Post-Pleistocene Mammals from Pratt Cave and their Environmental Significance. In *Biological Investigations in the Guadalupe Mountains National Park, Texas*, edited by Hugh H. Genoways and R. J. Baker, pp. 239–258. Proceedings and Transactions Series 4. National Park Service, Washington, DC.

1983. Climatic Implications of Late Pleistocene and Holocene Faunal Associations in Australia. *Alcheringa* 7:125–149.

1985. North American Pleistocene Mammals: Major Problems. *Acta Zoologici Fennica* 170:167–171.

1998. Development of Quaternary Vertebrate Paleontology in North America. In *Quaternary Paleozoology in the Northern Hemisphere*, edited by Jeffrey J. Saunders, Bonnie W. Styles, and Gennady F. Baryshnikov, pp. 235–248. Illinois State Museum Scientific Papers 27. Illinois State Museum, Springfield.

Luo, Zhe-Xi, Chong-Xi Yuan, Qing-Jin Meng, and Qiang Ji. 2010. A Jurassic Eutherian Mammal and Divergence of Marsupials and Placentals. *Nature* 476:442–445.

Lupo, Karen D. 2007. Evolutionary Foraging Models in Zooarchaeological Analysis: Recent Applications and Future Challenges. *Journal of Archaeological Research* 15: 143–189.

Lüthi, Dieter, Martine Le Floch, Bernhard Bereiter et al. 2008. High-Resolution Carbon Dioxide Concentration Record 650,000–800,000 Years before Present. *Nature* 453:379–382.

Lyell, Charles. 1832. *Principles of Geology, Being an Attempt to Explain Changes of the Earth's Surface by Reference to Causes Now in Operation*, vol. II. John Murray, London.

1833. *Principles of Geology, Being an Attempt to Explain Changes of the Earth's Surface by Reference to Causes Now in Operation*, vol. III. John Murray, London.

1863. *The Geological Evidences of the Antiquity of Man, with Remarks on Theories of the Origin of Species by Variation*. G. W. Childs, Philadelphia.

Lyman, R. Lee. 1986. On the Analysis and Interpretation of Species List Data in Zooarchaeology. *Journal of Ethnobiology* 6:67–81.

1987. On the Analysis of Vertebrate Mortality Profiles: Sample Size, Mortality Type, and Hunting Pressure. *American Antiquity* 52:125–142.

1994. *Vertebrate Taphonomy*. Cambridge University Press.

1995. Determining When Rare (Zoo)archaeological Phenomena are Truly Absent. *Journal of Archaeological Method and Theory* 2:369–424.

1996. Applied Zooarchaeology: The Relevance of Faunal Analysis to Wildlife Management. *World Archaeology* 28:110–125.

2000. Building Cultural Chronology in Eastern Washington: The Influence of Geochronology, Index Fossils, and Radiocarbon Dating. *Geoarchaeology* 15:609–648.

2002. Taxonomic Identification of Zooarchaeological Remains. *Review of Archaeology* 23:13–20.

2003. The Influence of Time Averaging and Space Averaging on Application of Foraging Theory in Archaeology. *Journal of Archaeological Science* 30:595–610.

2004. Late-Quaternary Diminution and Abundance of Prehistoric Bison (*Bison* sp.) in Eastern Washington State, USA. *Quaternary Research* 62:76–85.

2006a. Identifying Bilateral Pairs of Deer (*Odocoileus* sp.) Bones: How Symmetrical is Symmetrical Enough? *Journal of Archaeological Science* 33:1237–1255.

2006b. Paleozoology in the Service of Conservation Biology. *Evolutionary Anthropology* 15:11–19.

2008a. Estimating the Magnitude of Data Asymmetry in Paleozoological Biogeography. *International Journal of Osteoarchaeology* 18:85–94.

2008b. *Quantitative Paleozoology*. Cambridge University Press.

2010a. Paleozoology's Dependence on Natural History Collections. *Journal of Ethnobiology* 30:126–136.

2010b. Prehistoric Anthropogenic Impacts to Local and Regional Faunas are Not Ubiquitous. In *The Archaeology of Anthropogenic Environments*, edited by Rebecca M. Dean, pp. 204–224. Center for Archaeological Investigations Occasional Paper 37, Southern Illinois University Press, Carbondale.

2011. Paleoecological and Biogeographical Implications of Late Pleistocene Noble Marten (*Martes americana nobilis*) in Eastern Washington State, USA. *Quaternary Research* 75:176–182.

2012a. Biodiversity, Paleozoology, and Conservation Biology. In *Paleontology in Ecology and Conservation*, edited by Julien Louys, pp. 147–169. Springer, Heidelberg.

2012b. Human-Behavioral and Paleoecological Implications of Terminal Pleistocene Fox Remains at the Marmes Site (45FR50), Eastern Washington State, USA. *Quaternary Science Reviews* 41:39–48.

2012c. The Influence of Screen-Mesh Size, and Size and Shape of Rodent Teeth on Recovery. *Journal of Archaeological Science* 39:1854–1861.

2012d. Rodent-Prey Content in Long-Term Samples of Barn Owl (*Tyto alba*) Pellets from the Northwestern United States Reflects Local Agricultural Change. *American Midland Naturalist* 167:150–163.

2012e. A Warrant for Applied Paleozoology. *Biological Reviews* 87:513–525.

2013. Taxonomic Composition and Body-Mass Distribution in the Terminal Pleistocene Mammalian Fauna from the Marmes Site, Southeastern Washington State, USA. *Paleobiology* 39:345–359.

2014a. Paleoenvironmental Implications of Two Relative Indicator Rodent Taxa during the Pleistocene to Holocene Transition in Southeastern Washington State, USA. *Journal of Quaternary Science* 29:691–697.

2014b. Terminal Pleistocene Change in Mammal Communities in Southeastern Washington State, USA. *Quaternary Research* 81:295–304.

2015a. The History of "Laundry Lists" in North American Zooarchaeology. *Journal of Anthropological Archaeology* 39:42–50.

2015b. On the Variable Relationship between NISP and NTAXA in Bird Remains and in Mammal Remains. *Journal of Archaeological Science* 53:291–296.

2016. The Mutual Climatic Range Technique is (Usually) Not the Area of Sympatry Technique when Reconstructing Paleoenvironments Based on Faunal Remains. *Palaeogeography, Palaeoclimatology, Palaeoecology* 454:75–81.

2017. Paleoenvironmental Reconstruction from Faunal Remains: Ecological Basics and Analytical Assumptions. *Journal of Archaeological Research* 25:315–371.

2018a. A Critical Review of Four Efforts to Resurrect MNI in Zooarchaeology. *Journal of Archaeological Method and Theory*, doi.org/10.1007/s10816-018-9365-3.

2018b. The History of MNI in North American Zooarchaeology. In *Zooarchaeology in Practice: Case Studies in Methodology and Interpretation in Archaeofaunal Analysis*, edited by Christina M. Giovas and Michelle J. LeFebvre, pp. 13–33. Springer, New York.

2018c. Observations on the History of Zooarchaeological Quantitative Units: Why NISP, Then MNI, Then NISP Again? *Journal of Archaeological Science: Reports* 18:43–50.

Lyman, R. Lee, and Kenneth M. Ames. 2004. Sampling to Redundancy in Zooarchaeology: Lessons from the Portland Basin, Northwestern Oregon and Southwestern Washington. *Journal of Ethnobiology* 24:329–346.

2007. On the Use of Species-Area Curves to Detect the Effects of Sample Size. *Journal of Archaeological Science* 34:1985–1990.

Lyman, R. Lee, and Kenneth P. Cannon (editors). 2004. *Zooarchaeology and Conservation Biology*. University of Utah Press, Salt Lake City.

Lyman, R. Lee, and R. Jay Lyman. 2003. Lessons from Temporal Variation in the Mammalian Faunas from Two Collections of Owl Pellets in Columbia County, Washington. *International Journal of Osteoarchaeology* 13:150–156.

Lyman, R. Lee, and Michael J. O'Brien. 2000. Chronometers and Units in Early Archaeology and Paleontology. *American Antiquity* 65:691–707.

2005. Within-Taxon Morphological Diversity in Late-Quaternary *Neotoma* as a Paleoenvironmental Indicator, Bonneville Basin, Northwestern Utah, USA. *Quaternary Research* 63:274–282.

Lyman, R. Lee, and Todd L. VanPool. 2009. Metric Data in Archaeology: A Study of Intra-Analyst and Inter-Analyst Variation. *American Antiquity* 74:485–504.

Lynch, Michael, and Wilfried Gabriel. 1987. Environmental Tolerance. *American Naturalist* 129:283–303.

MacArthur, Robert H. 1964. Environmental Factors Affecting Bird Species Diversity. *American Naturalist* 98:387–397.

McCain, Christy M., and Sarah R. B. King. 2014. Body Size and Activity Times Mediate Mammalian Responses to Climate Change. *Global Change Biology* 20:1760–1769.

McCown, T. D. 1961. Animals, Climate and Palaeolithic Man. *Kroeber Anthropological Society Papers* 25:221–230.

McDonald, H. Gregory, and Reid A. Bryson. 2010. Modeling Pleistocene Local Climatic Parameters Using Macrophysical Climate Modeling and the Paleoecology of Pleistocene Megafauna. *Quaternary International* 217:131–137.

McDonald, Jerry N. 1981. *North American Bison: Their Classification and Evolution*. University of California Press, Berkeley.

MacFadden, Bruce J., Nikos Solounias, and Thure E. Cerling. 1999. Ancient Diets, Ecology, and Extinction of 5-Million-Year-Old Horses from Florida. *Science* 283:824–827.

McGill, Brian J., Rampal S. Etienne, John S. Gray et al. 2007. Species Abundance Distributions: Moving Beyond Single Prediction Theories to Integration within an Ecological Framework. *Ecology Letters* 10:995–1015.

McGuire, Jenny L. 2011. Identifying California *Microtus* Species Using Geometric Morphometrics Documents Quaternary Geographic Range Contractions. *Journal of Mammalogy* 92: 1383–1394.

McGuire, Jenny L., and Edward B. Davis. 2014. Conservation Paleobiogeography: The Past, Present and Future of Species Distributions. *Ecography* 37: 1092–1094.

McIntosh, Robert P. 1975. H. A. Gleason – "Individualistic Ecologist" 1882–1975: His Contributions to Ecological Theory. *Bulletin of the Torrey Botanical Club* 102:253–273.

1990. Henry Allan Gleason and the Individualistic Hypothesis: The Structure of a Botanist's Career. *Botanical Review* 56:91–161.

1995. H. A. Gleason's "Individualistic Concept" and Theory of Animal Communities: A Continuing Controversy. *Biological Reviews* 70:317–357.

1998. The Myth of Community as Organism. *Perspectives in Biology and Medicine* 41:426–438.

McNab, Brian K. 1971. On the Ecological Significance of Bergmann's Rule. *Ecology* 52:845–854.

2010. Geographic and Temporal Correlations of Mammalian Size Reconsidered: A Resource Rule. *Oecologia* 164:13–23.

Madsen, David B. (editor). 2000. *Late Quaternary Paleoecology in the Bonneville Basin.* Bulletin 130. Utah Geological Survey, Salt Lake City.

Madsen, David B., David Rhode, Donald K. Grayson et al. 2001. Late Quaternary Environmental Change in the Bonneville Basin, Western USA. *Palaeogeography, Palaeoclimatology, Palaeoecology* 167:243–271.

Magurran, Anne E. 1988. *Ecological Diversity and its Measurement.* Princeton University Press. 2004. *Measuring Biological Diversity.* Blackwell, Malden, MA.

Magurran, Anne E., and Peter A. Henderson. 2003. Explaining the Excess of Rare Species in Natural Species Abundance Distributions. *Nature* 422:714–716

Manley, Brian F. J. 2005. *Multivariate Statistical Methods: A Primer.* Chapman and Hall/CRC, Boca Raton, FL.

Mannion, Philip D., and Paul Upchurch. 2010. Completeness Metrics and the Quality of the Sauropodomorph Fossil Record through Geological and Historical Time. *Paleobiology* 36:283–302.

Marean, Curtis W. 1992. Implications of Late Quaternary Mammalian Fauna from Lukenya Hill (South-Central Kenya) for Paleoenvironmental Change and Faunal Extinctions. *Quaternary Research* 37:239–255.

Marean, Curtis W., and Naomi Cleghorn. 2003. Large Mammal Skeletal Element Transport: Applying Foraging Theory in a Complex Taphonomic System. *Journal of Taphonomy* 1:15–42.

Marean, Curtis W., Nina Mudida, and Kaye E. Reed. 1994. Holocene Paleoenvironmental Change in the Kenyan Central Rift as Indicated by Micromammals from Ekapune Ya Muto Rockshelter. *Quaternary Research* 41:376–389.

Marean, Curtis W., Haley C. Cawthra, Richard M. Cowling et al. 2014. Stone Age People in a Changing South African Greater Cape Floristic Region. In *Fynbos: Ecology, Evolution, and Conservation of a Megadiverse Region,* edited by Nicky Allsopp, Jonathan F. Colville and G. Anthony Verboom, pp. 164–199. Oxford University Press.

Mares, Michael A., and Michael R. Willig. 1994. Inferring Biome Associations of Recent Mammals from Samples of Temperate and Tropical Faunas: Paleoecological Considerations. *Historical Biology* 8:31–48.

Marshall, Charles R. 1990. Confidence Intervals on Stratigraphic Ranges. *Paleobiology* 16:1–10.

Marshall, Fiona, and Tom Pilgram. 1993. NISP vs. MNI in Quantification of Body-Part Representation. *American Antiquity* 58:261–269.

Martin, Paul S. 1958. Pleistocene Ecology and Biogeography of North America. In *Zoogeography*, edited by Carl L. Hubbs, pp. 375–420. Publication 51. American Association for the Advancement of Science, Washington, DC.

Martin, Robert A. 1984. The Evolution of Cotton Rat Body Mass. In *Contributions in Quaternary Vertebrate Paleontology: A Volume in Honor of John E. Guilday*, edited by Hugh H. Genoways and Mary R. Dawson, pp. 252–266. Special Publication 8. Carnegie Museum of Natural History, Pittsburgh.

1990. Estimating Body Mass and Correlated Variables in Extinct Mammals: Travels in the Fourth Dimension. In *Body Size in Mammalian Paleobiology: Estimation and Biological Implications*, edited by John Damuth and Bruce J. MacFadden, pp. 49–68. Cambridge University Press.

Martínez-Meyer, Enrique, A, Townsend Peterson, and William W. Hargrove. 2004. Ecological Niches as Stable Distributional Constraints on Mammal Species, with Implications for Pleistocene Extinctions and Climate Change Projections for Biodiversity. *Global Ecology and Biogeography* 13:305–314.

Maser, Chris, E. Wayne Hammer, and Stanley H. Anderson. 1970. Comparative Food Habits of Three Owl Species in Central Oregon. *The Murrelet* 51:29–33.

Matthews, Thalassa, Christiane Denys, and J. E. Parkington. 2005. The Palaeoecology of the Micromammals from the Late Middle Pleistocene Site of Hoedjiespunt 1 (Cape Province, South Africa). *Journal of Human Evolution* 49:432–451.

Matthews, Thalassa, Amy L. Rector, Zenobia Jacobs, Andy I. R. Herries, and Curtis W. Marean. 2011. Environmental Implications of Micromammals Accumulated Close to the MIS 6 to MIS 5 Transition at Pinnacle Point Cave 9 (Mossel Bay, Western Cape Province, South Africa). *Palaeogeography, Palaeoclimatology, Palaeoecology* 302:213–229.

Mayr, Ernst 1954. Change of Genetic Environment and Evolution. In *Evolution as a Process*, edited by Julian Huxley, A. C. Hardy and E. B. Ford, pp. 157–180. George Allen and Unwin, London.

1956. Geographical Character Gradients and Climatic Adaptations. *Evolution* 10:105–108.

1970. *Populations, Species, and Evolution*. Harvard University Press, Cambridge, MA.

Mead, Jim I. 1987. Quaternary Records of Pika, *Ochotona*, in North America. *Boreas* 16:165–171.

Mead, Jim I., and Frederick Grady. 1996. *Ochotona* (Lagomorpha) from Late Quaternary Cave Deposits in Eastern North America. *Quaternary Research* 45:93–101.

Mead, Jim I., and W. Geoffrey Spaulding. 1995. Pika (*Ochotona*) and Paleoecological Reconstructions of the Intermountain West, Nevada and Utah. In *Late Quaternary Environments and Deep History: A Tribute to Paul S. Martin*, edited by David W. Steadman and Jim I. Mead, pp. 165–186. Scientific Papers 3. Mammoth Site of Hot Springs, Hot Springs, SD.

Mead, Jim I., and Arthur E. Spiess. 2001. Reply to Russell Graham about *Mustela macrodon*. *Quaternary Research* 56:422–423.

Mead, Jim I., Christopher J. Bell, and Lyndon K. Murray. 1992. *Mictomys borealis* (Northern Bog Lemming) and the Wisconsin Paleoecology of the East-Central Great Basin. *Quaternary Research* 37:229–238.

Mead, Jim I., Arthur E. Spiess, and Kristin D. Sobolik. 2000. Skeleton of Extinct North American Sea Mink (*Mustela macrodon*). *Quaternary Research* 53:247–262.

Meadow, Richard H. 1999. The Use of Size Index Scaling Techniques for Research on Archaeozoological Collections from the Middle East. In *Historia Animalium ex Ossibus: Beiträge zur Paläoanatomie, Archäologie, Ägyptologie, Ethnologie und Geschichte der*

Tiermedizin, Festrchrift für Angela von den Driesch, edited by Cornelia Becker, Henriette Manhart, Joris Peters, and Jörg Schibler, pp. 285–300. Verlag Marie Leidorf, Rahden/ Westf.

Meiri, Shai. 2011. Bergmann's Rule – What's in a Name? *Global Ecology and Biogeography* 29:203–207.

Meiri, Shai, Natalie Cooper, and Andy Purvis. 2008. The Island Rule: Made to Be Broken? *Proceedings of the Royal Society B* 275:141–148.

Meiri, Shai, and Tamar Dayan. 2003. On the Validity of Bergmann's Rule. *Journal of Biogeography* 30:331–351.

Meiri, Shai, Tamar Dayan, and Daniel Simberloff. 2005. Area, Isolation and Body Size Evolution in Insular Carnivores. *Ecology Letters* 8:1211–1217.

Meloro, Carlo, and Kris Kovarovic. 2013. Spatial and Ecometric Analyses of the Plio-Pleistocene Large Mammal Communities of the Italian Peninsula. *Journal of Biogeography* 40:1451–1462.

Meltzer, David J. 2006. *Folsom: New Archaeological Investigations of a Classic Paleoindian Bison Kill.* University of California Press, Berkeley.

Mendoza, Manuel, and Paul Palmqvist. 2008. Hypsodonty in Ungulates: An Adaptation for Grass Consumption or for Foraging in Open Habitat? *Journal of Zoology* (London) 274:134–142.

Mendoza, Manuel, Christine M. Janis, and Paul Palmqvist. 2002. Characterizing Complex Craniodental Patterns Related to Feeding Behaviour in Ungulates: A Multivariate Approach. *Journal of Zoology* (London) 258:223–246.

——— 2005. Ecological Patterns in the Trophic-Size Structure of Large Mammal Communities: A "Taxon-Free" Characterization. *Evolutionary Ecology Research* 7:505–530.

Menge, Bruce A., and John P. Sutherland. 1987. Community Regulation: Variation in Disturbance, Competition, and Predation in Relation to Environmental Stress and Recruitment. *American Naturalist* 130:730–757.

Merceron, Gildas, and Peter Ungar. 2005. Dental Microwear and Paleoecology of Bovids from the Early Pliocene of Langebaanweg, Western Cape Province, South Africa. *South African Journal of Science* 101:365–370.

Merceron, Gildas, Cécile Blondel, Michel Brunet et al. 2004. The Late Miocene Paleoenvironment of Afghanistan as Inferred from Dental Microwear in Artiodactyls. *Palaeogeography, Palaeoclimatology, Palaeoecology* 207:143–163.

Merceron, Gildas, Louis de Bonis, Laurent Viriot, and Cécile Blondel. 2005a. Dental Microwear of Fossil Bovids from Northern Greece: Paleoenvironmental Conditions in the Eastern Mediterranean During the Messinian. *Palaeogeography, Palaeoclimatology, Palaeoecology* 217:173–185.

Merceron, Gildas, Cécile Blondel, Louis de Bonis, Georges D. Kuofos, and Laurent Viriot. 2005b. A New Method of Dental Microwear Analysis: Application to Extant Primates and *Ouranopithecus macedoniensis* (Late Miocene of Greece). *Palaios* 20:551–561.

Merriam, C. Hart. 1890. *Results of a Biological Survey of the San Francisco Mountain Region and Desert of the Little Colorado.* North American Fauna 3. Government Printing Office, Washington, DC.

——— 1892. The Geographical Distribution of Life in North America with Special Reference to the Mammalia. *Proceedings of the Biological Society of Washington* 7:1–64.

——— 1894. Laws of Temperature Control of the Geographic Distribution of Terrestrial Animals and Plants. *National Geographic Magazine* 6:229–238.

——— 1895. The Geographic Distribution of Animals and Plants in North America. *Yearbook of the US Department of Agriculture for 1894*, pp. 203–214.

1898. Life Zones and Crop Zones in the United States. *Bulletin of the Division of the Biological Survey, USDA* 10:1–79.

Mihlbachler, Matthew C., and Nikos Solounias. 2006. Coevolution of Tooth Crown Height and Diet in Oreodonts (Myerycoidodontidae, Artiodactyla) Examined with Phylogenetically Independent Contrasts. *Journal of Mammalian Evolution* 13:11–36.

Mihlbachler, Matthew C., Florent Rivals, Nikos Solounias, and Gina M. Semprebon. 2011. Dietary Change and Evolution of Horses in North America. *Science* 331:1178–1181.

Millar, J. S., and G. J. Hickling. 1990. Fasting Endurance and the Evolution of Mammalian Body Size. *Functional Ecology* 4:5–12.

Miller, Alden H. 1937. Biotic Associations and Life-Zones in Relation to the Pleistocene Birds of California. *Condor* 39:248–252.

Miller, G. H., Peter B. Beaumont, H. J. Deacon et al. 1999. Earliest Modern Humans in Southern Africa Dated by Isoleucine Epimerization in Ostrich Eggshell. *Quaternary Science Reviews* 18:1537–1548.

Miller, Joshua H. 2011. Ghosts of Yellowstone: Multi-Decadal Histories of Wildlife Populations Captured by Bones on a Modern Landscape. *PLoS One* 6(3):318057.

Miller, Joshua H., Anna K. Behrensmeyer, Andrew Du et al. 2014. Ecological Fidelity of Functional Traits Based on Presence–Absence in a Modern Mammalian Bone Assemblage (Amboseli, Kenya). *Paleobiology* 40:560–583.

Millien, Virginie, S. Kathleen Lyons, Link Olson et al. 2006. Ecotypic Variation in the Context of Global Climate Change: Revisiting the Rules. *Ecology Letters* 9:853–869.

Mills, J. R. E. 1955. Ideal Dental Occlusion in the Primates. *Dental Practice* 6:47–61.

Minagawa, Masao, and Eitaro Wada. 1984. Stepwise Enrichment of ^{15}N Along Food Chains: Further Evidence and the Relation between $\delta^{15}N$ and Animal Age. *Geochimica et Cosmochimica Acta* 48:1135–1140.

Mittelbach, Gary G., Christopher F. Steiner, Samuel M. Scheiner et al. 2001. What Is the Observed Relationship between Species Richness and Productivity? *Ecology* 82:2381–2396.

Mix, Alan C., Ann E. Morey, Nicklas G. Pisias, and Steven W. Hostetler. 1999. Foraminiferal Faunal Estimates of Paleotemperature: Circumventing the No-Analog Problem Yields Cool Ice Age Tropics. *Paleoceanography* 14:350–359.

Moffett, R. O., and H. J. Deacon. 1977. The Flora and Vegetation in the Surrounds of Boomplaas Cave: Cango Valley. *South African Archaeological Bulletin* 32:127–145.

Moine, Olivier, Denis-Didier Rousseau, Dominique Jolly, and Marc Vianey-Liaud. 2002. Paleoclimatic Reconstruction Using Mutual Climatic Range on Terrestrial Mollusks. *Quaternary Research* 57:162–172.

Monchot, Hervé, and Daniel Gendron. 2010. Disentangling Long Bones of Foxes (*Vulpes vulpes* and *Alopex lagopus*) from Arctic Archaeological Sites. *Journal of Archaeological Science* 37:799–806.

Montuire, Sophie, and Frederica Marcolini. 2002. Palaeoenvironmental Significance of the Mammalian Faunas of Italy since the Pliocene. *Journal of Quaternary Science* 17:87–96.

Moore, Jason R., David B. Norman, and Paul Upchurch. 2007. Assessing Relative Abundances in Fossil Assemblages. *Palaeogeography, Palaeoclimatology, Palaeoecology* 253:317–322.

Morales, Arturo, and Knud Rosenlund. 1979. *Fish Bone Measurements: An Attempt to Standardize the Measuring of Fish Bones from Archaeological Sites.* Steenstrupia, Copenhagen.

Morin, Eugène. 2012. *Reassessing Paleolithic Subsistence: The Neandertal and Modern Human Foragers of Saint-Césaire*. Cambridge University Press.

Morin, Eugène, Elspeth Ready, Arianne Boileau, Cédric Beauval, and Marie-Pierre Coumont. 2017a. Problems of Identification and Quantification in Archaeozoological Analysis, Part I: Insights from a Blind Test. *Journal of Archaeological Method and Theory* 24:886–937.

2017b. Problems of Identification and Quantification in Archaeozoological Analysis, Part II: Presentation of an Alternative Counting Method. *Journal of Archaeological Method and Theory* 24:938–973.

Morin, Xavier, and Martin J. Lechowicz. 2008. Contemporary Perspectives on the Niche that Can Improve Models of Species Range Shifts under Climate Change. *Biology Letters* 4:573–576.

Morlan, Richard E. 1984. Biostratigraphy and Biogeography of Quaternary Microtine Rodents from Northern Yukon Territory, Eastern Beringia. In *Contributions in Quaternary Vertebrate Paleontology: A Volume in Memorial to John E. Guilday*, edited by Hugh H. Genoways and Mary R. Dawson, pp. 184–199. Special Publication 8. Carnegie Museum of Natural History, Pittsburgh.

1991. Bison Carpal and Tarsal Measurements: Bulls versus Cows and Calves. *Plains Anthropologist* 36:215–227.

Morlot, A[dolphe von]. 1861. General Views on Archaeology. *Annual Report of the Smithsonian Institution for 1860*, pp. 284–343.

Mosbrugger, Volker. 2009. Nearest-Living-Relative Method. In *Encyclopedia of Paleoclimatology and Ancient Environments*, edited by Vivien Gornitz, pp. 607–609. Encyclopedia of Earth Sciences Series. Springer, Dordrecht.

Mota-Vargas, Claudio, and Octavio R. Rojas-Soto. 2016. Taxonomy and Ecological Niche Modeling: Implications for the Conservation of Wood Partridges (Genus *Dendrortyx*). *Journal for Nature Conservation* 29:1–13.

Munro, Natalie D. 2004. Zooarchaeological Measures of Hunting Pressure and Occupation Intensity in the Natufian: Implications for Agricultural Origins. *Current Anthropology* 45:S5–S33.

Murcia, Carolina, James Aronson, Gustavo H. Kattan et al. 2014. A Critique of the "Novel Ecosystem" Concept. *Trends in Ecology and Evolution* 29:548–553.

Nagaoka, Lisa. 2005. Differential Recovery of Pacific Island Fish Remains. *Journal of Archaeological Science* 32:941–955.

Nekola, Jeffrey C., and Peter S. White. 1999. The Distance Decay of Similarity in Biogeography and Ecology. *Journal of Biogeography* 26:867–878.

Nel, Thurid H., and Christopher S. Henshilwood. 2016. The Small Mammal Sequence from the C. 76–72 ka Still Bay Levels at Blombos Cave, South Africa: Taphonomic and Palaeoecological Implications for Human Behaviour. *PLoS ONE* 11:e0159817.

Nelson, Bruce K., Michael J. DeNiro, Margaret J. Schoeninger, Donald J. De Paolo, and P. E. Hare. 1986. Effects of Diagenesis on Strontium, Carbon, Nitrogen and Oxygen Concentration and Isotopic Composition of Bone. *Geochimica et Cosmochimica Acta* 50:1941–1949.

Nelson, Robert S., and Holmes A. Semken, Jr. 1970. Paleoecological and Stratigraphic Significance of the Muskrat in Pleistocene Deposits. *Geological Society of America Bulletin* 81:3733–3738.

Nesbit Evans, E. M., Judith H. van Couvering, and Peter Andrews. 1981. Palaeoecology of Miocene Sites in Western Kenya. *Journal of Human Evolution* 10:35–48.

Nikita, E. 2014. Estimation of the Original Number of Individuals Using Multiple Skeletal Elements. *International Journal of Osteoarchaeology* 24:660–664.

Nogués-Bravo, David. 2009. Predicting the Past Distribution of Species Climatic Niches. *Global Ecology and Biogeography* 18:521–531.

Nowak, Robert S., Cheryl L. Nowak, and Robin J. Tusch. 2000. Probability that a Fossil Absent from a Sample Is Also Absent from the Paleolandscape. *Quaternary Research* 54:144–154.

O'Connell, James F., Kristen Hawkes, and Nicholas Blurton-Jones. 1988. Hadza Hunting, Butchering, and Bone Transport and their Archaeological Implications. *Journal of Anthropological Research* 44:113–161.

O'Connor, Anne. 2007. *Finding Time for the Old Stone Age: A History of Palaeolithic Archaeology and Quaternary Geology in Britain, 1860–1960.* Oxford University Press.

O'Connor, Terry (editor). 2005. *Biosphere to Lithosphere: New Studies in Vertebrate Taphonomy.* Oxbow Books, Oxford.

Odling-Smee, F. John, Kevin N. Laland, and Marcus W. Feldman. 2003. *Niche Construction: The Neglected Process in Evolution.* Princeton University Press.

Odum, Eugene P. 1971. *Fundamentals of Ecology*, third edition. W. B. Saunders, Philadelphia.

Odum, Eugene P., and Gary W. Barrett. 2005. *Fundamentals of Ecology*, fifth edition. Thomson Brooks/Cole, Belmont, CA.

O'Gara, Bart W., and Jim D. Yoakum (editors). 2004. *Pronghorn: Ecology and Management.* University Press of Colorado, Boulder.

Olander, Heikki, Atte Korhola, and Tom Blom. 1997. Surface Sediment Chironomidae (Insecta: Diptera) Distributions along an Ecotonal Transect in Subarctic Fennoscandia: Developing a Tool for Palaeotemperature Reconstructions. *Journal of Paleolimnology* 18:45–59.

Olcott, Susan P., and Ronald E. Barry. 2000. Environmental Correlates of Geographic Variation in Body Size of the Eastern Cottontail (*Sylvilagus floridanus*). *Journal of Mammalogy* 81:986–998.

Olff, Han, M. E. Ritchie, and Herbert H. T. Prins. 2002. Global Environmental Controls of Diversity in Large Herbivores. *Nature* 415:901–905.

Olsen, John W. 1982. Prehistoric Environmental Reconstruction by Vertebrate Faunal Analysis. In *Multidisciplinary Research at Grasshopper Pueblo, Arizona*, edited by William A. Longacre, Sally J. Holbrook, and Michael W. Graves, pp. 63–72. Anthropological Papers 40. University of Arizona, Tucson.

Olson, D. M., E. Dinerstein, E. D. Wikramanayake et al. 2001. Terrestrial Ecoregions of the World: A New Map of Life on Earth. *BioScience* 51:933–938.

Olszewski, Thomas D. 1999. Taking Advantage of Time Averaging. *Paleobiology* 25:226–238.

O'Regan, Hannah J., and Alan Turner. 2004. The Interface between Conservation Biology, Palaeontology and Archaeozoology: Morphometrics and Population Viability Analysis. In *The Future from the Past*, edited by Roel C. G. M. Lauwerier and Ina Plug, pp. 90–96. Oxbow Books, Oxford.

Orians, Gordon H., and A. V. Milewski. 2007. Ecology of Australia: The Effects of Nutrient-Poor Soils and Intense Fires. *Biological Reviews* 82:393–423.

Orlando, Ludovic, and Alan Cooper. 2014. Using Ancient DNA to Understand Evolutionary and Ecological Processes. *Annual Review of Ecology, Evolution and Systematics* 45:573–598.

Ortiz, J. D., and A. C. Mix. 1997. Comparison of Imbrie-Kipp Transfer Function and Modern Analog Temperature Estimates Using Sediment Trap and Core Top Foraminiferal Faunas. *Paleoceanography* 12:175–190.

Orton, David C. 2014. Biometry in Zooarchaeology. In *Encyclopedia of Global Archaeology*, edited by Claire Smith, pp. 902–910. Springer, New York.

Overpeck, Jonathan T., Robert S. Webb, and Thompson Webb III. 1992. Mapping Eastern North American Vegetation Change of the Past 18 Ka: No-Analogs and the Future. *Geology* 20:1071–1074.

Owen, Pamela R., Christopher J. Bell, and Emilee M. Mead. 2000. Fossils, Diet, and Conservation of Black-Footed Ferrets (*Mustela nigripes*). *Journal of Mammalogy* 81:422–433.

Owen-Smith, R. Norman. 1988. *Megaherbivores: The Influence of Very Large Body Size on Ecology*. Cambridge University Press.

Palmqvist, Paul, Darren R. Gröcke, Alfonso Arribas, and Richard A. Fariña. 2003. Paleoecological Reconstruction of a Lower Pleistocene Large Mammal Community Using Biogeochemical (δ^{13}C, δ^{15}N, δ^{18}O, Sr:Zn) and Ecomorphological Approaches. *Paleobiology* 29:205–229.

Parmesan, Camille. 2006. Ecological and Evolutionary Responses to Recent Climate Change. *Annual Review of Ecology, Evolution, and Systematics* 37:637–669.

Pate, F. Donald. 1994. Bone Chemistry and Paleodiet. *Journal of Archaeological Method and Theory* 1:161–209.

Paterson, James D. 1990. Comment – Bergmann's Rule is Invalid: A Reply to V. Geist. *Canadian Journal of Zoology* 68:1610–1612.

Patton, J. L., U. F. J. Pardiñas, and G. D'Elía (editors). 2015. *Mammals of South America*, vol. II: *Rodents*. University of Chicago Press.

Patton, Thomas H. 1963. *Fossil Vertebrates from Miller's Cave, Llano County, Texas*. Texas Memorial Museum Bulletin 7. University of Texas, Austin.

Payne, Sebastian. 1969. A Metrical Distinction between Sheep and Goat Metacarpals. In *The Domestication and Exploitation of Plants and Animals*, edited by Peter J. Ucko and G. W. Dimbleby, pp. 295–305. Aldine Atherton, Chicago.

Peel, M. C., B. L. Finlayson, and T. A. McMahon. 2007. Updated World Map of the Köppen-Geiger Climate Classification. *Hydrology and Earth System Sciences* 11:1633–1644.

Peet, Robert K. 1974. The Measurement of Species Diversity. *Annual Review of Ecology and Systematics* 5:285–307.

Peet, Robert K., Robert G. Knox, J. Stephen Case, and R. B. Allen. 1988. Putting Things in Order: The Advantages of Detrended Correspondence Analysis. *American Naturalist* 131:924–934.

Peppe, Daniel J., Dana L. Royer, Bárbara Cariglino et al. 2011. Sensitivity of Leaf Size and Shape to Climate: Global Patterns and Paleoclimatic Applications. *New Phytologist* 190:724–739.

Pérez-Barbería, F. J., and I. J. Gordon. 2001. Relationships between Oral Morphology and Feeding Style in the Ungulata: A Phylogenetically Controlled Evaluation. *Proceedings of the Royal Society B* 268:1021–1030.

Pérez-Crespo, Víctor A., Christian R. Barrón-Ortiz, Joaquín Arroyo-Barales et al. 2016. Preliminary Data on the Diet and Habitat Preferences of *Capromeryx mexicana* (Mammalia: Antilocapridae) from the Late Pleistocene of Cedral, San Luis Potosí, Mexico. *Southwestern Naturalist* 61:152–155.

Peters, J., and James S. Brink. 1992. Comparative Postcranial Osteomorphology and Osteometry of Springbok, *Antidorcas marsupialis* (Zimmerman, 1780) and Grey Rhebok, *Pelea capreolus* (Forster, 1790) (Mammalia: Bovidae). *Navorsinge van die Nasionale Museum Bloemfontein* 8:161–206.

Peters, Robert H. 1983. *The Ecological Implications of Body Size*. Cambridge University Press.

Peterson, A. Townsend. 2011. Ecological Niche Conservatism: A Time-Structured Review of Evidence. *Journal of Biogeography* 38:817–827.

Peterson, A. Townsend, and Jorge Soberón. 2012. Species Distribution Modeling and Ecological Niche Modeling: Getting the Concepts Right. *Natureza & Conservação* 10:1–6.

Peterson, Charles H. 1977. The Paleoecological Significance of Undetected Short-Term Variability. *Journal of Paleontology* 51:976–981.

Petit, J. R., J. Jouzel, D. Raynaud et al. 1999. Climate and Atmospheric History of the Past 420,000 Years from the Vostok Ice Core, Antarctica. *Nature* 399:429–436.

Peyron, Odile, Joël Guiot, Rachid Cheddadi et al. 1998. Climatic Reconstruction in Europe for 18,000 Yr BP from Pollen Data. *Quaternary Research* 49:183–196.

Phillips, Arthur M., III, D. A. House, and B. G. Phillips. 1989. Expedition to the San Francisco Peaks: C. Hart Merriam and the Life Zone Concept. *Plateau* 60:19–30.

Pianka, E. R. 1978. *Evolutionary Ecology*, second edition. Harper and Row, New York.

    1988. *Evolutionary Ecology*, fourth edition. Harper and Row, New York.

Pickering, Travis, Kathy Schick, and Nicholas Toth (editors). 2007. *Breathing Life into Fossils: Taphonomic Studies in Honor of C. K. (Bob) Brain*. Stone Age Institute Publication Series no. 2. Stone Age Institute, Gosport, IN.

Pielou, E. C. 1966. The Measurement of Diversity in Different Types of Biological Collections. *Journal of Theoretical Biology* 13:131–144.

Pierce, Becky M., Vernon C. Bleich, Kevin L. Monteith, and R. Terry Bowyer. 2012. Top-Down Versus Bottom-Up Forcing: Evidence from Mountain Lions and Mule Deer. *Journal of Mammalogy* 93:977–988.

Pinto, C. Miguel, J. Angel Soto-Centeno et al. 2016. Archaeology, Biogeography, and Mammalogy Do Not Provide Evidence for Tarukas (Cervidae: *Hippocamelus antisensis*) in Ecuador. *Journal of Mammalogy* 97:41–53.

Pinto-Llona, Ana C. 2013. Macrowear and Occlusal Microwear on Teeth of Cave Bears *Ursus spelaeus* and Brown Bears *Ursus arctos*: Inferences Concerning Diet. *Palaeogeography, Palaeoclimatology, Palaeoecology* 370:41–50.

Plug, Ina. 2005. Osteomorphological Differences between Some Skeletal Elements of *Labeobarbus kimberleyensis*, *Labeobarbus aeneus* and *Labeo capensis* (Pisces: Cyprinidae). *Annals of the Transvaal Museum* 42:5–17.

Plummer, Thomas W., and Laura C. Bishop. 1994. Hominid Paleoecology at Olduvai Gorge, Tanzania, as Indicated by Antelope Remains. *Journal of Human Evolution* 27:47–75.

Plummer, Thomas W., Laura C. Bishop, and Fritz Hertel. 2008. Habitat Preference of Extant African Bovids Based on Astragalus Morphology: Operationalizing Ecomophology for Palaeoenvironmental Reconstruction. *Journal of Archaeological Science* 35:3016–3027.

Plummer, Thomas W., Laura C. Bishop, Peter Ditchfield et al. 2009. The Environmental Context of Oldowan Hominin Activities at Kanjera South, Kenya. In *Interdisciplinary Approaches to the Oldowan*, edited by Erella Hovers and David R. Braun, pp. 149–160. Springer, Dordrecht.

Plummer, Thomas W., Joseph V. Ferraro, Julien Louys et al. 2015. Bovid Ecomorphology and Hominin Paleoenvironments of the Shungura Formation, Lower Omo River Valley, Ethiopia. *Journal of Human Evolution* 88:108–126.

Pokines, James T. 1998. *The Paleoecology of Lower Magdalenian Cantabrian Spain*. British Archaeological Reports, International Series 713. BAR, Oxford.

Polley, H. Wayne. 1997. Implications of Rising Atmospheric Carbon Dioxide Concentrations for Rangeleands. *Journal of Range Management* 50:562–577.

Polley, H. Wayne, Herman S. Mayeux, Hyrum B. Johnson, and Charles R. Tischler. 1997. Viewpoint: Atmospheric CO_2, Soil Water and Shrub/Grass Ratios on Rangelands. *Journal of Range Management* 50:278–284.

Polly, P. David. 2010. Tiptoeing through the Trophics: Geographic Variation in Carnivoran Locomotor Ecomorphology in Relation to Environment. In *Carnivoran Evolution: New Views on Phylogeny, Form, and Function*, edited by Anjali Goswami and Anthony Friscia, pp. 374–410. Cambridge University Press.

Polly, P. David, and Jussi T. Eronen. 2011. Mammal Associations in the Pleistocene of Britain: Implications of Ecological Niche Modelling and a Method for Reconstructing Palaeoclimate. In *The Ancient Human Occupation of Britain*, edited by N. Ashton, S. Lewis, and C. Stringer, pp. 279–304. Developments in Quaternary Science 14. Elsevier, Amsterdam.

Polly, P. David, and Sana Sarwar. 2014. Extinction, Extirpation, and Exotics: Effects on the Correlation between Traits and Environment at the Continental Level. *Annales Zoologici Fennici* 51:209–226.

Polly, P. David, Jussi T. Eronen, Marianne Fred et al. 2011. History Matters: Ecometrics and Integrative Climate Change Biology. *Proceedings of the Royal Society B* 278:1131–1140.

Posadas, P., J. V. Crisci, and L. Katinas. 2006. Historical Biogeography: A Review of its Basic Concepts and Critical Issues. *Journal of Arid Environments* 66:389–403.

Potts, Richard. 1988. *Early Hominid Activities at Olduvai*. Aldine de Gruyter, New York.

Prentice, Colin I., S. P. Harrison, and P. J. Bartlein. 2011. Global Vegetation and Terrestrial Carbon Cycle Changes after the Last Ice Age. *New Phytologist* 189:988–998.

Preston, F. W. 1948. The Commonness, and Rarity, of Species. *Ecology* 29:254–283.

Price, Gilbert J., and Ian H. Sobbe. 2005. Pleistocene Palaeoecology and Environmental Change on the Darling Downs, Southeastern Queensland, Australia. *Memoirs of the Queensland Museum* 51:171–201.

Price, Gilbert J., Jian-xin Zhao, Yue-xing Feng, and Scott A. Hocknull. 2009. New Records of Plio-Pleistocene Koalas from Australia: Palaeoecological and Taxonomic Implications. *Records of the Australian Museum* 61:39–48.

Price, T. Douglas, Jennifer Blitz, James Burton, and Joseph A. Ezzo. 1992. Diagenesis in Prehistoric Bone: Problems and Solutions. *Journal of Archaeological Science* 19:513–529.

Prideaux, Gavin J., John A. Long, Linda K. Ayliffe et al. 2007. An Arid-Adapted Middle Pleistocene Vertebrate Fauna from South-Central Australia. *Nature* 445:422–425.

PRISM Climate Group. 2004. *PRISM Gridded Climate Data*. Oregon State University, http://prism.oregonstate.edu.

Purdue, James R. 1980. Clinal Variation of Some Mammals during the Holocene in Missouri. *Quaternary Research* 13:242–258.

——— 1986. The Size of White-Tailed Deer (*Odocoileus virginianus*) during the Archaic Period in Central Illinois. In *Foraging, Collecting, and Harvesting: Archaic Period Subsistence and Settlement in the Eastern Woodlands*, edited by Sarah W. Neusius, pp. 65–95. Occasional Paper 6, Center for Archaeological Investigations, Southern Illinois University, Carbondale.

——— 1987. Estimation of Body Weight of White-Tailed Deer from Bone Size. *Journal of Ethnobiology* 7:1–12.

——— 1989. Changes during the Holocene in Size of White-Tailed Deer (*Odocoileus virginianus*) from Central Illinois. *Quaternary Research* 32:307–316.

Qiao, Huijie, Jorge Soberón, and A. Townsend Peterson. 2015. No Silver Bullets in Correlative Ecological Niche Modeling: Insights from Testing among Many Potential Algorithms for Niche Estimation. *Methods in Ecology and Evolution* 6:1126–1136.

Quade, Jay, Thure E. Cerling, John C. Barry et al. 1992. A 16–Ma Record of Paleodiet Using Carbon and Oxygen Isotopes in Fossil Teeth from Pakistan. *Chemical Geology* 94:183–192.

Quirt-Booth, Tina, and Kathryn Cruz-Uribe. 1997. Analysis of Leporid Remains from Prehistoric Singagua Sites, Northern Arizona. *Journal of Archaeological Science* 24:945–960.

Rabenold, Diana, and Osbjorn M. Pearson. 2014. Scratching the Surface: A Critique of Lucas et al. (2013)'s Conclusion that Phytoliths Do Not Abrade Enamel. *Journal of Human Evolution* 74:130–133.

Raia, Pasquale, Francesco Carotenuto, Carlo Meloro, Paola Piras, and Diana Pushkina. 2010. The Shape of Contention: Adaptation, History, and Contingency in Ungulate Mandibles. *Evolution* 64:1489–1503.

Rainger, Ronald. 1981. The Continuation of the Morphological Tradition: American Paleontology, 1880–1910. *Journal of the History of Biology* 14:129–158.

1985. Paleontology and Philosophy: A Critique. *Journal of the History of Biology* 18:267–287.

1997. Everett C. Olson and the Development of Vertebrate Paleoecology and Taphonomy. *Archives of Natural History* 24:373–396.

Raper, Diana J., and Holli Zander. 2009. Paleoecology: An Untapped Resource for Teaching Environmental Change. *International Journal of Environmental and Science Education* 4:441–447.

Raup, David M. 1977. Stochastic Models in Evolutionary Paleobiology. In *Patterns of Evolution as Illustrated by the Fossil Record*, edited by Anthony Hallam, pp. 59–78. Elsevier, Amsterdam.

Raup, David M., and Steven M. Stanley. 1971. *Principles of Paleontology*. W. H. Freeman, San Francisco.

Rea, Amadeo M. 1986. Verification and Reverification: Problems in Archaeofaunal Studies. *Journal of Ethnobiology* 6:9–18.

Real, Leslie A., and James H. Brown (editors). 1991. *Foundations of Ecology: Classic Papers with Commentaries*. University of Chicago Press.

Réale, Denis, Andrew G. McAdam, Stan Boutin, and Dominique Berteaux. 2003. Genetic and Plastic Responses of a Northern Mammal to Climatic Change. *Proceedings of the Royal Society B* 270:591–596.

Rector, Amy L., and Kaye E. Reed. 2010. Middle and Late Pleistocene Faunas of Pinnacle Point and their Paleoecological Implications. *Journal of Human Evolution* 59:340–357.

Rector, Amy L., and Brian C. Verrelli. 2010. Glacial Cycling, Large Mammal Community Composition, and Trophic Adaptations in the Western Cape, South Africa. *Journal of Human Evolution* 58:90–102.

Redding, Richard W. 1978. Rodents and the Archaeological Paleoenvironment: Considerations, Problems, and the Future. In *Approaches to Faunal Analysis in the Middle East*, edited by Richard H. Meadow and Melinda A. Zeder, pp. 63–68. Peabody Museum of Archaeology and Ethnology Bulletin 2. Harvard University, Cambridge, MA.

Reed, Charles A. 1963. Osteo-archaeology. In *Science in Archaeology*, edited by D. Brothwell and E. Higgs, pp. 204–216. Basic Books, New York.

Reed, Charles A., and Robert J. Braidwood. 1960. Toward the Reconstruction of the Environmental Sequence of Northeastern Iraq. In *Prehistoric Investigations in Iraqi Kurdistan*, edited by Robert J. Braidwood and Bruce Howe, pp. 163–173. Studies in Ancient Oriental Civilization 31. University of Chicago Press.

Reed, D. N. 2007. Serengeti Micromammals and their Implications for Olduvai Paleoenvironments. In *Hominin Environments in the East African Pliocene: An Assessment of the Faunal Evidence*, edited by R[ené] Bobe, Z[eresenay] Alemseged, and A[nna] K. Behrensmeyer, pp. 217–255. Springer, Dordrecht.

Reed, Kaye E. 1997. Early Hominid Evolution and Ecological Change through the African Plio-Pleistocene. *Journal of Human Evolution* 32:289–322.

——— 1998. Using Large Mammal Communities to Examine Ecological and Taxonomic Structure and Predict Vegetation in Extant and Extinct Assemblages. *Paleobiology* 24:384–408.

——— 2008. Paleoecological Patterns at the Hadar Hominin Site, Afar Regional State, Ethiopia. *Journal of Human Evolution* 54:743–768.

——— 2013. Multiproxy Paleoecology: Reconstructing Evolutionary Context in Paleoanthropology. In *A Companion to Paleoanthropology*, edited by David R. Begun, pp. 204–225. Blackwell, Malden, MA.

Reed, Kaye E., Lillian M. Spencer, and Amy L. Rector. 2013. Faunal Approaches to Early Hominin Paleoecology. In *Early Hominin Paleoecology*, edited by Matt Sponheimer, Julia A. Lee-Thorp, Kaye E. Reed, and Peter S. Ungar, pp. 3–34. University Press of Colorado, Boulder.

Reimer, P. J., E. Bard, A. Bayliss et al. 2013. IntCal 13 and Marine 13 Radiocarbon Age Calibration Curves 0–50,000 Years cal BP. *Radiocarbon* 55:1869–1887.

Reitz, Elizabeth J., and Barbara Ruff. 1994. Morphometric Data for Cattle from North America and the Caribbean Prior to the 1850s. *Journal of Archaeological Science* 21:699–713.

Reitz, Elizabeth J., and Elizabeth S. Wing. 2008. *Zooarchaeology*, second edition. Cambridge University Press.

Rensberger, John M. 1978. Scanning Electron Microscopy of Wear and Occlusal Events in Some Small Herbivores. In *Development, Function, and Evolution of Teeth*, edited by Percy M. Butler and Kenneth A. Joysey, pp. 415–438. Academic Press, New York.

Rensch, Bernhard. 1938. Some Problems of Geographical Variation and Species Formation. *Proceedings of the Linnean Society of London* 150:275–285.

Reynolds, Sally C. 2007. Mammalian Body Size Changes and Plio-Pleistocene Environmental Shifts: Implications for Understanding Hominin Evolution in Eastern and Southern Africa. *Journal of Human Evolution* 53:528–548.

Rhoades, Robert E. 1978. Archaeological Use and Abuse of Ecological Concepts and Studies: The Ecotone Example. *American Antiquity* 43:608–614.

Rhodes, R. Sanders, II. 1984. *Paleoecology and Regional Paleoclimatic Implications of the Farmdalian Craigmile and Woodfordian Waubonsie Mammalian Local Faunas, Southwestern Iowa*. Reports of Investigations 40. Illinois State Museum, Springfield.

Rhodes, R. Sanders, II, and Holmes A. Semken, Jr. 1986. Quaternary Biostratigraphy and Paleoecology of Fossil Mammals from the Loess Hills Region of Western Iowa. *Proceedings of the Iowa Academy of Sciences* 93(3):94–139.

Richter, Kristine, Julie Wilson, Andrew K. G. Jones et al. 2011. Fish 'n Chips: ZooMS Peptide Mass Fingerprinting in a 96 Well Plate Format to Identify Fish Bone Fragments. *Journal of Archaeological Science* 38:1502–1510.

Ricklefs, Robert E., and Dolph Schluter (editors). 1993. *Species Diversity in Ecological Communities: Historical and Geographical Perspectives*. University of Chicago Press.

Ries, Leslie, Robert J. Fletcher, Jr., James Battin, and Thomas D. Sisk. 2004. Ecological Responses to Habitat Edges: Mechanisms, Models, and Variability Explained. *Annual Review of Ecology, Evolution and Systematics* 35:491–522.

Rivals, Florent, and Adrian M. Lister. 2016. Dietary Flexibility and Niche Partitioning of Large Herbivores through the Pleistocene of Britain. *Quaternary Science Reviews* 146:116–133.

Rivals, Florent, and Gina M. Semprebon. 2006. A Comparison of the Dietary Habits of a Large Sample of the Pleistocene Pronghorn *Stockoceros onusrosagris* from the Papago Springs Cave in Arizona to the Modern *Antilocapra americana*. *Journal of Vertebrate Paleontology* 26:495–500.

Rivals, Florent, Matthew C. Mihlbachler, and Nikos Solounias. 2007a. Effect of Ontogenetic-Age Distribution in Fossil and Modern Samples on the Interpretation of Ungulate Paleodiets Using the Mesowear Method. *Journal of Vertebrate Paleontology* 27:763–767.

Rivals, Florent, Nikos Solounias, and Matthew C. Mihlbachler. 2007b. Evidence for Geographic Variation in the Diets of Late Pleistocene and Early Holocene *Bison* in North America, and Differences from the Diets of Recent *Bison*. *Quaternary Research* 68:338–346.

Rivals, Florent, Nikos Solounias, and George B. Schaller. 2011. Diet of Mongolian Gazelles and Tibetan Antelopes from Steppe Habitats Using Premaxillary Shape, Tooth Mesowear and Microwear Analyses. *Mammalian Biology* 76:358–364.

Rivals, Florent, Gina M. Semprebon, and Adrian Lister. 2012. An Examination of Dietary Diversity Patterns in Pleistocene Proboscideans (*Mammuthus, Palaeoloxodon,* and *Mammut*) from Europe and North America as Revealed by Dental Microwear. *Quaternary International* 255:188–195.

Rivals, Florent, Marie-Anne Julien, Margot Kuitems et al. 2015. Investigation of Equid Paleodiet from Schöningen 13 II-4 through Dental Wear and Isotopic Analyses: Archaeological Implications. *Journal of Human Evolution* 89:129–137.

Roberts, D. L., M. D. Bateman, C. V. Murray-Wallace, A. S. Carr, and P. J. Holmes. 2009. West Coast Dune Plumes: Climate Driven Contrasts in Dunefield Morphogenesis along the Western and Southern South African Coasts. *Palaeogeography, Palaeoclimatology, Palaeoecology* 271:24–38.

Roberts, Linda J. 1982. The Formulation and Application of a Technique, Based on Phalanges, for Discriminating the Sex of Plains Bison (*Bison bison bison*). Unpublished Master of Arts thesis, Department of Anthropology, University of Manitoba, Winnipeg.

Roberts, Michael F. 1970. Late Glacial and Postglacial Environments in Southeastern Wyoming. *Palaeogeography, Palaeoclimatology, Palaeoecology* 8:5–17.

Rodríguez, Jesús. 1999. Use of Cenograms in Mammalian Palaeoecology: A Critical Review. *Lethaia* 32:331–347.

Rogers, Raymond R., David A. Eberth, and Anthony R. Fiorillo (editors). 2007. *Bonebeds: Genesis, Analysis, and Paleobiological Significance*. University of Chicago Press.

Romano, Marco. 2015. Reviewing the Term Uniformitarianism in Modern Earth Sciences. *Earth-Science Reviews* 148:65–76.

Romer, Alfred S. 1961. Palaeozoological Evidence of Climate: (I) Vertebrates. In *Descriptive Paleoclimatology*, edited by A. E. M. Nairn, pp. 183–206. Wiley-Interscience, New York.

Rosenzweig, Michael L. 1968. The Strategy of Body Size in Mammalian Carnivores. *American Midland Naturalist* 80:299–315.

1995. *Species Diversity in Space and Time*. Cambridge University Press.

Rosner, Hillary. 2015. Pine Beetle Epidemic: The Bug That's Eating the Woods. *National Geographic* (April).

Rosvold, Jørgen, Reidar Andersen, John D. C. Linnell, and Anne Karin Jufthammer. 2013. Cervids in a Dynamic Northern Landscape: Holocene Changes in the Relative Abundance of Moose and Red Deer at the Limits of their Distributions. *The Holocene* 23:1143–1150.

Rowan, John, J. Tyler Faith, Y. Gebru, and John G. Fleagle. 2015. Taxonomy and Paleoecology of Fossil Bovidae (Mammalia, Artiodactyla) from the Kibish Formation, Southern

Ethiopia: Implications for Dietary Change, Biogeography, and the Structure of Living Bovid Faunas of East Africa. *Palaeogeography, Palaeoclimatology, Palaeoecology* 420:210–222.

Rowe, Rebecca J., and Rebecca C. Terry. 2014. Small Mammal Responses to Environmental Change: Integrating Past and Present Dynamics. *Journal of Mammalogy* 95:1157–1174.

Roy, Kaustuv, James W. Valentine, David Jablonski, and Susan M. Kidwell. 1996. Scales of Climatic Variability and Time Averaging in Pleistocene Biotas: Implications for Ecology and Evolution. *Trends in Ecology and Evolution* 11:458–463.

Ruddiman, William F. 2013. The Anthropocene. *Annual Review of Earth and Planetary Sciences* 41:45–68.

Rudwick, Martin J. S. 1971. Uniformity and Progression: Reflections on the Structure of Biological Theory in the Age of Lyell. In *Perspectives in the History of Science and Technology*, edited by D. H. D. Roller, pp. 209–227. Oklahoma State University Press, Norman.

1978. Charles Lyell's Dream of a Statistical Paleontology. *Paleontology* 21:225–244.

1985. *The Meaning of Fossils: Episodes in the History of Palaeontology*, second edition. University of Chicago Press.

1996. Cuvier and Brongniart, William Smith, and the Reconstruction of Geohistory. *Earth Sciences History* 15:25–36.

1997. *Georges Cuvier, Fossil Bone, and Geological Catastrophes: New Translations and Interpretations of the Primary Texts*. University of Chicago Press.

Ruff, Christopher B. 1991. Climate and Body Shape in Hominid Evolution. *Journal of Human Evolution* 21:81–105.

Ruff, Christopher B. 1994. Morphological Adaptation to Climate in Modern and Fossil Hominids. *Yearbook of Physical Anthropology* 37:65–107.

Rull, Valentí. 2012. Palaeobiodiversity and Taxonomic Resolution: Linking Past Trends with Present Patterns. *Journal of Biogeography* 39:1005–1006.

Running, Steven W., R. Nemani Ramakrishna, Faith Ann Heinsch et al. 2004. A Continuous Satellite-Derived Measure of Global Terrestrial Primary Production. *BioScience* 54:547–560.

Ryan, Alan S. 1979. Wear Striation Direction on Primate Teeth: A Scanning Electron Microscope Examination. *American Journal of Physical Anthropology* 50:155–167.

Ryder, M. L. 1992. What Are We Measuring? *Circaea* 10(2):81–82.

Rymer, L. 1978. The Use of Uniformitarianism and Analogy in Palaeocology, Particularly Pollen Analysis. In *Biology and Quaternary Environments*, edited by D. Walker and J. C. Guppy, pp. 245–257. Australian Academy of Science, Canberra.

Sachs, Harvey Maurice, T[hompson] Webb III, and D. R. Clark. 1977. Paleoecological Transfer Functions. *Annual Review of Earth and Planetary Sciences* 5:159–178.

Sanchez, Julia L. 1996. A Re-evaluation of Mimbres Faunal Subsistence. *Kiva* 61:295–307.

Sand, Håkan, Göran Cederlund, and Kjell Danell. 1995. Geographical and Latitudinal Variation in Growth Patterns and Adult Body Size of Swedish Moose (*Alces alces*). *Oecologia* 102:433–442.

Sanders, Howard L. 1968. Marine Benthic Diversity: A Comparative Study. *American Naturalist* 102:243–282.

Sandford, Mary K. (editor). 1993. *Investigations of Ancient Human Tissue: Chemical Analyses in Anthropology*. Gordon and Breach, Langhorne, PA.

Sandweiss, Daniel H., and Alice R. Kelley. 2012. Archaeological Contributions to Climate Change Research: The Archaeological Record as a Paleoclimatic and Paleoenvironmental Archive. *Annual Review of Anthropology* 41:371–391.

Sankaran, Mahesh, Niall P. Hanan, Robert J. Scholes et al. 2005. Determinants of Woody Cover in African Savannas. *Nature* 438:846–849.

Sauer, John, Daniel Niven, James Hines et al. 2017. *The North American Breeding Bird Survey, Results and Analysis 1966–2015. Version 2.07.2017.* USGS Patuxent Wildlife Research Center, Laurel, MD.

Sayre, Nathan F. 2005. Ecological and Geographical Scale: Parallels and Potential for Integration. *Progress in Human Geography* 29:276–290.

Scheffer, Marten, Stephen R. Carpenter, Timothy M. Lenton et al. 2012. Anticipating Critical Transitions. *Science* 338:344–348.

Schmidt, Niels M., and Per M. Jensen. 2003. Changes in Mammalian Body Length over 175 Years: Adaptations to a Fragmented Landscape? *Conservation Ecology* 7:6 (online).

Schmidt, Niels M., and Per M. Jensen. 2005. Concomitant Patterns in Avian and Mammalian Body Length Changes in Denmark. *Ecology and Society* 10:5 (online).

Schmitt, Dave N., and Karen D. Lupo. 2012. The Bonneville Estates Rockshelter Rodent Fauna and Changes in Late Pleistocene–Middle Holocene Climates and Biogeography in the Northern Bonneville Basin, USA. *Quaternary Research* 78:95–102.

Schmitt, Dave N., David B. Madsen, and Karen D. Lupo. 2002. Small-Mammal Data on Early and Middle Holocene Climates and Biotic Communities in the Bonneville Basin, USA. *Quaternary Research* 58:255–260.

Schneider, David C. 2001. The Rise of the Concept of Scale in Ecology. *BioScience* 51:545–553.

Schoeninger, Margaret J. 1995. Stable Isotope Studies in Human Evolution. *Evolutionary Anthropology* 4:83–98.

Schoeninger, Margaret J., and Michael J. DeNiro. 1984. Nitrogen and Carbon Isotopic Composition of Bone Collagen from Marine and Terrestrial Animals. *Geochimica et Cosmochimica Acta* 48:625–639.

Scholtz, Anton. 1986. Palynological and Palaeobotanical Studies in the Southern Cape. Unpublished Master of Arts thesis, University of Stellenbosch, South Africa.

Schoonmaker, Peter K. 1998. Paleoecological Perspectives on Ecological Scale. In *Ecological Scale: Theory and Applications*, edited by David L. Peterson and V. Thomas Parker, pp. 79–103. Columbia University Press, New York.

Schoville, Benjamin J., and Erik Otárola-Castillo. 2014. A Model of Hunter-Gatherer Skeletal Element Transport: The Effect of Prey Body Size, Carriers, and Distance. *Journal of Human Evolution* 73:1–14.

Schubert, Blaine W., Peter S. Ungar, Matt Sponheimer, and Kaye E. Reed. 2006. Microwear Evidence for Plio-Pleistocene Bovid Diets from Makapansgat Limeworks Cave, South Africa. *Palaeogeography, Palaeoclimatology, Palaeoecology* 241:301–319.

Schultz, Gerald E. 1967. Four Superimposed Late-Pleistocene Vertebrate Faunas from Southwest Kansas. In *Pleistocene Extinctions: The Search for a Cause*, edited by P[aul] S. Martin and H. E. Wright, Jr., pp. 321–336. Yale University Press, New Haven, CT.

    1969. *Geology and Paleontology of a Late Pleistocene Basin in Southwest Kansas.* Special Paper 105. Geological Society of America, Boulder, CO.

    2010. Pleistocene (Irvingtonian, Cudahyan) Vertebrates from the Texas Panhandle, and their Geographic and Paleoecologic Significance. *Quaternary International* 217: 195–224.

Schweitzer, Franz R., and M. L. Wilson. 1982. Byneskranskop 1, a Late Quaternary Living Site in the Southern Cape Province, South Africa. *Annals of the South African Museum* 88:1–203.

Scott, G. H. 1963. Uniformitarianism, the Uniformity of Nature, and Paleoecology. *New Zealand Journal of Geology and Geophysics* 6:510–527.

Scott, Jessica R. 2012a. Dental Microwear Texture Analysis of Extant African Bovidae. *Mammalia* 76:157–174.

2012b. Dental Microwear Texture Analysis of Pliocene Bovids from Four Early Hominin Fossil Sites in Eastern Africa: Implications for Paleoenvironmental Dynamics and Human Evolution. Unpublished Ph.D. thesis, Department of Anthropology, University of Arkansas, Fayettefille.

Scott, Robert S., and W. Andrew Barr. 2014. Ecomorphology and Phylogenetic Risk: Implications for Habitat Reconstruction Using Fossil Bovids. *Journal of Human Evolution* 73:47–57.

Scott, Robert S., John Kappelman, and Jay Kelley. 1999. The Paleoenvironment of *Sivapithecus parvada*. *Journal of Human Evolution* 36:245–274.

Scott, Robert S., Peter S. Ungar, Torbjorn S. Bergstrom et al. 2005. Dental Microwear Texture Analysis Shows Within-Species Diet Variability in Fossil Hominins. *Nature* 436:693–695.

Scott, Robert S., Peter S. Ungar, Torbjorn S. Bergstrom et al. 2006. Dental Microwear Texture Analysis: Technical Considerations. *Journal of Human Evolution* 51:339–349.

Scott, Robert S., Mark F. Teaford, and Peter S. Ungar. 2012. Dental Microwear Texture and Anthropoid Diets. *American Journal of Physical Anthropology* 147:551–579.

Sealy, Judith, Julia Lee-Thorp, Emma Loftus, J. Tyler Faith, and Curtis W. Marean. 2016. Late Quaternary Environmental Change in the Southern Cape, South Africa, from Stable Carbon and Oxygen Isotopes in Faunal Tooth Enamel from Boomplaas Cave. *Journal of Quaternary Science* 31:919–927.

Secord, Ross, Jonathan I. Bloch, Stephen G. B. Chester et al. 2012. Evolution of the Earliest Horses Driven by Climate Change in the Paleocene–Eocene Thermal Maximum. *Science* 335:959–962.

Seddon, Alistair W. R., Anson W. Mackay, Ambroise G. Baker et al. 2014. Looking Forward through the Past: Identification of 50 Priority Research Questions in Palaeoecology. *Journal of Ecology* 102:256–267.

Semken, Holmes A., Jr. 1966. Stratigraphy and Paleontology of the McPherson Equus Beds (Sandahl Local Fauna), McPherson County, Kansas. *Contributions from the Museum of Paleontology* 20:121–178. University of Michigan, Ann Arbor.

1980. Holocene Climatic Reconstructions Derived from the Three Micromammal Bearing Cultural Horizons of the Cherokee Sewer Site, Northwestern Iowa. In *The Cherokee Excavations*, edited by Duane C. Anderson and Holmes A. Semken, Jr., pp. 67–99. Academic Press, New York.

1983. Holocene Mammalian Biogeography and Climatic Change in the Eastern and Central United States. In *Late-Quaternary Environments of the United States*, vol. II: *The Holocene*, edited by H. E. Wright, Jr., pp. 182–207. University of Minnesota Press, Minneapolis.

1988. Environmental Interpretations of the "Disharmonous" Late Wisconsinan Biome of Southeastern North America. In *Late Pleistocene and Early Holocene Paleoecology and Archeology of the Eastern Great Lakes Region*, edited by Richard S. Laub, Norton G. Miller, and David W. Steadman, pp. 185–194. Bulletin 33. Buffalo Society of Natural Sciences, Buffalo, NY.

Semken, Holmes A., Jr., and Russell W. Graham. 1987. Summary: Environmental Analysis and Plains Archaeology. In *Late Quaternary Mammalian Biogeography and Environments of the Great Plains and Prairies*, edited by Russell W. Graham, Holmes A. Semken, Jr., and Mary Ann Graham, pp. 474–480. Scientific Papers 22. Illinois State Museum, Springfield.

1996. Paleoecologic and Taphonomic Patterns Derived from Correspondence Analysis of Zooarchaeological and Paleontological Faunal Samples, a Case Study from the North American Prairie/Forest Ecotone. *Acta Zoologica Cracoviensia* 39:477–490.

Semken, Holmes A., Jr., and Steven C. Wallace. 2002. Key to Arvicoline ("Microtine" Rodents) and Arvicoline-Like Lower First Molars Recovered from Late Wisconsinan and Holocene Archaeological and Palaeontological Sites in Eastern North America. *Journal of Archaeological Science* 29:23–31.

Semken, Holmes A., Jr., Russell W. Graham, and Thomas W. Stafford, Jr. 2010. AMS ^{14}C Analysis of Late Pleistocene Non-Analog Faunal Components from 21 Cave Deposits in Southeastern North America. *Quaternary International* 217:240–255.

Semprebon, Gina M., Florent Rivals, Nikos Solounias, and Richard C. Hulbert, Jr. 2016a. Paleodietary Reconstruction of Fossil Horses from the Eocene through Pleistocene of North America. *Palaeogeography, Palaeoclimatology, Palaeoecology* 442:110–127.

Semprebon, Gina M., Deng Tao, Jelena Hasjanova, and Nikos Solounias. 2016b. An Examination of the Dietary Habits of *Platybelodon grangeri* from the Linxia Basin of China: Evidence from Dental Microwear of Molar Teeth and Tusks. *Palaeogeography, Palaeoclimatology, Palaeoecology* 457:109–116.

Sénégas, F., and J. F. Thackeray. 2008. Temperature Indices Based on Relative Abundances of Rodent Taxa in South Africa Plio-Pleistocene Assemblages. *Annals of the Transvaal Museum* 45:143–144.

Sepkoski, David. 2012. *Rereading the Fossil Record: The Growth of Paleobiology as an Evolutionary Discipline*. University of Chicago Press.

Sepkoski, J. John, Jr. 1988. Alpha, Beta, or Gamma: Where Does All the Diversity Go? *Paleobiology* 14:221–234.

Sesé Benito, Carmen. 1994. Paleoclimatical Interpretation of the Quaternary Small Mammals of Spain. *Geobios* 27:753–767.

Sexton, Jason P., Patrick J. McIntyre, Amy L. Angert, and Kevin J. Rice. 2009. Evolution and Ecology of Range Limits. *Annual Review of Ecology, Evolution and Systematics* 40:415–436.

Shabel, Alan B., Anthony D. Barnosky, Tonya Van Leuvan, Faysal Bibi, and Matthew H. Kaplane. 2004. Irvingtonian Mammals from the Badger Room in Porcupine Cave: Age, Taphonomy, Climate, and Ecology. In *Biodiversity Response to Climate Change in the Middle Pleistocene: The Porcupine Cave Fauna from Colorado*, edited by Anthony D. Barnosky, pp. 295–317. University of California Press, Berkeley.

Shaffer, Brian S., and Julia L. J. Sanchez. 1994. Comparison of 1/8"- and 1/4"- Mesh Recovery of Controlled Samples of Small-to-Medium-Sized Mammals. *American Antiquity* 59:525–530.

Shaffer, Brian S., and Christopher P. Schick. 1995. Environment and Animal Procurement by the Mogollon of the Southwest. *North American Archaeologist* 16:117–132.

Shapiro, Amy E., Virek V. Venkataraman, Nga Nguyen, and Peter J. Fashing. 2016. Dietary Ecology of Fossil *Theropithecus*: Inferences from Dental Microwear Textures of Extant Geladas from Ecologically Diverse Sites. *Journal of Human Evolution* 99:1–9.

Sharp, Zachary. 2007. *Principles of Stable Isotope Geochemistry*. Pearson Prentice Hall, Upper Saddle River, NJ.

Sheets, H. David., and Charles E. Mitchell. 2001. Why the Null Matters: Statistical Tests, Random Walks, and Evolution. *Genetica* 112–113:105–125.

Shelford, Victor E. 1913. *Animal Communities in Temperate America.* University of Chicago Press.
1931. Some Concepts of Bioecology. *Ecology* 12:455–467.

Sheridan, Jennifer A., and David Bickford. 2011. Shrinking Body Size as an Ecological Response to Climate Change. *Nature Climate Change* 1:401–406.

Shi, Guang R. 1993. Multivariate Data Analysis in Palaeoecology and Palaeobiogeography: A Review. *Palaeogeography, Palaeoclimatology, Palaeoecology* 105:199–234.

Shipman, Pat, and J. M. Harris. 1988. Habitat Preference and Paleoecology of *Australopithecus boisei* in Eastern Africa. In *Evolutionary History of the "Robust" Australopithecines*, edited by Frederick E. Grine, pp. 343–381. Aldine De Gruyter, New York.

Sholander, P. F. 1955. Evolution of Climatic Adaptation in Homeotherms. *Evolution* 9:15–26.

Shotwell, J. Arnold. 1955. An Approach to the Paleoecology of Mammals. *Ecology* 36:327–337.
1958. Inter-Community Relationships in Hemphillian (Mid-Pliocene) Mammals. *Ecology* 39:271–282.
1963. The Juntura Basin: Studies in Earth History and Paleoecology. *Transactions of the American Philosophical Society* 53:1–77.

Shurin, Jonathan B., and Emily G. Allen. 2001. Effects of Competition, Predation, and Dispersal on Species Richness at Local and Regional Scales. *American Naturalist* 158:624–637.

Simberloff, Daniel. 1998. Flagships, Umbrellas, and Keystones: Is Single-Species Management Passé in the Landscape Era? *Biological Conservation* 83:247–257.

Simpson, George G. 1936. Data on the Relationships of Local and Continental Mammalian Faunas. *Journal of Paleontology* 10:410–414.
1937. *The Fort Union of the Crazy Mountain Field, Montana, and its Mammalian Faunas.* United States National Museum Bulletin 169. Smithsonian Institution, Washington, DC.
1940. Mammals and Land Bridges. *Journal of the Washington Academy of Science* 30:137–163.
1942. The Beginnings of Vertebrate Paleontology in North America. *Proceedings of the American Philosophical Society* 81:130–188.
1943a. Criteria for Genera, Species, and Subspecies in Zoology and Paleozoology. *Annals of the New York Academy of Sciences* 44:145–178.
1943b. Mammals and the Nature of Continents. *American Journal of Science* 241:1–31.
1947. Holarctic Mammalian Faunas and Continental Relationship during the Cenozoic. *Geological Society of America Bulletin* 58:613–688.
1952. Probabilities of Dispersal in Geologic Time. *Bulletin of the American Museum of Natural History* 99:163–176.
1953. *Life of the Past: An Introduction to Paleontology.* Yale University Press, New Haven, CT.
1964. Species Density of North American Recent Mammals. *Systematic Zoology* 13:57–73.
1970. Uniformitarianism: An Inquiry into Principle, Theory, and Method in Geohistory and Biohistory. In *Essays in Evolution and Genetics*, edited by M. K. Hecht and W. C. Steere, pp. 43–96. Appleton-Century-Crofts, New York.

Simpson, George G., Anna Roe, and Richard C. Lewontin. 1960. *Quantitative Zoology,* revised edition. Harcourt, Brace, New York.

Sinninghe Damsté, Jaap S., Dirk Verschuren, Jort Osssebaar et al. 2011. A 25,000–Year Record of Climate-Induced Changes in Lowland Vegetation of Eastern Equatorial Africa Revealed by Stable Carbon-Isotopic Composition of Fossil Plant Leaf Waxes. *Earth and Planetary Science Letters* 302:236–246.

Skead, C. J. 2011. *Historical Incidence of the Large Land Mammals in the Broader Western and Northern Cape.* Centre for African Conservation Ecology, Nelson Mandela Metropolitan University, Port Elizabeth.

Skinner, J. D., and Christian T. Chimimba. 2005. *The Mammals of the Southern African Subregion*. Cambridge University Press.

Slaughter, Bob H. 1966. The Moore Pit Local Fauna: Pleistocene of Texas. *Journal of Paleontology* 40:78–91.

——— 1967. Animal Ranges as a Clue to Late-Pleistocene Extinction. In *Pleistocene Extinctions: The Search for a Cause*, edited by P[aul] S. Martin and H. E. Wright, Jr., pp. 155–167. Yale University Press, New Haven, CT.

Smith, Benjamin, and J. Bastow Wilson. 1996. A Consumer's Guide to Evenness Indices. *Oikos* 76:70–82.

Smith, C. Lavett. 1954. Pleistocene Fishes of the Berends Fauna of Beaver County, Oklahoma. *Copeia* 1954:282–289.

Smith, Douglas W., Rolf O. Peterson, and Douglas B. Houston. 2003. Yellowstone after Wolves. *BioScience* 53:330–340.

Smith, Felisa A., and Julio L. Betancourt. 1998. Response of Bushy-Tailed Woodrats (*Neotoma cinerea*) to Late Quaternary Climatic Change in the Colorado Plateau. *Quaternary Research* 50:1–11.

——— 2003. The Effect of Holocene Temperature Fluctuations on the Evolution and Ecology of *Neotoma* (Woodrats) in Idaho and Northwestern Utah. *Quaternary Research* 59:160–171.

Smith, Felisa A., and S. Kathleen Lyons. 2011. How Big Should a Mammal Be? A Macroecological Look at Mammalian Body Size over Space and Time. *Philosophical Transactions of the Royal Society B* 366:2364–2378.

Smith, Felisa A., Julio L. Betancourt, and James H. Brown. 1995. Evolution of Body Size in the Woodrat over the Past 25,000 Years of Climate Change. *Science* 270:2012–2014.

Smith, Felisa A., Hillary Browning, and Ursula L. Sheperd. 1998. The Influence of Climate Change on the Body Mass of Woodrats *Neotoma* in an Arid Region of New Mexico, USA. *Ecography* 21:140–148.

Smith, Felisa A., Dolly L. Crawford, Larisa E. Harding et al. 2009. A Tale of Two Species: Extirpation and Range Expansion during the Late Quaternary in an Extreme Environment. *Global and Planetary Change* 65:122–133.

Smith, Felisa A., John L. Gittleman, and James H. Brown (editors). 2014. *Foundations of Macroecology: Classic Papers with Commentaries*. University of Chicago Press.

Smith, James P. 1919. Climatic Relations of the Tertiary and Quaternary Faunas of the California Region. *Proceedings of the California Academy of Sciences* (fourth series) 9(4):123–173.

Smith, Philip W. 1957. An Analysis of Post-Wisconsin Biogeography of the Prairie Peninsula Region Based on Distributional Phenomena among Terrestrial Vertebrate Populations. *Ecology* 38:205–218.

Soberón, Jorge. 2007. Grinnellian and Eltonian Niches and Geographic Distributions of Species. *Ecology Letters* 10:1115–1123.

Soberón, Jorge, and B. Arroyo-Peña. 2017. Are Fundamental Niches Larger than the Realized? Testing a 50-Year-Old Prediction by Hutchinson. *PLoS One* 12(4): e0175138.

Soberón, Jorge, and Miguel Nakamura. 2009. Niches and Distributional Areas: Concepts, Methods, and Assumptions. *Proceedings of the National Academy of Sciences USA* 106 (Supplement 2):19644–19650.

Soberón, Jorge, and A. Townsend Peterson. 2005. Interpretation of Models of Fundamental Ecological Niches and Species' Distributional Areas. *Biodiversity Informatics* 2:1–10.

Socha, Pawel. 2014. Rodent Palaeofaunas from Bisnick Cave (Kraków–Czestochowa Upland, Poland): Palaeoecological, Palaeoclimatic and Biostratigraphic Reconstruction. *Quaternary International* 326–327:64–81.

Sokal, Robert R., and F. James Rohlf. 1995. *Biometry*, third edition. W. H. Freeman, New York.

Soligo, Christophe. 2002. Primatology, Paleoecology, and a New Method for Assessing Taphonomic Bias in Fossil Assemblages. *Evolutionary Anthropology* (Supplement) 1:24–27.

Soligo, Christophe and Peter Andrews. 2005. Taphonomic, Taxonomic and Historical Bias of Faunal Structure in Early Hominin Localities. *Journal of Human Evolution* 49:206–229.

Solounias, Nikos, and Sonja M. C. Moelleken. 1992a. Dietary Adaptations of Two Goat Ancestors and Evolutionary Considerations. *Geobios* 25:797–809.

———. 1992b. Tooth Microwear Analysis of *Eotragus sansaniensis* (Mammalia: Ruminantia), One of the Oldest Known Bovids. *Journal of Vertebrate Paleontology* 12:113–121.

Solounias, Nikos, Muhammad Tariq, Sukuan Hou, Melinda Danowitz, and Mary Harrison. 2014. A New Method of Tooth Mesowear and a Test of It on Domestic Goats. *Annales Zoologici Fennici* 51:111–118.

Southwood, T. R. E. 1987. The Concept and Nature of the Community. In *Organization of Communities: Past and Present*, edited by James H. R. Gee and Paul S. Giller, pp. 3–27. Blackwell Scientific, Oxford.

Sparks, B. W. 1961. The Ecological Interpretation of Quaternary Non-Marine Mollusca. *Proceedings of the Linnean Society of London* 172:71–80.

Spellerberg, Ian F., and Peter J. Fedor. 2003. A Tribute to Claude Shannon (1916–2001) and a Plea for More Rigorous Use of Species Richness, Species Diversity and the "Shannon-Wiener" Index. *Global Ecology and Biogeography* 12:177–179.

Spencer, Lillian M. 1995. Morphological Correlates of Dietary Resource Partitioning in the African Bovidae. *Journal of Mammalogy* 76:448–471.

———. 1997. Dietary Adaptations of Plio-Pleistocene Bovidae: Implications for Hominid Habitat Use. *Journal of Human Evolution* 32:201–228.

Sponheimer, Matt, and Julia A. Lee-Thorp. 2001. The Oxygen Isotope Composition of Mammalian Enamel Carbonate from Morea Estate, South Africa. *Oecologia* 126:153–157.

———. 2003. Using Carbon Isotope Data of Fossil Bovid Communities for Palaeoenvironmental Reconstruction. *South African Journal of Science* 99:273–275.

———. 2006. Enamel Diagenesis at South African Australopith Sites: Implications for Paleoecological Reconstruction with Trace Elements. *Geochimica et Cosmochimica Acta* 70:1644–1654.

Sponheimer, Matt, Kaye Reed, and Julia A. Lee-Thorp. 1999. Combining Isotopic and Ecomorphological Data to Refine Bovid Paleodietary Reconstruction: A Case Study from the Makapansgat Limeworks Hominin Locality. *Journal of Human Evolution* 36:705–718.

Sponheimer, Matt, Julia A. Lee-Thorp, Darryl J. DeRuiter et al. 2003. Diets of Southern African Bovidae: Stable Isotope Evidence. *Journal of Mammalogy* 84:471–479.

Staff, George M., Eric N. Powell, Robert J. Stanton, and Hays Cummins. 1985. Biomass – Is It a Useful Tool in Paleocommunity Reconstruction? *Lethaia* 18:209–232.

Stafford, Thomas W., Holmes A. Semken, Jr., Russell W. Graham et al. 1999. First Accelerator Mass Spectrometry [14]C Dates Documenting Contemporaneity of Nonanalog Species in Late Pleistocene Mammal Communities. *Geology* 27:903–906.

Stahl, Peter W. 2005. An Exploratory Osteological Study of the Muscovy Duck (*Cairina moschata*) (Aves: Anatidae) with Implications for Neotropical Archaeology. *Journal of Archaeological Science* 32:915–929.

Staver, A. Carla, Sally Archibald, and Simon A. Levin. 2011. The Global Extent and Determinants of Savanna and Forest as Alternative Biome States. *Science* 334:230–232.

Steele, Teresa E. 2003. Using Mortality Profiles to Infer Behavior in the Fossil Record. *Journal of Mammalogy* 84:418–430.

2005. Comparing Methods for Analyzing Mortality Profiles in Zooarchaeological and Paleontological Samples. *International Journal of Osteoarchaeology* 15:404–420.

Stegner, Mary Allison. 2015. The Mescal Cave Fauna (San Bernadino County, California) and Testing Assumptions of Habitat Fidelity in the Quaternary Fossil Record. *Quaternary Research* 83:582–587.

Steinhauser, F. 1979. *Climatic Atlas of North and Central America: I, Maps of Mean Temperature and Precipitation.* World Meteorological Organization, UNESCO, and Cartographia, Geneva.

Stephens, John J. 1960. Stratigraphy and Paleontology of a Late Pleistocene Basin, Harper County, Oklahoma. *Geological Society of America Bulletin* 71:1575–1702.

Sterling, K. B. 1989. Builders of the US Biological Survey, 1885–1930. *Journal of Forest History* 33:180–187.

Stewart, John R. 2005. The Use of Modern Geographical Ranges in the Identification of Archaeological Bird Remains. *Documenta Archaeobiologiae* 3:43–54.

2009. The Quaternary Fossil Record as a Source of Data for Evidence-Based Conservation: Is the Past the Key to the Future? In *Holocene Extinctions*, edited by Samuel T. Turvey, pp. 249–261. Oxford University Press.

Stiner, Mary C. 1990. The Use of Mortality Patterns in Archaeological Studies of Hominid Predatory Adaptations. *Journal of Anthropological Archaeology* 9:305–351.

Stiner, Mary C., Natalie D. Munro, and Todd A. Surovell. 2000. The Tortoise and the Hare: Small-Game Use, the Broad Spectrum Revolution, and Paleolithic Demography. *Current Anthropology* 41:39–73.

Stock, Chester. 1929. A Census of the Pleistocene Mammals of Rancho La Brea, Based on the Collections of the Los Angeles Museum. *Journal of Mammalogy* 10:281–289.

Stonehouse, Bernard. 1997. Animal Responses to Climate. In *Applied Climatology: Principles and Practice*, edited by Russell D. Thompson and Allen Parry, pp. 141–152. Routledge, London.

Storer, John E. 2003. Environments of Pleistocene Beringia: Analysis of Faunal Composition Using Cenograms. In *Advances in Mammoth Research*, edited by Jelle W. F. Reumer, John De Vos, and Dick Mol, pp. 405–414. Deinsea 9. Natural History Museum, Rotterdam.

Street-Perrott, F. Alayne, and R. A. Perrott. 1993. Holocene Vegetation, Lake Levels and Climate of Africa. In *Global Climates since the Last Glacial Maximum*, edited by H. E. Wright Jr., J. E. Kutzbach, T. Webb III, W. F. Ruddimann, F. A. Street-Perrott, and P. J. Bartlein, pp. 318–356. University of Minnesota Press, Minneapolis.

Stynder, Deano D. 2009. The Diets of Ungulates from the Hominid Fossil-Bearing Site of Elandsfontein, Western Cape, South Africa. *Quaternary Research* 71:62–70.

Su, Denise F., and Terry Harrison. 2007. The Paleoecology of the Upper Laetolil Beds at Laetoli: A Reconsideration of the Large Mammal Evidence. In *Hominin Environments in the East African Pliocene: An Assessment of the Faunal Evidence*, edited by R[ené] Bobe, Z[eresenay] Alemseged, and A[nna] K. Behrensmeyer, pp. 279–313. Springer, Dordrecht.

Sukselainen, Leena, Mikael Fortelius, and Terry Harrison. 2015. Co-occurrence of Pliopithecoid and Hominoid Primates in the Fossil Record: An Ecometric Analysis. *Journal of Human Evolution* 84:25–41.

Svenning, Jens-Christian, Wolf L. Eiserhardt, Signe Normand, Alejandro Ordonez, and Brody Sandel. 2015. The Influence of Paleoclimate on Present-Day Patterns in Biodiversity and Ecosystems. *Annual Review of Ecology, Evolution, and Systematics* 46:551–572.

Swihart, Robert K., Thomas M. Gehring, Mary Beth Kolozsvary, and Thomas E. Nupp. 2003. Responses of "Resistant" Vertebrates to Habitat Loss and Fragmentation: The Importance of Niche Breadth and Range Boundary. *Diversity and Distributions* 9:1–18.

Taber, Richard D., Kenneth Raedeke, and Donald A. McCaughran. 1982. Population Characteristics. In *Elk of North America: Ecology and Management*, edited by Jack W. Thomas and Dale E. Toweill, pp. 279–298. Stackpole Books, Harrisburg, PA.

Talma, A. S., and John C. Vogel. 1992. Late Quaternary Paleotemperatures Derived from a Speleothem from Cango Caves, Cape Province, South Africa. *Quaternary Research* 37:203–213

Taylor, Dwight W. 1965. The Study of Pleistocene Nonmarine Mollusks in North America. In *The Quaternary of the United States*, edited by H. E. Wright, Jr., and D. G. Frey, pp. 597–611. Princeton University Press.

Taylor, Lucy A., Thomas M. Kaiser, Christoph Schwitzer et al. 2013. Detecting Inter-Cusp and Inter-Tooth Wear Patterns in Rhinocerotids. *PLoS ONE* 8:e80921.

Tchernov, Eitan. 1968. *Succession of Rodent Faunas during the Upper Pleistocene of Israel*. Paul Parey, Hamburg.

   1975. Rodent Faunas and Environmental Changes in the Pleistocene of Israel. In *Rodents in Desert Environments*, edited by I. Prakash and P. K. Gosh, pp. 331–362. Junk, The Hague.

   1979. Polymorphism, Size Trends and Pleistocene Paleoclimatic Response of the Subgenus *Sylvaemus* (Mammalia: Rodentia) in Israel. *Israel Journal of Zoology* 28:131–159.

   1982. Faunal Responses to Environmental Changes in the Eastern Mediterranean during the Last 20,000 Years. In *Palaeoclimates, Palaeoenvironments and Human Communities in the Eastern Mediterranean Region in Later Prehistory*, edited by John L. Bintliff and Willem Van Zeist, pp. 105–129. British Archaeological Reports, International Series 133. BAR, Oxford.

Tchernov, Eitan, and Liora Kolska Horwitz. 1991. Body Size Diminution under Domestication: Unconscious Selection in Primeval Domesticates. *Journal of Anthropological Archaeology* 10:54–75.

Teaford, Mark F. 1991. Dental Microwear: What Can It Tell Us about Diet and Dental Function? In *Advances in Dental Anthropology*, edited by M. A. Kelley and C. S. Larsen, pp. 341–356. Wiley-Liss, New York.

   1994. Dental Microwear and Dental Function. *Evolutionary Anthropology* 3:17–30.

   2006. What Do We Know and Not Know about Dental Microwear and Diet? In *Evolution of the Human Diet: The Known, the Unknown, and the Unknowable*, edited by Peter Ungar, pp. 106–132. Oxford University Press.

Teaford, Mark F., and Alan Walker. 1984. Quantitative Differences in Dental Microwear between Primate Species with Different Diets and a Comment on the Presumed Diet of *Sivapithecus. American Journal of Physical Anthropology* 64:191–200.

Teaford, Mark F., and Ordean J. Oyen. 1989. *In Vivo* and *In Vitro* Turnover in Dental Microwear. *American Journal of Physical Anthropology* 80:447–460.

Teaford, Mark F., Peter Ungar, and Frederic E. Grine. 2013. Dental Microwear and Paleoecology. In *Early Hominin Paleoecology*, edited by Matt Sponheimer, Julia A. Lee-Thorp, Kaye E. Reed, and Peter Ungar, pp. 251–279. University Press of Colorado, Boulder.

Tedford, Richard H. 1970. Principles and Practices of Mammalian Geochronology in North America. In *Proceedings of the North American Paleontological Convention*, edited by Ellis L. Yochelson, pp. 666–703. Allen Press, Lawrence, KS.

Telford, R. J., C. Andersson, H. J. B. Birks, and S. Juggins. 2004. Biases in the Estimation of Transfer Function Prediction Errors. *Paleoceanography* 19:PA4014.

Telford, R. J., and H. J. B. Birks. 2009. Evaluation of Transfer Functions in Spatially Structured Environments. *Quaternary Science Reviews* 28:1309–1316.

Teplitsky, Céline, James A. Mills, Jussi S. Alho, John W. Yarall, and Juha Merilä. 2008. Bergmann's Rule and Climate Change Revisited: Disentangling Environmental and Genetic Responses in a Wild Bird Population. *Proceedings of the National Academy of Sciences USA* 105:13492–13496.

ter Braak, C. J. F. 1987. Ordination. In *Data Analysis in Community and Landscape Ecology*, edited by R. H. Jongman, C. J. F. ter Braak, and O. F. R. van Tongeren, pp. 91–173. Cambridge University Press.

ter Braak, C. J. F., S. Juggins, H. J. B. Birks, and H. Van der Voet. 1993. Weighted Averaging Partial Least Squares Regression (Wa-Pls): Definition and Comparison with Other Methods for Species–Environment Calibration. In *Multivariate Environmental Statistics*, edited by G. P. Patil and C. R. Rao, pp. 525–560. Elsevier, Amsterdam.

Terry, Rebecca C. 2007. Inferring Predator Identity from Skeletal Damage of Small-Mammal Prey Remains. *Evolutionary Ecology Research* 9:199–219.

——— 2008. Modeling the Effects of Predation, Prey Cycling, and Time Averaging on Relative Abundance in Raptor-Generated Small Mammal Death Assemblages. *Palaios* 23:402–410.

——— 2009. Paleoecology: Methods. *Encyclopedia of Life Sciences*, a0003274. John Wiley and Sons, Chichester.

——— 2010a. The Dead Do Not Lie: Using Skeletal Remains for Rapid Assessment of Historical Small-Mammal Community Baselines. *Proceedings of the Royal Society B* 277:1193–1201.

——— 2010b. On Raptors and Rodents: Testing the Ecological Fidelity and Spatiotemporal Resolution of Cave Death Assemblages. *Paleobiology* 36:137–160.

Terry, Rebecca C., Cheng (Lily) Li, and Elizabeth A. Hadly. 2011. Predicting Small-Mammal Responses to Climatic Warming: Autecology, Geographic Range, and the Holocene Fossil Record. *Global Change Biology* 17:3019–3034.

Terry, Rebecca C., and Mark Novak. 2015. Where Does the Time Go?: Mixing and the Depth-Dependent Distribution of Fossil Ages. *Geology* 43:487–490.

Terry, Rebecca C., and Rebecca J. Rowe. 2015. Energy Flow and Functional Compensation in Great Basin Small Mammals under Natural and Anthropogenic Environmental Change. *Proceedings of the National Academy of Sciences USA* 112:9656–9661.

Thackeray, J. Francis. 1987. Late Quaternary Environmental Changes Inferred from Small Mammalian Fauna, Southern Africa. *Climatic Change* 10:285–305.

——— 1990. Temperature Indices from Late Quaternary Sequences in South Africa: Comparisons with the Vostok Core. *South African Geographic Journal* 72:47–49.

——— 1992. Chronology of Late Pleistocene Deposits Associated with *Homo sapiens* at Klasies River Mouth, South Africa. *Palaeoecology of Africa and the Surrounding Islands* 23:177–191.

——— 2002. Palaeoenvironmental Change and Re-assessment of the Age of Late Pleistocene Deposits at Die Kelders Cave, South Africa. *Journal of Human Evolution* 43:749–753.

Thackeray, J. F[rancis], and D. M. Avery. 1990. A Comparison between Temperature Indices for Late Pleistocene Sequences at Klasies River and Border Cave, South Africa. *Palaeoecology of Africa* 21:311–316.

Thomas, Kenneth D., and Marcello A. Mannino. 2017. Making Numbers Count: Beyond Minimum Numbers of Individuals (MNI) for the Quantification of Mollusc Assemblages from Shell Matrix Sites. *Quaternary International* 427(A):47–58.

Thuiller, Wilfried, Sandra Lavorel, and Miguel B. Araújo. 2005. Niche Properties and Geographic Extent as Predictors of Species Sensitivity to Climate Change. *Global Ecology and Biogeography* 14:347–357.

Tiffney, Bruce H. 2008. Phylogeography, Fossils, and Northern Hemisphere Biogeography: The Role of Physiological Uniformitarianism. *Annals of the Missouri Botanical Garden* 95:135–143.

Tilman, David, and Stephen Pacala. 1993. The Maintenance of Species Richness in Plant Communities. In *Species Diversity in Ecological Communities*, edited by Robert E. Ricklefs and Dolph Schluter, pp. 13–25. University of Chicago Press.

Tipper, John C. 1979. Rarefaction and Rarefiction: The Use and Abuse of a Method in Paleoecology. *Paleobiology* 5:423–434.

Todd, Lawrence C. 1986. Determination of Sex of Bison Upper Forelimb Bones: The Humerus and Radius. *Wyoming Archaeologist* 29(1–3):109–124.

——— 1987. Taphonomy of the Horner II Bone Bed. In *The Horner Site: The Type Site of the Cody Cultural Complex*, edited by George C. Frison and Lawrence C. Todd, pp. 107–198. Academic Press, Orlando, FL.

Travouillon, K[enny] J., and S. Legendre. 2009. Using Cenograms to Investigate Gaps in Mammalian Body Mass Distributions in Australian Mammals. *Palaeogeography, Palaeoclimatology, Palaeoecology* 272:69–84.

Travouillon, Kenny J., Michael Archer, Serge Legendre, and Suzanne J. Hand. 2007. Finding the Minimum Sample Richness (MSR) for Multivariate Analyses: Implications for Palaeoecology. *Historical Biology* 19:315–320.

Travouillon, Kenny J., Gilles Escarguel, Serge Legendre, Michael Archer, and Suzanne J. Hand. 2011. The Use of MSR (Minimum Sample Richness) for Sample Assemblage Comparisons. *Paleobiology* 37:696–709.

Trinkaus, Erik. 1981. Neanderthal Limb Proportions and Cold Adaptation. In *Aspects of Human Evolution*, edited by C. B. Stringer, pp. 187–224. Taylor and Francis, London.

Tryon, Christian A., and J. Tyler Faith. 2016. A Demographic Perspective on the Middle to Later Stone Age Transition from Nasera Rockshelter, Tanzania. *Philosophical Transactions of the Royal Society B* 371:20150238.

Tryon, Christian A., J. Tyler Faith, Daniel J. Peppe et al. 2010. The Pleistocene Archaeology and Environments of the Wasiriya Beds, Rusinga Island, Kenya. *Journal of Human Evolution* 59:657–671.

Tsubamoto, Takehisa, Naoko Egi, Masanaru Takai, Chit Sein, and Maung Maung. 2005. Middle Eocene Ungulate Mammals from Myanmar: A Review with Description of New Specimens. *Acta Palaeontologica Polonica* 50:117–138.

Turvey, Samuel T., and Tim M. Blackburn. 2011. Determinants of Species Abundance in the Quaternary Vertebrate Fossil Record. *Paleobiology* 37:537–546.

Turvey, Samuel T., and Joanne H. Cooper. 2009. The Past is Another Country: Is Evidence for Prehistoric, Historical, and Present-Day Extinction Really Comparable? In *Holocene Extinctions*, edited by Samuel T. Turvey, pp. 193–212. Oxford University Press.

Ulbricht, Arlett, Lutz C. Maul, and Ellen Schulz. 2015. Can Mesowear Analysis Be Applied to Small Mammals? A Pilot-Study on Leporines and Murines. *Mammalian Biology* 80:14–20.

Ungar, Peter S. 2015. Mammalian Dental Function and Wear: A Review. *Biosurface and Biotribology* 1:25–41.

Ungar, Peter S., Christopher A. Brown, Torbjorn S. Bergstrom, and Alan Walker. 2003. Quantification of Dental Microwear by Tandem Scanning Confocal Microscopy and Scale-Sensitive Fractal Analyses. *Scanning* 25:183–193.

Ungar, Peter S., Gildas Merceron, and Robert S. Scott. 2007. Dental Microwear Texture Analysis of Varswater Bovids and Early Pliocene Paleoenvironments of Langebaanweg, Western Cape Province, South Africa. *Journal of Mammalian Evolution* 14:163–181.

Ungar, Peter S., Robert S. Scott, Jessica R. Scott, and Mark Teaford. 2008. Dental Microwear Analysis: Historical Perspectives and New Approaches. In *Technique and Application in Dental Anthropology*, edited by Joel D. Irish and Greg C. Nelson, pp. 389–425. Cambridge University Press.

Utescher, T., A. A. Bruch, B. Erdei et al. 2014. The Coexistence Approach: Theoretical Background and Practical Considerations of Using Plant Fossils for Climate Quantification. *Palaeogeography, Palaeoclimatology, Palaeoecology* 410:58–73.

Valverde, J. A. 1964. Remarques sur la Structure et l'Évolution des Communautés de Vertébrés Terrestres. I. Structure d'une Commumauté. II. Rapports entre Prédateurs et Proies. *Revue d'Écologie: La Terre et la Vie* 111:121–154.

Van Couvering, Judith A. 1980. Community Evolution in East Africa. In *Fossils in the Making: Vertebrate Taphonomy and Paleoecology*, edited by Anna K. Behrensmeyer and Andrew P. Hill, pp. 272–298. University of Chicago Press.

van der Klaauw, Cornelius J. 1948. Ecological Studies and Reviews: IV. Ecological Morphology. *Bibliotheca Biotheoretica* 4:27–111.

van der Meulen, Albert J., and Remmert Daams. 1992. Evolution of Early-Middle Miocene Rodent Faunas in Relation to Long-Term Palaeoenvironmental Changes. *Palaeogeography, Palaeoclimatology, Palaeoecology* 93:227–253.

VanPool, Todd L., and Robert D. Leonard. 2009. *Quantitative Archaeology*. Blackwell, Cambridge, MA.

Van Riper, A. Bowdoin. 1993. *Men among the Mammoths: Victorian Science and the Discovery of Human Prehistory*. University of Chicago Press.

Van Valen, L. 1973. Pattern and the Balance of Nature. *Evolutionary Theory* 1:31–49.

Van Valkenburgh, Blaire. 1987. Skeletal Indicators of Locomotor Behavior in Living and Extinct Carnivores. *Journal of Vertebrate Paleontology* 7:162–182.

    1988. Trophic Diversity in Past and Present Guilds of Large Predatory Mammals. *Paleobiology* 14:155–173.

Van Valkenburgh, Blaire, Mark F. Teaford, and Alan Walker. 1990. Molar Microwear and Diet in Large Carnivores: Inferences Concerning Diet in the Sabretooth Cat, *Smilodon fatalis*. *Journal of Zoology* (London) 222:319–340.

Varela, Sara, Jorge M. Lobo, and Joaquín Hortal. 2011. Using Species Distribution Models in Paleobiogeography: A Matter of Data, Predictors and Concepts. *Palaeogeography, Palaeoclimatology, Palaeoecology* 310:451–463.

Vartanyan, S. L., V. E. Garutt, and A. V. Sher. 1993. Holocene Dwarf Mammoths from Wrangel Island in the Siberian Arctic. *Nature* 362:337–340.

Verts, B. J., and Leslie N. Carraway. 1998. *Land Mammals of Oregon*. University of California Press, Berkeley.

Visher, S. S. 1954. *Climatic Atlas of the United States*. Harvard University Press, Cambridge, MA.

Vlok, Jan, and A. L. Schutte-Vlok. 2010. *Plants of the Klein Karoo*. Umdaus Press, Hatfield, South Africa.

Vogel, John C. 2001. Radiometric Dates for the Middle Stone Age in South Africa. In *Humanity from African Naissance to Coming Millennia: Colloquia in Human Biology and Paleoanthropology*, edited by Phillip V. Tobias, Michael A. Raath, Jacopo Maggi-Cecchi, and Gerald A. Doyle, pp. 261–268. Florence University Press.

von Humboldt, Alexander. 1850. *Views of Nature: Or Contemplations on the Sublime Phenomena of Creation; with Scientific Illustrations*. Translated by E. C. Otté and H. G. Bohn. H. G. Bohn, London.

Voorhies, Michael R. 1970. Sampling Difficulties in Reconstructing Late Tertiary Mammalian Communities. In *Proceedings of the North American Paleontological Convention*, edited by Ellis L. Yochelson, pp. 454–468. Allen Press, Lawrence, KS.

Vrba, E[lisabeth] S. 1974. Chronological and Ecological Implications of the Fossil Bovidae at the Sterkfontein Australopithecine Site. *Nature* 250:19–23.

——— 1975. Some Evidence of Chronology and Palaeoecology of Sterkfontein, Swartkrans and Kromdraai from the Fossil Bovidae. *Nature* 254:301–304.

——— 1980. The Significance of Bovid Remains as Indicators of Environment and Predation Patterns. In *Fossils in the Making: Vertebrate Taphonomy and Paleoecology*, edited by Anna K. Behrensmeyer and Andrew P. Hill, pp. 247–271. University of Chicago Press.

——— 1985. Environment and Evolution: Alternative Causes of the Temporal Distribution of Evolutionary Events. *South African Journal of Science* 81:229–236.

——— 1992. Mammals as a Key to Evolutionary Theory. *Journal of Mammalogy* 73:1–28.

——— 1995. The Fossil Record of African Antelopes (Mammalia, Bovidae) in Relation to Human Evolution and Paleoclimate. In *Paleoclimate and Evolution with Emphasis on Human Origins*, edited by Elisabeth S. Vrba, George H. Denton, Timothy C. Partridge, and Lloyd H. Burckle, pp. 385–424. Yale University Press, New Haven, CT.

Waide, R. B., M. R. Willig, C. F. Steiner et al. 1999. The Relationship between Productivity and Species Richness. *Annual Review of Ecology and Systematics* 30:257–300.

Wainwright, Peter C., and Stephen M. Reilly (editors). 1994. *Ecological Morphology: Integrative Organismal Biology*. University of Chicago.

Wake, David B., Elizabeth A. Hadly, and David D. Ackerly. 2009. Biogeography, Changing Climates, and Niche Evolution. *Proceedings of the National Academy of Sciences USA* 106:19631–19636.

Walde, Dale. 2004. Distinguishing Sex of *Bison bison bison* Using Discriminant Function Analysis. *Canadian Journal of Archaeology* 28:100–116.

Walker, Alan, H. N. Hoeck, and L. Perez. 1978. Microwear of Mammalian Teeth as an Indicator of Diet. *Science* 201:908–910.

Walker, D. 1978. Envoi. In *Biology and Quaternary Environments*, edited by D. Walker and J. C. Guppy, pp. 259–264. Australian Academy of Science, Canberra.

Walker, D., and J. C. Guppy (editors). 1978. *Biology and Quaternary Environments*. Australian Academy of Science, Canberra.

Walker, Danny N. 1982. Early Holocene Vertebrate Fauna. In *The Agate Basin Site: A Record of the Paleoindian Occupation of the Northwestern High Plains*, edited by George C. Frison and Dennis J. Stanford, pp. 274–308. Academic Press, New York.

——— 2007. Vertebrate Fauna. In *Medicine Lodge Creek: Holocene Archaeology of the Eastern Big Horn Basin, Wyoming*, vol. 1, edited by George C. Frison and Danny N. Walker, pp. 177–208. Clovis Press, Avondale, CO.

Walker, Danny N., and George C. Frison. 1982. Studies on Amerindian Dogs, 3: Prehistoric Wolf/Dog Hybrids from the Northwestern Plains. *Journal of Archaeological Science* 9:125–172.

Walter, H. 1970. *Vegetationszonen und Klima*. Eugen Ulmer, Stuttgart.

Wang, Yang, and Thure E. Cerling. 1994. A Model of Fossil Tooth and Bone Diagenesis: Implications for Paleodiet Reconstructions from Stable Isotopes. *Palaeogeography, Palaeoclimatology, Palaeoecology* 107:281–289.

Wartenburg, Daniel, Scott Ferson, and F. James Rohlf. 1987. Putting Things in Order: A Critique of Detrended Correspondence Analysis. *American Naturalist* 129:434–448.

Wasserman, David, and Donald J. Nash. 1979. Variation in Body Size, Hair Length, and Hair Density in the Deer Mouse *Peromyscus maniculatus* along an Altitudinal Gradient. *Holarctic Ecology* 2:115–118.

Watt, Cortney, Sean Mitchell, and Volker Salewski. 2010. Bergmann's Rule: A Concept Cluster? *Oikos* 119:89–100.

Weaver, Timothy D., Teresa E. Steele, and Richard G. Klein. 2011. The Abundance of Eland, Buffalo, and Wild Pigs in Middle and Later Stone Age Sites. *Journal of Human Evolution* 60:309–314.

Webb, Thompson, III, and Reid A. Bryson. 1972. Late- and Postglacial Climatic Change in the Northern Midwest, USA: Quantitative Estimates Derived from Fossil Pollen Spectra by Multivariate Statistical Analysis. *Quaternary Research* 2:70–115.

Webb, Thompson, III, and D. R. Clark. 1977. Calibrating Micropaleontological Data in Climatic Terms: A Critical Review. *Annals of the New York Academy of Sciences* 288:93–118.

Weinstock, Jaco. 1997. The Relationship between Body Size and Environment: The Case of Late Pleistocene Reindeer (*Rangifer tarandus*). *Archaeofauna* 6:123–135.

    2002. Reindeer Hunting in the Upper Paleolithic: Sex Ratios as a Reflection of Different Procurement Strategies. *Journal of Archaeological Science* 29:365–377.

Weisler, M. I. 1993. The Importance of Fish Otoliths in Pacific Island Archaeofaunal Analysis. *New Zealand Journal of Archaeology* 15:131–159.

Weissbrod, Lior, Dan Malkinson, Thomas Cucchi et al. 2014. Ancient Urban Ecology Reconstructed from Archaeozoological Remains of Small Mammals in the Near East. *Plos One* 9: e91795.

Wells, R. T. 1978. Fossil Mammals in the Reconstruction of Quaternary Environments with Examples from the Australian Fauna. In *Biology and Quaternary Environments*, edited by D. Walker and J. C. Guppy, pp. 103–124. Australian Academy of Science, Canberra.

West, Geoffrey B., James H. Brown, and Brian J. Enquis. 1997. A General Model for the Origin of Allometric Scaling Laws in Biology. *Science* 276:122–126.

Western, David, and Anna K. Behrensmeyer. 2009. Bone Assemblages Track Animal Community Structure over 40 Years in an African Savanna Ecosystem. *Science* 324:1061–1064.

White, Ethan P., S. K. Morgan Ernest, Andrew J. Kerkhoff, and Brian J. Enquist. 2007. Relationships between Body Size and Abundance in Ecology. *Trends in Ecology and Evolution* 22:323–330.

White, T. C. R. 2008. The Role of Food, Weather and Climate in Limiting the Abundance of Animals. *Biological Reviews* 83:227–248.

White, Theodore E. 1953. Studying Osteological Material. *Plains Archaeological Conference Newsletter* 6(1):58–67.

    1954. Preliminary Analysis of the Fossil Vertebrates of the Canyon Ferry Reservoir Area. *Proceedings of the United States National Museum* 103:395–438.

    1956. The Study of Osteological Material from the Plains. *American Antiquity* 21:401–404.

Whittaker, Robert J. 1975. *Communities and Ecosystems*, second edition. Macmillan, New York.

Whittaker, Robert J., Katherine J. Willis, and Richard Field. 2001. Scale and Species Richness: Towards a General Hierarchical Theory of Species Diversity. *Journal of Biogeography* 28:453–570.

Whittaker, Robert J., Brett R. Riddle, Bradford A. Hawkins, and Richard J. Ladle. 2013. The Geographical Distribution of Life and the Problem of Regionalization: 100 Years after Alfred Russell Wallace. *Journal of Biogeography* 40:2209–2214.

Whittington, H. B. 1964. Taxonomic Basis of Paleoecology. In *Approaches to Paleoecology*, edited by John Imbrie and Norman D. Newell, pp. 19–27. John Wiley and Sons, New York.

Widga, Chris. 2013. Evolution of the High Plains Paleoindian Landscape: The Paleoecology of Great Plains Faunal Assemblages. In *Paleoindian Lifeways of the Cody Complex*, edited by Edward J. Knell and Mark P. Muñiz, pp. 69–92. University of Utah Press, Salt Lake City.

Wiens, John A. 1989. Spatial Scaling in Ecology. *Functional Ecology* 3:385–397.

Wiens, John J., and Michael J. Donoghue. 2004. Historical Biogeography, Ecology and Species Richness. *Trends in Ecology and Evolution* 19:639–644.

Wiens, John J., David D. Ackerly, Andrew P. Allen et al. 2010. Niche Conservatism as an Emerging Principle in Ecology and Conservation Biology. *Ecology Letters* 13:1310–1324.

Wiens, John A., Gregory D. Hayward, Hugh D. Safford, and Catherine M. Giffen (editors). 2012a. *Historical Environmental Variation in Conservation and Natural Resource Management*. Wiley-Blackwell, Chichester.

Wiens, John A., Hugh D. Safford, Kevin McGarigal, William H. Romme, and Mary Manning. 2012b. What is the Scope of "History" in Historical Ecology? Issues of Scale in Management and Conservation. In *Historical Environmental Variation in Conservation and Natural Resource Management*, edited by John A. Wiens, Greogry D. Hayward, Hugh D. Safford, and Catherine M. Giffen, pp. 63–75. Wiley-Blackwell, Chichester.

Wiggington, John D., and F. Stephen Dobson. 1999. Environmental Influences on Geographic Variation in Body Size of Western Bobcats. *Canadian Journal of Zoology* 77:802–813.

Williams, John W., and Stephen T. Jackson. 2007. Novel Climates, No-Analog Communities, and Ecological Surprises. *Frontiers in Ecology and the Environment* 5:475–482.

Williams, John W., Bryan N. Shuman, and Thompson Webb III. 2001. Dissimilarity Analyses of Late-Quaternary Vegetation and Climate in Eastern North America. *Ecology* 82:3346–3362.

Williams, John W., Bryan Shuman, Patrick J. Bartlein, Noah S. Diffenbaugh, and Thompson Webb III. 2010. Rapid, Time-Transgressive, and Variable Responses to Early Holocene Midcontinental Drying in North America. *Geology* 38:135–138.

Williams, John W., Jessica L. Blois, Jacquelyn L. Gill et al. 2013. Model Systems for a No-Analog Future: Species Associations and Climates during the Last Deglaciation. *Annals of the New York Academy of Sciences* 1297:29–43.

Williams, Susan H., and Richard F. Kay. 2001. A Comparative Test of Adaptive Explanations for Hypsodonty in Ungulates and Rodents. *Journal of Mammalian Evolution* 8:207–229.

Willig, M. R., D. M. Kaufman, and R. D. Stevens. 2003. Latitudinal Gradients of Biodiversity: Pattern, Process, Scale, and Synthesis. *Annual Review of Ecology, Evolution, and Systematics* 34:273–309.

Wilson, Don E., and DeeAnn M. Reeder (editors) 2005. *Mammal Species of the World: A Taxonomic and Geographic Reference*, third edition. Johns Hopkins University Press, Baltimore.

Wilson, Michael [C]. 1973. The Early Historic Fauna of Southern Alberta: Some Steps to Interpretation. In *Historical Archaeology in Northwestern North America*, edited by Ronald M. Getty and Knut R. Fladmark, pp. 213–248. University of Calgary Archaeological Association, Calgary, Alberta.

1978. Archaeological Kill Site Populations and the Holocene Evolution of the Genus *Bison*. In *Bison Procurement and Utilization: A Symposium*, edited by Leslie B. Davis and Michael Wilson, pp. 9–22. *Plains Anthropologist* 23(82).

Wilson, M. V. H. 2001. Fossils as Environmental Indicators: Taphonomic Evidence. In *Paleobiology II*, edited by Derek E. G. Briggs and Peter R. Crowther, pp. 467–470. Blackwell Science, Oxford.

Wilson, R. L. 1968. Systematics and Faunal Analysis of a Lower Pliocene Vertebrate Assemblage from Trego County, Kansas. *Contributions from the Museum of Paleontology* 22:75–126. University of Michigan, Ann Arbor.

Wing, Scott L., Hans-Dieter Sues, Richard Potts, William A. DiMichele, and Anna K. Behrensmeyer. 1992. Evolutionary Paleoecology. In *Terrestrial Ecosystems through Time: Evolutionary Paleoecology of Terrestrial Plants and Animals*, edited by Anna K. Behrensmeyer, John D. Damuth, William A. DiMichele, Richard Potts, Hans-Dieter Sues, and Scott L. Wing, pp. 1–13. University of Chicago Press.

Winkler, Alisa J., and Wulf Gose. 2003. Mammalian Fauna and Paleomagnetics of the Middle Irvingtonian (Early Pleistocene) Fyllan Cave and Kitchen Door Localities, Travis County, Texas. In *Ice Age Cave Faunas of North America*, edited by Blaine W. Schubert, Jim I. Mead, and Russell W. Graham, pp. 215–261. Indiana University Press, Bloomington.

Wisz, M. S., J. Pottier, W. D. Kissling et al. 2013. The Role of Biotic Interactions in Shaping Distributions and Realised Assemblages of Species: Implications for Species Distribution Modeling. *Biological Reviews* 88:15–30.

Withnell, Charles B., and Peter S. Ungar. 2014. A Preliminary Analysis of Dental Microwear as a Proxy for Diet and Habitat in Shrews. *Mammalia* 78:409–415.

Wolff, Ronald G. 1973. Hydrodynamic Sorting and Ecology of a Pleistocene Mammalian Assemblage from California (USA). *Palaeogeography, Palaeoclimatology, Palaeoecology* 13:91–101.

——— 1975. Sampling and Sample Size in Ecological Analyses of Fossil Mammals. *Paleobiology* 1:195–204.

Wolverton, Steve. 2005. The Effects of the Hypsithermal on Prehistoric Foraging Efficiency in Missouri. *American Antiquity* 70:91–106.

——— 2008. Harvest Pressure and Environmental Carrying Capacity: An Ordinal-Scale Model of Effects on Unulate Prey. *American Antiquity* 73:179–199.

——— 2013. Data Quality in Zooarchaeological Faunal Identification. *Journal of Archaeological Method and Theory* 20:381–396.

Wolverton, Steve, and R. Lee Lyman (editors). 2012. *Conservation Biology and Applied Zooarchaeology*. University of Arizona Press, Tucson.

Wolverton, Steve, James H. Kennedy, and John D. Cornelius. 2007. A Paleozoological Perspective on White-Tailed Deer (*Odocoileus virginianus texana*) Population Density and Body Size in Central Texas. *Environmental Management* 39:545–552.

Wolverton, Steve, Michael A. Huston, James H. Kennedy, Kevin Cagle, and John D. Cornelius. 2009. Conformation to Bergmann's Rule in White-Tailed Deer Can Be Explained by Food Availability. *American Midland Naturalist* 162:403–417.

Wolverton, Steve, Lisa Nagaoka, and Torben C. Rick. 2016. *Applied Zooarchaeology: Five Case Studies*. Eliot Werner, Clinton Corners, NY.

Wood, Bernard A. 1979. An Analysis of Tooth and Body Size Relationships in Five Primate Taxa. *Folia Primatologica* 31:187–211.

Wood, David L., and Anthony D. Barnosky. 1994. Middle Pleistocene Climate Change in the Colorado Rocky Mountains Indicated by Fossil Mammals from Porcupine Cave. *Quaternary Research* 41:366–375.

Wood, David M., and Roger del Moral. 1987. Mechanisms of Early Primary Succession in Subalpine Habitats on Mount St. Helens. *Ecology* 68:780–790

Woodcock, D. W. 1992. Climate Reconstruction Based on Biological Indicators. *Quarterly Review of Biology* 67:457–477.

Woodring, W. P. 1951. Basic Assumption Underlying Paleoecology. *Science* 113:482–483.

Woodward, C. A., and J. Shulmeister. 2006. New Zealand Chironomids as Proxies for Human-Induced and Natural Environmental Change: Transfer Functions for Temperature and Lake Production (Chlorophyll *a*). *Journal of Paleolimnology* 36:407–429.

Wu, Jianguo. 2007. Scale and Scaling: A Cross-Disciplinary Perspective. In *Key Topics in Landscape Ecology*, edited by Jianguo Wu and Richard Hobbs, pp. 115–142. Cambridge University Press.

Yackulic, Charles B., Eric W. Sanderson, and María Uriarte. 2011. Anthropogenic and Environmental Drivers of Modern Range Loss in Large Mammals. *Proceedings of the National Academy of Sciences USA* 108:4024–4029.

Yalden, D. W. 2001. Mammals as Climatic Indicators. In *Handbook of Archaeological Sciences*, edited by D. R. Brothwell and A. M. Pollard, pp. 147–154. John Wiley and Sons, Chichester.

Yamada, Eisuke, Erl Hasumi, Nao Miyazato, Megumi Akahoshi, Mahito Watabe, and Hideo Nakaya. 2016. Mesowear Analyses of Sympatric Ungulates from the Late Miocene Maragheh, Iran. *Palaeobiodiversity and Palaeoenvironments* 96:445–452.

Yann, Lindsey T., Larisa R. G. DeSantis, Ryan J. Haupt et al. 2013. The Application of an Oxygen Isotope Aridity Index to Terrestrial Paleoenvironmental Reconstructions in Pleistocene North America. *Paleobiology* 39:576–590.

Yom-Tov, Yoram, and Eli Geffen. 2006. Geographic Variation in Body Size: The Effects of Ambient Temperature and Precipitation. *Oecologia* 148:213–218.

Yom-Tov, Yoram, and Henry Nix. 1986. Climatological Correlates for Body Size of Five Species of Australian Mammals. *Biological Journal of the Linnean Society* 29:245–262.

Yom-Tov, Yoram, and Jonathan Yom-Tov. 2005. Global Warming, Bergmann's Rule and Body Size in the Masked Shrew *Sorex cinereus* Kerr in Alaska. *Journal of Animal Ecology* 74:803–808.

Yom-Tov, Yoram, Shlomith Yom-Tov, Jonathan Wright, Chris J. R. Thorne, and Richard Du Feu. 2006. Recent Changes in Body Weight and Wing Length among Some British Passerine Birds. *Oikos* 112:91–101.

Yom-Tov, Yoram, Shlomith Yom-Tov, and Gordon Jarrell. 2008. Recent Increase in Body Size of the American Marten *Martes americana* in Alaska. *Biological Journal of the Linnean Society* 93:701–707.

Yom-Tov, Yoram, Noam Leader, Shlomith Yom-Tov, and Hans J. Baagøe. 2010. Temperature Trends and Recent Decline in Body Size of the Stone Marten *Martes foina* in Denmark. *Mammalian Biology* 75:146–150.

Young, Kenneth R. 2014. Biogeography of the Anthropocene: Novel Species Assemblages. *Progress in Physical Geography* 38:664–673.

Zeder, Melinda A. 2001. A Metrical Analysis of Modern Goats (*Capra hircus aegagrus* and *C. h. hircus*) from Iran and Iraq: Implications for the Study of Caprine Domestication. *Journal of Archaeological Science* 28:61–79.

Zeder, Melinda A., and Brian Hesse. 2000. The Initial Domestication of Goats (*Capra hircus*) in the Zagros Mountains 10,000 Years Ago. *Science* 287:2254–2257.

Zeuner, F. E. 1936. Climatic Research Based on the Association of Species in Fossil Faunas and Floras. In *Problems in Paleontology*, vol. 1, edited by A. Hartmann-Weinberg, pp. 200–216. Publications of the Laboratory of Paleontology, Moscow University.

———. 1961. Faunal Evidence for Pleistocene Climates. *Annals of the New York Academy of Science* 95(1):502–507.

Zhang, Hanwen, Yuan Wang, Christine M. Janis, Robert H. Goodall, and Mark A. Purnell. 2017. An Examination of Feeding Ecology in Pleistocene Proboscideans from

Southern China (*Sinomastodon, Stegodon, Elephas*), by Means of Dental Microwear Texture Analysis. *Quaternary International* 445:60–70.

Zhang, Yi Ge, Mark Pagani, Zhongui Liu, Steven M. Bohaty, and Robert DeConto. 2013. A 40-Million-Year History of Atmospheric CO_2. *Philosophical Transactions of the Royal Society A* 371:20130096.

Ziegler, Alan C. 1973. *Inference from Prehistoric Faunal Remains.* Addison-Wesley Module in Anthropology 43. Addison-Wesley, Reading, MA.

Zohar, Irit, and Miriam Belmaker. 2005. Size Does Matter: Methodological Comments on Sieve Size and Species Richness in Fishbone Assemblages. *Journal of Archaeological Science* 32:635–641.

INDEX

Note: Glossary terms appear in the index in *italic*.